Mathematical Economics

2 edition

Baldani
Bradfield
Turner

THOMSON

SOUTH-WESTERN

Australia · Canada · Mexico · Singapore · Spain · United Kingdom · United States

THOMSON

✦

SOUTH-WESTERN

Mathematical Economics, 2e
Jeffrey Baldani, James Bradfield, Robert W. Turner

VP/Editorial Director:
Jack W. Calhoun

VP/Editor-in-Chief:
Dave Shaut

Publisher:
Mike Mercier

Senior Acquisitions Editor:
Peter Adams

Developmental Editor:
Jennifer E. Baker

Production Editor:
Tamborah E. Moore

Marketing Manager:
John Carey

Senior Technology Project Editor:
Peggy Buskey

Technology Project Editor
Pam Wallace

Senior Manufacturing Coordinator:
Sandee Milewski

Production House:
Rozi Harris, Interactive Composition Corporation

Printer:
Courier Westford, MA

Art Director:
Chris A. Miller

Cover Image:
© Getty Images

Preface

This book is the outcome of many years of experience teaching courses that focus on how mathematics is used in economic analyses. We have found that although most economics students know the basics of calculus and economic theory, they rarely encounter interesting ways to apply mathematics to economics at the undergraduate level. In our teaching and in our book, we integrate mathematics and economics in a way that illustrates the insights that mathematics can bring to economic analysis. When we present mathematical procedures, we treat them as tools, emphasizing their applications and suggesting the various situations in which they are appropriate. By offering an intuitive understanding of why these procedures work, rather than going through detailed proofs, our method lets students focus on what is gained by applying mathematical tools to economic problems: establishing the generality of results, illustrating the roles that various assumptions and parameters play in establishing particular results, and finding the mathematical symmetry between different economic problems.

We decided to write this book because most existing textbooks in the field of mathematical economics are designed either for courses that teach mathematics to economics students or for courses that teach economic theory using mathematics. We wanted, instead, a book that focused on how mathematics is used in economic analysis. Finding none, we chose to write one ourselves. In so doing, we also sought to provide many examples of economic applications that are interesting in their own right.

The organization of this book reflects its purpose. We have structured this textbook in pairs of chapters, the first of which contains a brief description of certain mathematical procedures (or tools), while the second contains a collection of economic applications of those tools. This toolbox-applications approach is the way we prefer to teach. It allows the mathematics to be covered quickly and coherently without interrupting the discussion to set up and solve economic problems. Conversely, in the applications chapters, it is not necessary to digress into a detailed explanation of some new mathematical procedure. Our approach also provides an easy way for students to find and review the mathematical tools appropriate for particular economic applications.

In each theory chapter, we present the mathematics in abstract form, but we also motivate the discussion by particular problems in economics. Thus, we encourage students to view mathematical principles as the means to gain insight into specific economic applications. We have not tried to be rigorous in our presentation of the mathematics; there are other books that provide details and proofs of the mathematics we present. Instead, we have sought to present the mathematics in a practical way—one that will enable students to understand easily how to use mathematical tools in economic analysis. The applications chapters are designed for flexibility and utility. Each application in these chapters is self-contained, although many make reference to other applications in the same or previous chapters. They need not, therefore, be read from beginning to end. Rather, instructors may choose to cover any subset of the applications in each chapter. These chapters also emphasize the economic interpretation of results as well as the way in which mathematics is used to find those results. We have included a mix of applications— some that are familiar from intermediate economic theory and some that will stretch students' understanding of economics. Moreover, because most students learn more by doing problems than by reading textbooks or listening to lectures, our examples of economic applications are thorough enough to be used as models for students when they tackle the end-of-chapter problems. Except for Chapter 1, each chapter concludes with a set of problems. The problems for the theory chapters review the subject matter. The problems for the applications chapter, however, are intended to challenge students in a variety of ways: some extend the applications given in the text, while others introduce new, mathematically analogous applications. These problems are suitable for homework assignments and classroom examinations.

The organization of the book in pairs of chapters allows instructors great flexibility. Instructors who prefer the integrated approach can easily design a syllabus in which theory and applications chapters are read together and presented together in lectures. Alternatively, instructors teaching mathematics to economics students may use the set of theory chapters as the core of a course, employing a few of the applications as illustrations. Or some instructors may use the set of applications chapters for an applied economics course; in this case, the theory chapters can serve as refresher courses for students who should already know their contents. In addition, some professors may find this textbook suitable, either by itself or as a supplement, for some graduate-level courses.

We assume that all readers of this book will have taken at least one calculus course as well as one course each in microeconomic and macroeconomic theory. Many readers may also have taken an intermediate-level course in economic theory. Intended to address a variety of students having a range of preparation, this book includes a short review of the basics of differential calculus (see the appendix to Chapter 1), as well as challenging applications and exercises for the more-advanced students.

CHANGES TO THE SECOND EDITION

The second edition of this book retains the features that made the first edition popular with adopters, in particular our emphasis on using mathematics to gain insight into economic problems and our organization of paired chapters for each topic, and adds new material, providing even more flexibility to users. The main addition is a set of new chapters, with appendices, introducing dynamic analysis. We have also added new end-of-chapter problems and new economic applications, including more problems and applications dealing with macroeconomics and several investigating issues in environmental economics. In addition, we now integrate the discussion of the envelope theorem into all chapters dealing with optimization problems, while keeping the detailed description and analysis in a separate pair of chapters.

Users of the first edition will find the same organization of topics in the new edition, including our use of paired toolbox and application chapters for each topic. We, and other users of our book, continue to find this a successful way of teaching the use of mathematics in economic analysis. Each toolbox chapter introduces the basics of a mathematical method, with one or more economic applications being used as examples. Instructors can cover these chapters quickly or in more detail, depending on their needs and the backgrounds of their students. Then the applications chapters can focus on the use of mathematical analysis to learn more about economics, without long digressions about the details of the mathematical methods. The paired chapters also make it easy for instructors to pick and choose which applications they want to cover, because all the applications in a chapter use the same mathematical methods, which are covered separately.

In response to comments from users and reviewers of the first edition, we have added a pair of chapters introducing the use of difference and differential equations in economics; an appendix contains an introduction to dynamic programming. The new chapters are placed near the end of the book, but we have presented the material in such a way that instructors could choose to cover dynamics very early as well—as early as right after matrix theory is covered in Chapters 3 and 4. We believe most instructors who want to cover dynamics will choose to use the new chapters either after multivariate calculus is introduced in Chapters 5 and 6, or near the end of a semester-long course. We have written these chapters so that either option will work well.

Almost no semester-long course in mathematical economics will cover all the chapters of our book. Most instructors will want to cover all the material in Chapters 1–10 and then pick and choose among the remaining topics: inequality constraints, the details of value functions and the envelope theorem, an introduction to dynamic analysis, and an introduction to game theory. By adding the new chapters on dynamics, we have given instructors even more flexibility than was possible in the first edition, while still avoiding making the book too encyclopedic.

ACKNOWLEDGMENTS

We are indebted to many individuals for their support and contributions in completing the second edition. Many colleagues helped us by reviewing all or part of our manuscript and provided us with constructive critical comments that helped us improve our text.

Michael Caputo
University of California at Davis

Kamran M. Dadkhah
Northeastern University

Scott Fausti
South Dakota University

James Hartigan
University of Oklahoma

Mark R. Johnson
University of Alabama

Brian Kench
University of New Hampshire

Dean Kiefer
University of New Orleans

Janet Koscianski
Shippensburg University

Bruce Dean Larson
University of North Carolina, Asheville

Stephen Layson
University of North Carolina, Greenville

Bento Lobo
University of New Orleans

Kashif Mansori
Colby College

James J. Murphy
University of Massachusetts, Amherst

John Nachbar
Washington University

Babu Nahata
University of Louisville

John F. O'Connell
College of the Holy Cross

James Peach
New Mexico State University

Owen Phillips
University of Wyoming

Kevin Reffett
Florida State University

Allan Sleeman
Western Washington University

Thomas Rhoads
Towson University

Abu Wahid
Tennessee State University

Gilbert Skillman
Wesleyan University

Many individuals at Thomson Business and Professional Publishing contributed to this project, including Peter Adams, Senior Acquisitions Editor; Jennifer Baker, our Developmental Editor; Tamborah Moore, Production Editor; Chris Miller, who provided a beautiful and functional internal and cover design, and Rozi Harris, Lead Project Manager at Interactive Composition Corporation, who helped to see us through the copyediting and composition processes.

About the Authors

Jeffrey Baldani received his B.A. from the University of Kentucky and Ph.D. from Cornell University. He has taught economics at Colgate University since 1982. His teaching interests include mathematical economics, game theory, and applied microeconomics.

James Bradfield received his B.A. and Ph.D. degrees from the University of Rochester. He has taught economics at Hamilton College since 1976. His teaching interests include principles of economics, mathematical economics, microeconomics, and financial markets.

Robert W. Turner received his A.B. from Oberlin College and Ph.D. from the Massachusetts Institute of Technology. He has taught economics at Colgate University since 1983. His teaching interests include econometrics, mathematical economics, public and environmental economics, and principles of economics.

Brief Contents

Contents

To our families

An Introduction to Mathematical Economics

1.1 INTRODUCTION

Economics is often mathematical. As students with some background in economic analysis, you are probably already accustomed to working with geometric and algebraic economic models; these methods, while important and useful, are only part of an economist's mathematical tool kit. Professional economists use a wide variety of tools from higher mathematics. Knowledge of mathematical economics is, in fact, a prerequisite for reading and understanding many of the papers published in economic journals. This text will introduce some of the mathematical techniques that are fundamental to modern economic analysis and show you how the techniques are relevant and useful in analyzing economic issues.

This text focuses on how mathematical tools are used in economics. The approach is to first introduce mathematical concepts (with a few simple economic examples) and then present detailed economic applications. Throughout the text, odd-numbered chapters are the mathematics tool kits and even numbered chapters are the economics content.

1.2 THE CONCEPT OF A MATHEMATICAL ECONOMIC MODEL

1.2.1 Economic Models

Before embarking on an explanation of mathematical analysis, let's review the concept of an economic model. An economic model is an abstraction from reality that distills the essential features of an economic problem into a tractable and useful form. The typical economic model has five elements:

1. **Economic Agents**
   Economic agents include consumers, workers, firms, and governments. Agents are characterized by their ability to make decisions and to pursue goals.

2. **An Economic Environment**
   Agents do not pursue their goals in a vacuum: they face an economic environment encompassing the many factors that bear upon their choices. The term economic environment generally denotes economic factors that are important to an agent

but lie outside the direct control of the agent. Individual consumers, for example, have little or no control over the prices they pay for consumption goods. Elements in the economic environment are represented by **exogenous** variables—variables or **parameters** that are beyond the control of the agents or taken as externally given, rather than being explained by the model. We will discuss exogenous variables in more detail below.

3. **Choices**

   Choices by economic agents reflect judgments of how to best pursue goals in the face of their economic environment. The assumption that decisions are made rationally, that is, to achieve goals in the most effective manner, is very helpful in modeling economic issues mathematically. Some applications (such as consumer utility maximization) model choices explicitly whereas other applications (such as supply and demand or many macroeconomic models) model the results of choices.

4. **Equilibrium Solutions**

   Economic equilibrium is the outcome of agents' choices. A model is in equilibrium when there is no tendency for economic variables to change unless there are changes in the economic environment. As a general rule, an equilibrium specifies either how agents' optimal decisions are determined by the economic environment (such as what choices will be made given goals and given the factors outside the agents' control) or how economic variables (such as price and quantity in a supply and demand model or gross domestic product in a macroeconomic model) are determined by factors in the economic environment. The equilibrium values of economic variables are **endogenous**: they are determined within the model and depend on the values of the exogenous variables. Again, this is a term that we will discuss in more detail below.

5. **Analysis of How Changes in the Environment Affect Equilibrium**

   From an economist's point of view, the most important element of an economic model is the attempt to predict changes in economic behavior. By using a model to predict, economists are able to answer important economic questions. This is done through an analysis of how changes in the environment affect economic equilibrium. Consider, for example, questions about the effects of government policies. An economic analysis of how firms' and consumers' equilibrium choices change as a result of government policies can help in evaluating those policies. Most of the examples in the text focus on **static** changes, how one equilibrium compares to another, without worrying about the **dynamics**, how the model actually moves from the initial to the new equilibrium. Dynamics and the explicit model of the time-paths of changes in equilibrium are, however, covered in Chapters 15 and 16.

Throughout this text we will be building mathematical economic models that follow this general structure. To give a simple outline of an economic model consider a firm, an agent pursuing the goal of profit maximization. The firm's economic environment would include a production technology, supply schedules for its inputs, a demand curve for its output, and possibly taxes or government regulations. In this environment the firm may make many choices: input usage, output levels, research and development spending, and advertising spending are all possible choice variables for the firm. Equilibrium in this model would specify how the firm's decisions depend on its economic environment. Finally, once the model has been solved, the model's true value lies in its predictive power. For example, we may wish to know how an increase in payroll taxes will affect the firm's mix of labor and capital and its output level, or we might want to predict the effect of technological change on profitability.

Economic modeling is central to economic science. Because of this it is important to emphasize from the start that models are inherently incomplete and, in some sense, "unrealistic." No model can fully describe an underlying economic problem. A fully realistic model would collapse under its own weight: the model would be so complicated that it would be impossible to solve for equilibrium or to make predictions. An ideal model is simple enough to be solvable but at the same time complex enough to capture important economic features and to make predictions that can be verified (or refuted) by observations and statistical data.

As a practical matter, issues of model complexity are handled in a number of ways. One method is to make simplifying mathematical assumptions, such as assuming that the demand curve for a product is linear. Another method is to place certain elements in the environment instead of in the agents' choice sets. For example, a model might assume that product quality is technologically fixed rather than determined by a firm. A third method is to assume that some problems have already been solved for in another implicit model. For example, we might describe a firm's cost as a function of output by assuming that the firm will always adjust its inputs to minimize the cost of producing whatever level of output is chosen. A final method, which may include all of the previous elements, is to build up a model in steps. This means starting with very simple assumptions, solving the model, and then going back to relax the assumptions and re-solving. All of these methods are widely used in mathematical economics.

1.2.2   A Solvable Example: Linear Demand and Supply

We start with one of the most basic (and useful) economic models—the familiar demand and supply model. The agents here are consumers and firms. To keep our example simple we will assume that both the market demand and market supply curves are linear functions. Mathematically this means we can write the inverse demand and supply curves as

$$P = a - bQ_D \quad \text{(demand)} \tag{1.1}$$

and

$$P = c + dQ_S \quad \text{(supply)}, \tag{1.2}$$

where P is price, Q_D is quantity demanded, Q_S is quantity supplied, and a, b, c, and d are parameters.[1]

The parameters a, b, c, and d depend on many factors in the economic environment, such as consumer preferences and income (demand side) and technology and input prices (supply side). These parameters (and the underlying economic factors that determine them) are exogenous to our model because we do not explicitly model their determination or derivation. Similarly, we do not model the choice-making processes by consumers or firms; we simply assume that the optimal results of those processes are captured in the demand and supply schedules. In contrast, price and quantity will be determined within the solution to our model. These variables are therefore endogenous.

The value of using parameters, instead of specific numerical values, in a model is that parameters allow for a greater degree of generality. By this we mean that algebraic notation for the parameters automatically covers all cases of specific numerical values.

---

[1]Technically, when demand or supply curves are written with price as the dependent variable they are called the inverse demand and supply functions. This distinction occurs because economists violate standard mathematical practice by speaking of how quantity demanded (supplied) depends on price, but then graphing demand and supply curves with price on the vertical axis. In other words, quantity is the dependent variable—it depends on price— but contrary to normal conventions economists place this variable on the horizontal axis.

FIGURE 1.1

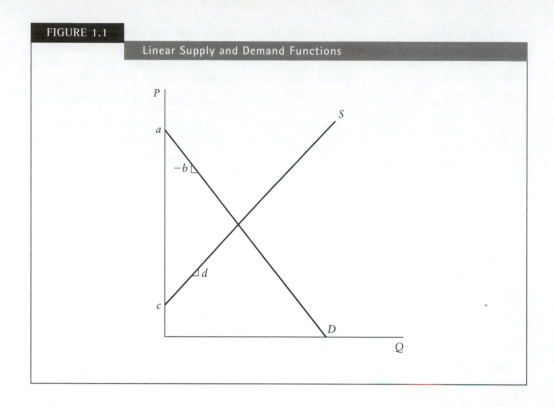

Once we solve the general model we need only make substitutions at the end to cover any particular numerical cases.

A greater degree of generality will prove to be most useful when we can supply an economic interpretation for parameters and/or use economic theory to put restrictions on the possible range of parameter values. In this example both possibilities apply. On the demand side the parameter a represents the demand curve intercept, and we know that $a > 0$, while b represents the demand slope and (since we have put a negative sign in front of b) we know that $b > 0$ also. For the supply curve, c and d are the intercept and slope, respectively. Again, economic theory tells us that $c > 0$ and $d > 0$.[2] Finally, for the model to have economic meaning, the demand curve intercept must be higher than the supply curve intercept, $a > c$. These restrictions and the graphs of the two curves are shown in Figure 1.1.

The solution to the model—the equilibrium price and quantity—lies at the point where quantity demanded equals quantity supplied. This is simply the intersection of the two curves and reflects the equilibrium condition that $Q_D = Q_S$. There are several methods for solving. One of the most efficient methods is through the use of matrix algebra, which will be introduced in Chapter 3. For now we can solve by simply setting quantity demanded equal to quantity supplied. Let Q be this equilibrium quantity; then demand equals supply when

$$a - bQ = c + dQ. \tag{1.3}$$

Rearranging yields

$$(b + d)Q = a - c, \tag{1.4}$$

---

[2]Technically, the supply curve could have a zero intercept or a zero slope (but not both at the same time) so our restrictions are $c \geq 0$, $d \geq 0$ and $c + d > 0$.

which solves as

$$Q^* = \frac{(a-c)}{(b+d)}. \tag{1.5}$$

Our assumptions $(a - c > 0, b > 0,$ and $d > 0)$ are sufficient to show that we have derived a positive solution for Q^*. (In this text we will often use an asterisk superscript on an endogenous variable to show that we have arrived at the final step of solving for the variable as a function that is dependent solely on the model's exogenous variable(s).) To solve for price we can substitute the solution for Q^* back into either the supply or demand equation. Using the supply equation, we get

$$P = c + d\frac{(a-c)}{(b+d)} \tag{1.6}$$

or

$$P = \frac{c(b+d) + d(a-c)}{(b+d)}. \tag{1.7}$$

Simplifying gives the equilibrium value of price as

$$P^* = \frac{(ad+bc)}{(b+d)}. \tag{1.8}$$

Note that these are equilibrium solutions for price and quantity in two senses. First, we have solved for a situation in which there is no tendency for market outcomes to change. Second, we have found the **reduced-form solutions** for the model, that is, we have expressed the values of the two endogenous variables solely as a function of the exogenous parameters. In the reduced form the endogenous variable is dependent on the exogenous variables. The mathematical solutions are shown graphically in Figure 1.2.

Once we have solved for an endogenous variable we are able to make predictions as to how changes in the environment will affect that variable. In a mathematical economic model this type of analysis is referred to as **comparative statics**: comparative because we are comparing one static equilibrium to the new one that would occur if parameters were to change; statics because we are abstracting from the description of the dynamics of how the equilibrium actually moves from one position to another. Mathematically, comparative statics involves taking the derivatives (or more generally the partial derivatives) of our solution functions with respect to the parameters. There are eight possible comparative static results in our model: changes in each of the four parameters can affect the two endogenous variables. To get comparative static results we must use the mathematical tool of partial derivatives. Partial derivatives are covered in detail in the Appendix, but the basic notion is fairly simple and can be explained here.

When an endogenous variable is a function of more than one exogenous variable we are often interested in how the endogenous variable changes when a single exogenous variable changes. To examine this we simply take the derivative with respect to the changing exogenous variable while treating the other exogenous variables as if they were constants.[3] This is a partial derivative in the sense that only a single change (out of several possible changes) is being considered. Later, in Chapter 5, we will learn how

---

[3]This technique is in fact used (without calling it partial differentiation) from the start in calculus courses. For example, consider the function $y = ax$. The derivative with respect to x implicitly treats a as a constant. One could just as easily take the derivative with respect to a while treating x as a constant.

FIGURE 1.2

Market Equilibrium

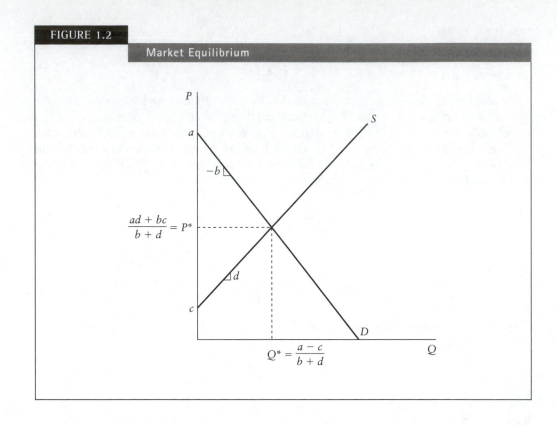

to show the simultaneous effects, on endogenous variables, of changes in more than one exogenous variable.

Rather than taking all eight possible partial derivatives of our model, let us focus on a couple of specific cases and their economic interpretations. First, let's look at the effects on quantity and price of an increase in the demand intercept, a. This shifts the demand curve up vertically. Such a change might be the result of an increased consumer preference for the product or of an increase in consumer income (if the good is normal with respect to changes in income). The partial derivatives showing the effect of an increase in the parameter a are

$$\frac{\partial Q^*}{\partial a} = \frac{1}{(b+d)} > 0 \quad \text{and} \quad \frac{\partial P^*}{\partial a} = \frac{d}{(b+d)} > 0. \tag{1.9}$$

Thus, an increase in the demand intercept leads to an increase in both price and quantity. This is shown graphically in Figure 1.3, where an increase in the intercept from a to a new value, a', leads to higher equilibrium values for P and Q.

Next let us examine the effect of a decrease in the supply curve slope, d. The decrease might result from such factors as improved production technology, lowered input prices, or entry of new firms into the industry, all of which increase the quantity supplied. Since the parameter d appears in both the numerator and denominator of the solutions we must use the quotient rule for taking derivatives. This rule (explained in the Appendix) is,

$$\text{if} \quad y = \frac{f(x)}{g(x)} \quad \text{then} \quad \frac{dy}{dx} = \frac{f'g - fg'}{g^2}. \tag{1.10}$$

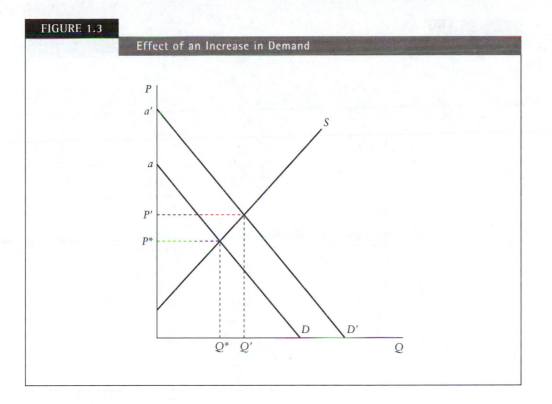

FIGURE 1.3

Effect of an Increase in Demand

Using this rule gives the derivatives of quantity and price with respect to the parameter d

$$\frac{\partial Q^*}{\partial d} = \frac{-(a-c)}{(b+d)^2} < 0 \tag{1.11}$$

and

$$\frac{\partial P^*}{\partial d} = \frac{a(b+d) - 1(ad+bc)}{(b+d)^2} = \frac{b(a-c)}{(b+d)^2} > 0. \tag{1.12}$$

Again, we are able to sign the derivatives by using the restrictions that economic theory places on the values of the parameters. The derivatives show that the change in P^* is in the same direction as a change in the parameter d and that the change in Q^* is opposite in sign to the direction of change in d. Thus, a decrease in d leads to an increase in equilibrium quantity but to a decline in equilibrium price. As an exercise the reader should check these results graphically.

In this model we have been able to derive qualitative comparative static predictions based on an economic interpretation of the parameters and their values. This is a common method for analyzing models and deriving results. There are also other methods that will be introduced later. The point to emphasize is that much of the power of mathematical economic analysis comes from the way it blends mathematical tools with economic analysis and theory.

Before leaving our demand-supply example, let us extend the model to include the effects of taxes. This extension will serve three purposes. First, it will illustrate the idea, that model building is often progressive. Economists start with simple models and then build upon them to derive additional results. Second, the extension will show how mathematical models can be used to examine public policy issues. Finally, in solving the extension we will examine how smart thinking can save time and effort in deriving mathematical results.

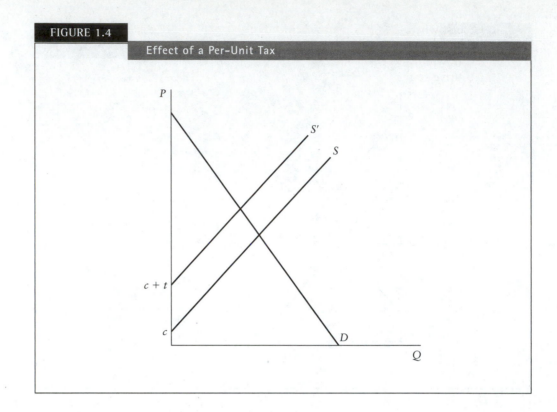

FIGURE 1.4

Effect of a Per-Unit Tax

The simplest case of government taxation is that of a per-unit tax on a product.[4] We will denote the tax rate by t/unit. As a modeling choice we might assume that firms are legally responsible for paying the tax.[5] From a firm's viewpoint the tax is equivalent to an additional cost of production. The supply schedule shifts up vertically by the amount of the tax: if a given quantity was supplied at a price P before the tax was imposed then the same quantity will now be supplied at a price of $P + t$. This is shown in Figure 1.4 and in the revised supply equation

$$P = (c + t) + dQ. \tag{1.13}$$

To solve for the new equilibrium we could simply repeat the substitution method used above to solve the original model. However, as will be true in many instances, a little bit of cleverness allows a shortcut that saves time and steps. Let $c' = c + t$ and let the supply equation be written as $P = c' + dQ$. It should then be obvious that the new solution will be the same as the original solution except that c' replaces c. Thus, the new solutions are

$$Q^* = \frac{(a - c')}{(b + d)} = \frac{(a - c - t)}{(b + d)} = \frac{(a - c)}{(b + d)} - \frac{t}{(b + d)}$$

$$P^* = \frac{(ad + bc')}{(b + d)} = \frac{(ad + bc + bt)}{(b + d)} = \frac{(ad + bc)}{(b + d)} + \frac{bt}{(b + d)}. \tag{1.14}$$

---

[4]The more common form of product taxation, a sales tax, is often identical in effect to a per-unit tax but is slightly more difficult to model.

[5]Microeconomic analysis can be used to show that the economic effects of the tax do not depend on who is legally responsible for the mechanics of sending the tax revenue to the government.

Such shortcuts, when used intelligently and correctly, can relieve some of the algebraic drudgery of working through mathematical models.

We can now conclude this modeling exercise by examining the impact of the tax. The comparative static partial derivatives of a change in the tax rate are[6]

$$\frac{\partial Q^*}{\partial t} = \frac{-1}{(b+d)} < 0$$

$$\frac{\partial P^*}{\partial t} = \frac{b}{(b+d)} > 0. \tag{1.15}$$

Our prediction therefore is that an increase in the tax rate will lead to a decrease in quantity and an increase in price. We can also see that the magnitude of the changes depends on the values of the other parameters (for example, Q decreases more if the demand and supply slopes are small). This latter type of result is often as of much interest as the signs of the derivatives themselves.[7]

1.2.3 A General Functional Form Example

The preceding example assumed specific (linear) functional forms. Although this is quite convenient, it is also quite restrictive. We now turn to an example in which we don't make any assumptions concerning specific functional forms. Let us consider a firm whose goal is to maximize profits. Mathematically we write the equation for profits as

$$\Pi = TR - TC, \tag{1.16}$$

where Π represents profits, TR represents total revenue, and TC represents total production costs. This equation, by itself, is simply an accounting identity that tells us little about how a firm would actually behave.

Next we specify the firm's economic environment. The modeling choices here might include whether the firm is in a competitive or monopolistic market, whether the firm spends money on research and development or advertising, and whether the firm is subject to government regulation or taxes. To keep our example simple let us assume the following:

1. The firm produces a single product and operates in a perfectly competitive market,
2. The firm's production technology is fixed,
3. The firm has (outside our model) already chosen the cost-minimizing mix of labor and capital for whatever level of output is to be produced,[8] and
4. The firm is not subject to taxes or regulations.

Next we translate these assumptions into a mathematical description. The first assumption means that the firm can sell any quantity, which we will denote by q, at the going market price, which we will denote by P. The next two assumptions let us write the firm's production cost as a function of its output. The final assumption simply allows us to escape any modeling of the government's impact on the firm. Mathematically, we

---

[6]These results could also be derived indirectly by using a chain rule, as in $(\partial Q^*/\partial c')(\partial c'/\partial t)$.

[7]Mathematically, this type of analysis would use cross-partial derivatives, which are covered in Chapter 5.

[8]Finding the cost-minimizing level of inputs is an example of a constrained optimization problem. See Chapter 10.

can now write

$$TR = Pq \quad TC = C(q)$$

$$\text{where } C'(q) > 0 \quad \text{and} \quad \Pi = Pq - C(q).$$

(1.17)

$C(q)$, read as "costs as a function of output," is a general functional form where we haven't chosen any specific mathematical description beyond the assumption that marginal cost, $C'(q)$, is positive.

In this example, price is an exogenous variable (beyond the firm's control), while output is endogenous (chosen by the firm in order to maximize profits), and profits are also endogenous since they are determined by the firm's output choice.

The distinction between exogenous and endogenous variables is crucial to understanding mathematical economics. Unfortunately, this distinction is often a cause of confusion and misunderstanding for students who are new to the subject. The main reason for this confusion is that there is no magic list of variables that are exogenous and variables that are endogenous. Instead the distinction is model-specific: it varies from one model to the next. The key point is that an economist, in constructing a model, chooses which variables to place in the economic environment and which variables to make endogenous. The distinction therefore depends on the specific context of the issues being modeled. Indeed, variables that are exogenous in one model (such as price in our example) may be endogenous in another model (such as the model of the overall competitive market above where price was endogenously determined by supply and demand).

The firm will pursue its goal of profit maximization by choosing an optimal output. By optimal output we simply mean the level of output, given the market price, that results in the highest-possible level of profit. Mathematically this type of solution would be derived using calculus. Putting aside, for now, the actual calculus-based derivation, the important point is that the solution, or equilibrium, is not a single level of output, but a function that specifies how the optimal output depends on the market price.

Let us write this solution as

$$q^* = f(P).$$

(1.18)

This supply function is the solution to our simple model. The central feature of equilibrium solutions is that they express the endogenous variable(s) as functions of the exogenous variable(s). In other words, we have solved for how the firm will behave or how choices depend on the environment.

In our simple example the only possible comparative static result is how output would change in response to a change in the market price.[9] The equation for this is

$$\frac{dq^*}{dP} = f'(P).$$

(1.19)

For the most part when we attempt to make predictions using comparative statics we are interested in qualitative results. Qualitative results involve determining the sign of the derivative (the direction of change in the endogenous variable) when there is a change in an exogenous variable. Only occasionally will we be interested in the quantitative result or the actual numerical magnitude of the change. The main reason for this is that mathematical models are abstractions and inherently lack the degree of concordance

---

[9]This is a very simple model. In most models there will be numerous parameters. For example, we might complicate the present model by adding parameters to the cost function.

with the real world that would allow us to place much confidence in predictions of numerical results.

To conclude this section we turn to the question of how we obtain qualitative comparative static predictions. In the context of our previous example, how can we determine the sign of the derivative or answer the economic question of whether an increase in price will lead to an increase or a decrease in output? At this point we have not yet developed the tools and methods that allow us to answer these questions.[10] As we progress, however, we will find that mathematical economists have several common methods for putting signs on derivatives and making qualitative predictions.

1.3 OPTIMIZATION

One of the most widely used tools in mathematical economics is optimization analysis. Economics, as the study of choices under scarcity, is fundamentally concerned with situations in which agents must maximize the achievement of some goal: consumers maximize utility, firms maximize profits, governments maximize welfare. In the next two subsections we will first review the calculus of single-variable optimization (multivariable optimization will be introduced in Chapter 7) and then examine a generic economic example of optimizing behavior.

1.3.1   Review of Calculus of Single-Variable Optimization

Introductory calculus courses emphasize the technique of using derivatives to find the extreme points of a function. Since a derivative of a function is simply the slope of that function, points where the derivative equals zero are potentially maximum points (top of a hill) or minimum points (bottom of a valley) of the original function.[11] If we write a function as

$$y = f(x), \tag{1.20}$$

then the derivative, or slope, of the function is written as

$$\frac{dy}{dx} = f'(x). \tag{1.21}$$

The notation $f'(x)$ indicates not only a derivative but also the idea that the value of the derivative may depend on the particular value of x at which the derivative is evaluated.

Examples of the graph of a function $f(x)$ and the graph of the function's derivative, $f'(x)$, are shown in Figures 1.5a and 1.5b. The function $f(x)$ has three local extreme points. The first is a maximum at x_0, the second is a minimum at x_1 and the third is a maximum at x_2. As can be seen from the graphs, the slope of the function and the value of its derivative are zero at all three of these points. This example illustrates two related mathematical points. First, there may be multiple local maxima or minima points for which $f'(x) = 0$. Second, a local maximum or minimum point need not be a global maximum or minimum. In fact, the graphed example is unbounded and has no global minimum. In most, but not all, economic applications—where there is usually a single

---

[10]If you guessed that an increase in price increases quantity, you're right. The proof, however, will have to wait until we develop a bit more mathematical sophistication.

[11]A derivative equal to zero may also correspond to an inflection point, i.e., a point that is neither a maximum nor a minimum. Such a case, however, is rarely of interest in economic applications.

FIGURE 1.5

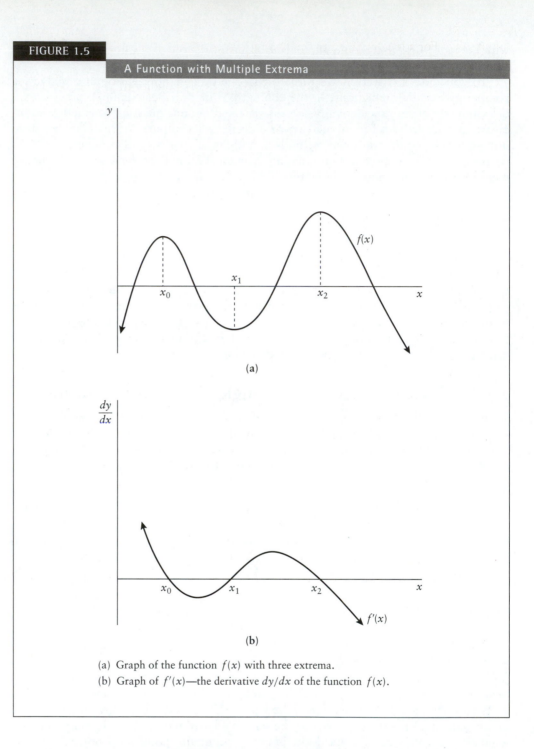

(a) Graph of the function $f(x)$ with three extrema.
(b) Graph of $f'(x)$—the derivative dy/dx of the function $f(x)$.

local extreme point that is also the global extreme point of the economic function that is being maximized or minimized—these complications will not arise.

In economic applications the condition that a first derivative should be zero for a local maximum or minimum is called a **first-order condition.** First-order conditions are necessary for a local extreme point but are not sufficient to indicate whether that extreme point is a minimum or maximum. To check whether a zero value of a derivative indicates a minimum or maximum we turn to the **second-order condition.** The second-order condition is an evaluation of the sign of the second derivative of the original function and is sufficient, given that the first-order condition holds, for a local maximum or minimum.[12] For a local maximum the two requirements are

$$\frac{dy}{dx} = f'(x) = 0 \quad \text{(first-order necessary condition)} \tag{1.22}$$

and

$$\frac{d^2y}{dx^2} = f''(x) < 0 \quad \text{(second-order sufficient condition).} \tag{1.23}$$

For a local minimum the requirements are

$$\frac{dy}{dx} = f'(x) = 0 \quad \text{(first-order necessary condition)} \tag{1.24}$$

and

$$\frac{d^2y}{dx^2} = f''(x) > 0 \quad \text{(second-order sufficient condition).} \tag{1.25}$$

The second-order conditions are checks on whether the function $f(x)$ is locally **strictly concave** or **strictly convex**. A single variable definition of these two terms is given in the Appendix. Multivariable definitions will be given later in Chapter 7. For the single variable case, a negative second derivative is sufficient to ensure that the function is locally strictly concave and a positive second derivative is sufficient to ensure that the function is locally strictly convex.[13] If the second derivative is single-signed for all values of x then the concavity or convexity property is global.

1.3.2   A Generic Example of Economic Maximization

It is only a slight exaggeration to claim that there is a single idea underlying all of microeconomics: agents maximize the net benefits of a course of action by setting marginal benefit equal to marginal cost. In this section we will work through this fundamental concept using calculus. To begin, let $x \geq 0$ be the level of some economic activity, such as output for a firm or consumption for a consumer. The agent's benefits and costs from the activity x can then be represented by the functions $B(x)$ and $C(x)$. The agent's overall gain, or net benefit, from the activity will be denoted by the variable y, which might be measured in dollars for a firm or utility units for a consumer. Let the value of y be given by the net benefit function $N(x)$,

$$y = N(x) = B(x) - C(x). \tag{1.26}$$

Maximizing net benefits requires satisfying the first- and second-order conditions for a maximum. These conditions are

$$\frac{dy}{dx} = N'(x) = B'(x) - C'(x) = 0 \tag{1.27}$$

---

[12]For the case where the second derivative equals zero we must check higher-order derivatives to determine whether the zero first derivative indicates a minimum, a maximum, or an inflection point.

[13]The converse is not true. It is possible that at a local maximum or minimum of a strictly concave or convex function the second derivative may be zero. This is why the second-order condition is sufficient, but not necessary.

and

$$\frac{d^2 y}{dx^2} = N''(x) = B''(x) - C''(x) < 0 \quad \text{or} \quad B''(x) < C''(x). \tag{1.28}$$

In mathematical economic models the economic interpretation of results is the primary aim of the model. Equations like those above have little significance or meaning until we interpret them through the lens of economic theory.

In economic terms the first-order condition states that the optimal level for the action x is where the marginal benefit of x, $B'(x)$, equals the marginal cost of x, $C'(x)$. This interpretation holds across the many different examples of agents and actions that the variable x might represent. The second-order condition states that the first-order condition will yield a maximum whenever (at the value of x that solves the first-order condition) the slope of the marginal benefit curve [$B'(x)$ is the curve and $B''(x)$ its slope] is less than the slope of the marginal cost curve [$C'(x)$ is the curve and $C''(x)$ its slope].

These conditions are illustrated in Figures 1.6a and 1.6b. Figure 1.6a shows a possible form for the benefit and cost curves. The net benefit is measured as the vertical distance between the two curves. Figure 1.6b shows the graphs of the marginal benefit and marginal cost curves. Note that marginal benefit equals marginal cost at two values of x, but that the first value of x, x_0, fails the second-order condition and is a minimum of net benefits. The second value, x_1, satisfies the second-order condition. The solution x_1 is therefore a local maximum.[14]

The point x_1 is not, however, the only maximum: $x = 0$ is also a local maximum. In our graph it is clear that the global maximum is achieved at x_1. Nonetheless, this raises the issue of **interior solutions** versus **corner solutions**. An interior solution occurs when the first- and second-order conditions yield a solution that is both a local and global extreme point. A corner solution occurs when the global extreme point occurs at the limit of economically permissible values for the x variable. First- and second-order conditions do not usually hold at corner solution extreme points. In our example the solution is interior because (1) the first- and second-order conditions hold at x_1 and (2) the value of net benefits is highest at x_1, as in $N(x_1) > N(0)$.

We now solve an example where we use specific functions. Suppose we assume that the benefit, cost, and net benefit functions have the following forms:

$$B(x) = ax - bx^2, \tag{1.29}$$

$$C(x) = cx^2 + f, \tag{1.30}$$

and

$$y = N(x) = B(x) - C(x) = ax - bx^2 - cx^2 - f, \tag{1.31}$$

where a, b, c, and f are positive parameters. Note that we are using the same parameter labels as in the supply and demand model, but the interpretation of the parameters is not the same. Here, the parameters a, b, and c are coefficients in the benefit and cost functions. We assume that they are all positive parameters in order to generate "normal looking" functions. The graphs of these functions are shown in Figure 1.7.

---

[14]Note that marginal cost exceeds marginal benefit for all activity levels between 0 and x_0, while marginal benefit exceeds marginal cost for all activity levels between x_0 and x_1.

FIGURE 1.6

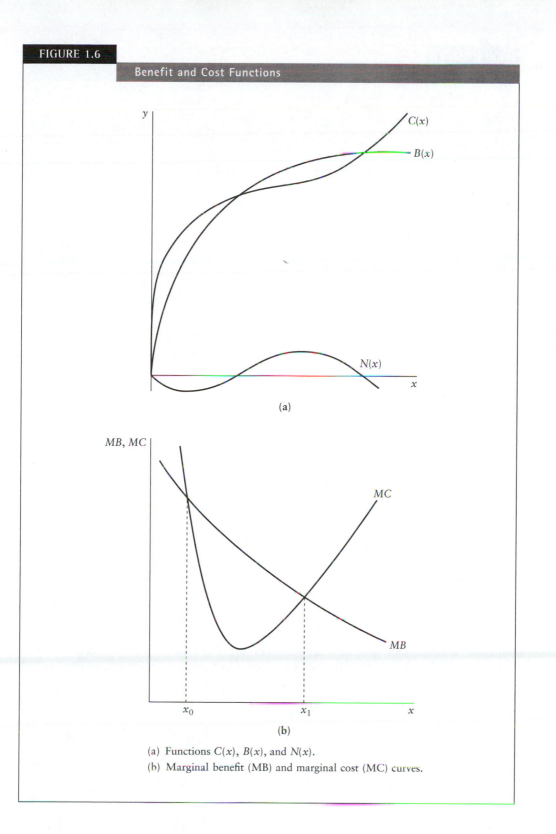

(a) Functions $C(x)$, $B(x)$, and $N(x)$.
(b) Marginal benefit (MB) and marginal cost (MC) curves.

FIGURE 1.7

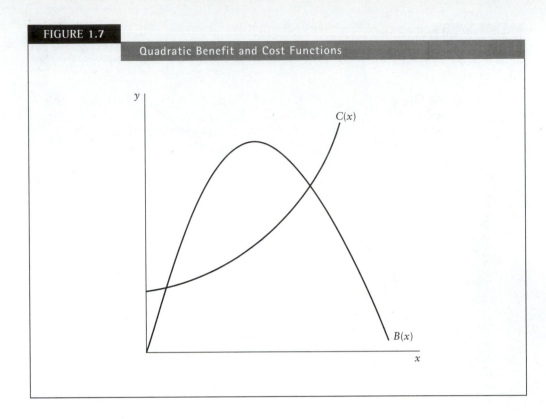

The first-order condition is

$$\frac{dy}{dx} = N'(x) = B'(x) - C'(x) = a - 2bx - 2cx = 0. \tag{1.32}$$

This equation can be solved for x:

$$x^* = \frac{a}{2(b+c)}. \tag{1.33}$$

To see whether x^* is a maximum we check the second-order condition and get

$$\frac{d^2y}{dx^2} = N''(x) = B''(x) - C''(x) = -2b - 2c = -2(b+c) < 0. \tag{1.34}$$

Since the second-order condition holds (remember that b and c are both positive parameters) the solution for x^* does correspond to a maximum. The resulting level of net benefits is found by substituting the solution for x^* into the function $N(x)$. This gives a level of net benefits of

$$y^* = ax^* - bx^{*2} - cx^{*2} - f. \tag{1.35}$$

Substituting in the solution for x^* from equation (1.33) yields

$$y^* = \frac{a^2}{2(b+c)} - b\left(\frac{a}{2(b+c)}\right)^2 - c\left(\frac{a}{2(b+c)}\right)^2 - f$$

or $\tag{1.36}$

$$y^* = \left(\frac{a^2}{4(b+c)^2}\right)(2(b+c) - b - c) - f.$$

The simplified reduced-form solution is

$$y^* = \frac{a^2}{4(b+c)} - f. \tag{1.37}$$

We conclude this modeling exercise with a discussion of the comparative statics of the solutions for x^* and y^*. There are four sets of possible comparative static derivatives: with respect to the parameters a, b, c, and f. We shall only go through the case of changes in the parameters f and c. The derivatives of x^* and y^* with respect to f are

$$\frac{\partial x^*}{\partial f} = 0 \tag{1.38}$$

and

$$\frac{\partial y^*}{\partial f} = -1. \tag{1.39}$$

Thus, increases in fixed cost directly reduce net benefits but have no effect on the economic agent's choice of activity level.

An increase in c represents an increase in the marginal cost of the economic activity.[15] The two comparative static partial derivatives for this case are derived using the quotient rule:

$$\frac{\partial x^*}{\partial c} = \frac{-a}{2(b+c)^2} < 0 \tag{1.40}$$

and

$$\frac{\partial y^*}{\partial c} = \frac{-a^2}{4(b+c)^2} < 0. \tag{1.41}$$

We therefore conclude that an increase in the marginal economic cost of an activity reduces both the level of the activity and the net benefit that results from that activity. Note that this mathematical economic result is moderately general: the result applies to any economic activity that has quadratic benefit and cost functions. The result does not, however, necessarily carry over to other benefit and cost functions. Later in the text we will develop tools that allow us to analyze more general specifications of benefits and costs.

1.4   VALUE FUNCTIONS AND THE ENVELOPE THEOREM

We conclude this chapter by presenting an introduction to value functions and the envelope theorem. Much of the analysis in this book is comparative statics; value functions and the envelope theorem facilitate comparative static analysis of optimization problems.

1.4.1   Three Kinds of Endogenous Variables

In optimization models there are three kinds of endogenous variables. The first kind is choice variables. A choice variable is a variable whose value is chosen directly by an economic agent. The second kind is the variables whose values the agent is trying to maximize or minimize. The third kind is other variables whose values depend on the values

---

[15]Since $MC = 2cx$, an increase in c will raise the height and slope of the marginal cost curve for any level of x.

of the choice variables. Economists use comparative static analysis to examine the effect of changes in the values of parameters on the equilibrium values of all three kinds of endogenous variables, but the first and second kinds are of particular importance.

Consider two examples that show up in various ways in later chapters. Firms choose technologies and combinations of quantities of inputs in those technologies to maximize profits. An endogenous choice variable, such as the quantity of labor to employ, will affect the level of profit, which is therefore also endogenous. A second example is a consumer allocating income among the purchases of several goods in order to maximize utility. Again, both the choices of quantities consumed and the resulting level of utility are endogenous.

Value functions and the envelope theorem facilitate comparative static analyses of the equilibrium values of variables that economic agents are trying to maximize or minimize. We will provide a detailed exposition and proof of this theorem in Chapter 13 and present several applications in Chapter 14. In the present chapter we will present an introduction to value functions and the envelope theorem. Throughout the book we will use value functions and the envelope theorem to conduct comparative static analyses in increasingly complex applications.

1.4.2 A Simple Value Function

A value function specifies the maximal (or minimal) value of an endogenous variable as a function of the parameters that determine the optimal values for an agent's choice variables.

Consider again the general maximization of net benefits in Section 1.3.2. First we rewrite equation (1.26) to include a parameter, θ, in addition to the choice variable, x, in the net benefit function:

$$y = N(x; \theta) = B(x; \theta) - C(x; \theta), \tag{1.42}$$

in which $N(x; \theta)$ is the net benefit that an agent obtains by choosing a value for x, given the value of a parameter, θ. The agent's *choice variable* is x. The agent's *objective* is to choose the value for x that will maximize the value of net benefit, $y = N(x; \theta)$. Accordingly, we call $N(x; \theta)$ in (1.26) the agent's **direct objective function:** an objective function that defines the value of an endogenous variable that an agent wants to maximize or minimize as a function of a variable (or variables) whose value(s) the agent chooses.

A **value function** specifies the *optimal* value for an agent's objective function in terms of the parameters that define the environment in which the agent makes choices. In the example in equation (1.42) the agent's choice of a value for x to maximize the net benefit, $N(x; \theta)$, depends on the value of θ. Let $x^* = x^*(\theta)$ be the value for the choice variable x that will maximize $N(x; \theta)$, given the value of θ. Substituting the function $x^*(\theta)$ for x in equation (1.42), we obtain the following value function:

$$y^*(\theta) = N(x^*(\theta); \theta) = B(x^*(\theta); \theta) - C(x^*(\theta); \theta). \tag{1.43}$$

The expression in (1.43) states directly in terms of θ the optimal (in this example, the maximal) value for the net benefit that the agent can obtain.

The value function in equation 1.43 is also called an **indirect objective function.** The arguments of the indirect objective function are the parameters of the agent's objective function. The agent's choice variables do not appear in the indirect objective function because the definition of that function requires that the choice variable is evaluated at its optimal value. Consequently, the indirect value function specifies the *maximal* value of the objective function in terms of the parameters.

Using the specific functional form example from Section 1.3.2, from equation (1.31) we may define

$$y = N(x; a, b, c, f) = ax - bx^2 - cx^2 - f \tag{1.44}$$

as the agent's *direct objective function* because it expresses the value of net benefit, N, as a direct function of the choice of a value for x. In equation (1.44), x is the agent's choice variable, and y is the variable that is the object of the agent's choice. The parameters are a, b, c, and f.

The agent's ultimate interest is in maximizing the value of the variable y, or equivalently the function N, which is done by choosing the optimal value of x based on the values of the parameters a, b, c, and f. Consequently, there is an indirect relationship between the values of the parameters and the maximal value that the agent can achieve for the variable y. We can specify this relationship by substituting for x in the agent's objective function the optimal value for x as a function of the parameters.

We know from our work in Section 1.3.2 that the value of x that will maximize the agent's net benefit is

$$x^* = a/2(b+c). \tag{1.45}$$

For the present example, only the parameters a, b, and c affect the agent's optimal choice of a value for x. If the values of those parameters change, the agent's choice of a value for x will change, and consequently the value for y will change. Although the value of the parameter f does not affect the agent's choice of a value for x, the value of f does affect the maximal value of y that the agent can achieve.

Substituting in the objective function (1.44) the optimal choice for x as defined by (1.45), we may write

$$y^* = N(x^*(a, b, c, f); a, b, c, f), \quad \text{or}$$
$$y^* = ax^* - bx^{*2} - cx^{*2} - f, \quad \text{or} \tag{1.46}$$
$$y^* = a[a/2(b+c)] - b[a/2(b+c)]^2 - c[a/2(b+c)]^2 - f, \quad \text{or,}$$

after simplification,

$$y^* = a^2/4(b+c) - f.$$

In equation (1.46) the variable y^* is the *maximal* value that the agent can obtain for N, given the values of the parameters a, b, c, and f.

The last line in (1.46) is the *indirect objective function* or *value function*. The arguments of the indirect objective function are the parameters of the agent's objective function. The agent's choice variables no longer appear in the indirect objective function because the definition of that function requires that the choice variable is evaluated at its optimal value and is substituted for by the solution that specifies x^* as a function of the parameters. Consequently, the indirect value function specifies the *maximal* value of the objective function in terms of the parameters.

A critical distinction between the direct objective function and the associated value function is this: the parameters of the direct objective function are the *arguments* of the value function. We can see this by examining the second and the fourth lines in equation (1.46). The second line is in the form of the direct objective function, whose argument is x, and whose parameters are a, b, c, and f. To obtain the fourth line, we substitute the optimal choice function, $x^* = a/2(b+c)$ for the argument x. The result is the indirect objective function in the fourth line. The arguments of the indirect objective function are the parameters a, b, c, and f; the choice variable x has been replaced by its optimal value as a function of those parameters.

1.4.3 The Envelope Theorem

The **envelope theorem** states that the marginal effect of a parameter on the *optimal* value of the agent's objective function is equal to the partial derivative of the agent's direct objective function with respect to that parameter, with that partial derivative evaluated at the optimal values of the choice variables. Even though a change in the value of a parameter will cause a change in the values of the choice variables, we can ignore those changes when evaluating the effect on the optimal value of the objective function. The power of the envelope theorem is that it enables us to ignore the fact that (in general) a change in the value of a parameter will cause the agent to change the value of the choice variable.

The optimal value that the agent can obtain for a general net benefit function $N(x; \theta)$ depends on the parameter θ in two ways: *The function N depends on θ directly, and N depends on the choice of x, which, in turn, depends on the value of θ.* In symbols, we have:

$$y^*(\theta) = N(x^*(\theta); \theta) = B(x^*(\theta); \theta) - C(x^*(\theta); \theta). \tag{1.47}$$

Suppose we want to calculate the marginal effect of the parameter θ on the maximal value that the agent can obtain for y. The direct effect is that θ is a parameter of the direct objective function and therefore has an effect on the function N. The indirect effect is that changes in θ affect the choice of x, which in turn affects the value of the function N. The total effect of a change in θ is simply the sum of the indirect and direct effects. Using the chain rule for the indirect effect, we obtain the total effect as:

$$\partial y^*/\partial \theta = (\partial N/\partial x^*)(\partial x^*/\partial \theta) + \partial N/\partial \theta, \tag{1.48}$$

in which all derivatives are evaluated at the value of $x = x^*$.

The term $\partial N/\partial x^*$ on the right side of equation (1.51) is equal to zero: we know this because $\partial N/\partial x = 0$ is the first-order condition for finding the value of x that gives a maximum of N. Therefore, we have:

$$\partial y^*/\partial \theta = \partial N/\partial \theta \quad \text{when } x = x^*. \tag{1.49}$$

Equation (1.49) is the **envelope theorem.** The theorem states that the marginal effect on y^*, which is the maximal value of N, of a change in a parameter is equal to the marginal effect of the parameter on the *direct objective function* N, ignoring the marginal effect of θ on the agent's optimal choice of the value for the decision variable, x. The first-order condition ensures that small changes in the value of the choice variable will not change the optimal value of the objective function and allows us to ignore the indirect effect.

To illustrate the envelope theorem and the relationship between direct and indirect objective function derivatives let us return to our specific function example:

$$y = N(x; a, b, c, f) = ax - bx^2 - cx^2 - f. \tag{1.50}$$

We found in equation (1.46) the indirect objective function by substituting in the solution for $x^* = a/2(b + c)$:

$$y^* = N(x^*(a, b, c), a, b, c, f) = a^2/4(b + c) - f. \tag{1.51}$$

The envelope theorem tells us that, for any parameter, the effect of the parameter on the optimal value of the variable that the agent is optimizing is equal to the partial derivative of the agent's direct objective function with respect to that parameter. Put another way, the derivative of the indirect objective function (1.51) with respect to the parameter has the same value as the partial derivative (holding x constant) of the direct objective function (1.50) with respect to the parameter. To illustrate, consider the effect of a change in the parameter c. The derivative of the direct objective function (1.50) with respect to c is

$$\partial y/\partial c = -x^2, \text{ which, when evaluated at } x = x^* = a/2(b+c), \text{ is}$$
$$\partial y/\partial c = -a^2/4(b+c)^2 < 0. \tag{1.52}$$

As stated by the envelope theorem we get the same result from the derivative of the indirect objective function (1.51):

$$\partial y^*/\partial c = -a^2/4(b+c)^2 < 0. \tag{1.53}$$

In cases where we can solve for x^* and y^*, the envelope theorem assures us that we can work with either the direct or indirect objective functions in evaluating comparative statics results. The envelope theorem, however, also allows us to find comparative statics results even when we are unable to solve for x^* and y^*. Consider a net benefit function

$$y = N(x; \theta) = B(x) - \theta C(x), \tag{1.54}$$

where θ is a parameter that affects costs. Since the functions $B(x)$ and $C(x)$ aren't specified, it is not possible to find explicit solutions for x^* and y^*. Nevertheless, by the envelope theorem we know that the derivatives of the direct and indirect objective functions are equal. Thus, $\partial y^*/\partial \theta = -C(x) < 0$ and increases in θ reduce the maximal value of y^*.

SUMMARY

This chapter has introduced the central elements of an economic model and how these elements can be expressed mathematically. We have also seen our first set of mathematical tools: the use of derivatives for maximization and of partial derivatives for making economic predictions. These fundamental tools, and the principles derived in the benefit-and-cost model, will show up repeatedly in economic contexts. Next, we will develop several economic applications using only the basic mathematical and economic concepts developed in this first chapter.

Appendix to Chapter 1: Calculus Review

1A.1   INTRODUCTION

Most formal mathematical economic analysis uses calculus. In this appendix we provide a brief review of the calculus tools that are used in this text. Students with no background in calculus will find basic rules and formulas in this appendix but may also want to consult an introductory calculus text for fuller explanations. Students with one or more semesters of work in calculus might use this appendix as a reference for formulas. All students should find this appendix useful as an introduction to the economic uses of calculus.

In this appendix we cover three mathematical topics from differential calculus. We start with first derivatives, or rates of change, of a function. We give both a general definition and a set of rules for finding derivatives for frequently encountered functions. We next extend this analysis to second- and higher-order derivatives of single variable functions. Our final mathematical topic is partial derivatives, or changes in a function with more than one right-hand-side variable.

We also show how these mathematical tools have immediate applications in economic analysis. We cover three economic applications of basic calculus: the concavity and convexity of functions, the concept of marginal economic analysis, and the elasticities of economic functions.

1A.2   FIRST DERIVATIVES

A **function** is a rule that specifies the relation between two variables, y and x. The set of possible values for the independent variable x is called the **domain,** and the set of possible values for the dependent variable y is called the **range.** A function $y = f(x)$ specifies a unique value in the range of y for each value in the domain of x. In economics the domain and range are most commonly real (as opposed to imaginary) valued.[1] The functional relation between x and y may also include parameters. We will use a, b, c, etc., to denote parameters that affect the functional relation between the variables x and y. Some examples of functions include:

$y = a$ (constant function)
$y = ax$ (linear function)
$y = ax^2 + bx + c$ (quadratic function)
$y = a_0 + a_1x + a_2x^2 + a_3x^3 + \cdots + a_nx^n$ (polynomial function of degree n)
$y = ae^x$ (exponential function)
$y = a \ln x$ (logarithmic function).

(1A.1)

---

[1] In many economic applications, variables, such as prices and quantities, must be nonnegative. Thus, in some cases the domain and range may be limited to nonnegative real numbers.

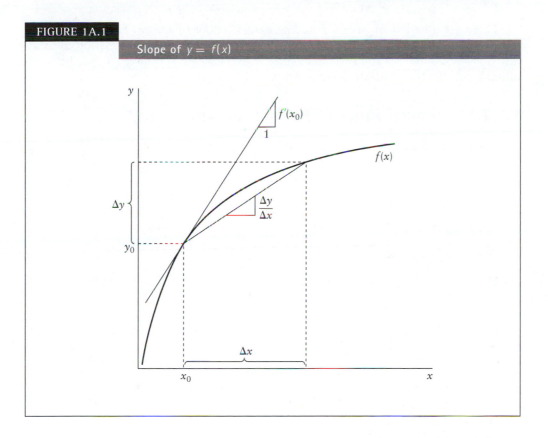

The slope of a function at a point $y_0 = f(x_0)$ is the slope of the line that is tangent to the function $y = f(x)$ at the point (x_0, y_0). Let Δy and Δx denote the changes in the variables y and x. The approximate value of the slope of a function at a point (x_0, y_0) is given by

$$\frac{\Delta y}{\Delta x} = \frac{f(x_0 + \Delta x) - f(x_0)}{\Delta x}. \tag{1A.2}$$

This slope is approximate because a tangent line slope is defined at a single point (x_0, y_0) (where x and y are fixed and unchanging), whereas the formula in (1A.2) allows x and y to vary away from the point (x_0, y_0).

Figure 1A.1 shows the difference between the slope, denoted by $f'(x_0)$, of a tangent line and the slope value given by equation (1A.2). The degree of discrepancy between the tangent line slope and the slope formula will depend on the magnitude of Δx. As Δx becomes smaller, the slope formula and the tangent line slope converge on the same value.[2]

We now define a derivative. The **derivative** of a function at a point $y_0 = f(x_0)$ is the rate of change, or slope of the tangent line, of $y = f(x)$ at the point (x_0, y_0). The value of the derivative is given by the limit of the slope function and is denoted by either dy/dx or $f'(x)$. We have

$$\frac{dy}{dx} \equiv f'(x) \equiv \lim_{\Delta x \to 0} \frac{f(x_0 + \Delta x) - f(x_0)}{\Delta x}. \tag{1A.3}$$

If this limit exists then the function is said to be **differentiable**.

---

[2]Only when the tangent line slope is the same for different values of x and y, that is, the function is linear, will the slope formula be exact.

There are a variety of rules for finding derivatives of functions. Some of these are general rules for finding derivatives when the function $f(x)$ is a combination of other functions and some are rules for finding derivatives when $f(x)$ takes a specific functional form. We start with the general rules.

1A.2.1   General Rules of Differentiation

1. A constant times a function

   Let $y = f(x) = a \cdot g(x)$, then $\dfrac{dy}{dx} = f'(x) = a \cdot g'(x)$.

2. Sum of two functions

   Let $y = f(x) = g(x) + h(x)$, then $\dfrac{dy}{dx} = f'(x) = g'(x) + h'(x)$.

3. Product of two functions

   Let $y = f(x) = g(x) \cdot h(x)$, then $\dfrac{dy}{dx} = f'(x) = g'(x) \cdot h(x) + g(x) \cdot h'(x)$.

4. Quotient of two functions

   Let $y = f(x) = \dfrac{g(x)}{h(x)}$, then $\dfrac{dy}{dx} = f'(x) = \dfrac{g'(x) \cdot h(x) - g(x) \cdot h'(x)}{[h(x)]^2}$.

5. Chain rule

   Let $y = g(z)$ and $z = h(x)$ so that $y = f(x) = g(h(x))$, then

   $$\frac{dy}{dx} = f'(x) = \frac{dy}{dz}\frac{dz}{dx} = g'(h(x)) \cdot h'(x).$$

   Alternatively, we can write the chain rule as the following:

   Let $y = f(g(x))$ then $\dfrac{dy}{dx} = f'(g(x)) \cdot g'(x)$.

1A.2.2   Rules of Differentiation for Specific Functional Forms

1. Constant function

   Let $y = f(x) = ax^0 = a$, then $\dfrac{dy}{dx} = f'(x) = 0$.

2. Power function

   Let $y = f(x) = ax^b$, where $b \neq 0$, then $\dfrac{dy}{dx} = f'(x) = abx^{b-1}$.

3. Polynomial functions

   Let $y = f(x) = a_0 + a_1x + a_2x^2 + a_3x^3 + \cdots + a_nx^n$, then (combining the rules for sums of functions, constant functions, and power functions)

   $$\frac{dy}{dx} = f'(x) = a_1 + 2a_2x + 3a_3x^2 + \cdots + na_nx^{n-1}.$$

4. Logarithmic functions

   a. Logarithm of base a

   Let $y = f(x) = \log_a g(x)$, then $\dfrac{dy}{dx} = f'(x) = \dfrac{g'(x)}{g(x)\ln a}$.

   b. Natural Logarithm

   Let $y = f(x) = \ln g(x)$, then $\dfrac{dy}{dx} = f'(x) = \dfrac{g'(x)}{g(x)}$.

5. Exponential functions

a. General

Let $y = f(x) = a^{g(x)}$, then $\dfrac{dy}{dx} = f'(x) = g'(x)a^{g(x)} \ln a$.

b. Base e

Let $y = f(x) = e^{g(x)}$, then $\dfrac{dy}{dx} = f'(x) = g'(x)e^{g(x)} \ln e = g'(x)e^{g(x)}$.

These rules (often in combination) cover all the derivatives that you will see in this text. Additional rules, such as for the derivatives of trigonometric functions (which are seldom used in economics) can be found in any introductory calculus text.

1A.2.3 Examples of Derivatives

Now let us present some examples of how to use the rules.

1. $y = f(x) = 5x^3 + 3x^2 + 2x + 1$
We use the polynomial rule to get $\dfrac{dy}{dx} = f'(x) = 15x^2 + 6x + 2$.

2. $y = f(x) = \ln(ax - b)$
We use the log rule to get $\dfrac{dy}{dx} = f'(x) = \dfrac{a}{ax - b}$.

3. $y = f(x) = e^{(3x^2 - 5)}$
We use the exponential rule to get $\dfrac{dy}{dx} = f'(x) = 6xe^{(3x^2 - 5)}$.

4. $y = f(x) = x^{0.5} \ln(x + 2)$
We use the power, log, and product rules to get
$$\frac{dy}{dx} = f'(x) = 0.5x^{-0.5} \ln(x + 2) + x^{0.5} \frac{1}{x + 2}.$$

5. $y = f(x) = \dfrac{x^2 + 1}{x + 3}$
We use the power and quotient rules to get
$$\frac{dy}{dx} = f'(x) = \frac{2x(x + 3) - 1(x^2 + 1)}{(x + 3)^2} = \frac{x^2 + 6x - 1}{(x + 3)^2}.$$

6. $y = f(z) = z^a$, $z = g(x) = 2x^2 - 3$
We use the power and chain rules to get
$$\frac{dy}{dx} = f'(z)g'(x) = az^{a-1}4x = 4a(2x^2 - 3)^{a-1}x.$$

You should notice that the value of the derivative of a function depends on both the value of the independent variable (the point at which the derivative is evaluated) and on the values of any parameters of the function. Further examples are given in the problems at the end of this appendix.

First derivatives play an important role in economic optimization problems. The use of first derivatives to find the maximum and minimum points of a function is discussed in Chapter 1.

1A.3 SECOND- AND HIGHER-ORDER DERIVATIVES

The first derivative of a function gives the slope of the function at a specific point. The **second derivative** of a function is the derivative of the first derivative. Thus, the second derivative of a function $f(x)$ can be interpreted as the slope of the function $f'(x)$ that is defined by the first derivative. Alternatively, the first derivative gives the slope of the original function and the second derivative is the rate of change in the slope of the original function.

We have already seen the notation that if $y = f(x)$ then the first derivative is $dy/dx \equiv f'(x)$. The second derivative of a function is the derivative of the first derivative and is written as

$$\frac{d}{dx}\left(\frac{dy}{dx}\right) \equiv \frac{d^2y}{dx^2} \equiv f''(x). \qquad \text{(1A.4)}$$

Finding second derivatives is relatively easy. Once the function, $f'(x)$, for the first derivative is known, we simply take the derivative of this function.

For example, suppose that $y = f(x) = ax^3 + bx^2 + cx + 3$. The first and second derivatives are

$$\frac{dy}{dx} = f'(x) = 3ax^2 + 2bx + c$$

$$\frac{d^2y}{dx^2} = \frac{d}{dx}\left(\frac{dy}{dx}\right) = f''(x) = 6ax + 2b. \qquad \text{(1A.5)}$$

Higher-order derivatives (which are seldom used in economics) are defined in an analogous fashion. The nth derivative of a function $y = f(x)$ is denoted by d^ny/dx^n and is found by taking the derivative of $f(x)$ n times.

1A.4 PARTIAL DERIVATIVES

When we defined functions and derivatives we wrote our results as if x were the only variable, yet many of our equations contained parameters, e.g., a, b, and c, that might take on different values. Since the parameters could vary (could be considered as variables) what we were really finding above were **partial derivatives**: the effect of a change in x when other "variables" were held constant. In this section we formally define partial derivatives.

Let $y = f(x, z)$ be a function of two variables. This function has two variables on the right-hand side. We can take the derivative with respect to either of these variables while treating the other variable as a constant. The notation for these partial derivatives is

$$\frac{\partial y}{\partial x} \equiv \frac{\partial f(x, z)}{\partial x} \equiv f_x(x, z)$$

$$\frac{\partial y}{\partial z} \equiv \frac{\partial f(x, z)}{\partial z} \equiv f_z(x, z) . \qquad \text{(1A.6)}$$

Note that: (1) for partial derivatives we use "∂" instead of "d," (2) we use a subscript instead of a "prime," because, since the function includes more than one variable, a prime would be ambiguous, and (3) the partial derivatives are generally functions of all of the original variables.

As an example of partial derivatives, suppose that we have a function $y = f(x, z) = 3x + 2z^3 - x^3z^2$. The partial derivatives of this function are

$$\frac{\partial f(x, z)}{\partial x} = f_x(x, z) = 3 - 3x^2z^2$$

$$\frac{\partial f(x, z)}{\partial z} = f_z(x, z) = 6z^2 - 2x^3z . \qquad \text{(1A.7)}$$

Each of the partials is simply a derivative taken with respect to one variable while holding the other variable constant. Although the other variable is held constant, the value of the other variable does affect the value of the partial derivative.

Just as we defined second- (and higher-) order derivatives for a function of one variable, we can also define second-partial or cross-partial derivatives. These higher-order partials measure the changes in the first partial derivatives when one of the variables changes (and the other is held constant). The higher-order partial derivatives of a function $y = f(x, z)$ are found by taking the partial derivatives of the first partial derivatives. Thus, we have

$$\frac{\partial^2 f(x, z)}{\partial x^2} = \frac{\partial}{\partial x}\left(\frac{\partial f(x, z)}{\partial x}\right) = f_{xx}(x, z)$$

$$\frac{\partial^2 f(x, z)}{\partial z^2} = \frac{\partial}{\partial z}\left(\frac{\partial f(x, z)}{\partial z}\right) = f_{zz}(x, z)$$

$$\frac{\partial^2 f(x, z)}{\partial x \partial z} = \frac{\partial}{\partial z}\left(\frac{\partial f(x, z)}{\partial x}\right) = f_{xz}(x, z) \tag{1A.8}$$

$$\frac{\partial^2 f(x, z)}{\partial z \partial x} = \frac{\partial}{\partial x}\left(\frac{\partial f(x, z)}{\partial z}\right) = f_{zx}(x, z).$$

In general, cross-partials are identical: the order in which cross-partials are taken does not affect the value of a cross-partial derivative.[3]

For the specific example of $y = f(x, z) = 3x + 2z^3 - x^3 z^2$ we have the following cross-partials:

$$\frac{\partial^2 f(x, z)}{\partial x^2} = \frac{\partial}{\partial x}\left(\frac{\partial f(x, z)}{\partial x}\right) = f_{xx}(x, z) = -6xz^2$$

$$\frac{\partial^2 f(x, z)}{\partial z^2} = \frac{\partial}{\partial z}\left(\frac{\partial f(x, z)}{\partial z}\right) = f_{zz}(x, z) = 12z - 2x^3$$

$$\frac{\partial^2 f(x, z)}{\partial x \partial z} = \frac{\partial}{\partial z}\left(\frac{\partial f(x, z)}{\partial x}\right) = f_{xz}(x, z) = -6x^2 z \tag{1A.9}$$

$$\frac{\partial^2 f(x, z)}{\partial z \partial x} = \frac{\partial}{\partial x}\left(\frac{\partial f(x, z)}{\partial z}\right) = f_{zx}(x, z) = -6x^2 z.$$

Again note that the values of the cross-partials generally depend on the levels of both variables.

Cross-partial derivatives for functions of more than two variables, such as $y = f(x_1, x_2, x_3, \ldots, x_m)$, are defined in an analogous manner. In economics, partial derivatives and cross-partial derivatives are used in multivariable optimization. Partial derivatives are also used when we wish to find the effects of changes in parameters on the equilibrium values of economic variables.

1A.5 ECONOMIC APPLICATIONS OF ELEMENTARY DIFFERENTIAL CALCULUS

In this section we apply differential calculus to economic analysis. We focus on three concepts that are often used in introductory economic courses: concavity and

---

[3]This is known as Young's Theorem. The theorem only requires that the second partials be continuous.

A Concave Production Possibility Frontier

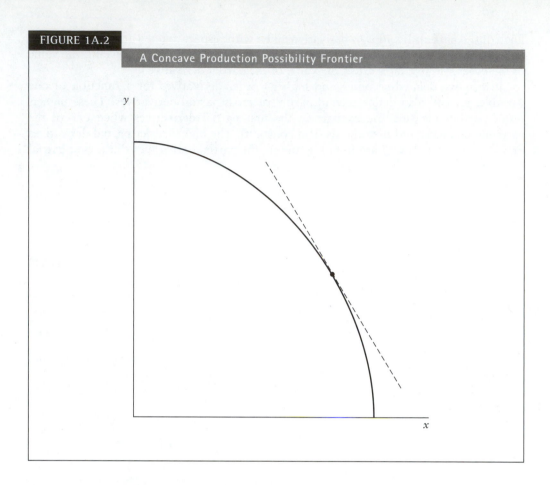

convexity, marginal analysis, and elasticities. We will see that even a brief introduction to calculus has useful economic applications.

1A.5.1 Concavity and Convexity

In economic analysis functions are often referred to as concave or convex. For example, production possibility frontiers are usually assumed to be concave, whereas indifference curves are assumed to be convex. We will first give definitions of concavity and convexity and then show how these definitions can be related to the derivatives of a function.

Let $y = f(x)$ be a function and (x_0, y_0) be any point on the graph of the function. The function is **strictly globally concave (convex)** if for every (x_0, y_0) the tangent line at that point lies everywhere above (below) the function.[4]

Figure 1A.2 shows a production possibility frontier. The frontier indicates the maximum combinations of goods x and y that can be produced given the resources and technology available to an economy, or the maximum production level of y for any given production level of x. The production frontier is strictly concave; all tangent lines lie strictly above the frontier. Figure 1A.3 shows a consumer's indifference curve (an indifference curve is all combinations of x and y that yield the same level of utility for a consumer) for the consumption goods x and y. Here, all tangent lines lie below the graph of the indifference curve, so the curve is convex.

---

[4]More formal definitions for concavity and convexity are given in Chapter 7. These definitions cover the multivariable case, as well as weak (as opposed to strict) and local (as opposed to global) concavity and convexity.

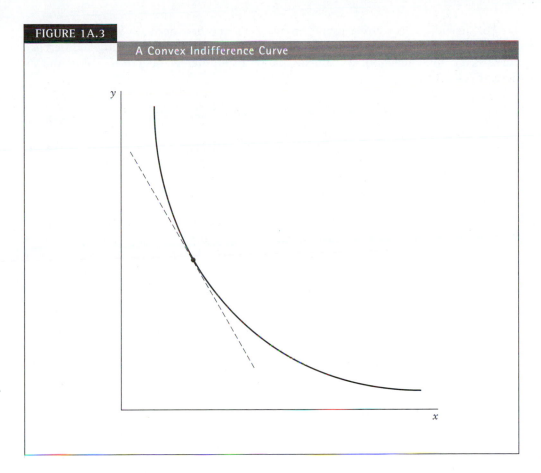

There is a straightforward relationship between the geometric definitions of concavity and convexity and the derivatives of a function. The function $y = f(x)$ is strictly globally concave (convex) if, for all values of x, the second derivative $f''(x)$ is negative (positive). Notice the wording of the definition: the sign of the second derivative is sufficient to determine concavity or convexity. The condition on the derivative is not, however, necessary. A function may have a zero second derivative at some points but still be strictly concave or convex.[5]

Let us apply the definitions to the production possibility frontier and indifference curve examples. Let the equation for the production frontier be $y = f(x)$. The first derivative is negative and increasing in absolute value. Thus, the second derivative is also negative and this is sufficient to establish the function's concavity. For example, the equation for a production possibility frontier in Figure 1A.2 and its first and second derivatives might be

$$y = f(x) = a - bx^2, \quad a > 0 \quad \text{and} \quad b > 0$$

$$\frac{dy}{dx} = f'(x) = -2bx \leq 0$$

$$\frac{d^2y}{dx^2} = f''(x) = -2b < 0.$$

(1A.10)

[5]Try graphing the (strictly convex) function $y = x^4$. Tangent lines always lie below the function, but the second derivative is zero when evaluated at $x = 0$.

For an indifference curve graph of $y = f(x)$, the first derivative is negative, but decreasing in absolute value. Thus, the second derivative is positive and the indifference curve is a convex function. As an example of an indifference curve equation and its derivatives consider

$$y = f(x) = \frac{a}{x}, a > 0$$

$$\frac{dy}{dx} = f'(x) = \frac{-a}{x^2} < 0 \qquad (1A.11)$$

$$\frac{d^2 y}{dx^2} = f''(x) = \frac{2a}{x^3} > 0.$$

Concavity and convexity are useful ways of characterizing economic relationships. We will also see (in Chapters 1 and 7) that these concepts are related to the conditions under which an extreme point of a function can be shown to be a maximum or a minimum.

1A.5.2 Marginal Analysis

Marginal analysis is central to microeconomics. In economics the term marginal refers to the change in one economic variable caused by the change in another economic variable. This is, of course, the definition of a derivative! We will see many examples of marginal analysis in this text. Let us work here with the example of a firm's marginal cost curve.

Let the variable x be a firm's output level and y be the firm's total cost of production. As a simple example let the function for total cost be

$$y = f(x; a, b, c) = ax^2 + bx + c, \qquad (1A.12)$$

where a, b, and c are positive parameters. These parameters summarize the effects of other economic factors, such as input prices and technology, that affect a firm's production costs. The equation uses the semicolon notation in $f(x; a, b, c)$ to draw a distinction between the economic variable (output) and the parameters (a, b, c) that affect the level of total cost.

The firm's marginal cost is the partial derivative of total cost with respect to output (with the parameters held constant). Marginal cost (MC) is

$$MC = \frac{\partial f(x; a, b, c)}{\partial x} = 2ax + b. \qquad (1A.13)$$

The firm's marginal cost depends on both its level of output and on the levels of the parameters a and b (total cost depends on c, but marginal cost does not).

In an economic application we might be interested in how the firm's marginal cost changes when output changes. We might also be interested in how marginal cost changes when the parameters change. To find these effects we would take the partial derivatives of marginal cost with respect to output and the parameters. First, with respect to output we find

$$\frac{\partial MC}{\partial x} = \frac{\partial^2 f(x; a, b, c)}{\partial x^2} = 2a. \qquad (1A.14)$$

Increases in output increase marginal cost: the MC curve slopes upward so that as output increases the marginal cost of production also increases.

For changes in the parameters we take the partial (cross-partial) derivatives of marginal (total) cost with respect to a, b, and c. These are

$$\frac{\partial MC}{\partial a} = \frac{\partial^2 f(x; a, b, c)}{\partial x \partial a} = 2x$$

$$\frac{\partial MC}{\partial b} = \frac{\partial^2 f(x; a, b, c)}{\partial x \partial b} = 1 \qquad \textbf{(1A.15)}$$

$$\frac{\partial MC}{\partial c} = \frac{\partial^2 f(x; a, b, c)}{\partial x \partial c} = 0.$$

These cross-partials have a somewhat different interpretation than the second partial with respect to output. Since we graph marginal cost with x on the horizontal axis, the partial of marginal cost with respect to x gives the slope of the marginal cost curve, while the partial with respect to a parameter gives the shift in the marginal cost curve when there is a change in the parameter.

This distinction between partials with respect to a variable and partials with respect to parameters is illustrated in Figure 1A.4. Consider an initial level a_0 of the parameter a and an initial output level x_0. At the initial output level a change in output (holding a, b, and c constant) will increase marginal cost by $2a_0$. This is just the slope of the MC curve. An increase in the parameter a to a new level a_1 (holding b and c constant) rotates the MC curve upward around the fixed vertical intercept. At the original output level x_0 the increase in marginal cost, for a one-unit change in a, is $2x_0$.

This example illustrates a general relation between mathematical and graphical analysis. The function $y = f(x; a, b, c)$ is five-dimensional with x, a, b, and c as independent variables and y as the (fifth dimension) dependent variable. Yet, since we can't

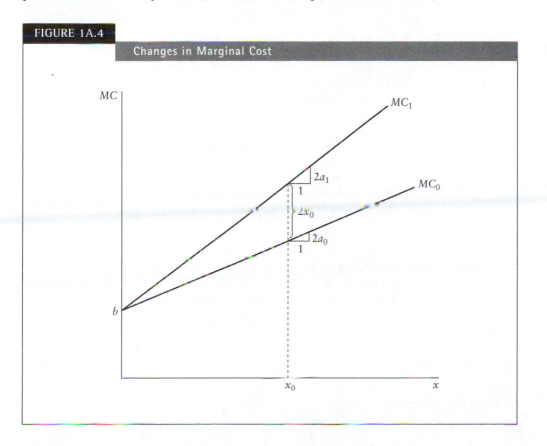

FIGURE 1A.4

Changes in Marginal Cost

graph this five-dimensional relationship, we graph the two-dimensional relation between the variables x and y holding the other right-hand-side terms constant. Whenever we graph a multidimensional function in two dimensions there will be a difference between a change in the horizontal axis variable, which causes a movement along the graphed curve, and a change in any other (held-constant) variable, which causes a shift in the graphed curve. In general, the magnitudes of both types of changes will depend on the levels of all of the right-hand-side variables.

1A.5.3   Elasticities

In many economic applications the rate of change of a function depends on the units in which variables are measured. Suppose, for example, an economic application looks at the quantity of corn demanded as a function of the price of corn. The slope of this demand curve will depend on whether quantity is measured in pounds or bushels and on whether price is measured in dollars or pennies. To avoid this dependence on units of measurement economists use elasticities. The **elasticity** of one variable, y, with respect to a second variable, x, is defined as the percentage change in y for a percentage change in x. Because elasticities are in terms of percentage changes they are invariant with respect to the actual units of measurement.

Let $y = f(x)$ specify the relationship between two economic variables. Then the elasticity (denoted by ε) of y with respect to x is defined as[6]

$$\varepsilon = \frac{dy/y}{dx/x} = \frac{dy}{dx}\frac{x}{y}, \tag{1A.16}$$

where the change in a variable divided by the initial level of the variable is, when multiplied by 100, the percentage change in the variable. Since the change in a variable and the level of a variable are measured in the same units, the actual units of measurement drop out of the formula.

In this section we will examine demand elasticity in three ways. First, we will derive the general relationship between marginal revenue and demand elasticity. Second, we will derive the elasticity formula for a linear demand function. Finally, we will show the relationship between changes in logarithms of variables and elasticity.

Consider a demand equation that specifies quantity demanded as a function of price. Let the quantity demanded, Q, be given by the function $Q = Q(P)$ where P is the price of a product. In this example we use the letter Q twice with two different meanings: first as the level of a variable and second as a label for a functional form that relates quantity demanded to price. This double usage is common in economics. The elasticity of quantity demanded with respect to price is given by

$$\varepsilon = \frac{dQ/Q}{dP/P} = \frac{dQ}{dP}\frac{P}{Q} = Q'(P)\frac{P}{Q}. \tag{1A.17}$$

Since $Q'(P) < 0$ (demand slopes down) the formula above yields a negative value. Economists often avoid this by defining demand elasticity as the absolute value of the formulas in equation (1A.17). Using this positive definition demand elasticity is characterized as inelastic ($|\varepsilon| < 1$), unit elastic ($|\varepsilon| = 1$), or elastic ($|\varepsilon| > 1$).

The elasticity categories can be related to the marginal revenue function. The demand function for a good is $Q(P)$ or how quantity demanded depends on price. Since, however, economists usually graph price on the vertical axis it is common to work with

---

[6]Another common symbol for elasticity is η.

the inverse demand function, $P(Q)$, or how price depends on quantity. Using the inverse demand function we can write total revenue, R, as a function of quantity:

$$R(Q) = P(Q)Q. \qquad \textbf{(1A.18)}$$

The product rule gives the marginal revenue for a change in quantity as

$$MR = R'(Q) = \frac{dP}{dQ}Q + P. \qquad \textbf{(1A.19)}$$

Note that there are two sources of change in the marginal revenue equation. The first term is the (negative) rate of change in price times the original quantity and the second term is the price at which the additional unit of output is sold.

We now relate marginal revenue to elasticity. Rearranging the terms in the marginal revenue equation yields

$$MR = P\left(\frac{dP}{dQ}\frac{Q}{P} + 1\right)$$
$$= P\left(\frac{1}{\varepsilon} + 1\right). \qquad \textbf{(1A.20)}$$

The elasticity in this equation is negative, so we get the immediate result that, for finite demand elasticities, marginal revenue is less than price. If we instead use the absolute value of elasticity we get

$$MR = P\left(1 - \frac{1}{|\varepsilon|}\right). \qquad \textbf{(1A.21)}$$

Thus, the marginal revenue from an extra unit of output is positive, zero, or negative as demand is elastic, unit elastic, or inelastic.

Our second example of elasticity uses a linear demand function. Suppose that the demand equation for a good is $Q = (a - P)/b$ so that the inverse demand equation for the good is $P = a - bQ$. The absolute value elasticity formula for this equation gives

$$|\varepsilon| = \left|\frac{dQ}{dP}\right|\frac{P}{Q} = \left|\frac{-1}{b}\right|\frac{P}{Q} = \frac{1}{b}\frac{P}{\left(\frac{a-P}{b}\right)} = \frac{P}{a - P}. \qquad \textbf{(1A.22)}$$

The demand elasticity therefore depends on the intercept of the inverse demand curve and on the price being charged. Note in particular that two linear demand curves, with different slopes but with the same price axis intercept will, at a given price, have the same demand elasticity. Figure 1A.5 shows the relationship between marginal revenue and elasticity for a linear demand curve. Since $TR = PQ = aQ - bQ^2$, $MR = a - 2bQ$. Thus, the MR curve has the same intercept as the demand curve but twice the slope. At the midpoint of the demand curve, where $P = a/2$, elasticity will equal one and MR will equal zero. Above this midpoint, demand is elastic and MR is positive. Below the midpoint, demand is inelastic and MR is negative.

For the last elasticity example we examine the relationship between changes in logarithms and elasticity. Suppose we have a function $u = \ln x$. We can write the changes in the variables and the function as

$$\frac{du}{dx} = \frac{d}{dx}(\ln x) = \frac{1}{x}. \qquad \textbf{(1A.23)}$$

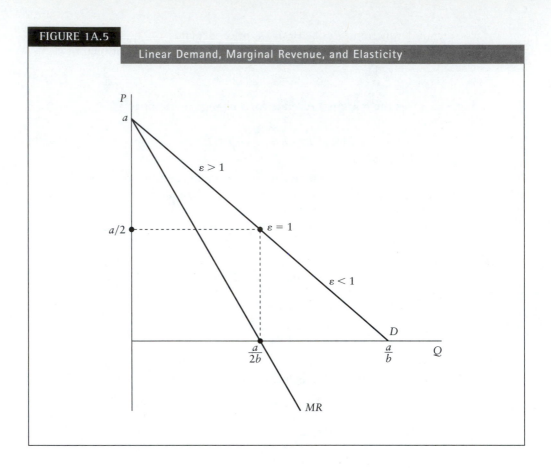

Linear Demand, Marginal Revenue, and Elasticity

Multiplying through by dx, the change in x, gives

$$du = d\ln x = \frac{dx}{x}. \qquad \textbf{(1A.24)}$$

Thus, we have the result that the percentage change in a variable is equal to the change in the logarithm of the variable.

Economists often work with economic relationships that are linear in the logarithms of variables, such as

$$\ln y = a \ln x. \qquad \textbf{(1A.25)}$$

Let $v = \ln y$ and $u = \ln x$. We then have

$$a = \frac{dv}{du} = \frac{d\ln y}{d\ln x} = \frac{dy/y}{dx/x} = \varepsilon. \qquad \textbf{(1A.26)}$$

Thus, if an economic equation is linear in logarithms then the elasticity of y with respect to x is the coefficient of $\ln x$.

This result is more useful than it might seem at first glance. Suppose we have the following equation:

$$y = 7x^3. \qquad \textbf{(1A.27)}$$

Since the equation can also be written as

$$\ln y = \ln(7x^3) = \ln 7 + \ln x^3 = \ln 7 + 3\ln x, \qquad \textbf{(1A.28)}$$

it is immediately apparent that the elasticity of y with respect to x is 3.

Problems

A.1 Find the first and second derivatives of the following functions:

(a) $y = f(x) = 3x^5 - 2x^{0.2}$

(b) $y = f(x) = \dfrac{\ln(x+1)}{x}$

(c) $y = f(x) = x^{-3} + ax^2 - e^{2x}$

(d) $y = f(x) = 2x^2(x^{-2} + ax^{0.5})$

(e) $y = f(x) = \dfrac{x^2 - 2}{4x}$

(f) $y = f(x) = (x^2 - ax)(x^3 - bx^c)$

(g) $y = f(g(x)) = a[g(x)]^2, g(x) = (x+1)$

(h) $y = f(g(x)) = a \ln[g(x)], g(x) = (x^2 + 1)^2$

(i) $y = f(x) = ax^b - cx^3$

(j) $y = f(x) = \dfrac{ax}{bx + c}$

(k) $y = f(x) = (ax^b + 3)(cx^3 - 2)$

(l) $y = f(x) = a(x+4)e^{2x}$

A.2 Find all first and second partial derivatives for the following functions:

(a) $y = f(x, z) = x^2 z^3 - \ln(x+1) - e^{3z}$

(b) $y = f(x, z) = (x+5)^a(z-b)^c$

(c) $y = f(x, z) = \ln(x+z) - (xz)^2$

(d) $y = f(x, z) = \dfrac{x^2}{z + x}$

(e) $y = f(x, z) = z^a x^b - 3xz$

(f) $y = f(x, z) = az \ln x + bx \ln z$

(g) $y = f(x, w, z) = x^a w^b z^c - 2x - 3w - 4z^2$

(h) $y = f(x, w, z) = \dfrac{x + w^2}{x + z}$

A.3 Check whether the following functions are concave or convex:

(a) $y = f(x) = x^2 + 3x$, where $x > 0$

(b) $y = f(x) = \ln(x + a)$, where $x > 0$ and $a > 0$

(c) $y = f(x) = -2x^2 - 3x$, where $x > 0$

(d) $y = f(x) = e^{2x+1}$, where $x > 0$

(e) $y = f(x) = x^a$, where $x > 0$ and $a > 0$

(f) $y = f(x) = ax^b$, where $x > 0, a > 0$, and $b < 0$

A.4 For the following cost functions find and graph marginal cost (assume that all parameters are positive):

(a) $y = f(x) = ax^{0.5} + b$

(b) $y = f(x) = a \ln x$

(c) $y = f(x) = ax^3 + bx^2 + cx$

(d) $y = f(x) = ae^{2x}$

(e) $y = f(x) = (x + a)^b$

A.5 For each total cost function in Problem A.4, find the effect of an increase in the parameter a on marginal cost.

A.6 For the following functions write the elasticity of y with respect to x as a function of x:

(a) $y = f(x) = ax^b$

(b) $y = f(x) = ax + b$

(c) $y = f(x) = \dfrac{a + x}{2x}$

(d) $y = f(x) = \ln(x + 1)$

(e) $y = f(x) = e^x$

An Introduction to Mathematical Economic Applications

2.1 INTRODUCTION

Throughout this text, chapters (such as Chapter 1) that introduce new mathematical tools will be followed by a chapter of economic applications. For the most part, the applications will be self-contained so that some applications may be skipped without a loss of continuity. The application chapters serve multiple purposes: The applications will reinforce your understanding of new mathematical tools and demonstrate how economists use mathematical tools and will (we hope) also be economically interesting in their own right. The primary theme for our first chapter of economic applications is the use of optimization and comparative statics in models of economic decision making. We start with microeconomics applications that draw on economic theory from the fields of industrial organization, labor economics, and public finance. We will first look at decisions made by a labor union, a monopoly supplier of labor to an industry, and examine the results of different possible union goals. The next set of microeconomic applications will cover four different market structures: perfect competition, monopoly, duopoly and oligopoly. We conclude the chapter with our first example of how mathematics is used in macroeconomics: a simple Keynesian model of national income determination.

2.2 LABOR UNIONS

In our first application we will focus on an example from labor economics. We will assume that an industry is completely unionized so that the union is a monopoly supplier of labor. Rather than examining a complex model of how the union would bargain with the firms in the industry, we will simply assume that the union picks a wage rate that firms then accept. Although there is no bargaining over the wage rate, firms do retain the power to choose how many union workers they wish to employ. Suppose that the industry's demand for labor is linear. We can write this linear demand function as

$$w = w_0 - bL, \qquad (2.1)$$

where w is the wage rate, $w_0 > 0$ is the demand intercept, $b > 0$ is the demand slope parameter, and L is the level of employment. The graph for this demand curve, with the

FIGURE 2.1

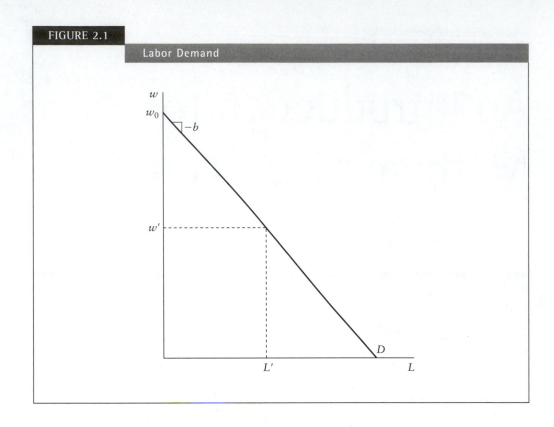

Chapter Two

Labor Demand

quantity of labor that firms wish to employ on the horizontal axis and the wage rate on the vertical axis, is shown in Figure 2.1. The way we interpret our model is that if the union names some wage, w', then the firms respond by employing labor at a level L'. This is shown in Figure 2.1.

The union knows the industry's demand curve for labor and can anticipate how firms will react to different wage demands. The question for the union is what wage level to ask for. To answer this question we must specify the union's goal. Unfortunately, economic theory does not provide any clear answer as to what goal a union might choose to pursue. To understand why unions' goals are ambiguous consider the case where employment is based on a seniority system. More senior workers would tend to favor high wage demands, because even though firms would reduce employment at high wages, senior workers would retain their jobs. Junior workers, on the other hand, would not be employed at high wages and would therefore tend to prefer a lower wage at which they would be employed. Throughout this analysis we will assume that union members who are not employed in the industry can earn an alternative wage of $\overline{w}$ either by finding a job in some other industry or through unemployment compensation.

Since union members have conflicting preferences, we cannot specify any single goal for the union. Instead we can examine what the union's wage demand would be for a range of possible goals. The four particular goals we will investigate are:

1. Maximizing the wage rate
2. Maximizing the level of employment
3. Maximizing the total income of employed workers
4. Maximizing the "rents" earned by employed workers

Maximizing the wage rate is a single-minded goal that ignores the level of employment. Simply maximizing a variable y equal to the wage rate w would give a first derivative of $dy/dw = 1$. A first-order condition that sets $1 = 0$ is obviously problematic. The problem is that the union would pick an infinitely high wage if it were not concerned about its members being employed. If the union needs to pick a wage at which employment would be positive then it would be constrained to a wage of $w_0 - \varepsilon$ (where ε is an arbitrarily small number) and accepts a level of employment approaching zero.

This is called a corner solution, i.e., a solution that lies at the limit of the possible range of values for an economic variable. Solutions that fall somewhere in between the maximum and minimum possible values for a variable are called interior solutions. Corner solutions occur when either (1) there is no solution to the first- and second-order conditions or (2) the first- and second-order conditions can be solved for a local maximum, but the global maximum is at the extreme point of economically feasible solutions. Here, the corner solution is an obviously unrealistic outcome. Nevertheless, corner solutions can be reasonable outcomes in a variety of other economic models.

A second possible goal is employment maximization. To analyze this case consider the level of employment when the wage declines from w_0. The quantity of labor demanded and employment will increase as the wage rate declines. There is a shift, however, when the wage rate reaches $\bar{w}$. Further declines in the wage rate will lead to increases in the quantity of labor demanded, but at wages below $\bar{w}$ union members would prefer to work elsewhere in alternative employment. In fact, at wages below $\bar{w}$ no labor will be supplied to the industry and the level of employment will be zero. The graph of employment as a function of the wage rate is shown in Figure 2.2. From the graph it is immediately evident that employment is maximized at a wage rate of $\bar{w}$. This outcome is also somewhat unrealistic because it implies that there is no wage premium

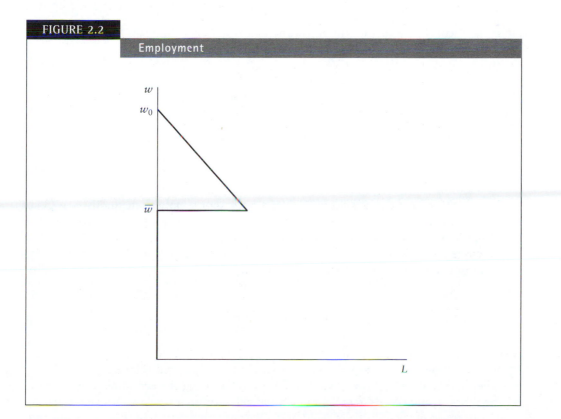

FIGURE 2.2

Employment

from being in the union, i.e., the union members earn the same wage as they could earn elsewhere.

The third possible goal is to maximize the total income of employed workers. Letting I stand for income the union's goal is to maximize

$$I = wL. \tag{2.2}$$

The union chooses the wage rate and firms respond by choosing the level of employment. By rewriting the labor demand curve in equation (2.1) we can represent the level of employment as a function of the wage rate:

$$L = \frac{w_0 - w}{b}. \tag{2.3}$$

This employment equation is the demand equation. It assumes that labor supply is positive, i.e., it assumes that w in equation (2.3) exceeds the alternative wage, $\overline{w}$. For wages below $\overline{w}$, the equation does not hold, instead $L = 0$. This will be important later.

The union's objective can then be written as maximize

$$I = wL = w \left(\frac{w_0 - w}{b} \right). \tag{2.4}$$

The first- and second-order conditions for a local maximum are

$$\frac{dI}{dw} = \frac{w_0 - 2w}{b} = 0$$
$$\frac{d^2 I}{dw^2} = \frac{-2}{b} < 0. \tag{2.5}$$

Since the second-order condition holds, the first-order condition can be solved for the local maximum. Call the wage rate that solves the first-order condition w_3 (since this is the third possible goal). We get

$$w_3 = \frac{w_0}{2}. \tag{2.6}$$

This solution for w_3 assumes

$$\frac{w_0}{2} > \overline{w}, \tag{2.7}$$

so that the wage exceeds the alternative wage, and labor supply to the industry is positive. If the inequality were reversed then labor supply and employment would be zero since union members would choose alternative employment at the wage $\overline{w} > w_3$. In this case income (and employment as well) would be maximized at the wage rate $\overline{w}$, where labor supply and employment are both positive. Formally, we would have to modify our solution for w_3 by writing

$$w_3 = \max \left(\frac{w_0}{2}, \overline{w} \right), \tag{2.8}$$

where the equation is read as w_3 equals the maximum of the two possible wage rates. In the rest of the analysis we will assume that the first argument is larger.

The final goal we will investigate is that of maximizing the economic rents earned by the union members. By rents we mean the value of the excess wages generated by being employed in the industry instead of in alternative employment. For any wage rate

w, each worker employed earns rents of r, where r is defined as

$$r = w - \overline{w}. \tag{2.9}$$

Total rents, R, earned by union members can be represented as

$$R = rL = (w - \overline{w})\left(\frac{w_0 - w}{b}\right) = \frac{(\overline{w} + w_0)w - w^2 - \overline{w}w_0}{b}. \tag{2.10}$$

The first- and second-order conditions for maximizing the rents earned by union workers are

$$\frac{dR}{dw} = \frac{\overline{w} + w_0 - 2w}{b} = 0$$

and

$$\frac{d^2R}{dw^2} = \frac{-2}{b} < 0. \tag{2.11}$$

The first-order condition can be solved for the wage rate, call it w_4, that maximizes rents. Solving gives

$$w_4 = \frac{w_0 + \overline{w}}{2} = w_3 + \frac{\overline{w}}{2}. \tag{2.12}$$

Thus, we can conclude that the wage rate that maximizes rents exceeds the wage rate that maximizes the total income of union members.

Figure 2.3 shows the outcomes for each of the four goals, where the subscript on the wage and labor solutions refers to the goal number. Our model is not determinate,

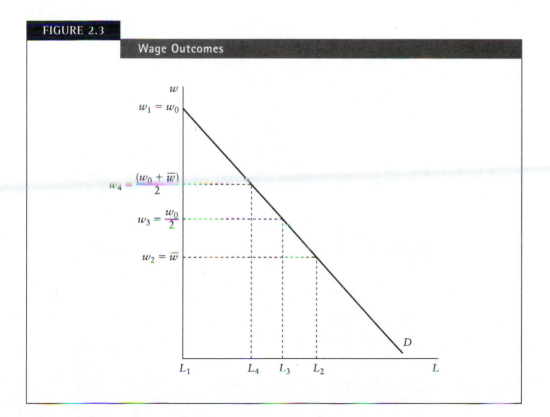

FIGURE 2.3

Wage Outcomes

in the sense that it yields a single solution, but the model does allow us to compare the goals in terms of their wage and employment outcomes. A fuller, but more complex, model might include a description of how union goals are determined, i.e., how the tensions between the conflicting preferences of senior and junior union members are resolved. A fuller model might also include a more realistic model of union and firm bargaining.[1]

2.3 PROFIT MAXIMIZATION: A COMPETITIVE FIRM

In this section we will consider a model of input and output choices by a firm in a perfectly competitive market. In a perfectly competitive market each firm is a price taker. By this we mean that the competitive firm has no control over the price at which it can sell its product, nor does it control the prices that must be paid for inputs. Instead, a competitive firm simply takes the going market prices as given, or exogenous, when making its decisions.

The general structure of this model is that the firm will choose inputs and output to maximize its profits. In our model there will be two inputs, labor and capital, denoted by the variables L and K. Inputs are linked to output, denoted by Q, via a production function. The production function can be represented mathematically as

$$Q = F(L, K).\tag{2.13}$$

For our example we will assume that the production technology takes the specific form of

$$Q = 2\sqrt{LK}$$

or

$$Q = 2L^{0.5}K^{0.5}.\tag{2.14}$$

In the long run the firm can choose all inputs, i.e., both labor and capital are endogenous variables. At this point, however, we don't have the mathematical techniques needed to handle the simultaneous choice of two inputs (these tools will be presented in Chapter 7). We will therefore model only the firm's short-run decision. By definition, the short run is a period of time in which at least one input is fixed. We will consider the usual case where capital is the fixed input. To indicate that capital is exogenously fixed in the short run we will follow a convention that puts a "zero" subscript on economically important exogenous variables.[2] The short-run production function is then written as

$$Q = 2K_0^{0.5}L^{0.5}.\tag{2.15}$$

All profit functions have the general form of total revenue minus total cost. In this case we can begin by writing

$$\Pi = PQ - wL - rK_0\tag{2.16}$$

where P is the price at which output can be sold, w is the wage rate paid for labor, and r is the price of capital. As we mentioned, competitive firms are price takers so P, w, and

---

[1]See the last chapter of this text for a game theory model of sequential bargaining.

[2]Another common convention is to put a bar over exogenous variables.

r are exogenous variables. The profit equation therefore has two endogenous variables, Q and L, and the four exogenous variables. The endogenous variables are not, however, independent. Once either L or Q is chosen the other variable is also chosen, i.e., specified by the production function.

Since there is only one independent endogenous variable, we can choose to write profits as a function of either L or Q. In this application we will write profits as a function of the labor input, and, after solving the model, we will examine how the firm's labor choice also specifies its output choice. Profits as a function of L are

$$\Pi = 2PK_0^{0.5}L^{0.5} - wL - rK_0. \tag{2.17}$$

To find a profit maximum we derive the first-order condition with respect to L:

$$\frac{d\Pi}{dL} = PK_0^{0.5}L^{-0.5} - w = 0. \tag{2.18}$$

This first-order condition is an example of the *marginal benefit equals marginal cost* rule. The first part of the equation indicates how much extra revenue the perfectly competitive firm gains from an increase in labor. This is called the value of the **marginal product of labor.** The w term in the equation represents the marginal cost of hiring an additional unit of labor. Thus, the first-order condition requires that labor be employed up until the point where the marginal revenue from an extra unit of labor equals the marginal cost of hiring that extra unit of labor.

We next check the second-order condition for a profit maximum:

$$\frac{d^2\Pi}{dL^2} = -0.5PK_0^{0.5}L^{-1.5} < 0. \tag{2.19}$$

Since the second-order condition holds we can go ahead and solve the first-order condition for the optimal level of labor. Rearranging the first-order condition gives

$$L^{0.5} = \frac{PK_0^{0.5}}{w}. \tag{2.20}$$

This gives the solution for L as a function of exogenous variables:

$$L^* = \frac{P^2K_0}{w^2}. \tag{2.21}$$

This solution is the firm's demand function for labor.

The comparative statics of the equilibrium level of labor are straightforward. You can easily check that the derivatives with respect to P and K_0 are positive and that the derivative with respect to the wage rate is negative. The economic interpretation is that increases in the product's price or the short-run capital stock will lead to increases in employment. The former variable increases the revenue from an additional unit of output, while the latter increases labor productivity. Finally, increases in the wage rate raise the cost of labor and output and will lead to lower levels of employment.

We can use the solution for labor to construct and analyze the corresponding output decision. To find the level of output as a function of exogenous parameters we substitute the solution L^* into the production function. This gives the firm's supply function

$$Q^* = 2K_0^{0.5}(L^*)^{0.5} = 2K_0^{0.5}\left(\frac{P^2K_0}{w^2}\right)^{0.5} = \frac{2PK_0}{w}. \tag{2.22}$$

Again the comparative static results are simple and straightforward. The firm's supply function is increasing in price and capital stock and decreasing in the wage rate. Even though the comparative static results confirm some obvious economic intuitions, the calculus-based solution for Q^* provides some insights that are not obvious from simple graphical analysis. For example, changes in the exogenous variables affect the slope of the supply curve rather than its (zero) intercept. In Chapter 8 we will return to this model to cover the case of the long run—where both labor and capital are variable inputs.

2.4 PROFIT MAXIMIZATION: MONOPOLY

Monopoly occurs when there is a single seller of a product. The salient point of any monopoly model is that the demand curve faced by a firm is identical to the market demand curve. In this application we will first explore a simple monopoly model with linear demand and constant marginal production cost. After we solve this first version of the model we will introduce government taxation.

2.4.1 Linear Demand and Costs

In most economic models of firm behavior, economists assume that a firm's goal is profit maximization. The firm wants to maximize a profit function, Π, of the form

$$\Pi = TR - TC. \tag{2.23}$$

where TR is total revenue and TC is total cost. Total revenue is (by definition) price times quantity, while total cost can also be written as a function of quantity. We will assume that the firm's inverse demand curve is linear so that we can write demand and total revenue as[3]

$$P = a - bQ \tag{2.24}$$

and

$$TR = PQ = (a - bQ)Q = aQ - bQ^2, \tag{2.25}$$

where a is the demand intercept, b is the demand slope, and both a and b are positive parameters. We will also assume that the firm's cost function has the linear form

$$TC = cQ + f, \tag{2.26}$$

where c represents marginal cost, f represents a fixed cost that is independent of output, and both c and f are positive parameters.

There are two endogenous variables in this model, Q and P. They cannot, however, be chosen independently: once the firm chooses the value of one of the variables, the other variable is determined by the demand curve. In this application we will eliminate the price variable by substitution and write the firm's profits as a function of the one endogenous variable, Q, which is chosen by the firm, and the four parameters:

$$\Pi = TR - TC = (aQ - bQ^2) - (cQ + f). \tag{2.27}$$

---

[3]In Chapter 6 we will present a monopoly model with more general demand and cost functions.

FIGURE 2.4

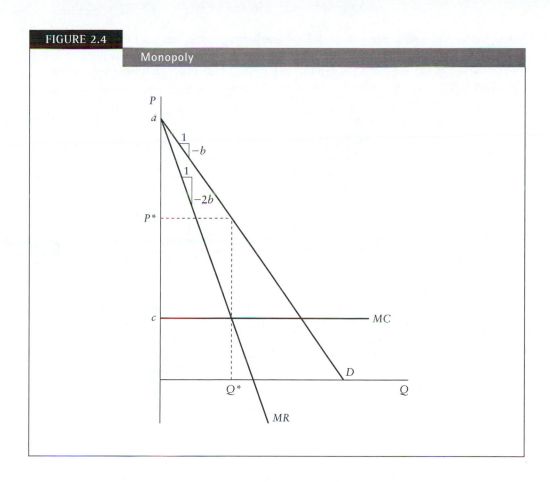

To maximize profits we take the first derivative of the profit function and set it equal to zero. This gives a first-order condition of

$$\frac{d\Pi}{dQ} = \frac{dTR}{dQ} - \frac{dTC}{dQ} = MR - MC = (a - 2bQ) - c = 0. \tag{2.28}$$

MR stands for *marginal revenue*, the change in total revenue for a change in quantity, and MC represents *marginal cost*, the change in total cost for a change in quantity. So the economic interpretation of the first-order condition is the familiar rule that the firm's profit maximum is at a quantity where marginal revenue equals marginal cost.

The equilibrium is shown graphically in Figure 2.4. Note that the graph for marginal revenue has the same intercept as the demand curve but twice the slope. Note also that the marginal cost curve, for our linear cost function, is flat and that the level of marginal cost is the parameter c.[4] The equilibrium values of price and quantity are labeled as P^* and Q^*. Before solving for these equilibrium values we must check to see that they correspond to a maximum of the profit function. To do this we find the second-order condition:

$$\frac{d^2\Pi}{dQ^2} = -2b < 0. \tag{2.29}$$

---

[4]A flat long-run marginal cost curve will occur if there are constant returns to scale in production. See Chapter 5 for a fuller exploration of the mathematics of returns to scale.

Since the second derivative is negative we know that our solutions for P^* and Q^* will yield the firm's maximum level of profits.

To solve for Q^* we simply solve the $MR = MC$ first-order condition. Rearranging this equation gives

$$Q^* = \frac{(a - c)}{2b}.$$ (2.30)

Note that for the quantity solution to be positive we must assume that $a > c$, i.e., that the demand intercept is higher than the level of marginal cost. To find price we substitute the equilibrium quantity back into the demand equation (2.24) and get

$$P^* = a - bQ^* = a - b\left(\frac{a - c}{2b}\right)$$

$$P^* = \frac{a + c}{2}.$$ (2.31)

It is worth noting that neither the quantity nor the price solutions depend on the firm's fixed cost; the derivatives of Q^* and P^* with respect to f are zero. This is the standard result (already seen in Section 1.3.2) that fixed costs are sunk and irrelevant to decisions at the margin. Fixed costs only matter if they are sufficiently large so that P^* and Q^* yield a maximum profit that is negative. In this case the firm should shut down in the long run.[5] The other comparative static results for changes in the demand and cost parameters are covered in the end-of-chapter exercises.

2.4.2   Taxation

We now introduce taxation into our model. The simplest case, as in the supply and demand example in Chapter 1, is a per-unit tax on the monopolist's output. With a tax rate of t dollars per unit the firm pays a total tax T of

$$T = tQ.$$ (2.32)

Net profits are given by the profit function defined above minus the amount paid in tax:

$$\Pi = TR - TC - T = PQ - cQ - tQ - f.$$ (2.33)

Substituting the demand equation into the profit equation gives

$$\Pi = aQ - bQ^2 - (c + t)Q - f.$$ (2.34)

The first- and second-order conditions for a profit maximum are found by taking derivatives with respect to Q. These conditions are essentially the same as the case without a tax. The only difference is that wherever the parameter c had appeared in the preceding equations, it is now replaced by $(c + t)$. Thus, the first- and second-order conditions are

$$\frac{d\Pi}{dQ} = a - 2bQ - (c + t) = 0$$ (2.35)

---

[5]A positive solution for Q^* is sufficient, given the assumptions of this particular model, to ensure that the firm covers more than its variable costs and stays open in the short run.

and

$$\frac{d^2\Pi}{dQ^2} = -2b < 0. \tag{2.36}$$

The interpretation of the first-order condition is that marginal revenue must equal marginal cost inclusive of the tax. The second-order condition is the same as before and is sufficient to ensure a profit maximum.

We can now solve for the new equilibrium values of quantity and price. Using the first-order condition, the solution for quantity is

$$Q^* = \frac{(a - c - t)}{2b}. \tag{2.37}$$

Substituting this solution into the demand equation (2.24) allows us to solve for price:

$$P^* = \frac{a + c + t}{2}. \tag{2.38}$$

Note that these solutions are the same as equations (2.30) and (2.31) except that the "new marginal cost" of $(c + t)$ replaces the old marginal cost of c.

Having solved for quantity and price as functions of the parameters and the tax rate, we can also solve for profits as a function of the same exogenous variables. Substituting Q^* and P^* into the profit function (2.33) gives

$$\Pi^* = (P^* - c - t)Q^* - f = \left(\frac{a + c + t}{2} - c - t\right)\left(\frac{a - c - t}{2b}\right) - f$$

or $\tag{2.39}$

$$\Pi^* = \frac{(a - c - t)^2}{4b} - f.$$

We can now solve for the comparative static effects of a change in the tax rate on the firm's output, price, and profits. These partial derivatives are

$$\frac{\partial Q^*}{\partial t} = -\frac{1}{2b} < 0,$$

$$\frac{\partial P^*}{\partial t} = \frac{1}{2} > 0, \tag{2.40}$$

and

$$\frac{\partial \Pi^*}{\partial t} = \frac{-(a - c - t)}{2b} < 0.$$

Thus, the model predicts that an increase in the tax rate will lead to a decrease in output, an increase in price, and a reduction in the firm's profits.

Do these results generalize? Equation (2.40) says that exactly half the tax gets passed on to consumers in the form of a higher price. This specific quantitative result is almost certainly due to the special assumptions of linear demand and constant marginal cost. We need more mathematical tools before we can see whether the qualitative result—that only part of the tax is passed on to consumers in the form of a higher price—is general.

What about the result that the tax lowers the firm's profits? We can use the envelope theorem to show that this result holds for any demand and cost functions. Recall, from Chapter 1, that the envelope theorem states that the derivative of the value function solution for Π^* (i.e., the indirect objective function derived in equation 2.39) with respect to a parameter will be identical to the derivative of the direct objective function, i.e., the profit function as first written in equation (2.33). If we examine equation (2.33), where $\Pi = TR - TC - tQ$, we immediately see that the partial derivative $\partial\Pi/\partial t = -Q$. This holds regardless of the specific functions for revenue and cost. Furthermore, note that the derivative of the value function, $\partial\Pi^*/\partial t$, in equation (2.40), while expressed in terms of parameters, is equal to our solution for Q^*.

Our next step is to extend the model by incorporating the government's decision on the level of the tax rate. To do this we will assume that the government is interested in maximizing the total tax revenue raised by this particular tax.[6] This extension of the model leads to a reinterpretation of the tax variable. In the original model the government's tax decision and the tax rate were exogenous, i.e., outside the control of the firm. In this extension, however, the government decision and the tax rate become endogenous to the model (although still outside the control of the firm).

Before delving into the mathematics of tax-revenue maximization let's start with some fairly obvious economic intuition about the extremes of possible tax rates. Clearly, a tax rate of zero will raise no revenue. At the other extreme there exists some tax rate so high that the firm's optimal output would be zero. From the equation for Q^*, tax rates of $t = (a - c)$ would result in zero output by the firm, and hence zero revenue for the government.[7] The revenue-maximizing tax rate must lie somewhere in between these two extremes.

To solve the model we first write tax revenues as a function of the tax rate and the parameters. Total tax revenue, T, is

$$T = tQ^* = t\left(\frac{a - c - t}{2b}\right). \tag{2.41}$$

The first- and second-order conditions for maximizing tax revenue are

$$\frac{dT}{dt} = \frac{a - c - 2t}{2b} = 0 \tag{2.42}$$

and

$$\frac{d^2T}{dt^2} = \frac{-1}{b} < 0. \tag{2.43}$$

Since the second-order condition holds we can solve the first-order condition for the revenue-maximizing rate, t^*. This gives

$$t^* = \frac{a - c}{2}. \tag{2.44}$$

---

[6]In a more complete model with many taxable goods we might make more realistic assumptions about the government's tax goal. For example, the government's goal might be to minimize the welfare losses from raising a fixed amount of revenue.

[7]There is an added complication that should be mentioned here. Since the fixed cost, f, is sunk in the short run, there are tax rates at which the firm would produce a positive output in the short run but would be losing money in the short run and would choose to shut down in the long run. We ignore this complication in the analysis in the text.

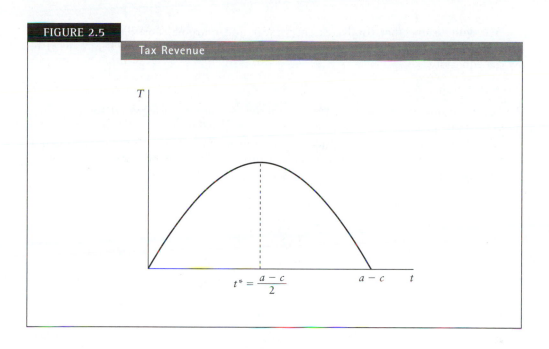

FIGURE 2.5

Tax Revenue

$t^* = \dfrac{a-c}{2}$

The tax revenue function and t^* are shown in Figure 2.5. This shows a function where tax rate increases increase total tax revenue for rates below t^* but decrease total tax revenue for rates above t^*. The revenue-maximizing tax rate itself is increasing in a, decreasing in c, and is independent of the demand slope parameter b.

2.5 PROFIT MAXIMIZATION: DUOPOLY

So far we have covered the extremes of possible market structures. Most real-world markets probably lie somewhere in between monopoly and perfect competition. Markets with a few firms are called oligopolies. The original oligopoly model by Antoine Cournot (1801–1877) was one of the first examples of mathematical economics.[8] Today, oligopolies are usually analyzed using a branch of applied mathematics called game theory, which is covered in this text in Chapters 17–20. In this section we will develop a mathematical model of a duopoly, a market in which two firms compete.

The duopoly model will use some of the same assumptions as in the monopoly model in Section 2.4. Market demand is assumed to be a linear function, i.e.,

$$P = a - bQ,\tag{2.45}$$

where Q represents the total combined output of the two firms. We will call the two firms firm 1 and firm 2 and use subscripts to indicate which firm we are referring to. Total output can then be written as

$$Q = q_1 + q_2\tag{2.46}$$

where q_i is the output of firm i. This way of writing total output and demand implicitly assumes that the two firms are producing identical products, and the market price therefore depends only on the total level of market output.

---

[8]For an English translation of Cournot's duopoly model see *Cournot Oligopoly: Characterizations and Applications*, Andrew Daughety, Ed., Cambridge University Press, New York: 1988.

We will assume that total costs are linear and marginal costs are constant. We will simplify by assuming that there are no fixed costs. Total costs for firm i are given by the equation

$$TC_i = cq_i. \tag{2.47}$$

The cost parameter c does not have a subscript: we are assuming that the two firms have the same production technology and the same level of marginal cost.

We can now write the profit function for a firm. We begin with the general definition of profits and then make successive substitutions to arrive at a usable form of the profit equation. Thus, for firm 1 we can develop the profit function

$$\Pi_1 = TR_1 - TC_1$$
$$\Pi_1 = Pq_1 - cq_1$$
$$\Pi_1 = (a - bQ)q_1 - cq_1 \tag{2.48}$$
$$\Pi_1 = (a - b(q_1 + q_2))q_1 - cq_1$$
$$\Pi_1 = aq_1 - bq_2q_1 - bq_1^2 - cq_1.$$

A similar derivation yields firm 2's profit function

$$\Pi_2 = aq_2 - bq_1q_2 - bq_2^2 - cq_2. \tag{2.49}$$

Before taking the first-order conditions, we need to carefully examine the endogeneity of the variables. For firm 1 q_1 is clearly an endogenous choice variable. The same is true for firm 2 and q_2. In terms of the entire model q_1, q_2, Q and P are all endogenously determined. The issue is whether firm 1 (or 2) views q_2 (or q_1) as endogenous or exogenous. Taking firm 1's point of view for the moment, q_2 would be seen as endogenous if firm 1 expected that changes in q_1 would predictably lead to changes in q_2. If this were the case then firm 1 would, in a sense, be able to control both q_1 and q_2 and would view both outputs as endogenous. Cournot's approach to this issue was to make the assumption that each firm treats the other firm's output as an exogenous variable. The *Cournot assumption* is that each firm chooses its own output, taking the other firm's output as exogenously fixed or given.

We can now turn to solving the two firms' profit-maximization problems. For firm 1 the first- and second-order conditions, treating q_2 as a constant, are

$$\frac{d\Pi_1}{dq_1} = (a - bq_2 - 2bq_1) - c = 0 \tag{2.50}$$

and

$$\frac{d^2\Pi_1}{dq_1^2} = -2b < 0. \tag{2.51}$$

For firm 2 these conditions, treating q_1 as a constant, are

$$\frac{d\Pi_2}{dq_2} = (a - bq_1 - 2bq_2) - c = 0 \tag{2.52}$$

and

$$\frac{d^2\Pi_2}{dq_2^2} = -2b < 0. \tag{2.53}$$

Both second-order conditions are satisfied, so the first-order conditions, when solved, will correspond to profit maxima. To interpret each of the first-order conditions note that the term in parentheses is the effect of a firm's quantity on the firm's total revenue, i.e., this is marginal revenue. The parameter c is marginal cost. So both first-order conditions embody the optimization rule that marginal revenue (or marginal benefit) must equal marginal cost.

The two first-order conditions can be solved respectively as

$$q_1 = \frac{(a - c)}{2b} - \frac{1}{2}q_2 \quad \text{and} \quad q_2 = \frac{(a - c)}{2b} - \frac{1}{2}q_1. \tag{2.54}$$

Note that we did not put "stars" on these solutions for q_1 and q_2. In the model as a whole, q_1 and q_2 are both endogenous, so we have not yet reached the stage where endogenous variables are expressed solely as a function of exogenous parameters. Instead, the two equations above express the q's in terms of each other.

Before proceeding, let us examine the economic intuition underlying the two equations. Taking the vantage point of firm 2 (similar logic would apply to firm 1) we can first explain and then graph the second equation in (2.54). The general interpretation of this equation is that it shows the profit maximizing level of q_2 for any given level of q_1. Put differently, the equation, called a *best-response function*, shows firm 2's best response to any level of q_1; the equation is also called firm 2's **reaction function**.

Now let us graph equation (2.54) with q_1 as the independent (horizontal axis) variable and q_2 as the dependent (vertical axis) variable. Since the equation is already in slope-intercept form we know that the vertical intercept is $(a - c)/2b$ and the slope is $-\frac{1}{2}$. This graph is shown in Figure 2.6. The intercept tells us the optimal value of q_2 when q_1 equals zero. If q_1 were to equal zero then under the Cournot Assumption firm 2 would believe that q_1 was exogenously fixed at this level. In this case firm 2 essentially believes that it is the only firm in the market, i.e., firm 2 should act as a monopolist. We can check this by looking at the results derived in the monopoly application in Section 2.4. There we found that the formula for the monopoly output, call it Q_M, was

$$Q_M = \frac{(a - c)}{2b}. \tag{2.55}$$

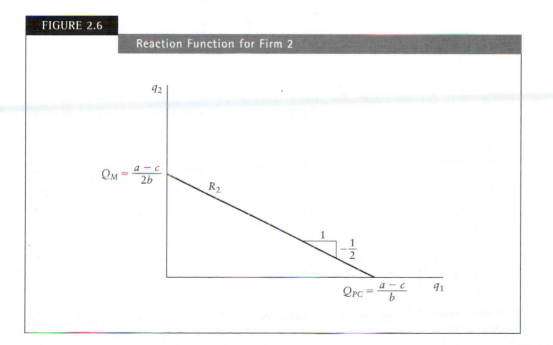

FIGURE 2.6

Reaction Function for Firm 2

FIGURE 2.7

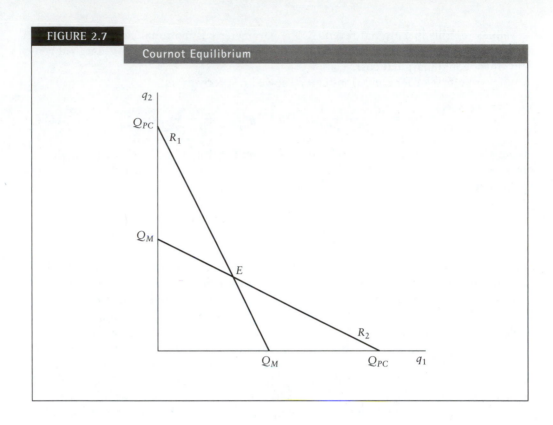

Thus, our first result is that the vertical intercept of firm 2's reaction function is the monopoly level of output.

With a vertical intercept of Q_M and a slope of $-\frac{1}{2}$, it is fairly easy to calculate that the horizontal intercept of the reaction function is $(a - c)/b$. To interpret this value, recall the linear supply and demand model of Chapter 1. There we found the output that would be produced by a perfectly competitive industry. Call this output Q_{PC}. The formula for the perfectly competitive output is[9]

$$Q_{PC} = \frac{(a - c)}{b}. \tag{2.56}$$

Why is the horizontal intercept of firm 2's reaction function equal to the perfectly competitive output? At the perfectly competitive output level price equals marginal cost, and profits in the market would be equal to zero. So, if $q_1 = Q_{PC}$ then firm 2 has no incentive to sell any output: any output by firm 2 would drive the market price below production cost and result in a negative profit for firm 2.

So far we have only explained firm 2's behavior. Firm 1 is in an analogous position, and we could explain firm 1's best reactions with reasons that parallel those given for firm 2. When $q_2 = 0$ firm 1's best output is $q_1 = Q_M$ and when $q_2 = Q_{PC}$ firm 1's best output is $q_1 = 0$. Figure 2.7 shows the two firms' best-response functions on the same graph, labeled as R_1 for firm 1 and R_2 for firm 2.

Now that we have set up the two response functions, we can find the equilibrium for our model. First, note that an individual firm is in equilibrium only when it is on its response function. By this we mean that if a quantity pair (q_1, q_2) is on a firm's response

---

[9]In Chapter 1 there was a parameter d that represented the slope of the marginal cost or supply curve. Here the marginal cost curve is flat so $d = 0$.

function, then the firm has chosen its best (profit-maximizing) output and has no incentive to change output. If, on the other hand, the quantity pair is not on a firm's response function then the firm has an incentive to change output in order to increase its profits. Finally, note that for the market to be in equilibrium both firms must be in equilibrium, i.e., market equilibrium occurs when neither firm has an incentive to change its output. This leads to the conclusion that the equilibrium for the model is the quantity pair that lies at the intersection of the two response functions. This is the point labeled E (for *Equilibrium*) in Figure 2.7.

Solving for point E requires finding the simultaneous solution to the two reaction functions. We can do this by substituting q_2 from firm 2's reaction function into firm 1's reaction function. Substituting gives

$$q_1 = \frac{(a-c)}{2b} - \frac{1}{2}q_2 = \frac{(a-c)}{2b} - \frac{1}{2}\left(\frac{(a-c)}{2b} - \frac{1}{2}q_1\right)$$

or   **(2.57)**

$$q_1 = \frac{(a-c)}{4b} + \frac{1}{4}q_1.$$

The solution for q_1 in terms of exogenous variables is

$$q_1^* = \frac{(a-c)}{3b}.$$ **(2.58)**

We can find q_2 by substituting the solution for q_1 into either reaction function. There is, however, a simpler way of solving. Since there is a mathematical symmetry between the two firms (the only difference between the firms is their numerical subscript), we can conclude that in equilibrium q_2 must equal q_1. The solution for firm 2 is

$$q_2^* = \frac{(a-c)}{3b}.$$ **(2.59)**

To finish the model we solve for the total market quantity and the market price. The market quantity is

$$Q^* = q_1^* + q_2^* = \frac{2(a-c)}{3b}.$$ **(2.60)**

The market price is

$$P^* = a - bQ^* = \frac{a+2c}{3}.$$ **(2.61)**

To illustrate the comparative statics of these solutions, let us examine the effect of an increase in production cost. We will first examine the effect graphically and then verify the graphical analysis by taking partial derivatives. Looking at the reaction function equations, we can see that an increase in marginal cost, the parameter c, will lead to an inward shift of each firm's reaction function. Since the reaction function slopes do not depend on c, the new reaction functions are parallel to the originals. Figure 2.8 shows the new reaction functions R_1' and R_2'. The new equilibrium point is labeled E'. Since in equilibrium firms produce identical outputs, both E and E' lie on the 45-degree line. It is clear in the graph that both firms' outputs are lower at the new equilibrium E' than at the original equilibrium E. We can therefore conclude that market output has fallen and market price has increased as a result of the rise in marginal cost.

FIGURE 2.8

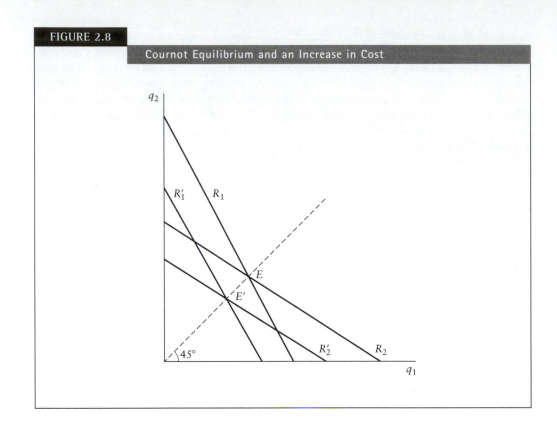

All of this can be verified by finding the partial derivatives of the solutions for the endogenous variables. The four partial derivatives with respect to marginal cost are

$$\frac{\partial q_1^*}{\partial c} = -\frac{1}{3b} < 0,$$

$$\frac{\partial q_2^*}{\partial c} = -\frac{1}{3b} < 0,$$

$$\frac{\partial Q^*}{\partial c} = -\frac{2}{3b} < 0,$$

(2.62)

and

$$\frac{\partial P^*}{\partial c} = \frac{2}{3} > 0.$$

This verifies the graphical results. Similar analysis could also be done for changes in the other parameters. In this particular model, graphical analysis and mathematical analysis are easily substitutable. In more complicated models graphical tools will not be sufficient and mathematical analysis will be our only method for deriving results. An example of a model that can only be solved mathematically will be presented next.

2.6   PROFIT MAXIMIZATION: OLIGOPOLY

In this section we consider the Cournot model of an oligopoly with an arbitrary number of firms. This example will allow us to make predictions as to how the market price and the optimal tax rate depend on market structure. These results will be derived by examining the comparative static effects of a change in the number of firms in the industry. Finally, we will also consider, as in the monopoly model, the effects of government taxation of the product.

Suppose that there are n firms producing identical products. Total market quantity, Q, can be written as:

$$Q = \sum_{i=1}^{n} q_i, \tag{2.63}$$

where q_i is the output of an individual firm. We will assume that demand is linear. The equation for price is

$$P = a - bQ = a - b\left(\sum_{i=1}^{n} q_i\right). \tag{2.64}$$

We will also assume that firms operate with constant marginal cost, zero fixed costs, and pay a per-unit tax at a rate t. Firm i's total cost can then be written as

$$TC_i = (c + t)q_i. \tag{2.65}$$

To work through the model let us focus on the decision making by the first firm. Firm 1's profit function can be derived in successive steps as

$$\Pi_1 = Pq_1 - (c + t)q_1$$

$$\Pi_1 = (a - bQ)q_1 - (c + t)q_1$$

$$\Pi_1 = \left(a - b\left(\sum_{i=1}^{n} q_i\right)\right)q_1 - (c + t)q_1 \tag{2.66}$$

$$\Pi_1 = aq_1 - b\left(\sum_{i=2}^{n} q_i\right)q_1 - bq_1^2 - (c + t)q_1.$$

The last step separates market quantity into two parts: the amount produced by firm 1 and the combined output of the remaining $n - 1$ firms, i.e., the last equation splits market quantity into the variable that is endogenous for firm 1 and the variables (other firms' outputs) that, by the Cournot assumption, are exogenous to firm 1's decision.

We can now derive the first- and second-order conditions for firm 1's profit maximization using the Cournot assumption that other firms' outputs are exogenous:

$$\frac{d\Pi_1}{dq_1} = a - b\left(\sum_{i=2}^{n} q_i\right) - 2bq_1 - (c + t) = 0$$

and $\tag{2.67}$

$$\frac{d^2\Pi_1}{dq_1^2} = -2b < 0.$$

Since the second-order condition is satisfied, the solution to the first-order condition will give a profit maximum. We can solve the first-order condition to derive the best-reponse function for firm 1. The profit-maximizing level of q_1 for any combination of other firms' outputs is

$$q_1 = \frac{(a - c - t)}{2b} - \frac{1}{2}\left(\sum_{i=2}^{n} q_i\right). \tag{2.68}$$

Note that the first term (the intercept) represents the monopoly output that firm 1 would produce if all the q_i's were zero. Also note that firm 1's output is a function of each of the other firms' outputs, but what matters is the combined total of other firms' outputs. The distribution of other firms' outputs, or which firm produces how much, does not matter.

For every other firm, 2 through n, there are similar first- and second-order conditions. Thus, to solve for outputs we need to solve n equations, the n first-order conditions, for n unknowns, the q_i's. Similarly, the n firms' reaction functions would be drawn in n dimensions: firm i's quantity as a function of the remaining $n-1$ quantities. Neither of these facts is of much practical use for representing or solving the model.

So how do we proceed? The answer is that we can resort to the mathematical symmetry of the firms to solve the model. In a mathematical model, variables are symmetric if they differ only by the ordering of subscripts. Furthermore, if the subscript ordering is arbitrary (as is usually true) then the solutions for the variables must be identical in form.

Here, since each firm has the same costs and the only difference between the firms is their identifying number, each firm produces the same equilibrium level of output. Let q without a subscript represent the common level of firm output in equilibrium. Then any firm's first-order condition can be represented as

$$a - b \left(\sum_{i=2}^{n} q \right) - 2bq - (c + t) = 0. \tag{2.69}$$

We can rewrite this equation and solve for q:

$$a - b(n-1)q - 2bq - (c + t) = 0$$

or

$$q^* = \frac{(a - c - t)}{(n+1)b}. \tag{2.70}$$

A word of caution is in order. Symmetry is often a powerful tool and a useful shortcut, but using it correctly requires careful thought. The solution method is to find the first-order conditions and then impose symmetry by assuming identical outputs. It is not valid to impose symmetry before taking the first-order conditions. In other words, imposing symmetry by putting identical q's in the profit function and then taking the first-order condition would not work. This is because by using symmetry too early we would be setting up the problem as if firm 1 could choose the other firms' outputs as well as its own. To emphasize: *symmetry only applies to the equilibrium values of endogenous variables; symmetry need not hold at nonequilibrium values of the endogenous variables.*

Now that we have solved for a typical firm's output we can also find the total market output and the market price. Market output is

$$Q^* = \sum_{i=1}^{n} q_i^* = nq^* = \left(\frac{n}{n+1} \right) \frac{(a - c - t)}{b}. \tag{2.71}$$

Using the demand equation (2.64) and rearranging gives the market price:

$$P^* = a - bQ^*$$

$$P^* = \left(\frac{n+1}{n+1} \right) a - \left(\frac{n}{n+1} \right) (a - c - t) \tag{2.72}$$

$$P^* = \left(\frac{a}{n+1} \right) + \left(\frac{n}{n+1} \right) (c + t).$$

Given the reduced-form solutions for the endogenous variables we can check to see how changes in market structure, i.e., changes in the number of firms, affect the market outcomes. Using the quotient rule, the derivatives of q^*, Q^*, and P^* with respect to n are

$$\frac{\partial q^*}{\partial n} = \frac{-(a - c - t)}{(n+1)^2 b} < 0,$$

$$\frac{\partial Q^*}{\partial n} = \frac{(a - c - t)(n+1) - n(a - c - t)}{(n+1)^2 b} = \frac{(a - c - t)}{(n+1)^2 b} > 0,$$

(2.73)

and

$$\frac{\partial P^*}{\partial n} = \frac{-a}{(n+1)^2} + \left(\frac{1}{(n+1)^2}\right)(c + t) = \frac{-(a - c - t)}{(n+1)^2} < 0.$$

All three functions are monotonic in n, i.e., the signs of these derivatives do not depend on the particular value of n. The derivative for q^* is negative. This means that an increase in the number of firms always leads to a decline in the level of output per firm. Despite this decline in output per firm, increases in n lead to increases in the total level of market output: the decline in output per firm is more than offset by the output increase from a greater number of active producers. Finally, since demand is downward sloping, increases in n, and hence Q^*, result in a lower market price.

In many instances in economics, mathematical analysis and graphical analysis complement each other quite nicely. In this model graphing Q^* and P^* as functions of n provides some interesting economic intuition concerning the mathematical results. In previous models we have already solved for the monopoly and competitive levels of market output and price. Let us use an M subscript for the monopoly outcomes and a PC subscript for the perfectly competitive outcomes. Figure 2.9 shows Q^* as a function

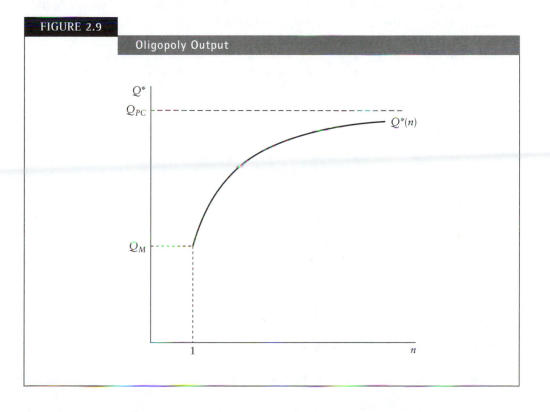

FIGURE 2.9

Oligopoly Output

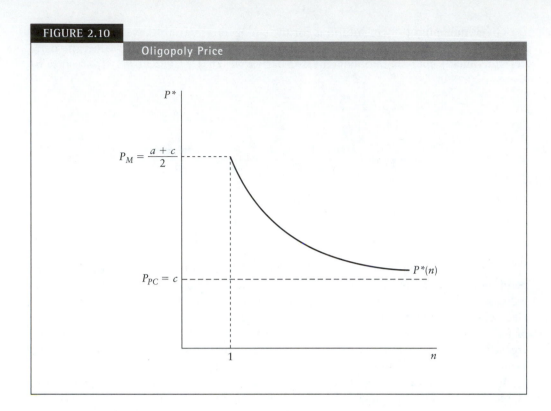

FIGURE 2.10

Oligopoly Price

of n. When n is 1 we get the monopoly outcome ($n = 2$ is the duopoly outcome). As n increases, market output approaches, in the limit, the competitive level. Figure 2.10 shows a similar, but declining, outcome for the market price. Thus, the model allows us to predict that more competitive market structures lead to lower prices. We might also extend the model to the case of free entry, so that n becomes an endogenous variable. This is covered in an exercise at the end of the chapter.

We finish this section by examining how market structure affects taxation. We will again assume, as in the monopoly section, that the government's tax goal is simply to maximize the tax revenue raised by the tax on this product. Total tax collections are

$$T = tQ^*$$

or

$$T = t\left(\frac{n}{n+1}\right)\frac{(a - c - t)}{b}.$$

(2.74)

The first- and second-order conditions for maximizing tax revenue are

$$\frac{dT}{dt} = \left(\frac{n}{n+1}\right)\frac{(a - c - 2t)}{b} = 0$$

and

(2.75)

$$\frac{d^2T}{dt^2} = \left(\frac{n}{n+1}\right)\left(\frac{-2}{b}\right) < 0.$$

We can solve the first-order condition for the revenue-maximizing tax rate

$$t^* = \frac{a - c}{2}.$$

(2.76)

This is the same result as in the monopoly model of Section 2.4.2. The answer to the question of how market structure affects the tax rate is that market structure doesn't matter! The revenue-maximizing tax rate is independent of the number of firms. This is a negative result, but the result is still interesting and useful since we had no way of knowing in advance that the tax rate would be independent of market structure.

This application has shown us how basic mathematical tools can be used to solve a simple economic model. We were able to derive comparative static results using one-variable calculus. We should be concerned, however, with whether these results can be generalized to more complex situations. For example, what if demand weren't linear, or what if marginal cost weren't constant? Even more generally, what if we don't know the equation for demand, or what if the government's tax goal were more complicated than simply maximizing revenue? Does our simple answer, that the tax rate is independent of market structure, still hold? To answer these questions we will need to expand our mathematical tool kit.

2.7   A SIMPLE MACROECONOMIC MODEL

The previous sections have all been from microeconomics. This section provides a first introduction to mathematical macroeconomic models. Both types of models use equilibrium analysis and comparative statics. The differences lie in the level of economic activity being analyzed and in the fact that microeconomic models are more likely to involve optimization analysis. In this section we will develop a simple Keynesian macroeconomic model, under the assumption that all elements of the model are linear.[10]

Our models of firm decision-making began with an accounting identity, profits equal revenue minus cost, and were given content by specifications of firm behavior, e.g., firms choose output to maximize profit. This macroeconomic model follows the same pattern. The accounting identity here is that gross domestic product (GDP) equals total spending. Mathematically, we have

$$Y = C + I + G, \tag{2.77}$$

where Y is GDP, C is consumption spending by consumers, I is investment spending by firms, and G is government spending.[11]

To give content to the model, we need to specify the determinants of the various components of GDP. To begin, consumption is a function of after-tax, or disposable, income. The linear consumption function is

$$C = C_0 + bY^d, \tag{2.78}$$

where C_0 and b are parameters and Y^d is disposable income. The parameter b is the marginal propensity to consume and we therefore assume

$$0 < b < 1. \tag{2.79}$$

To define disposable income we need to specify the level of taxes. We will assume that taxes are lump sum, which means that taxes are independent of the level of economic activity. In Chapter 4 we will relax this assumption and consider the case where

---

[10]Later, in Chapter 5, we will develop the tools needed to construct nonlinear macroeconomic models.

[11]This is the accounting definition of GDP for a closed economy without imports or exports. An open-economy model would add net exports to the definition of GDP.

taxes are proportional to income. Let T be the exogenous level of taxes; then disposable income is

$$Y^d = Y - T, \qquad (2.80)$$

and the consumption function can be rewritten as

$$C = C_0 + b(Y - T). \qquad (2.81)$$

Investment spending depends on the level of the interest rate, r. Because higher interest rates make investment more expensive, there is a negative relationship between investment and interest rates. We can write this relationship as

$$I = I_0 - er, \qquad (2.82)$$

where I_0 and e are positive parameters. In this simple Keynesian model the interest rate is exogenous. Later, we develop more realistic macroeconomic models, e.g., IS–LM models, that allow for an endogenous determination of the interest rate. Finally, we will assume that government spending is determined outside of the model, so that G is an exogenous variable.

We can now return to the definition of GDP and substitute the functions for the components of spending. This gives

$$Y = C_0 + b(Y - T) + I_0 - er + G. \qquad (2.83)$$

In this equation, the level of GDP is endogenous and all other variables in the equation are exogenous. Solving the equation gives the equilibrium level of GDP:

$$Y^* = \frac{C_0 - bT + I_0 - er + G}{1 - b}. \qquad (2.84)$$

The government spending multiplier is found by taking the partial derivative of Y^* with respect to G:

$$\frac{\partial Y^*}{\partial G} = \frac{1}{1 - b} > 1. \qquad (2.85)$$

Thus, because b, the marginal propensity to consume, is between zero and one, a one dollar increase in government spending leads to a greater than one dollar increase in the equilibrium level of GDP. With respect to an increase in taxes we find that

$$\frac{\partial Y^*}{\partial T} = \frac{-b}{1 - b} < 0. \qquad (2.86)$$

To conclude this section consider the combined effect of an increase in government spending that is matched by an identical increase in taxes. This is called the **balanced-budget multiplier**. Adding the partial effects of increases in G and T we get[12]

$$\frac{\partial Y^*}{\partial G} + \frac{\partial Y^*}{\partial T} = \frac{1}{1 - b} + \frac{-b}{1 - b} = \frac{1 - b}{1 - b} = 1. \qquad (2.87)$$

---

[12]This is less than rigorous mathematically. A more rigorous approach requires the multivariable calculus tools in Chapter 5.

Our result, which may be familiar, is that the balanced budget multiplier equals one: equal one dollar increases in G and T lead to a one dollar increase in Y^*.

In later chapters we will return to this macroeconomic model and extend it by allowing for more endogenous variables (e.g., the interest rate and the aggregate price level) and for nonlinear functional forms.

Problems

2.1 An economic agent engages in an activity x resulting in private benefits to the agent $B(x)$, private costs to the agent $C(x)$, and external costs $E(x)$. Because the costs are external, the agent maximizes private net benefits $PNB(x) = B(x) - C(x)$, but social welfare is maximized when social net benefits $SNB(x) = B(x) - C(x) - E(x)$ are maximized.

Suppose that $B(x) = bx(1 - x)$ for all $x \geq 0$, $C(x) = f + cx$ for all positive values of x and $C(x) = 0$ when $x = 0$, $E(x) = ex$ for all $x \geq 0$. Assume that the parameters b, f, c, and e are all positive.

(a) Assuming an interior solution,

    (i) Find the equilibrium solution $x^*(b, f, c)$ that maximizes private net benefits

    (ii) Show that the second-order condition is satisfied

    (iii) Find the equilibrium solution $PNB^*(b, f, c)$ for the magnitude of private net benefits when they are maximized

    (iv) Find and sign (if possible) all of the comparative static derivatives of x^* and PNB^*

    (v) Explain the economics of each of your results in part (4)

(b) Would the agent ever choose not to engage in this activity? (That is, would the corner solution $x = 0$ ever maximize private net benefits?) Explain both the mathematics and the economics.

(c) Again assuming an interior solution,

    (i) Find the equilibrium solution $x^{**}(b, f, c, e)$ that maximizes social net benefits

    (ii) Show that the second-order condition is satisfied

    (iii) Show that $x^{**} < x^*$ and explain the economics of this result

    (iv) Find the equilibrium solution $E^{**}(b, f, c, e)$ for the magnitude of external costs when social net benefits are maximized

    (v) Find and sign (if possible) all of the comparative static derivatives of x^{**} and E^{**}

    (vi) Explain the economics of each of your results in part (5)

(d) Would a social planner ever want to prohibit the agent from engaging in this activity? Explain both the mathematics and the economics.

2.2 For the labor union model in Section 2.2 find the comparative static effects on each of the four wage solutions when

(a) There is a reduction in the demand for union labor, w_0 decreases.

(b) There is a reduction in the overall demand for labor, $\bar{w}$ decreases.

2.3 Consider the labor union model in Section 2.2. Suppose that instead of pursuing a single goal the union maximizes a mixture of goals 3 and 4. Specifically, suppose that the union's objective is to maximize a utility function that depends on both total worker income and rents. Let U be given by $U = \alpha I + (1 - \alpha) R$, where I and R are total income and rents (as defined in Section 2.2) and $0 \le \alpha \le 1$.

(a) Is α endogenous or exogenous? How might α be interpreted?

(b) Find the optimal wage demand, w^*, as a function of the exogenous variables (you may assume that $w_0/2 > \bar{w}$).

(c) How does w^* depend on α? Interpret w^* when $\alpha = 0$ and when $\alpha = 1$.

2.4 Suppose that a firm in a perfectly competitive industry operates with the production function $Q = L^\alpha K^\beta$, where $0 \le \alpha \le 1$. Also suppose that the firm's capital stock is fixed in the short run.

(a) Solve for the firm's profit-maximizing levels of employment and output as a function of exogenous variables.

(b) Find the comparative static effects of increases in P, w, and r.

2.5 Demand elasticity (see the Appendix following Chapter 1) is defined as the absolute value of the percentage change in quantity demanded for a percentage change in price. Mathematically the percentage change in a variable, say x, is defined as dx/x. Thus demand elasticity, ε, is

$$\varepsilon = \left| \frac{dQ/Q}{dP/P} \right| = \left| \frac{dQ}{dP} \right| \left(\frac{P}{Q} \right)$$

(a) Find the formula for the labor demand elasticity for the labor demand curve in Section 2.2. What is the elasticity of demand at the wage rate that maximizes total worker income?

(b) Suppose that the demand curve for a product is given by the equation $P = Q^{-\alpha}$, where $\alpha > 0$

   (i) Find the elasticity of demand as a function of α. For what values of α is demand elastic, inelastic, and unit elastic? Why is this demand curve called a constant-elasticity curve?

   (ii) Graph the marginal revenue curve ($MR = dTR/dQ$) associated with this demand curve for the three cases of $\alpha < 1$, $\alpha = 1$, and $\alpha > 1$. How does the value of marginal revenue depend on the elasticity of demand?

2.6 Suppose that a monopolist operates with total costs of $TC = cQ$ and faces the constant elasticity demand curve $P = Q^{-\alpha}$.

(a) What are the first- and second-order conditions for a profit maximum? When does the second-order condition hold?

(b) Solve for the profit-maximizing levels of output and price. How do output and price change when marginal cost increases?

2.7 In Section 2.4 we chose to set up the monopolist's profits as a function of output. We could have, instead, set up profits as a function of price. Using the assumptions in Section 2.4 write the monopolist's profits with price as the endogenous variable. Solve for the profit-maximizing price and quantity. Are your answers the same as those in the text? Explain why.

2.8 Consider a monopolist that faces a linear demand curve and operates with a constant marginal cost of production (as in Section 2.4). Suppose that the monopolist must pay a sales tax on the product. A sales tax is a percentage of the product's price, so that if the sales tax rate is t, then the total tax owed by the monopolist, T, is t percent of total revenue, $T = tPQ$.

(a) Write the monopolist's profits as a function of output and the exogenous variables.

(b) Find the first- and second-order conditions for a profit maximum.

(c) Solve for the optimal levels of output and price as a function of the exogenous variables.

(d) Find the comparative static effects of a change in the sales tax rate.

(e) Suppose that the government chooses the sales tax rate to maximize tax revenue. Show how to solve for the revenue-maximizing tax rate. Does this relatively simple model yield a solution for the revenue-maximizing tax rate?

2.9 A monopolist faces a linear demand curve and operates with constant marginal cost (as in Section 2.4). The production of the monopolist's good creates a negative externality $E = eQ$. The monopolist has to pay a fine equal to the total external cost created (that is, the fine equals E).

(a) Find the equilibrium solutions for the profit-maximizing price and quantity.

(b) Find the equilibrium solution for the total external cost associated with the profit-maximizing quantity.

(c) If e increases, what will happen to the profit-maximizing output level? Explain the economics.

(d) Under what condition would the fine induce the monopolist to stop production? (That is, when would the monopolist produce a positive amount if it did not have to pay the fine, but choose to produce nothing when it does have to pay the fine?) Explain the economics, including a discussion of whether this possibility changes your answer to part c.

(e) Comparing this model to the model of Section 2.4.2, if the monopolist had to pay a per-unit tax instead of the fine, what tax rate would induce the monopolist to produce the same quantity as in part a? Explain the economics.

2.10 Using the Cournot duopoly model from Section 2.5, find the comparative static effects of increases in the demand intercept, a, and the demand slope, b.

2.11 Consider the Cournot duopoly model from Section 2.5. Suppose that demand is linear, but that each firm operates with a total cost function of $TC_i = cq_i^2$.

(a) Solve for the firms' outputs, the market output, and the market price.

(b) Find the comparative static effects of an increase in the parameter c.

2.12 Consider an industry with n firms. Suppose that each firm operates with a cost function of $TC_i = cq_i + f$, that total market demand is $P = a - bQ$, where $Q = \sum_{i=1}^{n} q_i$, and that output is taxed at a rate of $\$t$/unit.

(a) Assuming that each firm operates under the Cournot assumption with respect to other firms' outputs, solve for q_i, Q, and P.

(b) Now suppose that there is free entry into the industry, i.e., n is endogenous. Solve for the equilibrium number of firms in the industry (hints: (1) firms enter or exit so long as profits are nonzero, (2) you can treat n as a continuous variable). Re-solve for q_i, Q, and P using the equilibrium value of n. How do the variables depend on the level of fixed cost?

(c) Using your answers from part b find the tax rate that would maximize total tax revenue. Find the comparative static effect of a change in f on the revenue-maximizing tax rate.

2.13 Consider a single-good, two-country trade model. The good is produced in country Y and consumed in country X (the good is not consumed in Y nor is it produced in X). Suppose that there are n firms located in country Y, which export their output to country X. Also assume that each firm operates with zero production costs, that each firm operates under the Cournot assumption, and that exports are taxed at a rate of $\$t$ per unit. Finally assume that demand in country X is $P = a - Q$, where Q is the total level of imports into country X.

(a) Write the expression for a typical firm's profits. Find the first-order condition. Solve for the level of a typical firm's output, the level of total exports, and price.

(b) Suppose that welfare in country Y is $W =$ total profits $+$ tax revenue, and that the government of Y chooses t so as to maximize welfare. Solve for the optimal level of t. Is it true that t is chosen so as to generate a monopoly price for exports? Explain.

2.14 In the simple Keynesian model in Section 2.7, find the comparative static effects of changes in r and b.

2.15 Suppose that we take the simple Keynesian model in Section 2.7 and alter the assumption about government spending. Assume that the government adopts a countercyclical fiscal policy. When GDP falls below some target level, Y_0, government spending increases to avoid a recession, and when GDP exceeds Y_0 government spending decreases to avoid overheating the economy. Mathematically, one way of modeling this is by using an equation for government spending of $G = G_0 + g(Y_0 - Y)$, where G_0 and g are positive parameters.

(a) Solve for the equilibrium level of GDP. What assumptions about the value of g are needed to generate an economically meaningful solution to the model?

(b) Find the government spending multiplier. Is this larger or smaller than in the model where government spending is exogenous?

(c) Find the comparative static effect of an increase in g. Try to interpret your result.

2.16 Consider a model of optimal government policy with conflicting goals. The federal government would like to keep both unemployment and inflation low, but in the short run if it uses monetary or fiscal policy to lower one, the other will increase. We will model the government's problem as one of minimizing a loss function $L = u^2 + \pi^2$, where u is the unemployment rate and π is the inflation rate. The government can choose some monetary or fiscal policy action τ, which will affect both u and π. Let $u = A + a\tau$ and $\pi = B + b\tau$ where all parameters are positive.

(a) Write the loss function L as a function of τ.

(b) Find the first- and second-order conditions for the loss-minimization problem.

(c) Solve for the loss-minimizing value of τ.

(d) Find and sign the comparative static effect on τ of an increase in a.

(e) Find and sign the comparative static effect of an increase in a on the inflation rate that results when the government chooses τ to minimize its loss function. That is, find and sign $\partial \pi^* / \partial \tau$.

3 Matrix Theory

3.1 INTRODUCTION

Students of economics know that economics involves *systems* of relationships among many variables. These systems can be very intricate, involving many layers of feedback effects.

When two markets are related, a shift in a supply or demand function in one market can generate a sequence of feedback effects that reverberate through both markets. Suppose, for example, that consumers regard hot dogs and hamburgers as partial substitutes. An upward shift of the supply curve for hot dogs will raise the price of hot dogs and reduce the quantity demanded. Then, through a substitution effect, the demand curve for hamburgers will shift rightward. The resulting increase in the price of hamburgers will create a secondary substitution effect that will shift the demand curve for hot dogs rightward. This rightward shift in the demand curve for hot dogs will increase the quantity of hot dogs demanded. Thus, the net effect on the quantity of hot dogs demanded of the initial upward shift of the supply curve for hot dogs is ambiguous.

In Figure 3.1 we see the initial effects on the markets for hot dogs and hamburgers when the supply curve for hot dogs shifts upward. The initial equilibrium in the market for hot dogs is at point A' in Figure 3.1a. In the market for hamburgers the initial equilibrium is at point B'' in Figure 3.1b. First the supply function for hot dogs shifts upward to S''_{HD}, moving the equilibrium to point A''. The resulting increase in the price of hot dogs shifts the demand function for hamburgers rightward to D''_H. The new equilibrium in the hamburger market is at point B''. This increase in the price of hamburgers shifts the demand function for hot dogs rightward to D''_{HD}, moving the equilibrium in the hot dog market to point A'''. Whether the quantity of hot dogs supplied and demanded in the new equilibrium at point A''' is larger or smaller than the quantity at point A' depends on the relative values of the parameters.

We can see that the rightward shift of the demand curve for hot dogs creates a further increase in the price of hot dogs and that this secondary increase in the price of hot dogs sets off a sequence of feedback effects through both markets. How, then, are we to determine the net effect of an upward shift of the supply curve for hot dogs on the equilibrium quantity of hot dogs demanded?

FIGURE 3.1

Effects of an Upward Shift of the Supply Curve for One Good on the
Equilibria in Two Related Markets: Hot Dogs and Hamburgers

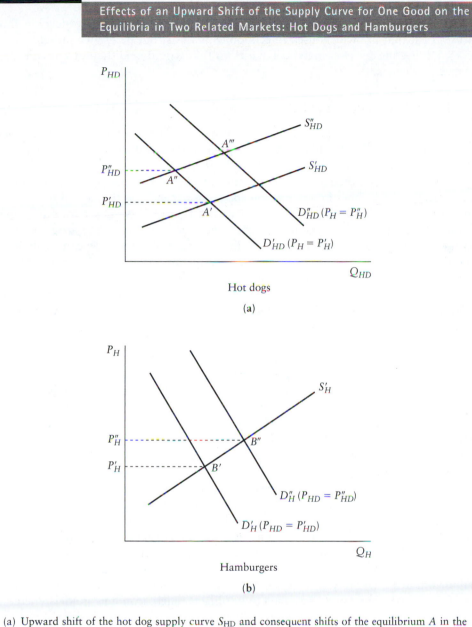

Hot dogs

(a)

Hamburgers

(b)

(a) Upward shift of the hot dog supply curve S_{HD} and consequent shifts of the equilibrium A in the hot dog market
(b) Shift of the equilibrium B in the hamburger market resulting from the upward shift of the hot dog supply curve

Comparative static analyses of related systems would be extremely difficult (if not impossible) if we had to rely on the kinds of graphical analysis presented in introductory economics courses. Fortunately, economists can often use matrix algebra to obtain comparative static results in complicated systems of economic relationships. For small changes in the parameters, we can approximate the consequent changes in the equilibrium values of the endogenous variables by creating systems of linear equations. Then using matrix algebra, we can solve these equations to obtain comparative static results.

The purpose of this chapter is to introduce matrix algebra and to develop those features of it that are widely used in economic analysis. But before we present matrix theory, let us consider two analytical questions that would be difficult to handle without matrix algebra.

3.1.1 Keynesian Systems

In the simplest Keynesian system we consider only one market, a market for goods and services. To establish equilibrium, we equate aggregate demand and supply, which gives us the familiar condition $Y = C(Y) + I + G$. In this equilibrating condition Y is the rate of national income (or national output), $C(Y)$ is the consumption function that determines how the rate of consumption spending depends on the rate of national income, I is the rate of spending on investment goods, and G is the rate of government spending for goods and services. In this model the variables Y and $C(Y)$ are endogenous and the variables I and G are exogenous. In such a simple model we can easily determine the effect of an exogenous change in governmental spending, G, on the equilibrium rate of national income, Y. We have no need for matrix algebra.

Suppose, however, we augment our simple Keynesian system by making investment spending depend on both the interest rate and the rate of national income. And suppose we add a market in which the supply and demand for monetary balances must be equal. Typically, the quantity of money supplied by the monetary authorities is exogenous in such a model.

In this augmented system it is no longer easy to determine the effect of an increase in government spending on the equilibrium rate of national income. Analytical difficulties arise because two or more interrelated markets will undergo a sequence of feedback effects like those we encountered in the markets for hot dogs and hamburgers. If we were to extend our Keynesian system still further by adding a market for labor as well as markets for imports, exports, and foreign currencies, the feedback relationships would become intricate indeed.

The world does, in fact, involve systems of many interrelated markets, and matrix algebra is well suited to sort through all these intricate feedback circuits. Using matrix algebra, we can quickly determine the net effect of a change in the value of an exogenous variable, such as the quantity of money or the rate of spending by the government, on endogenous variables, such as equilibrium national income.

In Chapter 4 we will show specifically how we use matrix algebra to analyze the equilibrium in a linear Keynesian system. Then in Chapter 6 we will see how the methods of matrix algebra can be extended to nonlinear macroeconomic models, including Keynesian models. We will use systems of linear equations to approximate the effects of small changes in the parameters on equilibrium values of the endogenous variables.

We conclude this section with a very simple, specific example.

3.1.2 A Competitive Market

Suppose that in a perfectly competitive market the quantities supplied Q_S and demanded Q_D of some good depend linearly on the price P as follows:

$$Q_D = a - bP$$
$$Q_S = c + dP, \tag{3.1}$$

where a, b, and d are positive constants, c is a constant of unrestricted sign, and $c < a$. To obtain the equilibrium value of the price for this good, we set $Q_S = Q_D$ and solve for P. The solution is $P^* = (a - c)/(b + d)$. The equilibrium value for P obviously depends on the numerical values assigned to the parameters a, b, c, and d.

Now notice that in the system of equations (3.1) the parameters a, b, c, and d form a rectangular array, with a in the upper left corner, b in the upper right corner, and so on. Clearly, the configuration in which the numerical values chosen for a, b, c, and d are distributed over the rectangular array can affect the value of P^*. For example, if the values assigned to a and d were interchanged, the value of P^* would change. Thus, the equilibrium *structure* of the market represented by equations (3.1) is captured by the particular *configuration* of the rectangular array of values for the parameters a, b, c, and d.

Matrix algebra is a mathematical technique that operates on a rectangular array as if that array were a single entity. Using matrix techniques, we can go quickly from the structure of the system to the equilibrium values of the endogenous variables. Furthermore, we can use matrix algebra to extract from the structure of a system of markets the effects of changes in the values of the exogenous variables or parameters on the equilibrium values of the endogenous variables.

3.2   SCALARS, VECTORS, AND MATRICES

The components of matrix algebra are scalars, vectors, and matrices together with a system of rules to define the algebraic operations of addition, subtraction, and multiplication. We need to know these algebraic operations in order to solve systems of equations that are much larger than the system in (3.1).

3.2.1   Scalars and Vectors

A **scalar** *is just a single number*. The scalar might be a known number, like 16, or it might be an unknown number, like the parameter b in equations (3.1).

A **vector** *is an ordered set of scalars*. There are two kinds of vectors: *row vectors* and *column vectors*. The distinction between row and column vectors is critical when we consider matrix multiplication, which is a necessary step in solving systems of equations.

A **row vector** *is an ordered set of scalars written horizontally*. The scalars that constitute a vector are known as the *elements* of that vector. The following is a row vector:

$$\mathbf{X}' = [10 \quad -7 \quad 4 \quad 6 \quad 21]. \tag{3.2}$$

Note that we designate row vectors by a prime ($'$), as do many authors. The row vector in (3.2) is read as "$\mathbf{X}$ prime."

A **column vector** *is an ordered set of scalars (or elements) written vertically*. The following is a column vector:

$$\mathbf{W} = \begin{bmatrix} 16 \\ 7 \\ -3 \\ 2 \end{bmatrix}. \tag{3.3}$$

Both row and column vectors have *dimension*. The dimension of a vector is the number of elements contained in that vector. The vector $\mathbf{X}'$ defined in equation (3.2) is a five-dimensional row vector, and the vector $\mathbf{W}$ defined in equation (3.3) is a four-dimensional column vector.

For both row and column vectors the order in which the elements are written is critical. For example, the row vector $\mathbf{V}' = [10 \; -7 \; 6 \; 4 \; 21]$ is not the same as the row vector $\mathbf{X}'$ defined in (3.2), even though both vectors contain the same elements. The fact that the elements appear in different orders in the vectors $\mathbf{X}'$ and $\mathbf{V}'$ makes these two vectors different.

Suppose, for example, that the elements of the vector $\mathbf{X}'$ specify the quantities of goods 1, 2, 3, 4, and 5 supplied as outputs or demanded as inputs by firm X. Positive values designate quantities supplied and negative values designate quantities demanded. Let the row vector $\mathbf{V}'$ have an analogous definition for firm V. Then firm X uses 7 units of good 2 and produces 10 units of good 1, 4 units of good 3, 6 units of good 4, and 21 units of good 5. Firms X and V differ in the amounts of goods 3 and 4 produced as outputs, even though both firms consume equal quantities of good 2 as an input.

3.2.2 Matrices

A **matrix** *is a rectangular array of scalars.* These scalars may be known constants, unknown constants, or variables. We will consider first the simplest case in which each scalar in the matrix is a known number.

Let the matrix $\mathbf{A}$ be defined by

$$\mathbf{A} = \begin{bmatrix} 10 & 6 & -2 \\ -9 & 0 & 8 \end{bmatrix}. \tag{3.4}$$

Each of the scalars $10, 6, -2, -9, 0$, and 8 is an element of the matrix $\mathbf{A}$. The position of each element in $\mathbf{A}$ is determined by specifying the row and the column in which that element appears. Rows are numbered from top to bottom and columns are numbered from left to right. Thus the element 6 is in row 1 and column 2; the element 8 is in row 2 and column 3.

In general, the typical element of a matrix $\mathbf{A}$ is designated as a_{ij}. The value of the first subscript i specifies a row, and the value of the second subscript j specifies a column. By assigning numerical values to the row and column subscripts, we identify a particular element in the matrix $\mathbf{A}$. For example, $a_{12} = 6$, $a_{23} = 8$, and $a_{21} = -9$. But for the matrix to have an element a_{31}, it would have to have three rows.

A matrix is like a spreadsheet, where numerical information is organized into rows and columns. This information is then manipulated arithmetically to conduct certain kinds of analyses. For example, we might sum the elements in each column. Or we might create a new row for the matrix in which each element is equal to 50% of the value of the corresponding element in row 1. Later we will see that much economic analysis can be done by manipulating rows and columns of matrices.

Economic applications of matrices require that we combine matrices using the algebraic operations of multiplication, addition, and subtraction. To perform these operations we must define the dimension of a matrix.

The **dimension of a matrix** is determined by the number of rows and the number of columns in that matrix. A matrix that has N rows and M columns has dimension $(N \times M)$. The matrix $\mathbf{A}$ defined in (3.4) has dimension (2×3).

Matrices are necessarily rectangular. A special case of a matrix is a *square matrix,* which is a matrix that has the same number of rows and columns $(N \times N)$. Most of the important applications of matrices in economics require that the structure of the system be defined by a square matrix.

It is often convenient to regard a matrix as a set of row vectors or as a set of column vectors. Analogously, a row vector of dimension N can be considered a $(1 \times N)$ matrix, and a column vector of dimension K can be treated as a $(K \times 1)$ matrix. By regarding vectors as special cases of matrices, and matrices as compilations of row and column vectors, we can easily develop the multiplication of vectors and matrices.

3.3 OPERATIONS ON VECTORS AND MATRICES

3.3.1 Multiplication of Vectors and Matrices by Scalars

The simplest operation in matrix algebra is the multiplication of a vector or a matrix by a scalar. To multiply a vector or a matrix by a scalar, we multiply every element of that vector or matrix by that scalar.

In the following examples of scalar multiplication we have multiplied the row vector $\mathbf{X}'$, the column vector $\mathbf{W}$, and the matrix $\mathbf{A}$, all defined below, by a scalar, t. The results are

$$t\mathbf{X}' = t[10 \quad -7 \quad 4 \quad 6 \quad 21] = [10t \quad -7t \quad 4t \quad 6t \quad 21t],$$

$$t\mathbf{W} = t\begin{bmatrix} 16 \\ 7 \\ -3 \\ 2 \end{bmatrix} = \begin{bmatrix} 16t \\ 7t \\ -3t \\ 2t \end{bmatrix}, \tag{3.5}$$

$$t\mathbf{A} = t\begin{bmatrix} 10 & 6 & -2 \\ -9 & 0 & 8 \end{bmatrix} = \begin{bmatrix} 10t & 6t & -2t \\ -9t & 0 & 8t \end{bmatrix}.$$

Scalar multiplication of a vector or a matrix can also be used to factor a constant out of the elements of that vector or matrix. In the following example we use scalar multiplication to factor the constant 2 out of the matrix $\mathbf{B}$:

$$\mathbf{B} = \begin{bmatrix} 16 & 20 \\ 4 & -6 \\ 0 & 14 \end{bmatrix} = \begin{bmatrix} 8(2) & 10(2) \\ 2(2) & -3(2) \\ 0(2) & 7(2) \end{bmatrix} = 2\begin{bmatrix} 8 & 10 \\ 2 & -3 \\ 0 & 7 \end{bmatrix}. \tag{3.6}$$

3.3.2 Addition and Subtraction of Vectors and Matrices

If $\mathbf{R}$ and $\mathbf{S}$ are two vectors that have the same dimension, the sum of $\mathbf{R}$ and $\mathbf{S}$ is a third vector whose dimension is the same as that of $\mathbf{R}$ and $\mathbf{S}$ and whose elements are the sums of the corresponding elements in $\mathbf{R}$ and $\mathbf{S}$. An analogous definition governs the subtraction of $\mathbf{S}$ from $\mathbf{R}$. We can see these operations in the following example. Let

$$\mathbf{R} = \begin{bmatrix} 6 \\ 7 \\ 13 \end{bmatrix} \quad \text{and} \quad \mathbf{S} = \begin{bmatrix} 10 \\ 3 \\ 12 \end{bmatrix}.$$

Then

$$\mathbf{R} + \mathbf{S} = \begin{bmatrix} 6 \\ 7 \\ 13 \end{bmatrix} + \begin{bmatrix} 10 \\ 3 \\ 12 \end{bmatrix} = \begin{bmatrix} 16 \\ 10 \\ 25 \end{bmatrix} \quad \text{and} \quad \mathbf{R} - \mathbf{S} = \begin{bmatrix} 6 \\ 7 \\ 13 \end{bmatrix} - \begin{bmatrix} 10 \\ 3 \\ 12 \end{bmatrix} = \begin{bmatrix} -4 \\ 4 \\ 1 \end{bmatrix}. \tag{3.7}$$

This example illustrates the addition and subtraction of column vectors. The addition and subtraction of row vectors is defined analogously.

Matrices can also be combined by addition or subtraction if the two matrices have the same dimensions. Addition or subtraction of two such matrices produces a third matrix whose dimension is the same as that of the first two and whose elements are generated by adding or subtracting corresponding elements from the first two matrices. Consider the following example.

Let

$$\mathbf{H} = \begin{bmatrix} 0 & 1 \\ -7 & 12 \\ 16 & -10 \end{bmatrix} \quad \text{and} \quad \mathbf{K} = \begin{bmatrix} 8 & 11 \\ 20 & -9 \\ 4 & 1 \end{bmatrix}.$$

Then

$$\mathbf{H} + \mathbf{K} = \begin{bmatrix} 8 & 12 \\ 13 & 3 \\ 20 & -9 \end{bmatrix} \quad \text{and} \quad \mathbf{H} - \mathbf{K} = \begin{bmatrix} -8 & -10 \\ -27 & 21 \\ 12 & -11 \end{bmatrix}. \tag{3.8}$$

3.3.3   Conformability and Transposition

Conformability Condition for Addition (Subtraction) of Matrices and Vectors

Conformability is a property that specifies the dimensions that vectors and matrices must have in order that certain algebraic operations can be performed on them. For example, we have just seen that the sum or difference of two vectors is defined if and only if both vectors have the same dimension. An analogous condition applies to matrices: If $\mathbf{A}$ is a matrix with dimension $(N \times K)$ and $\mathbf{B}$ is a second matrix, then both the sum $\mathbf{A} + \mathbf{B}$ and the difference $\mathbf{A} - \mathbf{B}$ are defined if and only if the dimension of $\mathbf{B}$ is also $(N \times K)$. (Of course, N and K could be equal; $\mathbf{A}$ and $\mathbf{B}$ would then be square matrices.)

If two vectors or two matrices have dimensions such that addition and subtraction are defined, we say that those vectors and matrices are *conformable* for addition and subtraction. Similarly, there is a conformability requirement that governs the multiplication of vectors and matrices. (That requirement is a bit more complicated. We will discuss it after we explain multiplication for vectors and matrices.)

Finally, the conformability requirement for the addition, subtraction, or multiplication of scalars is always satisfied because the dimension of any scalar is (1×1). Therefore, if a and b are any scalars, the sum $a + b$, the difference $a - b$, and the product ab are always defined.

Transposition

Solving systems of linear equations using matrix algebra often requires two kinds of transformations of vectors and matrices. One kind is to convert row vectors to column vectors. A second kind is to convert any matrix into a new one by using the rows of the first matrix to form the columns of the second. Both kinds of transformations are called **transpositions** and their results are known as **transposes.** We transpose vectors and matrices in order to perform matrix multiplication that would otherwise be impossible because conformability requirements would be violated.

Here are some simple examples of transpositions, where a prime (') denotes a transpose:

$$\text{If } \mathbf{R} = \begin{bmatrix} 6 \\ 7 \\ 13 \end{bmatrix}, \quad \text{then } \mathbf{R}' = [6 \quad 7 \quad 13].$$

$$\text{If } \mathbf{X}' = [10 \quad -7 \quad 4 \quad 6 \quad 21], \quad \text{then } (\mathbf{X}')' = \mathbf{X} = \begin{bmatrix} 10 \\ -7 \\ 4 \\ 6 \\ 21 \end{bmatrix}. \tag{3.9}$$

$$\text{If } \mathbf{B} = \begin{bmatrix} 16 & 20 \\ 4 & -6 \\ 0 & 14 \end{bmatrix}, \quad \text{then } \mathbf{B}' = \begin{bmatrix} 16 & 4 & 0 \\ 20 & -6 & 14 \end{bmatrix}.$$

In (3.9) the transpose of the column vector $\mathbf{R}$ is the row vector $\mathbf{R}'$, and the transpose of the row vector $\mathbf{X}'$ is the column vector $\mathbf{X}$. A superscript T may also be used to indicate a transpose. Thus, $\mathbf{R}^{\mathrm{T}}$ and $\mathbf{R}'$ are alternative ways to denote the transpose of the column vector $\mathbf{R}$.

Transposing the matrix $\mathbf{B}$ to the matrix $\mathbf{B}'$ is the result of sequentially transposing the column vectors of $\mathbf{B}$ to generate the corresponding row vectors of $\mathbf{B}'$. Equivalently, we could have transposed the row vectors of $\mathbf{B}$ to create the column vectors of $\mathbf{B}'$.

3.3.4 Multiplication of a Vector by a Vector

The multiplication of two vectors is more complicated than the multiplication of one scalar by another. If a and b are any two scalars (that is, if a and b are any numbers, including positive numbers, negative numbers, and zero), then the product of a times b is always defined. Moreover, scalar multiplication has the property of commutation; that is, the value of a times b is always the same as the value of b times a, regardless of the numerical values of a and b.

These two properties of scalar multiplication do not apply to the multiplication of vectors. If U and V are two vectors, either or both of the vector products UV VU may not be defined. Furthermore, vector multiplication is *not* commutative. Even if both vector products are defined, the product VU will not, in general, be the same thing as the product UV. If U is an N-dimensional row vector and V is an N-dimensional column vector, then the product UV is a scalar, but the product VU is an $(N \times N)$ matrix. The products UV and VU are identical *if and only if* $N = 1$. In that case the products are equal to the same scalar.

Unlike scalar multiplication, vector multiplication is subject to a conformability requirement. That is, two vectors are conformable for multiplication if and only if they have the same dimension. We shall consider first the *premultiplication* of an N-dimensional column vector by an N-dimensional row vector. The result will be a scalar. In the following section we shall examine *postmultiplication* of a column vector by a row vector. The result there will be a matrix.

Let $\mathbf{Q}' = [q_1 \ q_2 \ \cdots \ q_N]$ be an N-dimensional row vector and let $\mathbf{P} = [p_1 \ p_2 \ \cdots \ p_N]$ be an N-dimensional column vector. Then the vector product $E = \mathbf{Q}'\mathbf{P}$ is the scalar defined by

$$E = \mathbf{Q}'\mathbf{P} = [q_1 \quad q_2 \quad \cdots \quad q_N] \begin{bmatrix} p_1 \\ p_2 \\ \vdots \\ p_N \end{bmatrix} = \sum_{i=1}^{N} q_i p_i. \tag{3.10}$$

From (3.10) we see that the multiplication of one vector by another is defined if and only if the two vectors have the same dimension. The scalar produced by the multiplication of two vectors is a sum of N products, with each product formed by multiplying the corresponding elements from the two vectors.

For a numerical example of vector multiplication, let the four-dimensional row vector $\mathbf{Q}' = [20\ 16\ 0\ 13]$ be the quantities of four goods purchased each month by a consumer, and let the four-dimensional column vector $\mathbf{P} = [10\ 5\ 50\ 2]$ be the prices (in dollars) of those goods. Then the total monthly expenditure by this consumer is the vector product $\mathbf{Q}'\mathbf{P}$, whose value is

$$
\mathbf{Q}'\mathbf{P} = [20 \quad 16 \quad 0 \quad 13] \begin{bmatrix} 10 \\ 5 \\ 50 \\ 2 \end{bmatrix} = [(20)(10) + (16)(5) + (0)(50) + (13)(2)] = 306.
$$

(3.11)

For a visually convenient method of vector multiplication, imagine rotating the row vector $\mathbf{Q}'$ $90°$ clockwise (so that it becomes a column vector) and laying it beside the column vector $\mathbf{P}$ so that each element from $\mathbf{Q}'$ lies next to its corresponding element in $\mathbf{P}$. Next, compute the pairwise products of corresponding elements from $\mathbf{Q}'$ and $\mathbf{P}$, creating a column of products. Finally, "collapse" the column of products into their sum. Thus for the vector product $\mathbf{Q}'\mathbf{P}$ from equation (3.11), we have

$$
\mathbf{Q}'\mathbf{P} = [20 \quad 16 \quad 0 \quad 13] \begin{bmatrix} 10 \\ 5 \\ 50 \\ 2 \end{bmatrix} = \begin{bmatrix} (20)(10) \\ + \\ (16)(5) \\ + \\ (0)(50) \\ + \\ (13)(2) \end{bmatrix} = \begin{bmatrix} 200 \\ + \\ 80 \\ + \\ 0 \\ + \\ 26 \end{bmatrix} = 306. \quad \textbf{(3.12)}
$$

We need to treat one more issue. In equation (3.10) the vector product $\mathbf{Q}'\mathbf{P}$ is defined because $\mathbf{Q}'$ is an N-dimensional row vector and $\mathbf{P}$ is an N-dimensional column vector. The vector product $\mathbf{P}\mathbf{Q}'$ is *not* defined, however, because the order in which the multiplication is attempted is improper. Claiming that the vector product $\mathbf{P}\mathbf{Q}'$ is not defined, even though the vector product $\mathbf{Q}'\mathbf{P}$ is defined, might seem excessively fussy. If the product of two N-dimensional vectors is the sum of the pairwise products of their elements, why should we care whether the typical scalar product in that sum is calculated as $p_i q_i$ or as $q_i p_i$? Since scalar multiplication is commutative, the values of $\sum_{i=1}^{N} p_i q_i$ and $\sum_{i=1}^{N} q_i p_i$ are equal. We will see why we need to specify the order in which two vectors are multiplied when we consider matrix multiplication.

3.3.5 Multiplication of a Vector by a Matrix

Most of the applications of matrix algebra in economics require the multiplication of a vector by a matrix. We will develop this kind of multiplication by building on the example in the previous section, where we considered the multiplication of one vector by another vector. Once we complete this example, we will provide a systematic definition of multiplication between a matrix and a vector.

Suppose there are four consumers in the economy, each of whom purchases specified quantities of five goods each month. If we want to compute the total monthly expenditure of each consumer, we need to know the total amount spent each month by consumer 1, the total amount spent by consumer 2, and so on. Matrix multiplication of a vector is a convenient way to handle these calculations.

We begin by defining q_{ij} as the quantity of good j, where $j = 1, \ldots, 5$, purchased each month by consumer i, where $i = 1, \ldots, 4$. So we have 20 elements of the form q_{ij} (4 consumers times 5 goods). Note that it is critical to distinguish between the first and second subscripts.

We next arrange the 20 elements q_{ij} into a (4×5) matrix $\mathbf{Q}$ as follows:

$$\mathbf{Q} = \begin{bmatrix} q_{11} & q_{12} & q_{13} & q_{14} & q_{15} \\ q_{21} & q_{22} & q_{23} & q_{24} & q_{25} \\ q_{31} & q_{32} & q_{33} & q_{34} & q_{35} \\ q_{41} & q_{42} & q_{43} & q_{44} & q_{45} \end{bmatrix}. \tag{3.13}$$

Then let p_j, where $j = 1, \ldots, 5$, be the price of good j. We arrange this information on prices in a column vector $\mathbf{P}$ as follows:

$$\mathbf{P} = \begin{bmatrix} p_1 \\ p_2 \\ p_3 \\ p_4 \\ p_5 \end{bmatrix}. \tag{3.14}$$

Our objective is to calculate the total expenditure of each of the four consumers. Earlier, we stated that a matrix can be viewed as a set of row vectors, or as a set of column vectors. This concept will be handy now.

The total expenditure of consumer 1 is the product of the column vector $\mathbf{P}$ premultiplied by the first row vector of the matrix $\mathbf{Q}$. We know that this vector product is defined because the dimensions of the first row vector of $\mathbf{Q}$ and the column vector $\mathbf{P}$ are equal. In general, then, the total expenditure of any consumer i is the vector product of row i of matrix $\mathbf{Q}$ times the column vector $\mathbf{P}$.

We can now define the product of the column vector $\mathbf{P}$ premultiplied by the matrix $\mathbf{Q}$. If $\mathbf{Q}$ is a (4×5) matrix and $\mathbf{P}$ is a (5×1) (column) vector, then the product $\mathbf{QP}$ is a column vector $\mathbf{E}$ whose dimension is (4×1). The ith row of the (column) vector $\mathbf{E}$ is the scalar formed by the vector product of the ith row of the matrix $\mathbf{Q}$ times the column vector $\mathbf{P}$. Each row of the matrix $\mathbf{Q}$ generates a row in the product vector $\mathbf{E}$. In symbols, the product $\mathbf{QP} = \mathbf{E}$ is

$$\begin{matrix} \mathbf{Q} & \mathbf{P} & = & \mathbf{E} \end{matrix}$$

$$\begin{bmatrix} q_{11} & q_{12} & q_{13} & q_{14} & q_{15} \\ q_{21} & q_{22} & q_{23} & q_{24} & q_{25} \\ q_{31} & q_{32} & q_{33} & q_{34} & q_{35} \\ q_{41} & q_{42} & q_{43} & q_{44} & q_{45} \end{bmatrix} \begin{bmatrix} p_1 \\ p_2 \\ p_3 \\ p_4 \\ p_5 \end{bmatrix} = \begin{bmatrix} \sum_{j=1}^{5} q_{1j} p_j \\ \sum_{j=1}^{5} q_{2j} p_j \\ \sum_{j=1}^{5} q_{3j} p_j \\ \sum_{j=1}^{5} q_{4j} p_j \end{bmatrix}. \tag{3.15}$$

The premultiplication of a vector by a matrix is governed by rules of conformability that are analogous to the rules for the multiplication of a vector by a vector. The premultiplication of a vector by a matrix requires forming a set of vector products. Each of these products is generated by multiplying a row of the matrix $\mathbf{Q}$ against the column vector $\mathbf{P}$.

Now we can state the **conformability condition for the multiplication of a vector by a matrix.** If Q is a matrix and P is a column vector, the product QP is defined if and only if the number of columns in Q is equal to the number of rows in P.

We need this condition in order to compute the vector products formed by multiplying each row vector of Q by the column vector P. But notice that the definition of the product QP requires no condition on the number of rows of the matrix Q. We could, for example, add a fifth and a sixth consumer to our example. The effect would be to add fifth and sixth rows to the matrix Q, and fifth and sixth rows to the product vector E. Remember, each row of the matrix Q generates a row in the product vector E. Similarly, if we had only two consumers in our study, the matrix Q would have only two rows, and the product vector E would have only two rows. The number of rows in the matrix Q is irrelevant to the feasibility of multiplying the column vector P by the matrix Q. But the product QP is not defined unless the number of columns in Q is equal to the number of rows in P.

Let us gather what we have so far. If Q is a $(K \times N)$ matrix and P is an $(M \times 1)$ column vector, then the product QP is defined if and only if $N = M$. The result of the multiplication of P by Q will be a column vector E whose dimension is $(K \times 1)$.

Notice that we are always careful to speak of the *premultiplication* of the vector P by the matrix Q, never the other way around. Multiplication between matrices and vectors is not commutative. If the premultiplication QP of the vector P by the matrix Q is defined, that does not mean that the postmultiplication PQ of P by Q is defined. Even if both the products QP and PQ are defined, these products need not have the same value.

Suppose we were to attempt to compute the product PQ, in which P is a (5×1) (column) vector and Q is a (4×5) matrix. Using the rule for multiplication, we would attempt to form the product PQ by multiplying each row of P by a column of Q. Immediately, we have a problem: each row in P is one-dimensional but each column in Q is four-dimensional. The vector products we would need to calculate to multiply Q by P are not defined. We conclude that for this example, the premultiplication of P by Q is defined, but the postmultiplication of P by Q is not defined.

Let us now consider the economics of the matter. Suppose, for the moment, that we could compute the product PQ by multiplying the (column) vector P by any one of the columns in the matrix Q. (Arithmetically, this would be feasible if there were five consumers instead of four, so that the matrix Q contained five rows instead of four.) The elements in the vector P are prices of *different* goods. But the elements in any given column of Q are quantities of the *same* good. For example, the elements of the first column in Q specify the quantities of good 1 purchased each month by the various consumers. Even if it were arithmetically possible to form the vector product between the vector P and the first column of the matrix Q, it would make no sense economically, because all but one of the terms in that sum would consist of quantities of good 1 multiplied by prices of some other good.

Thus, even if the products QP and PQ are both defined mathematically, one or both of those products could be economic nonsense. When using matrix algebra, then, we must be careful about the order in which multiplication is considered.

3.3.6   Multiplication of a Matrix by a Matrix

We can now easily explain the multiplication of one matrix by another by combining what we have learned with the fact that any matrix can be considered either as a set of row vectors or as a set of column vectors. In fact, the multiplication of one matrix by another is an extension of the process of premultiplying a vector by a matrix.

Consider two matrices, F and G. Let the dimension of F be $(K \times L)$ and the dimension of G be $(L \times N)$. Note that the matrix F has the same number, L, of columns as the matrix **G** has rows. In this case the matrix product **FG** = **T** is defined, and **T** is a matrix that has K rows and N columns.

The product matrix **T** is formed as follows. Each element in the matrix **T** is obtained by multiplying one of the row vectors from **F** against one of the column vectors from **G**. In particular, the element in row i and column j in the matrix **T** is generated by multiplying row i from matrix **F** by column j from matrix **G**. For example, multiplying row 2 from matrix **F** against column 4 from matrix **G** yields the scalar element for row 2, column 4, in the product matrix **T**.

We know that multiplying any row of matrix **F** against any column of matrix **G** is feasible because the rows in **F** have the same dimension as the columns in **G**. (This is because the dimension of any row in **F** is equal to the number of columns in **F**, and the dimension of any column in **G** is equal to the number of rows in **G**.)

Confused? There is no need to be. Working through a simple numerical example will show you the logic in all this.

Let the matrices **F** and **G** and their product **T** be given by

$$\underset{\textbf{F}}{\begin{bmatrix} 6 & 0 & 5 \\ 1 & 8 & 4 \end{bmatrix}} \underset{\textbf{G}}{\begin{bmatrix} 2 & 3 & 0 & 1 \\ 3 & 1 & 1 & 10 \\ 0 & 0 & 5 & 2 \end{bmatrix}} = \underset{\textbf{T}}{\begin{bmatrix} 12 & 18 & 25 & 16 \\ 26 & 11 & 28 & 89 \end{bmatrix}}. \tag{3.16}$$

The four elements in the first row of the product matrix **T** are generated by multiplying the first row of **F** successively against each of the four columns of **G**. We can verify this by calculating the vector product of the first row of **F** and the first column of **G**:

$$\begin{bmatrix} 6 & 0 & 5 \end{bmatrix} \begin{bmatrix} 2 \\ 3 \\ 0 \end{bmatrix} = [(6 \times 2) + (0 \times 3) + (5 \times 0)] = 12. \tag{3.17}$$

Similarly, the element in row 2 and column 4 of the product matrix **T** is generated by the vector product of row 2 of **F** and column 4 of **G**. This product is $[(1 \times 1) + (8 \times 10) + (4 \times 2)] = 89$. (You should verify that the remaining elements in the matrix **T** are generated by vector multiplication of the appropriate rows from **F** and columns from **G**.)

From this example, we can see several properties of matrix multiplication in action.

1. Each row in the premultiplying matrix **F** generates a corresponding row in the product matrix **T**. For example, multiplying row 2 of **F** against column 1 from **G** produces the element for row 2, column 1 in **T**. Multiplying the same row 2 from **F** against column 2 from **G** produces the element for row 2, column 2 in **T**, and so on.

2. Similarly, each column in **G** generates a column in **T**. Each element in a particular column of **T** is produced by multiplying that column from **G** against one of the rows of **F**.

3. We also see the necessity of the conformability property. The elements for the product matrix **T** are generated by successive vector products of rows from **F** and columns from **G**. But these products are not defined unless the number of columns in **F** is equal to the number of rows in **G**.

Here is a good opportunity to use the visual technique for vector multiplication. This technique reinforces the fact that matrix multiplication is simply a systematic replication of vector multiplication. Let us try the technique on the matrix product **FG** = **T** in equation (3.16), where the first row of the matrix **F** is the row vector [6 0 5].

Step 1: To generate the first row of the product matrix **T**, multiply the first row of **F** successively across the columns of **G**.

Step 2: Rotate the first row of **F** clockwise through 90° and lay it along the first column of **G**.

Step 3: Sum the pairwise products, with each term containing one element from the row of **F** and the corresponding element from the column of **G**; that is, $[(6 \times 2) + (0 \times 3) + (5 \times 0)]$. This sum of products is equal to 12, which is the element in row 1, column 1 of **T**.

Step 4: Keeping the first row from **F** in its rotated (vertical) position, move it successively across the columns of **G**, and compute the vector products at each stage, thereby generating the entire first row of the product matrix **T**. Obviously, the four elements that constitute the second row of **T** are generated by rotating the second row of **F** and multiplying that row successively across the four columns of **G**.

We can also use our visual technique "backwards." Multiplying the columns of **G** against the rows of **F** will generate the elements of the product matrix **T**. Just as any row *i* of the matrix **F** generates the corresponding row *i* of the product matrix **T**, any column *j* of the matrix **G** will generate the corresponding columns of **T**.

4. If the matrix product **FG** = **T** is defined, then each row in **F** generates a row in **T** and each column in **G** generates a column in **T**. Therefore, for the matrix product **FG** to be defined, the dimension of the rows in **F** must be the same as the dimension of the columns in **G**. This is equivalent to the requirement that the number of columns in **F** be equal to the number of rows in **G**.

5. The matrix product **GF** is not defined because the number of columns in **G** is not equal to the number of rows in **F**. The matrices **G** and **F** are not conformable for multiplication in the order **GF**.

Notice what would happen if we were to attempt to compute the product **GF**. In equation (3.18) we have arranged the matrices **G** and **F** side by side as if we were going to compute the product **GF**:

$$\overset{\textbf{G}}{\begin{bmatrix} 2 & 3 & 0 & 1 \\ 3 & 1 & 1 & 10 \\ 0 & 0 & 5 & 2 \end{bmatrix}} \overset{\textbf{F}}{\begin{bmatrix} 6 & 0 & 5 \\ 1 & 8 & 4 \end{bmatrix}}. \tag{3.18}$$

To generate the elements for the product **GF**, we must compute the vector products formed by multiplying rows of **G** (since **G** is now the premultiplying matrix) against columns of **F** (since **F** is now the postmultiplying matrix). This is impossible because each row of **G** has four elements, while each column of **F** has only two elements. The conformability requirement is not met for multiplication in the order **GF**.

Finally, we can use the following simple procedure to determine whether a matrix product is defined and what dimensions the resulting product matrix (if defined) will take.

Write the sizes of the two matrices in the order in which you are considering multiplication. For example, if matrix **A** has 3 rows and 5 columns, and matrix **B** has 5 rows and 8 columns, the matrix product **AB** is represented as $[3 \times 5][5 \times 8]$. If the two "inside sizes" are the same (both equal to 5 in our example), then the matrix multiplication **AB** (in that order) is feasible. The resulting product matrix has the size [3 × 8], or 3 rows and 8 columns.

If the inside sizes are different, for example [3 × 5][6 × 3], the two matrices are not conformable for multiplication in this order. But if the outside sizes match, reversing the order of multiplication is possible. The resulting product matrix will have size [6 × 5].

We know that, even if the vector products $\mathbf{UV}$ and $\mathbf{VU}$ are both defined, one of these products will be a scalar and the other product will be a matrix (see Sections 3.3.4 and 3.3.5). Let $\mathbf{U}'$ be an N-dimensional row vector and $\mathbf{V}$ be an N-dimensional column vector. Then $\mathbf{U}'$ is a $(1 \times N)$ matrix and $\mathbf{V}$ is an $(N \times 1)$ matrix. Using our simple rule, the product $\mathbf{U}'\mathbf{V}$ can be represented as $[1 \times N][N \times 1]$, which will produce a (1×1) matrix, that is, a scalar. But the product $\mathbf{VU}'$ is the product $[N \times 1][1 \times N]$, which yields an $(N \times N)$ matrix.

For example, let $\mathbf{U}'$ be the row vector [2 4 0], and let $\mathbf{V}$ be the column vector [1 2 5]. Then the products $\mathbf{U}'\mathbf{V}$ and $\mathbf{VU}'$ are

$$\mathbf{U}'\mathbf{V} = [2 \quad 4 \quad 0]\begin{bmatrix} 1 \\ 2 \\ 5 \end{bmatrix} = 10,$$

$$\mathbf{VU}' = \begin{bmatrix} 1 \\ 2 \\ 5 \end{bmatrix}[2 \quad 4 \quad 0] = \begin{bmatrix} 2 & 4 & 0 \\ 4 & 8 & 0 \\ 10 & 20 & 0 \end{bmatrix}.$$

(3.19)

3.4 SYSTEMS OF EQUATIONS IN MATRIX FORM

3.4.1 Matrix Systems

Before we can use matrix algebra to obtain equilibrium values of endogenous variables as functions of exogenous variables or parameters, we need to express systems of economic relationships using matrix equations.[1] We can obtain many comparative static results by using linear approximations to describe relationships between small changes in values of parameters and the consequent changes in equilibrium values of endogenous variables.

Consider again the simple competitive market described in equations (3.1), which we write as a system:

$$\begin{aligned} Q_D &= a - bP \\ Q_S &= c + dP. \end{aligned}$$

(3.20)

Equilibrium in system (3.20) requires that $Q_S = Q_D$. Let Q be the unknown equilibrium quantity supplied and demanded. Substituting Q for Q_S and Q_D in (3.20) and transposing, we may rewrite equations (3.20) as

$$\begin{aligned} 1Q + bP &= a \\ 1Q - dP &= c. \end{aligned}$$

(3.21)

---

[1]Matrix algebra, also known as *linear algebra*, requires that the relationships to which it is applied be linear. Although matrix algebra is a powerful tool, we must pay for its power and convenience by restricting our analysis to linear relationships. But as economists, we know that life is full of tradeoffs.

Then, in matrix form, the system (3.21) is

$$
\begin{array}{ccc}
\mathbf{H} & \mathbf{K} & = \mathbf{S} \\
\begin{bmatrix} 1 & b \\ 1 & -d \end{bmatrix} & \begin{bmatrix} Q \\ P \end{bmatrix} & = \begin{bmatrix} a \\ c \end{bmatrix}.
\end{array}
\tag{3.22}
$$

Note that the matrix $\mathbf{H}$ and the column vector $\mathbf{S}$ contain the exogenous variables, and the endogenous variables are in the column vector $\mathbf{K}$. Note also that the matrices in (3.22) satisfy the conformability requirement for multiplication.

But what is the use of rewriting system (3.20) in the matrix form shown in (3.22)? Will matrix notation help us get a solution for the equilibrium values of Q and P? Our answer is yes, but we need just a few more tools first.

3.4.2   The Concept of a Solution to a Matrix System

A good way to understand both the nature of a solution to a matrix system of equations and the procedure for obtaining that solution is to study an analogy based on a single equation.

Consider the equation $ax = b$. In this equation a and b are scalar parameters and x is a scalar variable. This single equation is a very simple system of equations: namely, a system containing just one equation and one variable. Nevertheless, the concept of a solution to this equation and the procedure for finding that solution involve the same principles used to solve matrix systems that contain many equations and variables.

If the value of a is nonzero, we can easily solve the equation $ax = b$ for x by dividing both sides by a. That is, the value $x = b/a$ will satisfy the equation. More formally: The equation $ax = b$ has a solution for x if and only if the coefficient a has a multiplicative inverse.

The multiplicative inverse of a scalar a is a number a^{-1}, which has the following property:

$$ a(a^{-1}) = (a^{-1})a = 1. $$

If we either pre- or postmultiply a number by its multiplicative inverse, we always get the number 1. Mathematicians define the multiplicative inverse of any scalar a as the scalar $1/a$. Division may therefore be defined as multiplication by a multiplicative inverse. For example, the division of the scalar 6 by the scalar 3 would be regarded formally as multiplying the scalar 6 by the scalar 1/3.

We now apply this concept of multiplication by a multiplicative inverse to the process of finding a solution to the equation $ax = b$. Remember that a, x, and b are all scalars.

If a is nonzero, then a multiplicative inverse for a will exist. Call this multiplicative inverse a^{-1}. Now use these steps:

$$
\begin{aligned}
ax &= b \\
(a^{-1})ax &= (a^{-1})b \\
(1)x &= \left(\frac{1}{a}\right)b \\
x &= \frac{b}{a}.
\end{aligned}
\tag{3.23}
$$

This procedure is tedious and unneccessary when our system of equations consists of a single equation in just one variable, but what happens when we use this procedure for solving a matrix system that contains several equations and several variables?

Suppose we have the system $\mathbf{AX} = \mathbf{B}$, in which $\mathbf{A}$ is a matrix having K rows and L columns, $\mathbf{X}$ is a column vector having L rows, and $\mathbf{B}$ is a column vector having K rows. Written out, the matrix equation is

$$
\overset{\mathbf{A}}{\begin{bmatrix} a_{11} & a_{12} & \cdots & a_{1L} \\ a_{21} & a_{22} & \cdots & a_{2L} \\ \cdot & \cdot & \cdots & \cdot \\ \cdot & \cdot & \cdots & \cdot \\ \cdot & \cdot & \cdots & \cdot \\ \cdot & \cdot & \cdots & \cdot \\ \cdot & \cdot & \cdots & \cdot \\ a_{K1} & a_{K2} & \cdots & a_{KL} \end{bmatrix}} \overset{\mathbf{X}}{\begin{bmatrix} x_1 \\ x_2 \\ \cdot \\ \cdot \\ \cdot \\ x_L \end{bmatrix}} = \overset{\mathbf{B}}{\begin{bmatrix} b_1 \\ b_2 \\ \cdot \\ \cdot \\ \cdot \\ \cdot \\ \cdot \\ b_K \end{bmatrix}} \tag{3.24}
$$

To find a solution to the simple system, $ax = b$, we premultiplied both sides of the equation by a special number, the inverse a^{-1}. The result was to remove the coefficient a from the left side and to transform the right side to the number b/a, which is the solution for x. By analogy, then, to find a solution to the matrix equation $\mathbf{AX} = \mathbf{B}$, we need to find a special matrix, which we call $\mathbf{A}^{-1}$ (read as "$\mathbf{A}$ inverse") that will act as a multiplicative inverse for the coefficient matrix $\mathbf{A}$.

The matrix $\mathbf{A}^{-1}$ must have two properties. First, we must be able to premultiply both sides of the matrix equation $\mathbf{AX} = \mathbf{B}$ by the matrix $\mathbf{A}^{-1}$. Premultiplying both sides of the equation by the same matrix preserves the relationships among $\mathbf{A}$, $\mathbf{X}$, and $\mathbf{B}$ defined by the equation $\mathbf{AX} = \mathbf{B}$. The second property is that the premultiplication by $\mathbf{A}^{-1}$ must remove the coefficient matrix $\mathbf{A}$ from the left side of $\mathbf{AX} = \mathbf{B}$, leaving only the vector of variables on the left side. The elements remaining on the right side will be the solution values for the elements of vector $\mathbf{X}$.

3.5 THE IDENTITY MATRIX AND THE INVERSE OF A MATRIX

3.5.1 The Identity Matrix

An **identity matrix** is a square matrix in which every element on the diagonal is a 1 and every element off the diagonal is a 0. The *order* of an identity matrix is the number of rows (or equivalently, the number of columns) that it contains. In (3.25), for example, we have an identity matrix $\mathbf{I}$ of order 4:

$$
\mathbf{I} = \begin{bmatrix} 1 & 0 & 0 & 0 \\ 0 & 1 & 0 & 0 \\ 0 & 0 & 1 & 0 \\ 0 & 0 & 0 & 1 \end{bmatrix}. \tag{3.25}
$$

In matrix algebra, an identity matrix $\mathbf{I}$ has a multiplicative property that is (almost) analogous to the multiplicative property of a 1 in scalar algebra. We know that if s is any scalar, then $1s = s1 = s$. In matrix algebra, if $\mathbf{X}$ is a K-dimensional column vector and $\mathbf{I}$ is a $(K \times K)$-dimensional identity matrix, then $\mathbf{IX} = \mathbf{X}$. That is, premultiplying the K-dimensional column vector $\mathbf{X}$ by the K-dimensional identity matrix $\mathbf{I}$ will reproduce the column vector $\mathbf{X}$. For example,

$$\begin{array}{ccc} \mathbf{I} & \mathbf{X} = & \mathbf{X} \\ \begin{bmatrix} 1 & 0 & 0 \\ 0 & 1 & 0 \\ 0 & 0 & 1 \end{bmatrix} \begin{bmatrix} 16 \\ -4 \\ 8 \end{bmatrix} & = & \begin{bmatrix} 16 \\ -4 \\ 8 \end{bmatrix}. \end{array} \tag{3.26}$$

Multiplying the first row of $\mathbf{I}$ against the first (and only) column of $\mathbf{X}$ produces the sum of products $[(1)(16) + (0)(-4) + (0)(8)] = 16$, which is the entry for the first row, first (and only) column in the matrix on the right side. Similarly, the second and third rows of the identity matrix, multiplied against the column vector $\mathbf{X}$, generate the second and third rows of the column vector on the right side.

What about reversing the order of the multiplication to compute the matrix product $\mathbf{XI}$? The product $\mathbf{XI}$ is not defined because the conformability requirement is not satisfied. Because $\mathbf{X}$ is a column vector, the expression $\mathbf{XI}$ requires us to premultiply the $(K \times K)$ matrix $\mathbf{I}$ by the $(K \times 1)$ column vector $\mathbf{X}$. This operation is not feasible because there is only one column in the vector $\mathbf{X}$, while there are K rows in the identity matrix $\mathbf{I}$. If two matrices are to be multiplied, the first matrix must have as many columns as the second matrix has rows.

But there is a way to preserve the analogy between matrix algebra and scalar algebra as far as multiplying by the identity is concerned. The transpose of the K-dimensional column vector $\mathbf{X}$ is the K-dimensional row vector $\mathbf{X}'$. Now the matrix product $\mathbf{X}'\mathbf{I}$ is defined. Premultiplication of any column vector by an identity matrix of proper size will reproduce that column vector. Similarly, any row vector can be reproduced by postmultiplying it by an identity matrix of proper size. Finally, any matrix can be reproduced by pre- or postmultiplying that matrix by identity matrices of proper size. (Unless the matrix to be reproduced is square, different sizes of identity matrices are required for pre- and postmultiplication.)

3.5.2 The Inverse of a Matrix

Let $\mathbf{A}$ be a matrix whose dimension is $(N \times K)$. If the matrix $\mathbf{A}$ has an **inverse matrix $\mathbf{A}^{-1}$**, then

$$\mathbf{A}^{-1}\mathbf{A} = \mathbf{A}\mathbf{A}^{-1} = \mathbf{I}. \tag{3.27}$$

That is, by definition (3.27), multiplication between matrix $\mathbf{A}$ and its inverse $\mathbf{A}^{-1}$ must be feasible in either order and the product of either order of multiplication must be the identity matrix $\mathbf{I}$.

As an exercise, we can demonstrate the fact that if a matrix $\mathbf{A}$ has an inverse, then $\mathbf{A}$ must be square ($N = K$). Squareness, however, is only a *necessary* condition for a matrix to have an inverse. It is not a *sufficient* condition. There are some square matrices that have no inverses. In the following numerical example of a (3×3) matrix $\mathbf{A}$ and its inverse $\mathbf{A}^{-1}$, you should be able to verify that $\mathbf{A}\mathbf{A}^{-1} = \mathbf{A}^{-1}\mathbf{A} = \mathbf{I}$, where $\mathbf{I}$ is the (3×3) identity matrix.

$$\mathbf{A} = \begin{bmatrix} -2 & 0 & 10 \\ 4 & 1 & -1 \\ 0 & 5 & 8 \end{bmatrix}, \quad \mathbf{A}^{-1} = \frac{1}{174} \begin{bmatrix} 13 & 50 & -10 \\ -32 & -16 & 38 \\ 20 & 10 & -2 \end{bmatrix}. \tag{3.28}$$

In (3.28) we have, for convenience, factored out the constant $(1/174)$ from $\mathbf{A}^{-1}$. The elements in the first row of $\mathbf{A}^{-1}$ are therefore $13/174$, $50/174$, and $-10/174$. An analogous statement applies to the second and third rows. Remember that factoring a constant out of a matrix is an application of scalar multiplication (see Section 3.3.1).

Summarizing what we have established so far about identity matrices and inverse matrices, we know that an identity matrix is a square matrix whose diagonal elements are all equal to 1 and whose off-diagonal elements are all equal to 0. The dimension of an identity matrix is the number of rows (or, equivalently, the number of columns) that it contains. Assuming that an identity matrix is of the appropriate dimension, premultiplying a column vector by the identity matrix, or postmultiplying a row vector by the identity matrix, will reproduce that column vector or that row vector. An analogous conclusion holds for pre- or postmultiplying a square matrix by an identity matrix.

Some square matrices are special in that they have inverse matrices associated with them. If the (square) matrix $\mathbf{A}$ has an inverse $\mathbf{A}^{-1}$, then either pre- or postmultiplying $\mathbf{A}$ by its inverse $\mathbf{A}^{-1}$ will produce an identity matrix that has the same square dimensions as those of $\mathbf{A}$ and $\mathbf{A}^{-1}$. But note that only certain square matrices have inverses. A matrix that has an inverse is called a **nonsingular matrix**. Square matrices that do not have inverses are called **singular matrices**.

3.6 DETERMINANTS

Associated with each square matrix is a unique scalar called the **determinant** of that matrix. Economists use the value of that determinant for many analytical purposes, such as calculating the inverse of a matrix. A rigorous, general definition of a determinant is very intricate. Luckily, we do not need to understand either that rigorous definition or its implications to use determinants in most areas of economic analysis. We therefore constrain our discussion to those properties of determinants that are necessary for reading most economics literature.[2]

We shall use a general (2×2) matrix to construct a pragmatic definition of a determinant. We will then explain how this pragmatic definition can be combined with a procedure called the expansion of a determinant by its cofactors to calculate the determinant of a square matrix of any size.

3.6.1 The Determinant of a (2×2) Matrix

Consider the (2×2) matrix $\mathbf{A}$:

$$\mathbf{A} = \begin{bmatrix} a_{11} & a_{12} \\ a_{21} & a_{22} \end{bmatrix}. \tag{3.29}$$

Since $\mathbf{A}$ is a square matrix, it has a determinant. The determinant of $\mathbf{A}$ is the scalar

$$\det \mathbf{A} = |\mathbf{A}| = (a_{11})(a_{22}) - (a_{21})(a_{12}). \tag{3.30}$$

By convention, the expression "det" in front of a matrix, or vertical bars enclosing a matrix, denotes the determinant of that matrix. (Note that we use vertical lines instead of square brackets to designate the determinant of a matrix.) The determinant of any (2×2) matrix is the difference of two terms. The first term is the product of the elements on the diagonal that runs from the upper left corner of the matrix to the lower right corner. This is the term $(a_{11})(a_{22})$. The second term in the determinant, which is subtracted from the first, is the product of the elements along the other diagonal. This is the term $(a_{21})(a_{12})$.

---

[2]Students interested in a rigorous treatment of determinants might consult Anton, Howard, *Elementary Linear Algebra*, 7th edition, John Wiley and Sons, Inc., 1994; or Andrilli, S., and Hecker, D., *Elementary Linear Algebra*, PWS–Kent Publishing Company, Boston, 1993.

Consider the following numerical examples of three determinants:

$$\begin{vmatrix} 16 & 5 \\ 2 & -1 \end{vmatrix} = (16)(-1) - (5)(2) = -26,$$

$$\begin{vmatrix} 8 & -3 \\ 2 & 4 \end{vmatrix} = (8)(4) - (-3)(2) = 38, \tag{3.31}$$

$$\begin{vmatrix} 4 & 2 \\ 12 & 6 \end{vmatrix} = (4)(6) - (2)(12) = 0.$$

The three examples in (3.31) show that the determinant can be positive, negative, or zero. We see also that it is important not to confuse a negative sign attached to an element with the fact that the determinant is the difference between two products. For example, the determinant of the second matrix is $(8)(4) - (-3)(2)$, which equals 38, NOT $(8)(4) - (3)(2)$, which equals 26.

Finally, notice that the determinant of the third matrix is zero. A theorem in matrix algebra states that if a square matrix has any row that is a multiple of any other row, or any column that is a multiple of any other column, then the determinant of that matrix must be zero.

We conclude this section by discussing the relationship of the numerical value of the determinant to the solution of a system of linear equations. A necessary and sufficient condition for a system of N linear equations to have a unique solution is that the determinant of the matrix of coefficients be nonzero.

Consider the following system of two linear equations in two variables, x and y:

$$\begin{matrix} ax + by = c \\ dx + ey = f \end{matrix} \quad \text{or} \quad \begin{bmatrix} a & b \\ d & e \end{bmatrix} \begin{bmatrix} x \\ y \end{bmatrix} = \begin{bmatrix} c \\ f \end{bmatrix}. \tag{3.32}$$

A necessary and sufficient condition for this pair of equations to have a unique solution is that the graphs of the two equations intersect. Since the graph of each equation is a straight line, the condition for a unique solution is that the slopes of the two equations be different.

With y plotted vertically and x plotted horizontally, the slope of the first equation is $-a/b$. The slope of the second equation is $-d/e$. Assume that neither b nor e is zero. Then both equations have finite slopes; that is, neither equation is a vertical line. If the two slopes are different, then the expression

$$-\frac{a}{b} - \left(-\frac{d}{e}\right) = -\left(\frac{a}{b} - \frac{d}{e}\right) = -\frac{1}{be}(ae - bd) \tag{3.33}$$

must be nonzero. If neither b nor e is zero, then a necessary and sufficient condition for the slopes to differ is that the quantity $ae - bd$ be nonzero. But $ae - bd$ is the determinant of the matrix of coefficients for the system defined in (3.32).

We conclude that a necessary and sufficient condition for a system of linear equations to have a unique solution, assuming that the graphs of all of the equations have finite slopes, is that the determinant of the matrix of coefficients be nonzero.

3.6.2 Determinants of Larger Matrices

Expansion of a Determinant by Cofactors

In matrix algebra there are several theorems that enable us to calculate the determinant of a square matrix of any size by constructing a weighted sum of the determinants of (2×2) matrices. This procedure is called the *expansion of a determinant by its cofactors*.

The first step is to define *minors* and *cofactors*. Minors and cofactors are special kinds of determinants. Every square matrix has both a minor and a cofactor associated with each of its elements.

Let A be any square matrix of dimension $(K \times K)$ and let a_{ij} denote the typical element of A. The **minor** of the element a_{ij} is the determinant of the submatrix created by deleting row i and column j of the matrix A. Let M_{ij} denote the minor associated with the element a_{ij} of the matrix A. Thus for a matrix A whose determinant is

$$\det A = \begin{vmatrix} a_{11} & a_{12} & a_{13} \\ a_{21} & a_{22} & a_{23} \\ a_{31} & a_{32} & a_{33} \end{vmatrix},$$

the minor M_{11} of element a_{11} is the determinant of the submatrix created by deleting row 1 and column 1 from A; that is,

$$M_{11} = \begin{vmatrix} \cancel{a_{11}} & \cancel{a_{12}} & \cancel{a_{13}} \\ \cancel{a_{21}} & a_{22} & a_{23} \\ \cancel{a_{31}} & a_{32} & a_{33} \end{vmatrix} = \begin{vmatrix} a_{22} & a_{23} \\ a_{32} & a_{33} \end{vmatrix}.$$

Notice that the minor, sometimes called a *subdeterminant*, is a scalar, just like any other determinant.

The **cofactor**, C_{ij}, associated with an element a_{ij} is defined as

$$C_{ij} = (-1)^{i+j} M_{ij}. \tag{3.34}$$

That is, cofactors are minors whose signs are modified, depending on the position of the element with which the minor is associated. Thus the cofactor associated with the element a_{11} is

$$C_{11} = (-1)^{1+1} M_{11} = (-1)^2 M_{11} = (+1) M_{11}$$

but the cofactor of the element a_{12} is

$$C_{12} = (-1)^{1+2} M_{12} = (-1)^3 M_{12} = (-1) M_{12}.$$

Cofactors are scalars, of course, because minors are scalars.

Let us calculate the cofactors of the second row of the matrix A defined in (3.28). For convenience, we repeat that matrix here and write the three minors associated with the elements of the second row:

$$A = \begin{bmatrix} -2 & 0 & 10 \\ 4 & 1 & -1 \\ 0 & 5 & 8 \end{bmatrix} \tag{3.35}$$

$$M_{21} = \begin{vmatrix} 0 & 10 \\ 5 & 8 \end{vmatrix}, \quad M_{22} = \begin{vmatrix} -2 & 10 \\ 0 & 8 \end{vmatrix}, \quad M_{23} = \begin{vmatrix} -2 & 0 \\ 0 & 5 \end{vmatrix}.$$

Then, using the definition (3.34), the cofactors of the elements of the second row of A are

$$C_{21} = (-1)^{2+1} M_{21} = (-1)[(0)(8) - (10)(5)] = +50$$
$$C_{22} = (-1)^{2+2} M_{22} = (+1)[(-2)(8) - (10)(0)] = -16 \tag{3.36}$$
$$C_{23} = (-1)^{2+3} M_{23} = (-1)[(-2)(5) - (0)(0)] = +10.$$

To check your understanding, you should verify that the values of the cofactors of the first and the third rows are

$$C_{11} = +13, \quad C_{12} = -32, \quad C_{13} = +20,$$
$$C_{31} = -10, \quad C_{32} = +38, \quad C_{33} = -2. \tag{3.37}$$

THEOREM

Expansion of a Determinant by Cofactors

The determinant of any square matrix is equal to the sum of the elements of any row, with each element multiplied by its cofactor. Similarly, the determinant is equal to the sum of the elements of any column, with each element multiplied by its cofactor.

To state this theorem in symbols, let $\mathbf{A}$ be any square matrix of dimension N, let a_{ij} be the element in row i and column j, and let C_{ij} be the cofactor of that element. Then

$$\det \mathbf{A} = |\mathbf{A}| = \sum_{j=1}^{N} a_{ij} C_{ij}, \qquad \text{for any } i, \quad i = 1, \ldots, N$$

$$= \sum_{i=1}^{N} a_{ij} C_{ij}, \qquad \text{for any } j, \quad j = 1, \ldots, N. \tag{3.38}$$

COROLLARY

Effect on the Determinant of Multiplying Any Row or Any Column of the Matrix by a Scalar.

If the matrix $\mathbf{B}$ is formed by multiplying all the elements in any single row or any single column of the square matrix $\mathbf{A}$ by the scalar h, then $|\mathbf{B}| = h|\mathbf{A}|$.

Proof

Let $a_{i,j}$ be the element in row i and column j of $\mathbf{A}$, and let $b_{i,j}$ be the element in row i and column j of $\mathbf{B}$. Suppose that $\mathbf{A}$ and $\mathbf{B}$ are identical except for row k. Each element in row k of $\mathbf{B}$ is h times the corresponding element in $\mathbf{A}$. Then $b_{i,j} = a_{i,j}$ for $i \neq j$, and $b_{k,j} = ha_{k,j}$. Let $C_{k,j}$ be the cofactor of the element $a_{k,j}$ in $\mathbf{A}$. By definition, $C_{k,j} = (-1)^{k+j}$ times the determinant of the submatrix of $\mathbf{A}$ created by deleting row k and column j. Since $b_{i,j} = a_{i,j}$ for $i \neq j$, $C_{k,j}$ is also the cofactor of the element $b_{k,j}$ in $\mathbf{B}$. By using the preceding theorem to expand the determinant of $\mathbf{B}$ by the cofactors of its kth row, and using the fact that $b_{k,j} = ha_{k,j}$, we have:

$$|\mathbf{B}| = \sum_{j=1}^{n} b_{k,j} C_{k,j}$$

$$= \sum_{j=1}^{n} ha_{k,j} C_{k,j} \tag{3.39}$$

$$= h \sum_{j=1}^{n} a_{k,j} C_{k,j}$$

$$= h|\mathbf{A}|,$$

which proves the corollary with respect to forming **B** by multiplying any single row of **A** by the scalar h. The proof with respect to a single column is analogous.

The process of calculating the determinant of a matrix by summing the elements of any row, with each element weighted by its cofactor, is called expanding the determinant by cofactors of a row. Similarly, we can expand the determinant by summing the products of elements and their cofactors from any column.

Let us try this technique on the (3×3) matrix **A** defined in (3.28). Since we can expand the determinant by cofactors of any row or column, it is convenient to choose a row or a column with 0s or 1s in it. Expanding the determinant of **A** by cofactors of its second column, we have

$$\begin{aligned} |\mathbf{A}| &= (0)C_{12} + (1)C_{22} + (5)C_{32} \\ &= (0)(-32) + (1)(-16) + (5)(38) \\ &= 0 - 16 + 190 \\ &= 174. \end{aligned} \tag{3.40}$$

According to the theorem, we will get the same result if we expand the determinant of **A** by any row or any column. Checking this by expanding the determinant of **A** using the cofactors of its third row, we have

$$\begin{aligned} |\mathbf{A}| &= (0)C_{31} + (5)C_{32} + (8)C_{33} \\ &= (0)(-10) + (5)(38) + (8)(-2) \\ &= 0 + 190 - 16 \\ &= 174, \end{aligned} \tag{3.41}$$

as required.

In the last example we constructed the determinant of a (3×3) matrix as a weighted sum of determinants of (2×2) submatrices. Because we know the procedure for obtaining the determinant of a (2×2) matrix, we encountered no problem. But what if we want the determinant of a matrix larger than (3×3)? After a moment's thought, we can see that successive application of the theorem we just applied to a (3×3) matrix lets us calculate the determinant of a square matrix of any size. This calculation is possible because the cofactors of the elements of any matrix are signed subdeterminants of that matrix, so we can expand each of those subdeterminants in terms of the cofactors of that submatrix.

Suppose, for example, we want the determinant of a (4×4) matrix. In a (4×4) matrix the cofactor of the element in any row i and column j will be $(-1)^{i+j}$ times the determinant of the (3×3) submatrix formed by deleting row i and column j from the original matrix. But the determinant of the (3×3) submatrix can itself be calculated by an expansion using the cofactors of any of *its* rows or columns.

In the following numerical example we use cofactors to expand the determinant of (4×4) matrix **H**. To minimize the number of calculations required, we expand the determinant in terms of the cofactors of that row or column having the largest number of 0s in it. Both the third row and the fourth column are candidates; we choose the former.

Let **H** be the matrix

$$\mathbf{H} = \begin{bmatrix} 6 & -1 & 10 & 5 \\ 9 & 2 & -3 & 0 \\ 0 & 4 & 2 & 0 \\ 8 & 1 & -1 & 2 \end{bmatrix}. \tag{3.42}$$

To expand the determinant |**H**| by using cofactors of the third row, we need only the minors M_{32} and M_{33} for the elements h_{32} and h_{33}, respectively. The minors for the elements h_{31} and h_{34} are unnecessary because both of those elements are zero.

The minor M_{32} is the determinant of the submatrix obtained by deleting row 3 and column 2 from the matrix **H**, namely

$$M_{32} = \begin{vmatrix} 6 & 10 & 5 \\ 9 & -3 & 0 \\ 8 & -1 & 2 \end{vmatrix}. \tag{3.43}$$

We can calculate the minor M_{32} by using the cofactors of the third column of the relevant submatrix. The result is

$$\begin{aligned} M_{32} &= 5(-1)^{1+3}[(9)(-1) - (8)(-3)] \\ &\quad + 0 \\ &\quad + 2(-1)^{3+3}[(6)(-3) - (9)(10)] \\ &= 5(15) + 2(-108) \\ &= -141. \end{aligned} \tag{3.44}$$

Using an analogous procedure, the minor M_{33} can be calculated as follows:

$$\begin{aligned} M_{33} &= \begin{vmatrix} 6 & -1 & 5 \\ 9 & 2 & 0 \\ 8 & 1 & 2 \end{vmatrix} \\ &= 5(-1)^{1+3}[(9)(1) - (8)(2)] \\ &\quad + 0 \\ &\quad + 2(-1)^{3+3}[(6)(2) - (9)(-1)] \\ &= 5(-7) + 2(21) \\ &= 7. \end{aligned} \tag{3.45}$$

Returning to the original matrix **H**, we find that the cofactors of the second and third elements of the third row are

$$\begin{aligned} C_{32} &= (-1)^{3+2} M_{32} = (-1)(-141) = 141 \\ C_{33} &= (-1)^{3+3} M_{33} = (+1)(7) = 7. \end{aligned} \tag{3.46}$$

Finally, we have the data required to expand |**H**| by using the cofactors of its third row. The calculation is

$$\begin{aligned} |\mathbf{H}| &= 0C_{31} + 4C_{32} + 2C_{33} + 0C_{34} \\ &= 4(141) + 2(7) \\ &= 578. \end{aligned} \tag{3.47}$$

To obtain the determinant of a (4×4) matrix, we used a concatenated application of the expansion of a determinant by cofactors. We could apply this process to find the determinant of any square matrix, no matter how large.[3]

---

[3]Only masochists calculate determinants for large matrices by hand. Computers easily calculate determinants. We have explained the calculation of determinants because the understanding of some of the economics literature requires a knowledge of how determinants are formed. The same comment applies to the calculation of adjoints and inverses, which we discuss in the next section.

3.7   CONSTRUCTING THE INVERSE OF A MATRIX

The final step required to construct the inverse of a square matrix is to define the adjoint of a matrix.

Let **A** be a square $(N \times N)$ matrix in which a_{ij} denotes the element in row i and column j. Let **C** be the $(N \times N)$ matrix of cofactors associated with the matrix **A**, so that in the matrix **C**, the typical element c_{ij} is the cofactor of the element a_{ij} from the matrix **A**. The **adjoint matrix** for the matrix **A** is the transpose **C'** of its matrix of cofactors, **C**. The adjoint matrix for a matrix **A** is designated as adj **A**. Using this notation, we may write the matrix of cofactors as

$$\mathbf{C} = \begin{bmatrix} c_{11} & c_{12} & \cdots & c_{1N} \\ c_{21} & c_{22} & \cdots & c_{2N} \\ \cdots & & & \\ \cdots & & & \\ c_{N1} & c_{N2} & \cdots & c_{NN} \end{bmatrix}$$

so that

$$\mathbf{C'} = \begin{bmatrix} c_{11} & c_{21} & c_{31} & \cdots & c_{N1} \\ c_{12} & c_{22} & c_{32} & \cdots & c_{N2} \\ & & \cdots & & \\ & & \cdots & & \\ c_{1N} & c_{2N} & c_{3N} & \cdots & c_{NN} \end{bmatrix} = \text{adj } \mathbf{A}. \tag{3.48}$$

Recall that each column j in the transposed matrix is row j in the original matrix and each row i in the transposed matrix is column i in the original matrix. In (3.48) we see that the element in row i and column j in the adjoint matrix is the cofactor of the element of column i and row j in the original matrix **A**. The element in row i and column j of the adjoint matrix **C'** is c_{ji}, and c_{ji} is the cofactor of the element in row j and column i of the original matrix **A**.

We can now (at long last!) construct the inverse of any square matrix. Let **A** be a square matrix. Then $\mathbf{A}^{-1}$, the inverse of matrix **A**, is defined by

$$\mathbf{A}^{-1} = \frac{1}{|\mathbf{A}|} \text{ adj } \mathbf{A}. \tag{3.49}$$

Recall that $|\mathbf{A}|$, which is the determinant of the matrix **A**, is a scalar. Therefore, $1/|\mathbf{A}|$ is also a scalar, if $|\mathbf{A}|$ is *not* equal to zero.

Suppose that $|\mathbf{A}|$ is nonzero. Then definition (3.49) states that the element in row i and column j of $\mathbf{A}^{-1}$ is the element in row i and column j of the adjoint matrix, divided by the scalar, $|\mathbf{A}|$. But the adjoint matrix is the transpose of the cofactor matrix. Therefore, to determine the value of the element in row i and column j of the inverse matrix $\mathbf{A}^{-1}$, we divide the cofactor of the element in row j and column i of the original matrix **A** by the determinant of that matrix.

For example, we will construct the inverse of the matrix **K**, defined below, by using the principle stated in (3.49). Let

$$\mathbf{K} = \begin{bmatrix} 2 & 3 \\ 7 & 16 \end{bmatrix}. \tag{3.50}$$

Then the determinant of **K** is

$$|\mathbf{K}| = [(2)(16) - (3)(7)] = 11. \tag{3.51}$$

Now let **C** be the matrix of cofactors for the matrix **K**. Using the results of Sections 3.6.1, we have

$$C = \begin{bmatrix} 16 & -7 \\ -3 & 2 \end{bmatrix}.$$ (3.52)

Then the adjoint matrix for the matrix **K** is

$$\text{adj } K = C' = \begin{bmatrix} 16 & -3 \\ -7 & 2 \end{bmatrix}.$$ (3.53)

Finally, using the rule stated by definition (3.49) and the information from (3.51) and (3.53), we conclude that the inverse matrix for **K** is

$$K^{-1} = \frac{1}{11} \begin{bmatrix} 16 & -3 \\ -7 & 2 \end{bmatrix} = \begin{bmatrix} 16/11 & -3/11 \\ -7/11 & 2/11 \end{bmatrix}.$$ (3.54)

By the definition of an inverse matrix, multiplying the matrix **K** on either side by its inverse K^{-1} must produce the (2×2) identity matrix. The following calculation demonstrates that postmultiplication of **K** by K^{-1} satisfies this condition:

$$
\begin{aligned}
KK^{-1} &= \begin{bmatrix} 2 & 3 \\ 7 & 16 \end{bmatrix} \begin{bmatrix} 16/11 & -3/11 \\ -7/11 & 2/11 \end{bmatrix} \\
&= \begin{bmatrix} [32/11 - 21/11] & [-6/11 + 6/11] \\ [(7)(16/11) + (16)(-7/11)] & [(7)(-3/11) + (16)(2/11)] \end{bmatrix} \\
&= \begin{bmatrix} 11/11 & 0/11 \\ 0/11 & 11/11 \end{bmatrix} \\
&= \begin{bmatrix} 1 & 0 \\ 0 & 1 \end{bmatrix}.
\end{aligned}
$$ (3.55)

We could show that premultiplication of **K** by its inverse will produce the identity matrix as well.

The computations involving the matrix **K** and its inverse K^{-1} are fairly simple because these matrices are of dimension (2×2). In particular, this small dimension makes the calculation of cofactors and adjoints trivial. To make certain you understand the application of definition (3.48) to the construction of inverse matrices, try constructing the inverse of the (3×3) matrix **A** defined by (3.28).

Note that definition (3.48) establishes that a necessary and sufficient condition for any square matrix to have an inverse matrix is that $|A|$ be nonzero. (At this point it is useful to recall that to solve a single linear equation, or a system of linear equations, we need an inverse to remove the coefficients from the left side of the equations [see Section 3.4.2].) If $|A|$ were equal to zero, then definition (3.49) would require division of each element in the adjoint matrix by zero. But this operation is not defined. Therefore, if the square matrix **A** is to have an inverse, it is necessary that $|A|$ be nonzero.

To see that a nonzero determinant is a sufficient condition for **A** to have an inverse, observe from (3.49) that the value of any specific element of the inverse matrix is found by dividing the corresponding element of the adjoint matrix by the determinant of **A**. Every element of the adjoint matrix is a (transposed) cofactor of an element of **A**. Every element of **A** has a cofactor defined for it because a cofactor is just a determinant of a square submatrix; and every square matrix (or submatrix) has a determinant, even if that determinant has a value of zero. Therefore, the division required by the rule in (3.49) to form the elements of A^{-1} is feasible if and only if $|A|$ is nonzero.

3.8 A NUMERICAL EXAMPLE SOLVED BY MATRIX INVERSION

Consider the following system of three equations in three variables:

$$3x + 5y - 4z = 12$$
$$10y = 6x - 9 \tag{3.56}$$
$$8x = 7y - 3z + 10.$$

There are three ways to solve this system of equations.

One method is to use elementary algebra. We could solve the second equation for y in terms of x, use the resulting expression for y to eliminate that variable from the first and third equations, then solve those two equations simultaneously for x and z. We could then find the solution for y by substituting in any of the three equations the values that we found for x and z. This method is manageable if the number of equations and variables is not large. (Imagine how tedious this process would be if the system were (10×10) rather than (3×3)!)

A second method is to use matrix inversion. To do this, we begin by rewriting equations (3.56) so that all the variables are on the left-hand side and in the same order within each equation. We also want all the constant terms on the right-hand side. The rewritten system is

$$3x + 5y - 4z = 12$$
$$-6x + 10y + 0z = -9 \tag{3.57}$$
$$8x - 7y + 3z = 10.$$

Now define $\mathbf{A}$ as the (3×3) matrix of coefficients, $\mathbf{V}$ as the column vector of variables $[x \; y \; z]$, and $\mathbf{B}$ as the column vector of constant terms $[12 \; -9 \; 10]$. Then equations (3.57) can be rewritten in matrix form as

$$\overset{\mathbf{A}}{\begin{bmatrix} 3 & 5 & -4 \\ -6 & 10 & 0 \\ 8 & -7 & 3 \end{bmatrix}} \overset{\mathbf{V}}{\begin{bmatrix} x \\ y \\ z \end{bmatrix}} = \overset{\mathbf{B}}{\begin{bmatrix} 12 \\ -9 \\ 10 \end{bmatrix}}. \tag{3.58}$$

To use matrix inversion, we must calculate the elements of the inverse matrix $\mathbf{A}^{-1}$. Recalling Section 3.7, we need the cofactors of each element of $\mathbf{A}$, which are

$$c_{11} = (+1)[(10)(3) - (-7)(0)] = 30$$
$$c_{12} = (-1)[(-6)(3) - (8)(0)] = 18$$
$$c_{13} = (+1)[(-6)(-7) - (8)(10)] = -38$$
$$c_{21} = (-1)[(5)(3) - (-7)(-4)] = 13$$
$$c_{22} = (+1)[(3)(3) - (8)(-4)] = 41 \tag{3.59}$$
$$c_{23} = (-1)[(3)(-7) - (8)(5)] = 61$$
$$c_{31} = (+1)[(5)(0) - (10)(-4)] = 40$$
$$c_{32} = (-1)[(3)(0) - (-6)(-4)] = 24$$
$$c_{33} = (+1)[(3)(10) - (-6)(5)] = 60.$$

Then the matrix of cofactors is

$$\mathbf{C} = \begin{bmatrix} 30 & 18 & -38 \\ 13 & 41 & 61 \\ 40 & 24 & 60 \end{bmatrix}. \tag{3.60}$$

Transposing the cofactor matrix gives us the adjoint matrix:

$$\text{adj } \mathbf{A} = \mathbf{C}' = \begin{bmatrix} 30 & 13 & 40 \\ 18 & 41 & 24 \\ -38 & 61 & 60 \end{bmatrix}. \tag{3.61}$$

To obtain the inverse $\mathbf{A}^{-1}$, we divide each element of the adjoint by $|\mathbf{A}|$. Expanding $|\mathbf{A}|$ by cofactors of the second row of $\mathbf{A}$, we have

$$\begin{aligned} |\mathbf{A}| &= -6c_{21} + 10c_{22} + 0c_{23} \\ &= -6(13) + 10(41) = 332. \end{aligned} \tag{3.62}$$

Finally, using definition (3.49), we have

$$\mathbf{A}^{-1} = \frac{1}{|\mathbf{A}|} \text{ adj } \mathbf{A} = \frac{1}{332} \begin{bmatrix} 30 & 13 & 40 \\ 18 & 41 & 24 \\ -38 & 61 & 60 \end{bmatrix}. \tag{3.63}$$

To verify that the matrix in (3.63) is the inverse of $\mathbf{A}$, we perform the following calculation:

$$\begin{aligned} \mathbf{A}^{-1}\mathbf{A} &= \frac{1}{332} \begin{bmatrix} 30 & 13 & 40 \\ 18 & 41 & 24 \\ -38 & 61 & 60 \end{bmatrix} \begin{bmatrix} 3 & 5 & -4 \\ -6 & 10 & 0 \\ 8 & -7 & 3 \end{bmatrix} \\ &= \frac{1}{332} \begin{bmatrix} 332 & 0 & 0 \\ 0 & 332 & 0 \\ 0 & 0 & 332 \end{bmatrix} \\ &= \begin{bmatrix} 1 & 0 & 0 \\ 0 & 1 & 0 \\ 0 & 0 & 1 \end{bmatrix}. \end{aligned} \tag{3.64}$$

With the elements of $\mathbf{A}^{-1}$ calculated, we can now solve equations (3.56) using matrix inversion as follows:

$$\mathbf{AV} = \mathbf{B}$$
$$\mathbf{A}^{-1}\mathbf{AV} = \mathbf{A}^{-1}\mathbf{B}.$$

Then

$$\mathbf{IV} = \mathbf{A}^{-1}\mathbf{B}$$

where

$$\mathbf{V} = \frac{1}{332} \begin{bmatrix} 30 & 13 & 40 \\ 18 & 41 & 24 \\ -38 & 61 & 60 \end{bmatrix} \begin{bmatrix} 12 \\ -9 \\ 10 \end{bmatrix},$$

so that

$$\begin{bmatrix} x \\ y \\ z \end{bmatrix} = \frac{1}{332} \begin{bmatrix} (30)(12) - (13)(9) + (40)(10) \\ (18)(12) - (41)(9) + (24)(10) \\ (-38)(12) - (61)(9) + (60)(10) \end{bmatrix} = \frac{1}{332} \begin{bmatrix} 643 \\ 87 \\ -405 \end{bmatrix} = \begin{bmatrix} 1.94 \\ 0.26 \\ -1.22 \end{bmatrix}. \tag{3.65}$$

We leave you to verify that these solutions for x, y, and z satisfy equations (3.56), subject to a rounding error.

The third method of solving equations (3.56) is to use Cramer's rule.

3.9   CRAMER'S RULE

3.9.1   Definition of Cramer's Rule

Cramer's rule is a quick method that is widely used in economics to solve systems of linear equations.

CRAMER'S RULE

If X is a column vector of variables in the matrix equation $AX = B$, then the solution value for the element in the ith row of X, namely the element x_i, is the quotient of two determinants. The determinant in the denominator is $|A|$. The determinant in the numerator is the determinant of the matrix obtained by replacing the ith column of matrix A with the column vector B of constants from the right side of the equation.

Let us examine Cramer's rule in symbols. The terms of the matrix equation $AX = B$ are

$$\begin{bmatrix} a_{11} & a_{12} & \cdots & a_{1i} & \cdots & a_{1N} \\ a_{21} & a_{22} & \cdots & a_{2i} & \cdots & a_{2N} \\ . & . & \cdots & . & \cdots & . \\ a_{i1} & a_{i2} & \cdots & a_{ii} & \cdots & a_{iN} \\ . & . & \cdots & . & \cdots & . \\ a_{N1} & a_{N2} & \cdots & a_{Ni} & \cdots & a_{NN} \end{bmatrix} \begin{bmatrix} x_1 \\ x_2 \\ . \\ x_i \\ . \\ x_N \end{bmatrix} = \begin{bmatrix} b_1 \\ b_2 \\ . \\ b_i \\ . \\ b_N \end{bmatrix}. \tag{3.66}$$

We define $A(i)$ to be the matrix obtained by replacing column i of the matrix A with the column vector B:

$$A(i) = \begin{bmatrix} a_{11} & a_{12} & \cdots & b_1 & \cdots & a_{1N} \\ a_{21} & a_{22} & \cdots & b_2 & \cdots & a_{2N} \\ . & . & \cdots & . & \cdots & . \\ . & . & \cdots & . & \cdots & . \\ a_{N1} & a_{N2} & \cdots & b_N & \cdots & a_{NN} \end{bmatrix}. \tag{3.67}$$

Then by Cramer's rule, the solution value for x_i, the ith element of the vector X, is

$$x_i = \frac{|A(i)|}{|A|}. \tag{3.68}$$

3.9.2   Applying Cramer's Rule

To solve equations (3.56) by Cramer's rule, we calculate three ratios of determinants.

Let $A(i)$ be the matrix obtained by replacing column i of matrix A with the column vector B of constants from the right-hand side of equations (3.56). Then the solution value for the first variable x is

$$\begin{aligned} x &= \frac{|A(1)|}{|A|} \\ &= \frac{1}{332} \begin{vmatrix} 12 & 5 & -4 \\ -9 & 10 & 0 \\ 10 & -7 & 3 \end{vmatrix} \\ &= \frac{1}{332}\{(-9)(-1)[(5)(3) - (-7)(-4)] + 10(+1)[(12)(3) - (10)(-4)]\} \\ &= \frac{1}{332}[9(15 - 28) + 10(36 + 40)] \\ &= 1.94. \end{aligned} \tag{3.69}$$

Notice that in equations (3.69) we calculated $|\mathbf{A}(1)|$ by using the cofactors of the second row of the matrix $\mathbf{A}(1)$.

The solution for y is found by

$$
\begin{aligned}
y &= \frac{|\mathbf{A}(2)|}{|\mathbf{A}|} \\
&= \frac{1}{332} \begin{vmatrix} 3 & 12 & -4 \\ -6 & -9 & 0 \\ 8 & 10 & 3 \end{vmatrix} \\
&= \frac{1}{332}\{(-6)(-1)[(12)(3)-(10)(-4)]-(9)(+1)[(3)(3)-(8)(-4)]\} \\
&= \frac{1}{332}[6(36+40)-9(9+32)] = \frac{456-369}{332} \\
&= 0.26.
\end{aligned}
\tag{3.70}
$$

Finally, the solution for z is

$$
\begin{aligned}
z &= \frac{|\mathbf{A}(3)|}{|\mathbf{A}|} \\
&= \frac{1}{332} \begin{vmatrix} 3 & 5 & 12 \\ -6 & 10 & -9 \\ 8 & -7 & 10 \end{vmatrix} \\
&= \frac{1}{332}\{3(+1)[100-(-7)(-9)]+5(-1)[(-6)(10)-(8)(-9)] \\
&\quad + 12(+1)[(-6)(-7)-(8)(10)]\} \\
&= \frac{1}{332}\{3[37]-5[-60+72]+12[42-80]\} = -\frac{405}{332} \\
&= -1.22.
\end{aligned}
\tag{3.71}
$$

Problems

3.1   Given $\mathbf{A} = \begin{bmatrix} 8 & 3 & 5 \\ 2 & 5 & 9 \end{bmatrix}.$

    (a)   Write the elements a_{23}, a_{22}, a_{13} of matrix $\mathbf{A}$.

    (b)   Write all (1×3) row vectors and all (2×1) column vectors of matrix $\mathbf{A}$.

3.2   Given $\mathbf{A}' = [3 \ -4 \ 9 \ 12]$ and $\mathbf{B}' = [-1 \ 0 \ 4 \ 5]$.

    (a)   Compute $3\mathbf{A}$ and $5\mathbf{B}$.

    (b)   Compute $2\mathbf{A} - 3\mathbf{B}$ and $\mathbf{A} + \mathbf{B}$.

    (c)   Compute $\mathbf{A}'\mathbf{B}$.

3.3   Given $\mathbf{A} = \begin{bmatrix} 3 & 7 \\ -2 & 4 \\ 5 & 8 \end{bmatrix}$ and $\mathbf{B} = \begin{bmatrix} -1 & 5 \\ 0 & 7 \\ -2 & 9 \end{bmatrix}.$

    (a)   Compute $5\mathbf{A} + \mathbf{B}$.

(b) Compute $\mathbf{A'}$, $\mathbf{B'}$, $\mathbf{A'} - \mathbf{B'}$, and $\mathbf{A'} + \mathbf{B'}$.

(c) Compute $\mathbf{AB'}$ and $\mathbf{BA'}$.

3.4 Given $\mathbf{A} = \begin{bmatrix} -1 & -2 \\ 0 & 1 \\ 2 & 15 \\ 3 & 9 \end{bmatrix}$ and $\mathbf{B} = \begin{bmatrix} 3 \\ 5 \end{bmatrix}$.

(a) Calculate $\mathbf{AB}$.

(b) Is $\mathbf{BA}$ defined? Explain.

3.5 Write the following system of equations in matrix form:

$$Q - 5P = 100$$
$$2Q - 3P = 80$$

3.6 Pre- and postmultiply the following matrices with appropriate identity matrices and show that in each case you reproduce the original matrix.

$$\mathbf{A} = \begin{bmatrix} 1 & 3 \\ 3 & 6 \\ 5 & 7 \end{bmatrix} \quad \text{and} \quad \mathbf{B} = \begin{bmatrix} 1 & -2 & 4 \\ 5 & 2 & 8 \\ 10 & 6 & 7 \end{bmatrix}$$

3.7 Calculate the determinants of the following matrices:

$$\mathbf{A} = \begin{bmatrix} -3 & 6 \\ -1 & 2 \end{bmatrix} \quad \mathbf{B} = \begin{bmatrix} 8 & 6 \\ 8 & 7 \end{bmatrix} \quad \mathbf{F} = \begin{bmatrix} 12 & 5 \\ 16 & 3 \end{bmatrix}$$

3.8 Given

$$\mathbf{A} = \begin{bmatrix} -3 & 0 & 7 \\ 2 & 5 & 1 \\ -1 & 0 & 5 \end{bmatrix}, \quad \mathbf{B} = \begin{bmatrix} 4 & -1 & 1 & 6 \\ 0 & 0 & -3 & 3 \\ 4 & 1 & 0 & 14 \\ 3 & 1 & 3 & 2 \end{bmatrix}, \quad \text{and} \quad \mathbf{G} = \begin{bmatrix} 3 & 0 & 7 \\ -2 & 5 & 1 \\ 1 & 0 & 5 \end{bmatrix}.$$

For any matrix let c_{ij} be the cofactor of the element in row i and column j.

(a) For matrix $\mathbf{A}$, calculate c_{13}, c_{32}, c_{22}.

(b) For matrix $\mathbf{B}$, calculate c_{24}, c_{32}.

3.9 Use expansion by cofactors to calculate the determinants of matrices $\mathbf{A}$, $\mathbf{B}$, and $\mathbf{G}$ in Problem 3.8.

3.10 Using matrix $\mathbf{A}$ in Problem 3.8, calculate (a) adj $\mathbf{A}$ and (b) $\mathbf{A}^{-1}$.

3.11 Solve the following system of equations using matrix inversion:

$$4x + 5y = 2$$
$$11x + y + 2z = 3$$
$$x + 5y + 2z = 1$$

3.12 Solve the system of equations in Problem 3.11 using Cramer's rule.

4 Applications of Matrix Theory to Linear Models

4.1 INTRODUCTION

In this chapter we demonstrate several applications of matrix theory, choosing examples from both microeconomics and macroeconomics that are graduated in complexity. In each example we will first obtain *reduced-form solutions,* which define equilibrium values of the endogenous variables as functions of the exogenous variables. We will then conduct comparative static analyses of these solutions.

We begin by using matrix algebra to reformulate the simple model of a single competitive market first analyzed in Chapter 1. Then we turn to a system with two competitive markets, using the power of matrix algebra to examine the interactions that arise between markets when goods are substitutes or complements. We consider substitutability and complementarity both for consumers and producers.

Our next two examples come from the model of duopoly presented in Chapter 2. We begin with the simplest version of duopoly, then relax the Cournot assumption to consider the effect of nonzero conjectural variations.

We conclude the chapter by analyzing two macroeconomic models. The first is the simple Keynesian model in which both the interest rate and the rate of government spending are exogenous and investment spending depends on the interest rate. The second model is an IS–LM model. The endogenous variables are the rate of national income, the interest rate, and the rates of spending for consumption and investment. The money supply and the rate of government spending are exogenous.

In the models that we examine in this chapter all relationships are linear. This is necessary if we are to use matrix algebra. In Chapters 5 and 6 we will use matrix algebra to do comparative static analyses for linear approximations of nonlinear systems.

4.2 A SINGLE COMPETITIVE MARKET

In the model of a competitive market for a single good that we examined in Chapter 1 the inverse demand and supply functions are

$$P = a - bQ_D \quad \text{(demand)}$$
$$P = c + dQ_S \quad \text{(supply)}.$$

(4.1)

The endogenous variables in this system are P, Q_D, and Q_S. The parameters are a, b, c, and d.

We will use matrix algebra to determine the equilibrium values of the endogenous variables in terms of the parameters. Our first step is to impose the equilibrium condition $Q_D = Q_S$. If we let Q be the (common) equilibrium value of Q_D and Q_S, we can rewrite equations (4.1) as the system

$$P + bQ = a$$
$$P - dQ = c. \tag{4.2}$$

In system (4.2) the endogenous variables P and Q are on the left side and the parameters a and c are on the right side. This reformulation leads easily to the matrix form

$$\begin{array}{ccc} A & Q & = B \end{array}$$
$$\begin{bmatrix} 1 & b \\ 1 & -d \end{bmatrix} \begin{bmatrix} P \\ Q \end{bmatrix} = \begin{bmatrix} a \\ c \end{bmatrix}. \tag{4.3}$$

Note that the matrix system (4.3) is *equivalent* to the system in equations (4.1); that is, both systems express the same relationships among variables. If two systems of equations are equivalent, any solution to either system is a solution to the other system.

Let P^* and Q^* be the values of P and Q that will satisfy system (4.3). The values P^* and Q^* are, of course, *equilibrating* (market-clearing) values for price and quantity in the market represented by system (4.3). We will use Cramer's rule to find P^* and Q^*.

According to Cramer's rule, the value of the ith variable in a system of linear equations is equal to a quotient of two determinants. The denominator is the determinant of the coefficient matrix A. The numerator is the determinant of the matrix formed by replacing the ith column of the matrix A by the column of constants on the right side of the matrix equation (4.3).

Let $|A|$ be the determinant of the original coefficient matrix A and let $|A(i)|$ be the determinant of the matrix formed by replacing the ith column of A by the column vector of constants from the right side of equation (4.3). These determinants are

$$|A| = [(1)(-d) - (b)(1)] = -(d + b)$$

$$|A(1)| = [(a)(-d) - (b)(c)] = -(ad + bc) \tag{4.4}$$

$$|A(2)| = [(1)(c) - (a)(1)] = c - a.$$

Using Cramer's rule, the solution to (4.3) is

$$P^* = \frac{|A(1)|}{|A|} = \frac{ad + bc}{d + b}$$

$$\tag{4.5}$$

$$Q^* = \frac{|A(2)|}{|A|} = \frac{a - c}{d + b}.$$

Note that the values for P^* and Q^* obtained in equations (4.5) by Cramer's rule are identical to the values that we obtained in Chapter 1, where we used the conventional method to solve a system of two linear equations.

In Chapter 1 we also considered the effect of a per-unit tax on the equilibrium values for price and quantity. We did this by replacing the parameter c by $c + t$, and defining t as the unit tax rate. As you know, the imposition of a unit tax t shifts each firm's marginal cost curve upward by the amount of the tax. Consequently, the competitive industry's supply curve shifts upward by the amount of the tax. Clearly, we can conduct the same kind of analysis using matrix algebra. We simply replace the parameter c with a new parameter, $c + t$, in the column vector of constants on the right side of system (4.3).

4.3 TWO COMPETITIVE MARKETS: SUBSTITUTABILITY AND COMPLEMENTARITY

In the preceding section we had only two variables—equilibrium price and quantity for a single good. With only two variables, and hence only two equations, it would have been easy to obtain the equilibrating values P^* and Q^* by using ordinary algebra. Using matrix algebra to solve a (2×2) system is like building an entire railroad to haul just one ton of coal. It's more trouble than it's worth! In this section, however, we will analyze the equilibrium of two related competitive markets in which we have FOUR variables and FOUR equations—a supply equation and a demand equation for each of the two markets. In this case, matrix algebra is worth the trouble; in fact, the larger the system under analysis, the greater the advantage of using matrix algebra. Solving a system of four or more simultaneous equations using ordinary algebra is not fun.

Suppose there are perfectly competitive markets for goods 1 and 2. Suppose also that the quantity demanded of each good depends linearly on the prices of both goods. This structure of dependence will enable us to consider cases in which consumers regard the two goods as substitutes, complements, or neither. We can now construct a model that can handle the kinds of questions we raised at the outset of Chapter 3 about interrelated markets for goods like hot dogs and hamburgers.

Let the demand functions for the two goods be

$$
\begin{aligned}
Q_1^D &= d_1 + d_{11} P_1 + d_{12} P_2 \\
Q_2^D &= d_2 + d_{21} P_1 + d_{22} P_2.
\end{aligned}
\tag{4.6}
$$

The signs of d_{12} and d_{21} depend on the relationship between goods 1 and 2. If the two goods are *substitutes*, so that an increase in the price of one good will increase the quantity demanded of the other good, the signs of d_{12} and d_{21} will be positive. If goods 1 and 2 are *complements*, so that an increase in the price of one good will decrease the quantity demanded of the other good, the signs of d_{12} and d_{21} will be negative. If there is *no relationship* between the quantity demanded of one good and the price of the other good, both d_{12} and d_{21} will be zero.[1]

---

[1]If goods 1 and 2 are complements or substitutes in consumption, then the quantity demanded of good 1 will depend on both the price and quantity demanded of good 2, as well as on the price of good 1. Similarly, the quantity demanded of good 2 will depend on the quantity demanded of good 1 and on both prices. For example, automotive gasoline and oil are complements. The quantity demanded of either will depend on the quantity demanded of the other.

To show this interdependence of quantities demanded, we might rewrite equations (4.6) as

$$
\begin{aligned}
Q_1^D &= d_1' + d_{11}' P_1 + d_{12}' P_2 + e_{12} Q_2^D \\
Q_2^D &= d_2' + d_{21}' P_1 + d_{22}' P_2 + e_{21} Q_1^D.
\end{aligned}
\tag{4.6a}
$$

If goods 1 and 2 are complements, then e_{12} and e_{21} will be positive.

Systems (4.6) and (4.6a) are equivalent. In (4.6a) we use the second equation to substitute for Q_2^D in the first equation. Solving the first equation for Q_1^D will produce the first equation in (4.6) if we define $d_1 = (d_1' + e_{12} d_2')/(1 - e_{12} e_{21})$, $d_{11} = (d_{11}' + e_{12} d_{21}')/(1 - e_{12} e_{21})$, and $d_{12} = (d_{12}' + e_{12} d_{22}')/(1 - e_{12} e_{21})$. An analogous manipulation will produce the second equation in (4.6).

4.3.1 Graphical Illustration

There is a graphical way to exhibit the relationships of substitutes and complements. If we solve the demand function for good 1 to obtain P_1, we have the inverse demand function $P_1 = Q_1^D/d_{11} - (d_1 + d_{12}P_2)/d_{11}$. We plot this demand function on a two-dimensional graph, with P_1 measured vertically and Q_1^D measured horizontally. The vertical intercept will be the term $-(d_1 + d_{12}P_2)/d_{11}$. If we make the usual assumption that the quantity demanded of any good is inversely related to its price, then d_{11} (and d_{22}) are negative.[2] Consequently, an increase in P_2 will shift the demand function for good 1 upward and rightward if d_{12} is positive, and downward and leftward if d_{12} is negative. If goods 1 and 2 are substitutes, an increase in the price of good 2 should shift the demand function for good 1 rightward and upward. In Figure 4.1 an upward shift of the supply function for good 2 in the right panel increases the price of that good. If goods 1 and 2 are substitutes, the sign of d_{12} must be positive in order to shift the demand function for good 1 rightward and upward in the left panel. By symmetry, both d_{12} and d_{21} must be positive if goods 1 and 2 are substitutes. If goods 1 and 2 are complements, d_{12} (and d_{21}) must be negative.

We now consider the supply side of this competitive market. The phenomena of substitutability and complementarity apply to producers as well as to consumers. Let the two linear supply functions be

$$Q_1^S = s_1 + s_{11}P_1 + s_{12}P_2$$
$$Q_2^S = s_2 + s_{21}P_1 + s_{22}P_2. \tag{4.7}$$

Note that each of the supply functions contains a term for the price of the other good. The signs of the coefficients s_{12} and s_{21} for the other prices P_2 and P_1 are determined by whether the two goods are substitutes or complements in production.[3]

Suppose, for example, that good 1 is oats and good 2 is wheat. Since both crops can be grown on the same type of farm, these two goods are substitutes in production; that is, oats and wheat compete for the use of the farmer's land, equipment, and time. The inverse supply function for good 1, oats, is

$$P_1 = \frac{Q_1^S}{s_{11}} - \frac{s_1 + s_{12}P_2}{s_{11}}. \tag{4.8}$$

If the quantity of good 1 supplied is plotted horizontally and the price of good 1 is plotted vertically, then equation (4.8) provides the (inverse) supply function for good 1 with the price of good 2 held constant. The vertical intercept of this function is the term $-(s_1 + s_{12}P_2)/s_{11}$.

The usual assumption is that s_{11} is nonnegative. We assume that s_{11} is positive and finite.

Suppose that P_2, the price of wheat, increases. This means that for any given price of oats, it is now profitable for the farmer to increase the percentage of land allocated to wheat, and reduce the percentage allocated to oats. The (inverse) supply function for oats will shift leftward and upward in response to an increase in the price of wheat.

---

[2] If d_{11} were equal to zero, the demand for good 1 would be perfectly inelastic (quantity demanded of good 1 does not depend on its price) and the inverse demand function would be vertical. If the demand for good 1 were perfectly elastic, the value of d_{11} would be infinitely large and negative.

[3] The quantity supplied of each good could depend on the quantity supplied of the other good, as well as on the two prices. The comment in footnote 1 about demand functions applies to supply functions also.

FIGURE 4.1

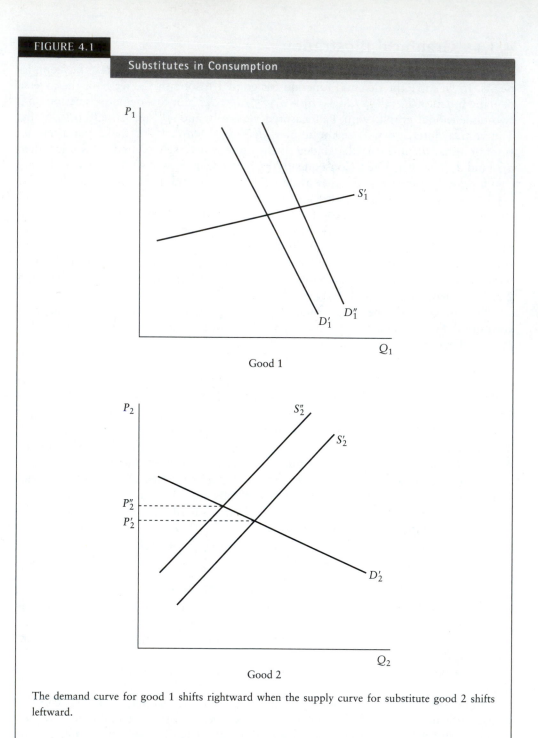

Substitutes in Consumption

The demand curve for good 1 shifts rightward when the supply curve for substitute good 2 shifts leftward.

Therefore, if wheat and oats are substitutes in production, the sign of s_{12} (and s_{21}) must be negative.

In Figure 4.2, we see how a change in the price of wheat P_2 shifts the supply function of oats S_1, assuming these two goods are substitutes in production. The prices for wheat and oats are initially in equilibrium where S_2' intersects D_2' at point A' for wheat, and where S_1' intersects D_1' at point B' for oats. If the demand function for wheat shifts

FIGURE 4.2

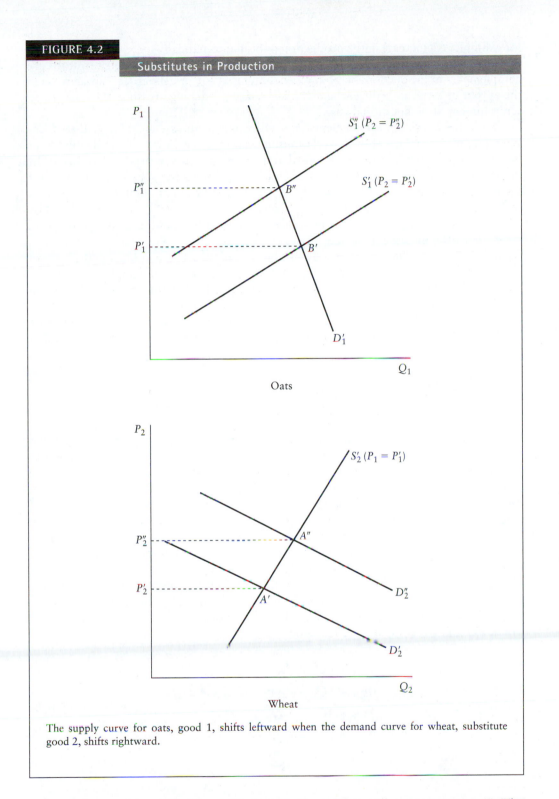

Oats

Wheat

The supply curve for oats, good 1, shifts leftward when the demand curve for wheat, substitute good 2, shifts rightward.

rightward to position D_2'', the equilibrium in that market will move to point A''. The higher price of wheat will shift the supply function for oats leftward and upward to position S_1'', resulting in a higher price P_1'' of oats with equilibrium at point B''.

Two goods are complements in production if an increase in the output of one good reduces the (marginal) cost of producing the other good. Consider beef and cowhides, for example. An increase in the production of beef automatically increases (at zero marginal cost) the output of cowhides. Similarly, pressboard and wooden furniture are complements in production. Scrap lumber is an input for pressboard, so an increase in the output of furniture reduces the marginal cost of producing pressboard by providing scrap lumber at no additional cost.[4]

If goods 1 and 2 are complements in production, s_{12} and s_{21} must be positive. Consider Figure 4.3, which shows a relationship between the (inverse) supply functions (marginal cost functions) of two goods that are complements in production. Let P_2' be the initial equilibrium price for good 2. Figure 4.3a shows the supply function S_1' for good 1 when the initial price of good 2 is P_2' with equilibrium at point A'. Suppose that the demand function for good 2 shifts rightward from D_2' to D_2'', as shown in Figure 4.3b. Producers of good 2 will then increase their rates of output from Q_1' to Q_2'' in response to the higher equilibrium price, P_2'', for that good. The increased output of good 2 will increase the output of a by-product of good 2, thereby reducing the price of that by-product. If that by-product is an input for good 1, the supply function for good 1 will shift downward and rightward from S_1' to S_1''. For this to occur, the sign of s_{12} must be positive.[5]

Finally, if goods 1 and 2 are unrelated in production, in the sense that an increase in the output of either good has no effect on the marginal costs of producing the other good, then the coefficients s_{12} and s_{21} will be zero.

4.3.2 Matrix Algebra Methods

The changes we have just described represent only the first round of changes if two goods are substitutes or complements in consumption or production. A shift of the supply (demand) function for good 1 causes its price to change, leading to a shift in the supply (demand) function for good 2. Then the price of good 2 changes, leading to a further shift in the supply (demand) functions for good 1. Thus, a second round of reverberations between the two markets is touched off. We cannot easily determine the final equilibrium prices using ordinary algebraic methods, but we can use matrix algebra. As we do this, however, it is important to keep the economics in prominent view. Failure to do so can allow the mathematics to obfuscate the economics.

Let us now rewrite the system of two demand and two supply functions specified in equations (4.6) and (4.7) in a form convenient for matrix algebra. Define Q_1 and Q_2 as the equilibrium quantities of goods 1 and 2, respectively. In equilibrium, $Q_1 = Q_1^D = Q_1^S$ and $Q_2 = Q_2^D = Q_2^S$. Therefore

$$
\begin{aligned}
1Q_1 + 0Q_2 - d_{11}P_1 - d_{12}P_2 &= d_1 \\
0Q_1 + 1Q_2 - d_{21}P_1 - d_{22}P_2 &= d_2 \\
1Q_1 + 0Q_2 - s_{11}P_1 - s_{12}P_2 &= s_1 \\
0Q_1 + 1Q_2 - s_{21}P_1 - s_{22}P_2 &= s_2.
\end{aligned}
$$

(4.9)

---

[4]Reducing the cost of an input for pressboard will shift the marginal cost for pressboard downward. Increasing the output of furniture reduces to zero the marginal cost for scrap lumber incurred by pressboard operation. Consequently, the marginal cost for producing pressboard shifts downward when the output of furniture increases.

[5]This is a good opportunity to check your understanding of the difference between a shift of a curve and a movement along the curve.

FIGURE 4.3

Complements in Production

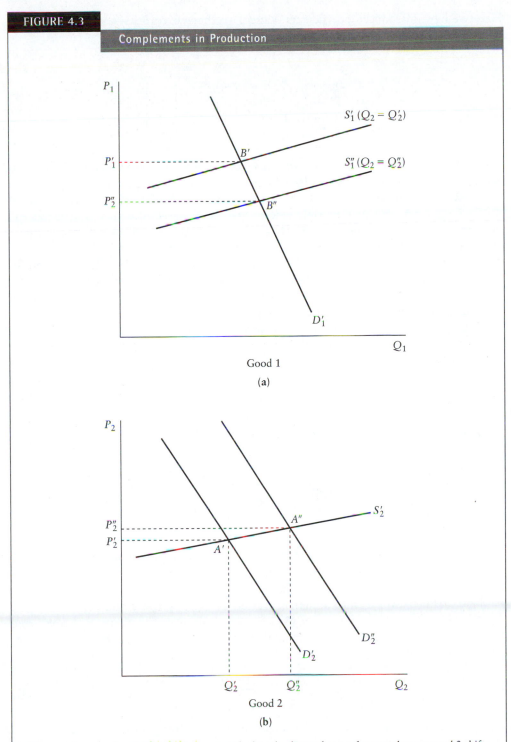

Good 1

(a)

Good 2

(b)

The supply curve for good 1 shifts downward when the demand curve for complement good 2 shifts rightward.

Here we have introduced coefficients 0 and 1 so that we can easily create the matrix formulation

$$
\begin{matrix}
\mathbf{A} & \mathbf{X} & = & \mathbf{B}
\end{matrix}
$$

$$
\begin{bmatrix}
1 & 0 & -d_{11} & -d_{12} \\
0 & 1 & -d_{21} & -d_{22} \\
1 & 0 & -s_{11} & -s_{12} \\
0 & 1 & -s_{21} & -s_{22}
\end{bmatrix}
\begin{bmatrix}
Q_1 \\ Q_2 \\ P_1 \\ P_2
\end{bmatrix}
=
\begin{bmatrix}
d_1 \\ d_2 \\ s_1 \\ s_2
\end{bmatrix},
\tag{4.10}
$$

where $\mathbf{A}$ is a (4×4) matrix of coefficients, $\mathbf{X}$ is a (4×1) column vector of the four endogenous variables, and $\mathbf{B}$ is a (4×1) column vector of constants. Multiplying the first row of the matrix $\mathbf{A}$ against the column vector $\mathbf{X}$ will produce the left side of the first equation in (4.9). The right side of that equation is, of course, the element in the first row of the column vector $\mathbf{B}$. [Note that the matrix equation in (4.10) is equivalent to the systems in (4.6) and (4.7), as well as to the system (4.9).]

The equilibrium values of the two prices and the two quantities will depend on the structure of the system. This structure is defined by the matrix of coefficients $\mathbf{A}$ and the parameters in $\mathbf{B}$. We will use Cramer's rule to obtain the reduced-form expression for P_1^*, which is the equilibrium value for the price of good 1.

Let $\mathbf{A}(j)$ be the matrix obtained by replacing column j of matrix $\mathbf{A}$ with the column vector $\mathbf{B}$. The price of good 1 is the third element of the column vector $\mathbf{X}$. Then by Cramer's rule

$$
P_1^* = \frac{|\mathbf{A}(3)|}{|\mathbf{A}|}.
\tag{4.11}
$$

The numerator in equation (4.11) is the determinant

$$
|\mathbf{A}(3)| =
\begin{vmatrix}
1 & 0 & d_1 & -d_{12} \\
0 & 1 & d_2 & -d_{22} \\
1 & 0 & s_1 & -s_{12} \\
0 & 1 & s_2 & -s_{22}
\end{vmatrix}.
\tag{4.12}
$$

We can expand the determinant of $\mathbf{A}(3)$ by using the cofactors of any row or column. We choose the first column because the 0s in that column will eliminate two calculations and the 1s will simplify the remaining two.

First, define $\mathbf{A}(3)_{ij}$ as the submatrix created by deleting row i and column j of the matrix $\mathbf{A}(3)$. Then the determinant $|\mathbf{A}(3)_{ij}|$ is the minor of the element in row i and column j of the matrix $\mathbf{A}(3)$. Expanding the determinant of $\mathbf{A}(3)$ by the cofactors of its first column produces

$$
\begin{aligned}
|\mathbf{A}(3)| &= 1(-1)^{1+1}|\mathbf{A}(3)_{11}| + 0 + 1(-1)^{3+1}|\mathbf{A}(3)_{31}| + 0 \\
&= |\mathbf{A}(3)_{11}| + |\mathbf{A}(3)_{31}|.
\end{aligned}
\tag{4.13}
$$

To verify equation (4.13), recall that the cofactor of the element located in row i and column j of any (square) matrix is the signed minor of that element. The sign is determined by $(-1)^{i+j}$.

Now the submatrix $\mathbf{A}(3)_{11}$ is

$$
\mathbf{A}(3)_{11} =
\begin{bmatrix}
1 & d_2 & -d_{22} \\
0 & s_1 & -s_{12} \\
1 & s_2 & -s_{22}
\end{bmatrix}.
\tag{4.14}
$$

Expanding the determinant of $\mathbf{A}(3)_{11}$ by the cofactors of its first column, we have

$$
\begin{aligned}
|\mathbf{A}(3)_{11}| &= 1(-1)^{1+1}[(s_1)(-s_{22}) - (-s_{12})(s_2)] \\
&\quad + 0 \\
&\quad + 1(-1)^{3+1}[(d_2)(-s_{12}) - (-d_{22})(s_1)] \\
&= s_{12}(s_2 - d_2) + s_1(d_{22} - s_{22}).
\end{aligned}
\tag{4.15}
$$

The submatrix $\mathbf{A}(3)_{31}$ is

$$
\mathbf{A}(3)_{31} = \begin{bmatrix} 0 & d_1 & -d_{12} \\ 1 & d_2 & -d_{22} \\ 1 & s_2 & -s_{22} \end{bmatrix}.
\tag{4.16}
$$

Expanding the determinant of $\mathbf{A}(3)_{31}$ by the cofactors of its first column yields

$$
\begin{aligned}
|\mathbf{A}(3)_{31}| &= 0 \\
&\quad + 1(-1)^{2+1}[(d_1)(-s_{22}) - (-d_{12})(s_2)] \\
&\quad + 1(-1)^{3+1}[(d_1)(-d_{22}) - (-d_{12})(d_2)] \\
&= d_{12}(d_2 - s_2) + d_1(s_{22} - d_{22}).
\end{aligned}
\tag{4.17}
$$

We now have the result that the numerator of P_1^* is

$$
\begin{aligned}
|\mathbf{A}(3)| &= |\mathbf{A}(3)_{11}| + |\mathbf{A}(3)_{31}| \\
&= s_{12}(s_2 - d_2) + s_1(d_{22} - s_{22}) + d_{12}(d_2 - s_2) + d_1(s_{22} - d_{22}) \\
&= (s_2 - d_2)(s_{12} - d_{12}) + (s_{22} - d_{22})(d_1 - s_1).
\end{aligned}
\tag{4.18}
$$

The denominator of P_1^* is $|\mathbf{A}|$. The matrix $\mathbf{A}$ is

$$
\mathbf{A} = \begin{bmatrix} 1 & 0 & -d_{11} & -d_{12} \\ 0 & 1 & -d_{21} & -d_{22} \\ 1 & 0 & -s_{11} & -s_{12} \\ 0 & 1 & -s_{21} & -s_{22} \end{bmatrix}.
\tag{4.19}
$$

Expanding $|\mathbf{A}|$ by the cofactors of its first column produces

$$
\begin{aligned}
|\mathbf{A}| &= 1(-1)^{1+1}|\mathbf{A}_{11}| + 0 + 1(-1)^{3+1}|\mathbf{A}_{31}| + 0 \\
&= |\mathbf{A}_{11}| + |\mathbf{A}_{31}|.
\end{aligned}
\tag{4.20}
$$

Expanding the two minors by the cofactors of the first columns of the submatrices $\mathbf{A}_{11}$ and $\mathbf{A}_{31}$, we have

$$
\begin{aligned}
|\mathbf{A}_{11}| &= \begin{vmatrix} 1 & -d_{21} & -d_{22} \\ 0 & -s_{11} & -s_{12} \\ 1 & -s_{21} & -s_{22} \end{vmatrix} \\
&= 1(-1)^{1+1}[(-s_{11})(-s_{22}) - (-s_{12})(-s_{21})] + 0 \\
&\quad + 1(-1)^{3+1}[(-d_{21})(-s_{12}) - (-d_{22})(-s_{11})] \\
&= (s_{11}s_{22} - s_{12}s_{21}) + (d_{21}s_{12} - d_{22}s_{11}) \\
&= s_{12}(d_{21} - s_{21}) + s_{11}(s_{22} - d_{22})
\end{aligned}
\tag{4.21}
$$

and

$$|\mathbf{A}_{31}| = \begin{vmatrix} 0 & -d_{11} & -d_{12} \\ 1 & -d_{21} & -d_{22} \\ 1 & -s_{21} & -s_{22} \end{vmatrix}$$
$$= 0 + 1(-1)^{2+1}[(-d_{11})(-s_{22}) - (-d_{12})(-s_{21})]$$
$$+ 1(-1)^{3+1}[(-d_{11})(-d_{22}) - (-d_{12})(-d_{21})]$$
$$= -(d_{11}s_{22} - d_{12}s_{21}) + (d_{11}d_{22} - d_{12}d_{21})$$
$$= d_{11}(d_{22} - s_{22}) + d_{12}(s_{21} - d_{21}).$$

$$\text{(4.22)}$$

Using the values of these two minors, we have from (4.20)

$$|\mathbf{A}| = (s_{22} - d_{22})(s_{11} - d_{11}) + (d_{21} - s_{21})(s_{12} - d_{12}). \qquad \text{(4.23)}$$

We can now write the equilibrium price of good 1 in terms of the parameters of the four equations for supply and demand in equations (4.6) and (4.7). Using Cramer's rule and the determinants calculated in equations (4.15), (4.17), and (4.23), we have

$$P_1^* = \frac{|\mathbf{A}(3)|}{|\mathbf{A}|}$$
$$= \frac{(s_2 - d_2)(s_{12} - d_{12}) + (s_{22} - d_{22})(d_1 - s_1)}{(s_{22} - d_{22})(s_{11} - d_{11}) + (d_{21} - s_{21})(s_{12} - d_{12})}. \qquad \text{(4.24)}$$

The expression for P_1^* is complicated, but no more so than the system of four linear equations from which it was derived. The dependence of the equilibrium price for good 1 on the interrelated structure of the markets for goods 1 and 2 is fully specified by equation (4.24).

To derive expressions for the equilibrium price of good 2, and the equilibrium quantities of goods 1 and 2, we would use the column vector $\mathbf{B}$ to replace columns 1, 2, and 4 of the matrix $\mathbf{A}$, and thus obtain the determinants for the numerators of expressions analogous to equation (4.24). By Cramer's rule, the denominators are always $|\mathbf{A}|$.

We can conduct various kinds of comparative static analyses using equation (4.24). Suppose, for instance, we want to examine the effects of unit taxes on the equilibrium values of the prices and quantities of both goods. We know that placing a unit tax on producers will shift their (inverse) supply functions upward by the tax rate. The vertical intercept of the (inverse) supply function for good 1 is $-(s_1 + s_{12}P_2)/s_{11}$. A unit tax at the rate t_1 on good 1 changes this intercept to $-(s_1 + s_{12}P_2)/s_{11} + t_1$. The new inverse supply function for good 1 will be

$$P_1 = \frac{Q_1^S}{s_{11}} - \frac{s_1 + s_{12}P_2}{s_{11}} + t_1. \qquad \text{(4.25)}$$

Similarly, the new inverse supply function for good 2 will be

$$P_2 = \frac{Q_2^S}{s_{22}} - \frac{s_2 + s_{21}P_1}{s_{22}} + t_2. \qquad \text{(4.26)}$$

If we rewrite equations (4.25) and (4.26) so that quantities supplied are functions of the prices and the unit tax rates, equations (4.7) become

$$Q_1^S = s_1 + s_{11}P_1 + s_{12}P_2 - s_{11}t_1,$$
$$Q_2^S = s_2 + s_{21}P_1 + s_{22}P_2 - s_{22}t_2. \qquad \text{(4.27)}$$

Now if we replace the term s_1 with the term $s_1 - s_{11}t_1$ and the term s_2 with the term $s_2 - s_{22}t_2$ in equations (4.9) and (4.10), we can proceed as we did in equations (4.11) through (4.24). The result is that in the formula for P_1^* given in equation (4.24), s_1 and s_2 will be replaced by $s_1 - s_{11}t_1$ and $s_2 - s_{22}t_2$. Then the effect on the equilibrium price of good 1 of an increase in the unit tax rate imposed on producers of good 2 can be obtained by differentiating P_1^* with respect to t_2. (*Note:* Our analysis of the effect of unit taxes on two interrelated markets is analogous to what we did for a single market in Section 1.2.3.)

Although the technique we used to analyze a system of two interrelated markets can be used to study any number of markets, the larger the number of markets, the more burdensome the task of computing all the determinants required to apply Cramer's rule. Sometimes this burden can be reduced by imposing restrictions on the system. For example, if some pairs of goods are unrelated either on the demand side, the supply side, or both, then the *interaction coefficients*, like d_{12} and s_{21} in the system that we just studied, will be 0. The more 0s there are in the matrix of coefficients, the easier it is to calculate determinants. If we were studying an empirical system, we would have numerical estimates of all the elements in the coefficient matrix and the elements of the vector on the right-hand side. Computers are adept at solving matrix systems of large size, either by using Cramer's rule or by other methods.

4.4 TWO FIRMS WITH DIFFERENTIATED OUTPUTS

4.4.1 Obtaining Equilibrium Values Using Matrix Inversion

Let us now use matrix inversion to determine the equilibrium prices and quantities for a market in which two firms sell differentiated products. Then we will use the form of this solution to obtain some comparative static results.

The inverse demand functions for firms 1 and 2 are

$$P_1 = a_1 - b_1 q_1 - q_2$$
$$P_2 = a_2 - q_1 - b_2 q_2, \qquad (4.28)$$

in which a_1, a_2, b_1, and b_2 are positive constants. Notice that an increase in the output q of either firm shifts the demand function for the other firm downward and leftward. An easy way to see this for firm 1 is to write its inverse demand function in the form: $P_1 = (a - q_2) - b_1 q_1$. The vertical intercept of this demand function for firm 1 depends negatively on the output q_2 of firm 2.

Let $C(q_i) = c_i q_i$ be the total cost for firm $i (i = 1, 2)$, and assume that each c_i, is a positive constant. For each firm, total cost is proportional to output, but the proportionality need not be the same for the two firms.

Let $\pi_i(q_1, q_2)$ be the rate of profit for firm i. The profit functions are

$$\pi_1 = P_1 q_1 - c_1 q_1 = (a_1 - b_1 q_1 - q_2)q_1 - c_1 q_1$$
$$\pi_2 = P_2 q_2 - c_2 q_2 = (a_2 - q_1 - b_2 q_2)q_2 - c_2 q_2. \qquad (4.29)$$

Thus each firm's own rate of profit depends in part on the rates of output chosen by both firms.

To obtain the first-order conditions for maximizing profits, we set the first partial derivative of each firm's profit function with respect to its own output equal to zero. Although each firm's profit depends on the output levels of both firms, each firm controls only its own level of output. Therefore, the first-order conditions for maximum profits do not depend on the cross-partial derivatives $\partial \pi_1 / \partial q_2$ and $\partial \pi_2 / \partial q_1$.[6] Rather, differentiating each equation in (4.29) with respect to each firm's *own level* of output only, we have

$$\frac{\partial \pi_1}{\partial q_1} = a_1 - 2b_1 q_1 - q_2 - c_1$$

$$\frac{\partial \pi_2}{\partial q_2} = a_2 - q_1 - 2b_2 q_2 - c_2. \tag{4.30}$$

Equating these derivatives to zero and transposing, we can write the conditions for profit maximization in matrix form as

$$\begin{matrix} \mathbf{A} & \mathbf{Q} & = & \mathbf{B} \\ \begin{bmatrix} 2b_1 & 1 \\ 1 & 2b_2 \end{bmatrix} \begin{bmatrix} q_1 \\ q_2 \end{bmatrix} & = & \begin{bmatrix} a_1 - c_1 \\ a_2 - c_2 \end{bmatrix}. \end{matrix} \tag{4.31}$$

To solve this system of equations by matrix inversion, we need to construct the inverse $\mathbf{A}^{-1}$ for the matrix of coefficients $\mathbf{A}$. Since the cofactor matrix for $\mathbf{A}$, $\mathbf{C}(\mathbf{A})$, is symmetric, it is also the adjoint matrix. Thus

$$\mathbf{C}(\mathbf{A}) = \begin{bmatrix} 2b_2 & -1 \\ -1 & 2b_1 \end{bmatrix} = \text{adj } \mathbf{A}. \tag{4.32}$$

The determinant of $\mathbf{A}$ is

$$|\mathbf{A}| = 4b_1 b_2 - 1. \tag{4.33}$$

Then the inverse of $\mathbf{A}$ is

$$\mathbf{A}^{-1} = \frac{1}{|\mathbf{A}|} \begin{bmatrix} 2b_2 & -1 \\ -1 & 2b_1 \end{bmatrix} = \frac{1}{4b_1 b_2 - 1} \begin{bmatrix} 2b_2 & -1 \\ -1 & 2b_1 \end{bmatrix}. \tag{4.34}$$

Returning to equation (4.31) and multiplying through by $\mathbf{A}^{-1}$ from equation (4.34), we have

$$\mathbf{AQ} = \mathbf{B},$$
$$\mathbf{A}^{-1}\mathbf{AQ} = \mathbf{A}^{-1}\mathbf{B},$$
$$\mathbf{Q} = \mathbf{A}^{-1}\mathbf{B},$$

so that

$$\mathbf{Q} = \begin{bmatrix} q_1 \\ q_2 \end{bmatrix} = \frac{1}{4b_1 b_2 - 1} \begin{bmatrix} 2b_2 & -1 \\ -1 & 2b_1 \end{bmatrix} \begin{bmatrix} a_1 - c_1 \\ a_2 - c_2 \end{bmatrix}. \tag{4.35}$$

The equilibrium rates of output are therefore

$$q_1^* = \frac{2b_2(a_1 - c_1) - (a_2 - c_2)}{4b_1 b_2 - 1} \quad \text{and} \quad q_2^* = \frac{2b_1(a_2 - c_2) - (a_1 - c_1)}{4b_1 b_2 - 1}. \tag{4.36}$$

---

[6]Later in this chapter we will consider a version of the duopoly model in which each firm recognizes that a change in the level of its own output will induce the other firm to change its level of output. In this sense, each firm has some control over the quantity produced by the other firm. When this is true, the conditions for profit maximization do involve cross-partial derivatives.

Equations (4.36) are the reduced-form solutions for the rates of output q_1 and q_2. Each expression specifies an equilibrium value for an endogenous variable solely in terms of the parameters $a_1, a_2, b_1, b_2, c_1,$ and c_2. If this model contained other parameters, their values would also appear on the right sides of equations (4.36). The reduced-form solutions for the two prices can be obtained by substituting the values for q_1^* and q_2^* defined by equations (4.36) for the values of q_1 and q_2 in the inverse demand functions (4.28).

We derived equations (4.36) with no constraints on the signs of the equilibrium levels of output. Without conditions on the relative values of the parameters, the solutions in (4.36) could specify negative levels of output, which are impossible.

4.4.2 Interpreting the Solution

This system of two interrelated markets served by firms 1 and 2 is complex enough so that no simple relationship exists among the parameter values that will ensure nonnegative outputs. We therefore suggest a simple version of this system that will ensure positive outputs for each firm.

Suppose that the product $b_1 b_2 > 1/4$, so that the denominators in (4.36) are positive. Next, suppose that both $(a_1 - c_1)$ and $(a_2 - c_2)$ are positive, so that each firm's demand function lies above its marginal (and average) cost function at the vertical axis, assuming that the other firm produces no output. Finally, suppose that the numerators in equations (4.36) are positive.

We now consider the economic significance of these assumptions in more detail. In so doing we will provide an economic rationale that explains why some of the parameters appear as they do in the reduced-form solutions for the optimal outputs in equations (4.36).

Assume for the moment that $q_2 = 0$. Then for firm 1, the term $(a_1 - c_1)$ is the distance between the demand function and the average (and marginal) cost function at the vertical axis. If $(a_1 - c_1)$ is nonpositive, firm 1 has no levels of output for which price covers average cost. In this case it is impossible for firm 1 to operate profitably, even if there is no competition from firm 2. But if $(a_1 - c_1) > 0$, firm 1 has a range of positive levels of output that are profitable.

The greater the distance between firm 1's demand function and its average cost function, and the less steep its demand function, the larger is firm 1's profit-maximizing rate of output. When firm 2 produces no output, the distance between the vertical intercepts of the demand and average cost functions for firm 1 is $(a_1 - c_1)$. The slope of its demand function is b_1.

In Figure 4.4 we plot the demand (D), marginal revenue (MR), and marginal cost (MC) functions for firm 1, assuming a fixed rate of output for firm 2. An increase in the value of a_1 will shift the demand and marginal revenue functions rightward. As a result, the intersection between the marginal revenue and marginal cost functions will move rightward. Thus, an increase in the value of a_1 increases the optimal rate of output for firm 1. Similarly, a decrease in the value of c_1 will increase the optimal rate of output. Finally, a decrease in the value of b_1 will cause both the demand and marginal revenue functions to become less steep. (Both functions rotate counterclockwise about their common vertical intercept.) Then the optimal rate of output for firm 1 varies inversely with the value of b_1.

The foregoing discussion explains why the term $(a_1 - c_1)$ appears positively in the numerator of the expression for q_1^* in equation (4.36). We also see why b_1 appears positively in the denominator of that expression.

Now remove the assumption that $q_2 = 0$. The greater the distance between the vertical intercepts of the demand and average cost functions for firm 2, measured by

FIGURE 4.4

A Duopolist

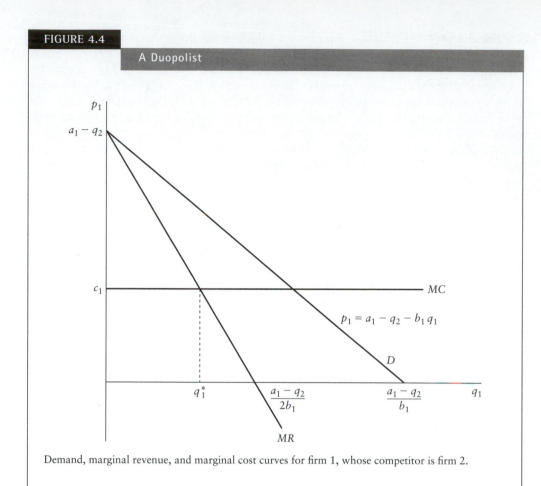

Demand, marginal revenue, and marginal cost curves for firm 1, whose competitor is firm 2.

$(a_2 - c_2)$, the larger will be its profit-maximizing rate of output. But as firm 2 produces more output, the demand function for firm 1 shifts farther leftward and downward, because the vertical intercept of that demand function is $(a_1 - q_2)$. As firm 1's demand function shifts leftward, that firm reduces its rate of output. We have now explained why the term $(a_2 - c_2)$ should appear with a negative sign in the numerator of the expression for q_1^* in equation (4.36).

Let us now consider the effect of b_2 in the expression for q_1^*. The partial derivative of q_1^* with respect to b_2 is the quantity $2[2b_1(a_2 - c_2) - (a_1 - c_1)]/(4b_1b_2 - 1)^2$. The term in square brackets in the numerator is positive by assumption. The denominator is a square, so the partial derivative of q_1^* with respect to b_2 is positive. It is easy to understand why this is so. An increase in b_2 makes firm 2's demand function steeper. This causes firm 2 to reduce its rate of output. The reduction in q_2^* shifts firm 1's demand function rightward, causing that firm to increase its rate of output.

There are several more questions of comparative statics that we can easily handle. Suppose that firm 1 makes some expenditures on advertising, with the result that its demand function shifts outward. This shift would increase the intercept term a_1 in firm 1's (inverse) demand function. From equation (4.36) we have

$$\frac{\partial q_1^*}{\partial a_1} = \frac{2b_2}{4b_1b_2 - 1} > 0 \quad \text{and} \quad \frac{\partial q_2^*}{\partial a_1} = \frac{-1}{4b_1b_2 - 1} < 0. \tag{4.37}$$

A rightward and upward shift in firm 1's demand function causes firm 1 to increase output and firm 2 to decrease output. Consumers have strengthened their preferences for the product of firm 1 relative to its competitor's product.

We can interpret the economics of this result by considering the inverse demand functions (4.28) and Figure 4.4. Hold q_2 fixed for a moment. Then an increase in a_1 will shift firm 1's demand and marginal revenue functions rightward, increasing that firm's optimal rate of output. This increase in q_1^* will shift firm 2's demand and marginal revenue functions leftward, reducing the value for q_2^*. The lower value for q_2^* will shift firm 1's demand and marginal revenue functions further rightward. The value for q_1^* will increase further, generating another round of adjustments in the rates of output for both firms. The new equilibrium values for q_1^* and q_2^* associated with the higher value for a_1 can be obtained from equations (4.36).

We can also determine the effect on both P_1^* and P_2^* of a rightward shift of firm 1's demand function. From equations (4.28), the marginal effects on the optimal prices of a rightward shift in firm 1's demand function are

$$\frac{\partial P_1^*}{\partial a_1} = 1 - b_1 \frac{\partial q_1^*}{\partial a_1} - \frac{\partial q_2^*}{\partial a_1} = \frac{2b_1 b_2}{4b_1 b_2 - 1} > 0$$

$$\frac{\partial P_2^*}{\partial a_1} = -\frac{\partial q_1^*}{\partial a_1} - b_2 \frac{\partial q_2^*}{\partial a_1} = \frac{-b_2}{4b_1 b_2 - 1} < 0.$$

(4.38)

To derive equations (4.38) we use the chain rule because each price depends on a_1 indirectly through q_1^* and q_2^*. As an alternative, we could use equations (4.36) to substitute the expressions for q_1^* and q_2^* into the inverse demand functions (4.28). This would create reduced-form expressions showing the optimal prices P_1^* and P_2^* as functions of parameters only. We could then differentiate the optimal prices directly with respect to a_1 (or any parameter).

What would happen to equilibrium prices and rates of output of both firms if a unit tax were imposed only on firm 2? If the unit tax rate were t_2, then firm 2's cost function would become $C_2(q_2) = (c_2 + t_2)q_2$. To obtain the effects of the tax on the rates of output simply replace c_2 by $(c_2 + t_2)$ in equations (4.36) and differentiate with respect to t_2. (Imposing a unit tax on firm 2 is equivalent to increasing the unit tax rate t_2 from zero to some positive value.) The effects of the tax on the prices can be obtained by using the chain rule as we did in obtaining equations (4.38). The results are

$$\frac{\partial q_1^*}{\partial t_2} = \frac{1}{4b_1 b_2 - 1} > 0 \quad \text{and} \quad \frac{\partial q_2^*}{\partial t_2} = \frac{-2b_1}{4b_1 b_2 - 1} < 0$$

so that

(4.39)

$$\frac{\partial P_1^*}{\partial t_2} = \frac{b_1}{4b_1 b_2 - 1} > 0 \quad \text{and} \quad \frac{\partial P_2^*}{\partial t_2} = \frac{2b_1 b_2 - 1}{4b_1 b_2 - 1}.$$

To interpret these results, recall from equations (4.28) that the vertical intercept of each firm's (inverse) demand function depends negatively on its competitor's rate of output. An increase in the unit tax imposed on firm 2 will cause firm 1 to increase both its rate of output and its price. This is possible because firm 1's demand function shifts rightward as a consequence of firm 2's reducing its rate of output.

The effect of the tax on firm 2's price is ambiguous. This is because firm 2's demand function shifts leftward.

Suppose for a moment that the position of firm 2's demand function did not depend on its competitor's rate of output. Then from equations (4.28) the demand function for firm 2 would degenerate to $P_2 = a_2 - b_2 q_2$. The marginal effect of t_2 on P_2^* would then be $-b_2$ times the partial derivative of q_2^* with respect to t_2. From the second equation in the first line in (4.39), this effect is $-b_2[-2b_1/(4b_1 b_2 - 1)]$, which is positive. But if firm 1's rate of output does affect the location of firm 2's demand function, then the marginal effect of t_2 on P_2 is $-b_2[-2b_1/(4b_1 b_2 - 1)] - [1/(4b_1 b_2 - 1)]$, whose sign is ambiguous. The first term is the increase in P_2^* caused by moving along an unchanged demand function. The second term is the decrease in P_2^* caused by the leftward (and downward) shift of the demand function. The net effect on P_2^* of the change in t_2 depends on the relative values of the slopes of the two firms' demand functions.

4.5   A SIMPLE MODEL OF DUOPOLY

Let us now use matrix algebra to replicate the analysis of the simple model of duopoly presented in Section 2.5. We begin with that model to gain practice using matrix algebra. Then in Sections 4.6 through 4.8 we will examine more complicated versions of the model.

Two firms produce identical output goods (coal of the same grade available at the same location, for example). Firm 1 produces q_1 and firm 2 produces q_2 units per year; $Q = q_1 + q_2$ is the annual rate of output for the industry. The demand function for the industry is

$$P = a - bQ = a - b(q_1 + q_2),\qquad\qquad \textbf{(4.40)}$$

in which both a and b are positive constants. Each firm's total cost is equal to the same constant c times its rate of output.

The price at which either firm can sell its product depends both on its own rate of output and on the rate of output chosen by its competitor. In view of this interdependence, what is each firm's optimal rate of output?

To answer this question, we must make assumptions about how each firm will adjust its own rate of output in response to a change in its competitor's rate of output. Cournot called these assumptions *conjectural variations*.

Let $\partial q_i / \partial q_j$ be firm j's conjecture about the rate at which its competitor, firm i, will adjust its rate of output in response to a change in firm j's rate of output. In the simplest form of the duopoly model $\partial q_1 / \partial q_2 = \partial q_2 / \partial q_1 = 0$, which is the Cournot assumption. Each firm treats its competitor's rate of output as a constant.

We want to determine the equilibrium price, rates of output, and rates of profit for these duopolists. Let $\pi_i(q_1, q_2)$ denote the profit for firm i, $i = 1, 2$. Then

$$\pi_i(q_1, q_2) = Pq_i - cq_i = [a - b(q_1 + q_2)]q_i - cq_i, \quad i = 1, 2. \qquad \textbf{(4.41)}$$

Using our assumption that the conjectural variations $\partial q_1 / \partial q_2$ and $\partial q_2 / \partial q_1$ are both zero, and recognizing that $\partial q_i / \partial q_i = 1$ for $i = 1, 2$, the first-order conditions to maximize profit are as follows:
For firm 1

$$\frac{\partial \pi_1}{\partial q_1} = [a - b(q_1 + q_2)] - b\left(\frac{\partial q_1}{\partial q_1} + \frac{\partial q_2}{\partial q_1}\right)q_1 - c = 0,$$

$$\text{or}\quad a - bq_2 - 2bq_1 = c. \qquad\qquad \textbf{(4.42a)}$$

For firm 2

$$\frac{\partial \pi_2}{\partial q_2} = [a - b(q_1 + q_2)] - b\left(\frac{\partial q_1}{\partial q_2} + \frac{\partial q_2}{\partial q_2}\right)q_2 - c = 0,$$

$$\text{or}\quad a - bq_1 - 2bq_2 = c. \qquad\qquad \textbf{(4.42b)}$$

Note that each first-order condition in equations (4.42) is the familiar requirement that marginal revenue equals marginal cost. Marginal cost for each firm is the same constant, c. Each firm's marginal revenue function is a linear function of its own rate of output, with the same vertical intercept as its demand function and a slope twice as steep. This relationship between the marginal revenue and demand functions is a consequence of our assumption that each firm behaves as if its competitor's rate of output were constant. The common vertical intercept of firm 1's demand and marginal revenue functions is $a - bq_2$; the analogous vertical intercept for firm 2 is $a - bq_1$.

Solving the two conditions (4.42a) and (4.42b) simultaneously establishes that the optimal output levels q_1^* and q_2^* are equal to each other.[7] Using this fact and equations (4.42), we have

$$q_1^* = q_2^* = \frac{a - c}{3b}. \tag{4.43}$$

Then from equations (4.40) and (4.41), the equilibrium price and rates of profit are

$$P^* = \frac{a + 2c}{3}$$

$$\pi_1^* = \pi_2^* = \frac{1}{b}\left(\frac{a - c}{3}\right)^2. \tag{4.44}$$

Using matrix algebra to analyze the preceding problem is too cumbersome because the analysis is so simple. We now consider a more complicated version of the duopoly problem.

4.6   DUOPOLY WITH NONZERO CONJECTURAL VARIATIONS

We retain all the assumptions of the preceding section, except that we will use the following values for the conjectural variations:

$$\frac{\partial q_2}{\partial q_1} = 1 \quad \text{and} \quad \frac{\partial q_1}{\partial q_2} = 0.5. \tag{4.45}$$

Firm 1 now assumes that whenever it changes its own rate of output, firm 2 will match that change one for one. Firm 2 expects that any change in its own rate of output will cause firm 1 to match only half of that change. The numerical values in equations (4.45) are hypothetical; we did not derive them from prior assumptions.

The two first-order conditions for maximal profits are now
For firm 1:

$$\frac{\partial \pi_1}{\partial q_1} = [a - b(q_1 + q_2)] - b\left(\frac{\partial q_1}{\partial q_1} + \frac{\partial q_2}{\partial q_1}\right)q_1 - c = 0,$$

$$\text{or} \quad a - b(3q_1 + q_2) = c. \tag{4.46a}$$

For firm 2:

$$\frac{\partial \pi_2}{\partial q_2} = [a - b(q_1 + q_2)] - b\left(\frac{\partial q_1}{\partial q_2} + \frac{\partial q_2}{\partial q_2}\right)q_2 - c = 0,$$

$$\text{or} \quad a - b(q_1 + 2.5q_2) = c. \tag{4.46b}$$

---

[7]As an alternative to solving the two conditions (4.42a) and (4.42b) simultaneously, we could recognize that the two conditions are symmetric and the numbering of the firms is arbitrary. Then the common equilibrium rate of output is the solution to $a - 3bq = c$, or $q^* = (a - c)/3b$.

The equilibrium price and rates of output can be found by solving simultaneously the demand function (4.40) and the two first-order conditions (4.46a) and (4.46b). In matrix form these three conditions are

$$
\begin{array}{ccc}
\mathbf{A} & \mathbf{V} = & \mathbf{B} \\
\begin{bmatrix} 1 & b & b \\ 0 & -3b & -b \\ 0 & -b & -2.5b \end{bmatrix} \begin{bmatrix} P \\ q_1 \\ q_2 \end{bmatrix} = & \begin{bmatrix} a \\ c & - & a \\ c & - & a \end{bmatrix}.
\end{array}
\tag{4.47}
$$

As before, define $\mathbf{A}(j)$ as the matrix created by replacing column j in matrix $\mathbf{A}$ with the column vector $\mathbf{B}$. Using Cramer's rule, the solution to equation (4.47) is

$$
\begin{aligned}
P^* &= \frac{|\mathbf{A}(1)|}{|\mathbf{A}|} = \frac{3a + 3.5c}{6.5} \\
q_1^* &= \frac{|\mathbf{A}(2)|}{|\mathbf{A}|} = \frac{1.5(a - c)}{6.5b} \\
q_2^* &= \frac{|\mathbf{A}(3)|}{|\mathbf{A}|} = \frac{2(a - c)}{6.5b}.
\end{aligned}
\tag{4.48}
$$

The quantity $(a - c)$ in the numerators of the expressions for the optimal rates of output q_1^* and q_2^* is the distance between the vertical intercepts of the demand and marginal cost functions. If this distance is nonpositive, neither firm can obtain a positive profit at any rate of output. Both firms would shut down, and the industry would disappear. We assume that $a > c$, so that each firm's optimal rate of output is positive.

The economic interpretation of the roles of the parameters in the reduced-form solutions for the optimal price and rates of output is fairly transparent. In some ways, this duopoly behaves as a monopoly would. The parameter a is the vertical intercept of the duopoly's (inverse) demand function. An increase in a would shift the demand function upward and rightward; the vertical component of the shift is the change in the value of a. A monopolist in this situation would increase both output and price; the larger output restricts the increase in the price to something less than the change in a. The joint behavior of the duopolists produces the same result. Thus, $\partial q_1^*/\partial a > 0$, $\partial q_2^*/\partial a > 0$, and $0 < \partial P^*/\partial a < 1$. Similarly, the parameter c measures the height of the (constant) marginal cost function facing each firm. An increase in marginal cost causes each duopolist to reduce output and increase price, just as a monopolist would.

Note that $q_2^* > q_1^*$. The fact that firm 2 will produce more output than firm 1, even though both firms have the same cost structure, is a result of our assumptions about conjectural variations. From equations (4.45) we see that firm 2 assumes that when it changes its own rate of output, firm 1 will respond by changing its output by only half as much. Firm 1, on the other hand, assumes that firm 2 will fully meet every change in firm 1's output. This means that firm 2 regards the industry demand function as having a slope equal to $-b(1 + 0.5)$ with respect to changes in its own rate of output. Firm 1 acts as if the slope were equal to $-b(1 + 1)$ with respect to its rate of output. We know that the steeper the slope of the demand function, the smaller the optimal rate of output. Firm 1 chooses a smaller rate of output than that of firm 2 because firm 1 perceives the industry demand function to be steeper than firm 2 does.

We can easily obtain from equations (4.48) the effect of imposing a unit tax at rate t on each firm. We simply replace the parameter c with $c + t$. We could also easily consider a more general version of this linear model of duopoly by letting the marginal cost for firm i be the constant c_i, so that costs differ between the two firms. The terms for the conjectural variations, $\partial q_1/\partial q_2$ and $\partial q_2/\partial q_1$, could be left unspecified, with the condition that each of these terms is a constant. Without this condition the first-order conditions (4.46a) and (4.46b) will be nonlinear in q_1 and q_2. Matrix algebra would then be inapplicable.

After making these modifications to the terms for costs and conjectural variations, the element a_{22} in matrix A in equation (4.47) will be $-(2 + \partial q_2/\partial q_1)b$, and the element a_{33} will become $-(2 + \partial q_1/\partial q_2)b$. On the right side of equation (4.47) the second element of the vector B will become $c_1 - a$, and the third element $c_2 - a$. We leave for you the task of determining the new expressions for the equilibrium price and quantities in (4.48).

4.7 A SIMPLE KEYNESIAN MODEL

In this section we will use matrix algebra to analyze the equilibrium level of national income in an economy that has only two endogenous variables. In this economy national income Y is the sum of the annual rates of spending by consumers, C, by investors, I, and by the government, G. The consumption function is $C = C_0 + b(1 - t)Y$, in which C_0, b, and t are constants and $0 < b(1 - t) < 1$. Autonomous spending for consumption is C_0, the marginal propensity to consume out of disposable income is b, and the marginal tax rate is t. Investment spending is constant at I_0, and spending by the government is constant at G_0. Both I_0 and G_0 are positive.

The equations specifying the structure of the economy are

$$Y = C + I_0 + G_0$$
$$C = C_0 + b(1 - t)Y. \tag{4.49}$$

The endogenous variables are Y and C, and the parameters are C_0, I_0, G_0, b, and t. To translate the system defined by equations (4.49) into a matrix equation, we rewrite those two equations as a system in which the endogenous variables and their coefficients are on the left-hand sides and the parameters are on the right-hand sides:

$$Y - C = I_0 + G_0$$
$$b(1 - t)Y - C = -C_0 \tag{4.50}$$

or in matrix form

$$
\begin{array}{ccc}
A & V & = & B
\end{array}
$$
$$
\begin{bmatrix} 1 & -1 \\ b(1-t) & -1 \end{bmatrix} \begin{bmatrix} Y \\ C \end{bmatrix} = \begin{bmatrix} I_0 + G_0 \\ -C_0 \end{bmatrix}. \tag{4.51}
$$

We now obtain an expression for the equilibrium rate of national income Y by using Cramer's rule. The determinant of A is

$$|A| = -1 - [(-1)b(1 - t)] = -1 + b(1 - t). \tag{4.52}$$

By Cramer's rule, the equilibrium rate of national income is

$$Y = \frac{\begin{vmatrix} I_0 + G_0 & -1 \\ -C_0 & -1 \end{vmatrix}}{|A|} = \frac{(I_0 + G_0)(-1) - (-1)(-C_0)}{-1 - (-1)[b(1-t)]} = \frac{C_0 + I_0 + G_0}{1 - b(1-t)}. \tag{4.53}$$

The equilibrium rate of national income depends positively on the autonomous component of consumption spending, C_0, and on the autonomous rates of spending by investors, I_0, and the government, G_0. An increase in the marginal tax rate, t, will reduce national income because an increase in t will reduce the denominator. The standard Keynesian multiplier can be derived by differentiating equation (4.53) with respect to any of the three components of autonomous spending. The result is

$$\frac{\partial Y}{\partial C_0} = \frac{\partial Y}{\partial I_0} = \frac{\partial Y}{\partial G_0} = \frac{1}{1 - b(1 - t)} > 0. \tag{4.54}$$

Notice that the autonomous spending multiplier takes the familiar form $1/[1 - b(1 - t)]$. The relevant marginal propensity to consume is $b(1 - t)$, rather than simply b, because it is necessary to convert national income to disposable income. If the tax rate were zero, the expression in equation (4.54) would be $1/(1 - b)$, which is the multiplier in its simplest form.

To determine the effect on equilibrium income of a change in the tax rate, we differentiate equation (4.53) with respect to t:

$$\frac{\partial Y}{\partial t} = \frac{-b[C_0 + I_0 + G_0]}{[1 - b(1 - t)]^2} < 0. \tag{4.55}$$

Now suppose the government operated on a balanced budget. Then $G_0 = tY$. Substituting this relationship into equation (4.53) and solving for Y yields

$$Y_B = \frac{C_0 + I_0}{1 - t - b(1 - t)}, \tag{4.56}$$

in which Y_B is the equilibrium rate of national income under the constraint that the government's budget is balanced. The autonomous spending multiplier for an economy constrained by the condition $G_0 = tY$ is

$$\frac{\partial Y_B}{\partial C_0} = \frac{\partial Y_B}{\partial I_0} = \frac{1}{1 - t - b(1 - t)}. \tag{4.57}$$

Comparing equations (4.57) and (4.54), we see that the multiplier in an economy in which the government must balance its budget is larger than the multiplier in an economy in which the government may run a deficit or a surplus. The reason is not difficult to understand. If the government must balance its budget, then $G_0 = tY$. The aggregate demand function becomes $C + I + G = C_0 + b(1 - t)Y + I_0 + tY$. Collecting terms in Y, the aggregate demand *is* $C_0 + [t + b(1 - t)]Y + I_0$. The aggregate marginal propensity to spend out of national income has increased from $b(1 - t)$ to $t + b(1 - t)$. Whenever national income rises, the government must spend the additional tax receipts, tY. The larger the (aggregate) marginal propensity to spend, the larger is the multiplier.

Note that the multiplier for the economy constrained by the condition $G_0 = tY$ is not the same as the "balanced budget multiplier" presented in most introductory macroeconomic courses and in Chapter 2. In the introductory course both government spending and tax collections are treated as exogenous constants. In particular, tax collections are not a proportion of national income. The balanced budget multiplier in that introductory model does not require that the budget be balanced, only that the *changes* in tax collections and government spending be equal. Consequently, the balanced budget multiplier of the introductory model is equal to unity. In the model that leads to equation (4.56) it is the *totals* of government spending and tax collections, not just the changes, that must be equal.

4.8   AN IS–LM MODEL

In the preceding section our macroeconomic model had only one market, a market for goods. Consumers, investors, and the government competed for goods in that market. Equilibrium required that the quantity of goods supplied, which is national income Y, be equal to the aggregate demand for goods, which is $C + I + G$.

We now augment this model by creating a market for money. The quantity of money supplied, M_0, is exogenous. The quantity of money demanded is $M_D = fY - gr$, in which f and g are positive constants, Y is national income, and r is the interest rate. Both Y and r are endogenous. Money is a *stock variable*, unlike national income, which is a *flow variable*. The usual assumption is that the quantity of monetary balances people want to hold depends on the rate of economic activity and on the cost of holding wealth in the form of money. A proxy for the rate of economic activity is Y. The cost of holding money is r, the rate of interest foregone by holding money rather than financial assets. Requiring f and g to be positive makes our demand for money compatible with the usual assumptions.

Equilibrium in the augmented model requires that the markets for goods and money be cleared simultaneously. Equilibrium will therefore be defined by a pair of values for Y and r such that the quantities of goods supplied and demanded are equal, and the quantities of money supplied and demanded are also equal. The condition for equilibrium in the money market is

$$fY - gr = M_0. \tag{4.58}$$

We will allow investment spending to be endogenous through dependence on the interest rate. Investment spending is now determined by $I = I_0 - er$, in which e is a positive constant.

The economy is now described by the system of equations

$$\begin{aligned} Y &= C + I + G \\ C &= C_0 + b(1 - t)Y \\ I &= I_0 - er \\ G &= G_0 \\ \hline fY - gr &= M_0. \end{aligned} \tag{4.59}$$

The endogenous variables in system (4.59) are Y, C, I, and r. The exogenous variables are the three levels of autonomous spending, C_0, I_0, and G_0, and the constant stock of money M_0.

A simultaneous solution of the first four equations in system (4.59) will define a set of combinations of values for Y and r that will establish equilibrium in the goods market. The locus of these combinations is known as the **IS curve**. (Equilibrium in the goods market requires that planned *investment* and planned *saving* be equal. Hence "IS" is the name of the locus of combinations of Y and r that will clear the goods market.)

The fifth equation in system (4.59) defines a set of combinations of values for Y and r that will establish equilibrium in the money market. The locus of these combinations is known as the **LM curve** ("L" for *liquidity preference*, as the demand for money is sometimes known; "M" for the quantity of money supplied).

We can rewrite system (4.59) as

$$Y - C - I + 0r = G_0$$
$$b(1-t)Y - C + 0I + 0r = -C_0$$
$$0Y + 0C + I + er = I_0$$
$$fY + 0C + 0I - gr = M_0$$

(4.60)

or in matrix form as

$$
\overset{\mathbf{A}}{\begin{bmatrix} 1 & -1 & -1 & 0 \\ b(1-t) & -1 & 0 & 0 \\ 0 & 0 & 1 & e \\ f & 0 & 0 & -g \end{bmatrix}}
\overset{\mathbf{X}}{\begin{bmatrix} Y \\ C \\ I \\ r \end{bmatrix}} =
\overset{\mathbf{B}}{\begin{bmatrix} G_0 \\ -C_0 \\ I_0 \\ M_0 \end{bmatrix}}.
$$

(4.61)

We will now work through the explicit steps required to obtain the equilibrium value of national income Y^* by use of Cramer's rule. The interested reader can obtain the equilibrium values of C, I, and r.

The first step is to obtain the determinant of the coefficient matrix $\mathbf{A}$. Using cofactors of its third column, we write the determinant of $\mathbf{A}$ as

$$
|\mathbf{A}| = (-1)(-1)^{1+3}\begin{vmatrix} b(1-t) & -1 & 0 \\ 0 & 0 & e \\ f & 0 & -g \end{vmatrix} + (1)(-1)^{3+3}\begin{vmatrix} 1 & -1 & 0 \\ b(1-t) & -1 & 0 \\ f & 0 & -g \end{vmatrix}
$$

(4.62)

$$
= -\begin{vmatrix} b(1-t) & -1 & 0 \\ 0 & 0 & e \\ f & 0 & -g \end{vmatrix} + \begin{vmatrix} 1 & -1 & 0 \\ b(1-t) & -1 & 0 \\ f & 0 & -g \end{vmatrix}.
$$

Expanding the first determinant in equation (4.62) by the cofactors of the second row of its matrix produces

$$
-\begin{vmatrix} b(1-t) & -1 & 0 \\ 0 & 0 & e \\ f & 0 & -g \end{vmatrix} = -e(-1)^{2+3}[b(1-t)(0) - f(-1)]
$$

(4.63)

$$
= -e(-1)(f) = ef.
$$

Expanding the second determinant in equation (4.62) by the cofactors of the third column of its matrix, we have

$$
\begin{vmatrix} 1 & -1 & 0 \\ b(1-t) & -1 & 0 \\ f & 0 & -g \end{vmatrix} = -g(-1)^{3+3}[1(-1) - b(1-t)(-1)]
$$

(4.64)

$$
= -g[-1 + b(1-t)]
$$

$$
= g[1 - b(1-t)].
$$

Then using equations (4.62), (4.63), and (4.64), we conclude that the determinant of matrix $\mathbf{A}$ is

$$
|\mathbf{A}| = ef + g[1 - b(1-t)].
$$

(4.65)

Applications of Matrix Theory to Linear Models

Next, we see in equation (4.61) that the endogenous variable Y is the first variable in the column vector $\mathbf{X}$ of endogenous variables. By Cramer's rule, the equilibrium value of Y is

$$Y^* = \frac{|\mathbf{A}(1)|}{|\mathbf{A}|}. \tag{4.66}$$

The numerator is the determinant of the matrix obtained by replacing the first column of $\mathbf{A}$ with the column vector of constants $\mathbf{B}$ from the right side of equation (4.61).

We now evaluate the two determinants in equation (4.66) by expansion with cofactors. The numerator of equation (4.66) is

$$|\mathbf{A}(1)| = \begin{vmatrix} G_0 & -1 & -1 & 0 \\ -C_0 & -1 & 0 & 0 \\ I_0 & 0 & 1 & e \\ M_0 & 0 & 0 & -g \end{vmatrix} \tag{4.67}$$

Expanding this determinant by the cofactors of its second column produces two (3×3) determinants, as follows:

$$|\mathbf{A}(1)| = (-1)(-1)^{1+2} \begin{vmatrix} -C_0 & 0 & 0 \\ I_0 & 1 & e \\ M_0 & 0 & -g \end{vmatrix} + (-1)(-1)^{2+2} \begin{vmatrix} G_0 & -1 & 0 \\ I_0 & 1 & e \\ M_0 & 0 & -g \end{vmatrix}$$

$$= \begin{vmatrix} -C_0 & 0 & 0 \\ I_0 & 1 & e \\ M_0 & 0 & -g \end{vmatrix} - \begin{vmatrix} G_0 & -1 & 0 \\ I_0 & 1 & e \\ M_0 & 0 & -g \end{vmatrix}. \tag{4.68}$$

Expanding the first (3×3) determinant by cofactors of the first row of its matrix, and expanding the second (3×3) determinant by cofactors of the second column of its matrix, we have the final expression for the numerator in equation (4.66):

$$\begin{aligned} |\mathbf{A}(1)| &= -C_0(-1)^{1+1}[(1)(-g) - (0)(e)] \\ &\quad - \{(-1)(-1)^{1+2}[I_0(-g) - M_0 e] + (1)(-1)^{2+2}[G_0(-g) - M_0(0)]\} \\ &= C_0 g - [I_0(-g) - M_0 e] - [G_0(-g)] \\ &= M_0 e + g(C_0 + I_0 + G_0). \end{aligned} \tag{4.69}$$

Finally, using equations (4.65), (4.66), and (4.69), we have the expression for the equilibrium rate of national income:

$$Y^* = \frac{|\mathbf{A}(1)|}{|\mathbf{A}|} = \frac{M_0 e + g(C_0 + I_0 + G_0)}{ef + g[1 - b(1 - t)]}. \tag{4.70}$$

By comparing equations (4.53) and (4.70), we can see how it complicates the system to add a market for money and to make the interest rate and investment spending endogenous variables. From equation (4.70), the multiplier for the three sources of autonomous spending is

$$\frac{\partial Y^*}{\partial C_0} = \frac{\partial Y^*}{\partial I_0} = \frac{\partial Y^*}{\partial G_0} = \frac{g}{ef + g[1 - b(1 - t)]}. \tag{4.71}$$

And the effect of an exogenous increase in the quantity of money on the equilibrium rate of income is

$$\frac{\partial Y^*}{\partial M_0} = \frac{e}{ef + g[1 - b(1 - t)]}.$$

(4.72)

Problems

4.1 Using the model of two competitive markets described in Section 4.3, derive expressions that can be used to show the effects on P_1^* and Q_2^* of an increase in a unit subsidy granted to producers of good 1. Provide economic interpretations for the cases in which goods 1 and 2 are complements, substitutes, or neithers.

4.2 In equation (4.36) assume that $b_1 b_2 > 1/4$. Then specify conditions for the remaining parameters so that the equilibrium levels of outputs will be positive. Provide economic interpretations for your conditions.

4.3 Using the model of Section 4.4, determine the unit tax or subsidy for firm 1 that will ensure that both firms have the same equilibrium price. To simplify the calculations, assume $b_1 = b_2$. Express the required unit tax or subsidy in terms of the cost and demand parameters of the problem.

4.4 For the model of duopoly in Section 4.6, assume that the conjectural variations satisfy $dq_2/dq_1 = 3(dq_1/dq)$, in which $dq_1/dq_2 = h$, a positive constant. Further assume that $c_1 = 2c_2$, and that c_2 is a positive constant. Derive the expressions for P^*, q_1^*, and q_2^* that are analogous to equation (4.48). Provide an economic interpretation of why your results differ from equation (4.48).

4.5 Suppose that in the simple Keynesian model of Section 4.7 we allow investment I to be an endogenous variable governed by $I = I_0 + aY$, in which a is a positive constant. Using matrix algebra, derive expressions for

(a) Equilibrium national income. Specify the (necessary and sufficient) relationship among the parameters a, b, and t for equilibrium to be positive. Give an economic interpretation of this relationship among the parameters.

(b) The multipliers for the three components (C_0, I_0, and G_0) of autonomous spending. Explain how and why the parameter a changes the multiplier from the result in equation (4.54).

(c) Equilibrium national income when the government must balance its budget.

(d) The multipliers for autonomous consumption and investment spending when the government must balance its budget.

Provide economic interpretations for the relationship between your answers and the results derived in Section 4.7.

4.6 Using the IS–LM model of Section 4.8:

(a) Derive expressions for the equilibrium values of Y, C, and r subject to the constraint that $G = tY$.

(b) Derive an expression for the money multiplier (i.e., $\delta Y^*/\delta M_0$) subject to the constraint that $G = tY$.

5

Multivariate Calculus: Theory

5.1 INTRODUCTION

So far we have dealt only with linear functions, which are simple but relatively unrealistic. But it is important to know whether the results derived with linear functions change qualitatively when more realistic, nonlinear functions are used. For example, introductory textbooks usually derive the simple Keynesian multiplier by assuming that the consumption function is linear, as we did in Chapters 2 and 4. Then the equilibrium condition $Y = C + I + G$ can be solved for the equilibrium value of Y as a function of the exogenous variables and parameters. If the consumption function is $C = a + bY$, then $Y = (a + bY) + I + G$; so equilibrium output is

$$Y^* = \left(\frac{1}{1 - b} \right) (a + I + G) \tag{5.1}$$

and the government expenditure multiplier formula follows directly:

$$\frac{\partial Y^*}{\partial G} = \frac{1}{1 - b}. \tag{5.2}$$

A linear consumption function is a restrictive assumption, however. What would the multiplier equal if we had a more general consumption function $C(Y)$? The equilibrium condition is now

$$Y = C(Y) + I + G \tag{5.3}$$

and, without knowing the functional form of the consumption function, we cannot solve explicitly for the equilibrium value of Y. The equilibrium condition (5.3) does, however, *implicitly* define the equilibrium value of Y as a function of the exogenous variables I and G. That is, in equilibrium the value of Y must be such that equation (5.3) holds. If either I or G (or both) changes, the equilibrium value of Y must change in such a way that the equilibrium condition continues to hold. Condition (5.3) does not *explicitly* define Y as a function of exogenous variables because the endogenous variable Y still appears on the right-hand side. But the equilibrium value of Y is implicitly defined by

$$Y^*(I,G) = C(Y^*(I,G)) + I + G. \tag{5.4}$$

In this chapter we will see how to use equilibrium conditions that implicitly define equilibrium values of endogenous variables. We will examine the implicit function theorem, which shows under what conditions explicit functions (such as $Y^*(I,G)$ in our example) exist that give equilibrium values of endogenous variables as functions of exogenous variables and parameters. We will also see how to derive comparative static results even if the explicit forms of the equilibrium functions (like $Y^*(I,G)$) are not known.

In addition, we will discover the method of finding out how the equilibrium values of endogenous variables change when two or more exogenous variables change simultaneously. Last, we will discuss two other concepts regarding functions of several variables that are used extensively in economics: level curves and homogeneity. Level curves play a particularly important role in constrained optimization problems, which we discuss in several chapters beginning with Chapter 9. Homogeneous functions appear often in consumer theory and the theory of the firm.

5.2 PARTIAL AND TOTAL DERIVATIVES

Partial derivatives were introduced in Chapter 1. They measure the change in a function caused by a change in one of its arguments, holding other arguments constant. The notation for the partial derivative of a function $f(x, y, z)$ with respect to, say, x is f_x or $\partial f / \partial x$. Using the symbol ∂ rather than d shows that the function has several arguments. When a function has only a single argument, there is no other argument to hold constant, so the partial derivative notation is unnecessary. That is, the notation for the derivative with respect to x of the function $g(x)$ is either dg/dx or $g'(x)$. We can use the prime notation, such as $g'(x)$, in turn, only for functions of one variable because the prime notation is ambiguous about which argument the derivative is being taken with respect to if there are many arguments to the function.

Second (and higher-order) partial derivatives are obtained in a manner analogous to those of single-variable functions, except that the second derivative can be with respect to a different variable than the first. So, for instance, $f_{xy} = \partial f^2 / \partial x \, \partial y$ refers to the second derivative of the function f, taking first the derivative with respect to x (holding all other arguments constant) and then with respect to y (holding all other arguments constant). This kind of second derivative is called a **cross-partial derivative**. In symbols,

$$f_{xy} = \frac{\partial^2 f}{\partial x \, \partial y} = \frac{\partial \, (\partial f / \partial x)}{\partial y}.$$

This second derivative measures the rate at which the value of the first partial derivative with respect to x, f_x, changes when the value of y is increased incrementally. The rate at which the value of f_x changes when the value of x changes is also a second derivative: $f_{xx} = \partial^2 f / \partial x^2$ refers to the second derivative of the function f with respect to x, holding all other arguments constant. Higher-order derivatives can be taken as well. For example, $f_{xyz} = \partial^3 f / \partial x \, \partial y \, \partial z$ refers to the third derivative of f, where the order of differentiation is with respect to x, then y, then z.

As an example, consider the function $f(x, y, z) = x^2y^2z^2$. For this function,

$$\frac{\partial f}{\partial x} = 2xy^2z^2, \quad \frac{\partial^2 f}{\partial x^2} = 2y^2z^2, \quad \frac{\partial^2 f}{\partial x\,\partial y} = 4xyz^2, \quad \frac{\partial^3 f}{\partial x\,\partial y\,\partial z} = 8xyz.$$

It is very important to remember that, just as in the case of functions of one variable, partial derivatives are functions of all of the arguments of the parent function. If the value of one argument, say x, changes, the value of the function f changes; but in addition, the values of all the derivatives of the function f will change as well, even derivatives with respect to other variables.

A very useful theorem says that the order of differentiation does not matter; that is, the second derivative of a function f with respect to first x, then y is the same as the second derivative with respect to first y, then x.

YOUNG'S THEOREM

For any function f(x,y) with continuous second derivatives,

$$\frac{\partial^2 f}{\partial x\,\partial y} = \frac{\partial^2 f}{\partial y\,\partial x}.$$

This result extends to any second partial of a function of many variables and to higher-order derivatives as well.

Now suppose that x, y, and z are all in turn functions of some parameter, θ. The total derivative of the function $f(x, y, z)$ with respect to θ is

$$\frac{df(x(\theta), y(\theta), z(\theta))}{d\theta} = \frac{\partial f}{\partial x}\frac{dx}{d\theta} + \frac{\partial f}{\partial y}\frac{dy}{d\theta} + \frac{\partial f}{\partial z}\frac{dz}{d\theta}. \tag{5.5}$$

This is nothing more than the chain rule of differentiation applied to a function of many variables. As an example of finding a total derivative, suppose that $f(x, y, z) = 3x + 2y + z$ and that $x(\theta) = \theta^2$, $y(\theta) = \theta$, and $z(\theta) = 0.5\theta$. Then

$$\frac{df}{d\theta} = 3(2\theta) + 2(1) + 1(0.5) = 2.5 + 6\theta. \tag{5.6}$$

A somewhat more confusing case arises when some of the arguments of the function, say x and y, are endogenous variables while others, say z, are exogenous. In equilibrium, the endogenous variables will in general be functions ($x^*(z)$ and $y^*(z)$) of the exogenous variables. Thus the function might be written as $f(x^*(z), y^*(z), z)$ showing that, while f is a function of x, y, and z, the equilibrium values of x and y will change if the value of z changes. The partial derivative of f with respect to z, $\partial f/\partial z$, is the change in the value of the function when z changes but x and y are held constant. The total derivative of f with respect to z takes into account the changes in x and y that will be caused by the change in z:

$$\frac{df}{dz} = \frac{\partial f}{\partial x}\frac{dx}{dz} + \frac{\partial f}{\partial y}\frac{dy}{dz} + \frac{\partial f}{\partial z}. \tag{5.7}$$

For example, if $f(x, y, z) = 3x + 2y + z$ and $y = 4z$ while $x = 2z$, then

$$\frac{df}{dz} = 3(2) + 2(4) + 1 = 15. \tag{5.8}$$

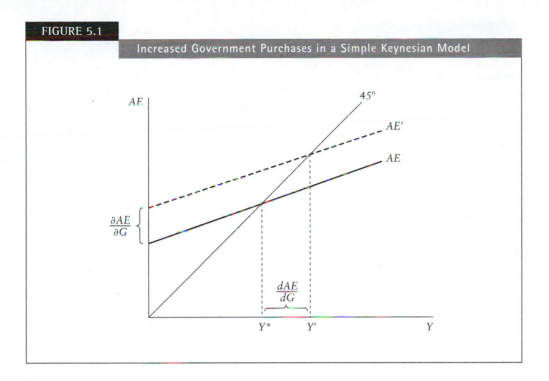

FIGURE 5.1

Increased Government Purchases in a Simple Keynesian Model

For an economics example, consider the effect on aggregate expenditures (AE) of a change in government purchases. In our simple, fixed-price Keynesian model, aggregate expenditures are given by $AE(C(Y), I, G) = C(Y) + I + G$. We know that the equilibrium value of output is a function of all the exogenous variables in the system; these in our simple case are I and G and the parameters of the consumption function. Thus we can write equilibrium output as $Y^*(I, G)$. Aggregate expenditures can now be written as $AE(C(Y^*), I, G) = C(Y^*(I, G)) + I + G$. The partial derivative of AE with respect to G represents the effect of G on aggregate expenditures, holding consumption and investment constant, and in our example equals 1. Graphically this represents the vertical shift in the AE curve in the typical Keynesian cross diagram, shown in Figure 5.1. The total change in *equilibrium* aggregate expenditures caused by an increase in G is the total derivative with respect to G,

$$\frac{dAE}{dG} = \frac{\partial C}{\partial Y}\frac{\partial Y^*}{\partial G} + \frac{\partial I}{\partial G} + \frac{\partial G}{\partial G} = \frac{\partial C}{\partial Y}\frac{\partial Y}{\partial G} + 1.$$

Note that $\partial I/\partial G = 0$ because in this model investment is exogenous.

In other words, aggregate expenditures go up because of the autonomous increase in G (the shift of the AE curve in Figure 5.1) but also because of the induced increase in consumption (the movement along the new AE curve to the new equilibrium). Graphically, dAE/dG is the horizontal distance (or equivalently the vertical distance, since both equilibria are on the 45° line) between the two equilibrium points.

5.3 DIFFERENTIALS

Differentials are defined in the context of functions. Consider two variables, y and x, related by a function: $y = f(x)$. Figure 5.2 shows the graph of a function and its tangent at a particular point. The differentials dy and dx of y and x are the related changes

Multivariate Calculus: Theory

FIGURE 5.2

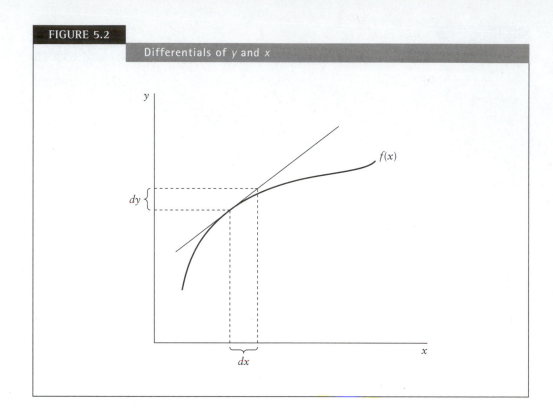

Differentials of y and x

in the two variables along the tangent to the function. That is, $dy = (dy/dx)\,dx$ or $dy = f'(x)\,dx$. Loosely translated into words, this equation says that the change in y is equal to the amount by which y changes when x changes, times the change in x. The differential of y, dy, is thus an approximation to the change in the value of the function f when the argument of the function, x, changes by dx. It is only an approximation because it measures the movement along a tangent to the function f rather than a change in the function f itself; obviously, the smaller the differential of the argument of the function, the better the approximation will be.

For a slightly more complicated example, let $y = f(u)$ and $u = g(x)$. Then, using the chain rule,

$$dy = \frac{df}{du}\frac{du}{dx}\,dx \quad \text{or} \quad dy = f'(u(x))u'(x)\,dx. \tag{5.9}$$

The definition and interpretation of differentials extend in an obvious manner to functions of several variables. Let $y = f(x_1, x_2, \ldots, x_n)$. Then the differentials of y and each x_i are defined by

$$dy = \frac{\partial f}{\partial x_1}\,dx_1 + \frac{\partial f}{\partial x_2}\,dx_2 + \cdots + \frac{\partial f}{\partial x_n}\,dx_n \tag{5.10}$$

which, again loosely speaking, says that the change in y is equal to the sum of the changes in each x_i times the changes in the value of the function caused by the change in each argument (holding the others constant).

The right-hand side of equation (5.10) is called the **total differential** of the function $f(x_1, x_2, \ldots, x_n)$. We will use differentials for the most part to find the total effect on an endogenous variable when more than one exogenous variable or parameter is changing simultaneously. For example, market demand curves are functions of the price of the good (p_x), prices of other goods (for instance, p_y), and income (I). In a two-good model, the demand for good x is given by $x = f(p_x, p_y, I)$. We often assume that only one of these factors changes at a time, but in the real world all three may be changing simultaneously. We can find the shift in the demand curve for x resulting from simultaneous increases in the price of good y and income by holding the price of x constant and taking the total differential of the demand curve; that is,

$$dx = \frac{\partial f}{\partial p_x} dp_x + \frac{\partial f}{\partial p_y} dp_y + \frac{\partial f}{\partial I} dI \qquad \text{(5.11)}$$

which, since $dp_x = 0$ because p_x is being held constant, simplifies to

$$dx = \frac{\partial f}{\partial p_y} dp_y + \frac{\partial f}{\partial I} dI. \qquad \text{(5.12)}$$

As another example, we could add taxes to our simple Keynesian model and use differentials to find the effect on equilibrium output of simultaneous increases in government purchases and taxes. The solution of this problem requires an understanding of implicit functions, to which we now turn.

5.4 IMPLICIT FUNCTIONS

An implicit function is one that is implied by an equation of the form $f(x_1, x_2, \ldots, x_n) = 0$. We will begin by considering the implicit function theorem in the context of a single equation and then we will extend it to many equations.

5.4.1 The Implicit Function Theorem and Implicit Differentiation for One Equation

Consider the equation $f(x, y) = 0$. Since the function always equals zero, df will always equal zero as well. Taking the total differential of the equation, we therefore obtain

$$\frac{\partial f(x, y)}{\partial x} dx + \frac{\partial f(x, y)}{\partial y} dy = 0. \qquad \text{(5.13)}$$

Rearranging terms, assuming that $\partial f/\partial y \neq 0$, yields

$$\frac{dy}{dx} = -\frac{\partial f/\partial x}{\partial f/\partial y} \qquad \text{(5.14)}$$

which illustrates the implicit function theorem:

IMPLICIT FUNCTION THEOREM (ONE EQUATION)

If $f(x, y) = 0$ and $\partial f/\partial x$ and $\partial f/\partial y$ both exist (at least in the vicinity of particular values of x and y, say x_0 and y_0) and $\partial f/\partial y \neq 0$, then the function y(x) exists (again, in that vicinity), because the derivative dy/dx exists (at least locally). Furthermore, the **implicit function rule** *is that $dy/dx = -(\partial f/\partial x)/(\partial f/\partial y)$ (evaluated at x_0 and y_0).*

The implicit function theorem and rule extend to any pair of arguments of a function of many variables, if all other arguments are held constant.

Said another way, the equation $f(x, y) = 0$ implicitly defines the function $y(x)$. This suggests another method, called implicit differentiation, of finding dy/dx from an equation implicitly defining the relationship between y and x—a method that we often use to find comparative static derivatives.

The equation $f(x, y) = 0$ implicitly defines y as a function of x in that, if x changes, y must change to keep the value of $f(x, y)$ equal to zero. If we assert that the function $y(x)$ exists, then $f(x, y(x))$ must identically equal zero. That is, when x changes, the function $y(x)$ automatically adjusts the value of y to make the value of f equal to zero. Since $f(x, y(x))$ identically equals zero, the total derivative of f with respect to x must also equal zero:

$$\frac{df}{dx} = \frac{\partial f}{\partial x} + \frac{\partial f}{\partial y}\frac{dy}{dx} = 0. \tag{5.15}$$

Differentiating through the equation $f(x, y) = 0$, where y is treated as a function of x, is called **implicit differentiation**. Rearranging terms yields the same formula for dy/dx that we derived by using differentials. We will be able to solve for dy/dx, that is to say that dy/dx exists, as long as the partial derivatives of f exist and $\partial f/\partial y \neq 0$. Thus if we can solve for dy/dx we know that our assertion that $y(x)$ exists was justified, at least locally.

Let us look at two simple examples before illustrating the use of implicit differentiation in economics. Consider first the function $f(x, y) = y - ax^2 = 0$. The implicit function rule says that

$$\frac{dy}{dx} = -\frac{\partial f/\partial x}{\partial f/\partial y} = -\frac{-2ax}{1} = 2ax.$$

This can be confirmed by solving for y as an explicit function of x and then differentiating: $y = ax^2$ so $dy/dx = 2ax$. For a slightly more complicated example, consider the function $f(x, y) = y^3 - x^2 = 0$. The implicit function rule says that

$$\frac{dy}{dx} = -\frac{\partial f/\partial x}{\partial f/\partial y} = -\frac{-2x}{3y^2} = \frac{2x}{3y^2}. \tag{5.16}$$

At first glance, solving for y as an explicit function of x and then differentiating seems to give a different formula for the derivative: $y = x^{2/3}$, so $dy/dx = 2x^{-1/3}/3$. But the two formulas are in fact the same, as can be seen by substituting in for y:

$$\frac{dy}{dx} = \frac{2x}{3y^2} = \frac{2x}{3(x^{2/3})^2} = \frac{2x}{3x^{4/3}} = \frac{2}{3}x^{-1/3}. \tag{5.17}$$

As an example of implicit differentiation, the simple Keynesian expenditure multiplier can now be derived. The equation $Y = C(Y) + I + G$ implicitly defines equilibrium output as a function of I and G (and the parameters of the consumption function). Asserting that the function $Y^*(I, G)$ exists and plugging it into the equilibrium condition yields

$$Y^*(I, G) = C(Y^*(I, G)) + I + G. \tag{5.18}$$

Substituting the function $Y^*(I, G)$ for Y turns the equilibrium condition (an equation that holds only in equilibrium) into an identity (an equation that always holds), since the function $Y^*(I, G)$ ensures that, for any values of I and G, the value of Y will change in a way that makes the equation hold. Since it is an identity, we can differentiate both sides with respect to G to get

$$\frac{\partial Y^*}{\partial G} = C'(Y^*(I,G))\frac{\partial Y^*}{\partial G} + 1$$

$$(1 - C')\frac{\partial Y^*}{\partial G} = 1 \tag{5.19}$$

$$\frac{\partial Y^*}{\partial G} = \frac{1}{1 - C'}.$$

As long as the marginal propensity to consume, C', is less than one, we can calculate the effect on equilibrium output of a change in government purchases, holding investment constant. That is, we can find the government expenditure multiplier, which equals $1/(1 - C')$.

The multiplier could have been solved for directly using the implicit function rule by considering the equation $f(Y,I,G) = Y - C(Y) - I - G = 0$. Using the implicit function rule,

$$\frac{\partial Y^*}{\partial G} = -\frac{\partial f/\partial G}{\partial f/\partial Y} = -\frac{-1}{1 - C'} = \frac{1}{1 - C'}. \tag{5.20}$$

5.4.2 The Implicit Function Theorem and Implicit Differentiation for Multiple Equations

The implicit function theorem extends to cases where several equations together implicitly define one set of variables as functions of another set of variables. Consider the two-equation system

$$f(x_1, x_2, \alpha) = 0$$
$$g(x_1, x_2, \alpha) = 0 \tag{5.21}$$

where each x_i is an endogenous variable and α is an exogenous variable or parameter. Taking total differentials of the two equations,

$$\frac{\partial f}{\partial x_1} dx_1 + \frac{\partial f}{\partial x_2} dx_2 + \frac{\partial f}{\partial \alpha} d\alpha = 0$$

$$\frac{\partial g}{\partial x_1} dx_1 + \frac{\partial g}{\partial x_2} dx_2 + \frac{\partial g}{\partial \alpha} d\alpha = 0, \tag{5.22}$$

we get two equations in which the variables are the differentials dx_1 and dx_2 rather than x_1 and x_2. Written in matrix notation, these two equations in the two unknowns dx_1 and dx_2 are

$$\begin{bmatrix} \dfrac{\partial f}{\partial x_1} & \dfrac{\partial f}{\partial x_2} \\ \dfrac{\partial g}{\partial x_1} & \dfrac{\partial g}{\partial x_2} \end{bmatrix} \begin{bmatrix} dx_1 \\ dx_2 \end{bmatrix} = \begin{bmatrix} -\dfrac{\partial f}{\partial \alpha} d\alpha \\ -\dfrac{\partial g}{\partial \alpha} d\alpha \end{bmatrix}. \tag{5.23}$$

Using Cramer's rule to solve for dx_1,

$$dx_1 = \frac{\begin{vmatrix} -\dfrac{\partial f}{\partial \alpha} d\alpha & \dfrac{\partial f}{\partial x_2} \\ -\dfrac{\partial g}{\partial \alpha} d\alpha & \dfrac{\partial g}{\partial x_2} \end{vmatrix}}{\begin{vmatrix} \dfrac{\partial f}{\partial x_1} & \dfrac{\partial f}{\partial x_2} \\ \dfrac{\partial g}{\partial x_1} & \dfrac{\partial g}{\partial x_2} \end{vmatrix}} = \frac{-\begin{vmatrix} \dfrac{\partial f}{\partial \alpha} & \dfrac{\partial f}{\partial x_2} \\ \dfrac{\partial g}{\partial \alpha} & \dfrac{\partial g}{\partial x_2} \end{vmatrix} d\alpha}{\begin{vmatrix} \dfrac{\partial f}{\partial x_1} & \dfrac{\partial f}{\partial x_2} \\ \dfrac{\partial g}{\partial x_1} & \dfrac{\partial g}{\partial x_2} \end{vmatrix}}. \tag{5.24}$$

The denominator of the right-hand side is the determinant of a matrix formed by taking the first derivatives of a system of equations with respect to the endogenous variables. These kinds of determinants are called **Jacobian determinants,** or **Jacobians,** and are frequently denoted by $|J|$. With this notation and a small rearrangement of terms, the formula for the comparative static derivative $dx_1/d\alpha$ is

$$\frac{dx_1}{d\alpha} = \frac{-\begin{vmatrix} \frac{\partial f}{\partial \alpha} & \frac{\partial f}{\partial x_2} \\ \frac{\partial g}{\partial \alpha} & \frac{\partial g}{\partial x_2} \end{vmatrix}}{|J|}. \tag{5.25}$$

The derivative $dx_1/d\alpha$ exists and can be solved for as long as the partial derivatives in the numerator exist and the Jacobian is nonzero. This illustrates the implicit function theorem for a system of equations:

IMPLICIT FUNCTION THEOREM AND RULE (MULTIPLE EQUATIONS)

If $f(x_1, x_2, \boldsymbol{\alpha}) = 0$ and $g(x_1, x_2, \boldsymbol{\alpha}) = 0$, where x_1 and x_2 are endogenous variables and $\boldsymbol{\alpha}$ is a vector of exogenous variables or parameters; and if all partial derivatives of f and g exist (at least in the vicinity of particular values of the variables); and if the Jacobian of the system of equations is nonzero (evaluated at those particular values); then the functions $x_1^*(\boldsymbol{\alpha})$ and $x_2^*(\boldsymbol{\alpha})$ exist (at least locally), since the derivatives $\partial x_1^*/\partial \alpha_i$ and $\partial x_2^*/\partial \alpha_i$ both exist (locally) for each parameter. The generalization of the implicit function rule is that

$$\frac{\partial x_1^*}{\partial \alpha_i} = -\frac{\begin{vmatrix} \frac{\partial f}{\partial \alpha_i} & \frac{\partial f}{\partial x_2} \\ \frac{\partial g}{\partial \alpha_i} & \frac{\partial g}{\partial x_2} \end{vmatrix}}{|J|} \quad and \quad \frac{\partial x_2^*}{\partial \alpha_i} = -\frac{\begin{vmatrix} \frac{\partial f}{\partial x_1} & \frac{\partial f}{\partial \alpha_i} \\ \frac{\partial g}{\partial x_1} & \frac{\partial g}{\partial \alpha_i} \end{vmatrix}}{|J|}.$$

The theorem and rule extend to systems of many equations with many endogenous variables.

As was true in the case of a single equation, the formula for $\partial x_1^*/\partial \alpha_i$ can also be derived by asserting that the functions $x_1^*(\boldsymbol{\alpha})$ and $x_2^*(\boldsymbol{\alpha})$ exist, substituting them into the two equations to get the identities $f(x_1^*(\boldsymbol{\alpha}), x_2^*(\boldsymbol{\alpha}), \boldsymbol{\alpha}) \equiv 0$ and $g(x_1^*(\boldsymbol{\alpha}), x_2^*(\boldsymbol{\alpha}), \boldsymbol{\alpha}) \equiv 0$, and totally differentiating the two equations with respect to α_i.

For example, if $f(x, y, \alpha) = y - \alpha x^2 = 0$ and $g(x, y, \alpha) = y^2 + \alpha x = 0$, then the conditions of the implicit function theorem are met: both f and g are differentiable with respect to both x and y, and the Jacobian of the system,

$$|J| = \begin{vmatrix} \partial f/\partial x & \partial f/\partial y \\ \partial g/\partial x & \partial g/\partial y \end{vmatrix} = \begin{vmatrix} -2\alpha x & 1 \\ \alpha & 2y \end{vmatrix} = -4\alpha xy - \alpha, \tag{5.26}$$

is nonzero. Thus the functions $x^*(\alpha)$ and $y^*(\alpha)$ exist. Using the implicit function rule, the derivatives[1] of these functions are

$$\frac{dx^*}{d\alpha} = -\frac{\begin{vmatrix} \partial f/\partial \alpha & \partial f/\partial x \\ \partial g/\partial \alpha & \partial g/\partial x \end{vmatrix}}{|J|} = -\frac{\begin{vmatrix} -x^2 & 1 \\ x & 2y \end{vmatrix}}{-4\alpha xy - \alpha} = -\frac{-2x^2 y - x}{-4\alpha xy - \alpha} = -\frac{x(2xy+1)}{\alpha(4xy+1)} \tag{5.27}$$

---

[1]Since each function has only one argument, the partial derivative notation is not necessary.

and

$$\frac{dy^*}{d\alpha} = -\frac{\begin{vmatrix} \partial f/\partial x & \partial f/\partial \alpha \\ \partial g/\partial x & \partial g/\partial \alpha \end{vmatrix}}{|J|} = -\frac{\begin{vmatrix} -2\alpha x & -x^2 \\ \alpha & x \end{vmatrix}}{-4\alpha xy - \alpha}$$

$$= -\frac{-2\alpha x^2 + \alpha x^2}{-4\alpha xy - \alpha} = -\frac{-\alpha x^2}{-\alpha(4xy+1)} = -\frac{x^2}{4xy+1}.$$

(5.28)

To illustrate the use of the implicit function theorem for more than one equation, we will derive the government expenditure multiplier in a Keynesian model with money. (This model is referred to as the IS–LM model.) We will continue to use a closed-economy model for simplicity, but now in addition to the goods market we have a money market in which equilibrium is defined by equality between money supply and money demand (sometimes called the demand for liquidity). Money supply (M) is considered to be exogenous but money demand (L) is endogenous: it is an increasing function of income and a decreasing function of the interest rate. Also, investment demand (I) is now a decreasing function of the interest rate (r) instead of being exogenous. Thus the equilibrium of the system is the solution of the two-equation system

$$Y = C(Y) + I(r) + G \qquad \text{(goods market equilibrium)}$$
$$M = L(Y, r) \qquad \text{(money market equilibrium)}$$

(5.29)

where $G = G_0$ and $M = M_0$. We can use the implicit function theorem and rule if we rewrite these equations as

$$Y - C(Y) - I(r) - G = f(Y, r, G) = 0$$
$$M - L(Y, r) = g(Y, r, M) = 0$$

(5.30)

where

$$\frac{\partial f}{\partial Y} = 1 - \frac{\partial C}{\partial Y} = 1 - C'(Y) > 0, \qquad \frac{\partial f}{\partial r} = -\frac{\partial I}{\partial r} = -I'(r) > 0, \qquad \frac{\partial f}{\partial G} = -1,$$
$$\frac{\partial g}{\partial Y} = -\frac{\partial L}{\partial Y} = -L_Y < 0, \qquad \frac{\partial g}{\partial r} = -\frac{\partial L}{\partial r} = -L_r > 0, \quad \text{and} \quad \frac{\partial g}{\partial M} = 1.$$

The implicit function rule says that the multiplier, $\partial Y^*/\partial G$, is

$$\frac{\partial Y^*}{\partial G} = -\frac{\begin{vmatrix} \partial f/\partial G & \partial f/\partial r \\ \partial g/\partial G & \partial g/\partial r \end{vmatrix}}{|J|} = -\frac{\begin{vmatrix} \partial f/\partial G & \partial f/\partial r \\ \partial g/\partial G & \partial g/\partial r \end{vmatrix}}{\begin{vmatrix} \partial f/\partial Y & \partial f/\partial r \\ \partial g/\partial Y & \partial g/\partial r \end{vmatrix}}$$

(5.31)

$$= -\frac{\begin{vmatrix} -1 & -I'(r) \\ 0 & -L_r \end{vmatrix}}{\begin{vmatrix} 1 - C'(Y) & -I'(r) \\ -L_Y & -L_r \end{vmatrix}} = -\frac{L_r}{-L_r(1 - C'(Y)) - L_Y I'(r)}.$$

The difference between this and the simple multiplier (equation (5.20)) can be seen most easily by rewriting the new formula as

$$\frac{\partial Y^*}{\partial G} = \frac{1}{(1 - C'(Y)) + (L_Y I'(r)/L_r)}.$$

(5.32)

FIGURE 5.3

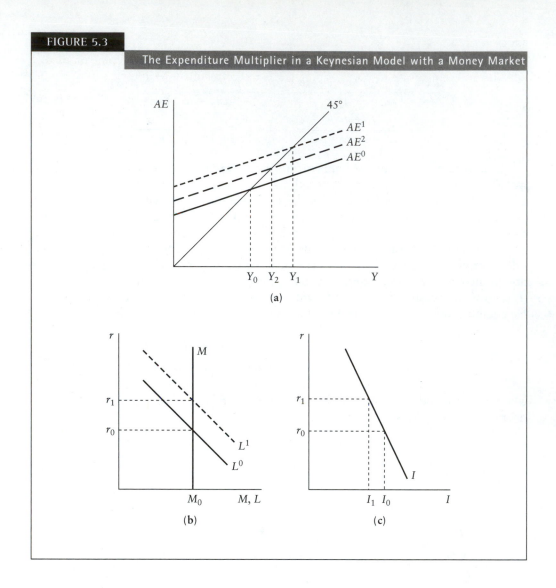

(a)

(b)   (c)

The second term in parentheses in the denominator is positive, making the denominator larger and the multiplier smaller.

The economic explanation of this is illustrated in Figure 5.3.[2] The higher government expenditure leads to higher income Y_1, as shown in Figure 5.3a. This in turn increases the demand for money, as shown in Figure 5.3b, raising the interest rate to r_1. Figure 5.3c shows the resulting crowding out of private investment I, which shifts the aggregate expenditure curve in Figure 5.3a back down to AE^2, resulting in a smaller increase in equilibrium income than in the simple Keynesian case where there is no money market and therefore no crowding out. The graphical analysis is incomplete in the sense that the reduction in income caused by lower investment will lead to a new sequence of events: as money demand decreases, the equilibrium interest rate falls, and investment increases, leading to yet a new sequence. These rounds of the adjustment process eventually peter out, but this process is essentially impossible to show graphically. The mathematical formula for the multiplier shows (1) that the process does indeed converge,

[2]IS and LM curves, discussed in Chapter 6, could also be used.

(2) what the resulting change in equilibrium income equals, and (3) the three factors that influence the extent of the crowding out of investment—the effect of income on money demand and the sensitivities of money demand and investment to interest rates.

5.5   LEVEL CURVES

Another common use of the implicit function theorem and rule in economics is in the context of level curves. Consider the equation $y_0 = f(x_1, x_2)$. Written in the form appearing in the implicit function theorem, the equation $g(x_1, x_2, y_0) = f(x_1, x_2) - y_0 = 0$ implicitly defines x_1 as a function of x_2 and y_0 (or x_2 as a function of x_1 and y_0). If this function exists, its graph is called a **level curve** of the function f because it shows all the combinations of x_1 and x_2 that yield a particular value (or level), y_0, of the function. The implicit function theorem tells us whether $x_1^*(x_2, y_0)$ exists (it does as long as $\partial g/\partial x_1 \neq 0$ and the function g is differentiable with respect to y_0 and both x's) and the implicit function rule tells us the slope of the level curve:

$$\frac{\partial x_1^*(x_2, y_0)}{\partial x_2} = -\frac{\partial g(x_1, x_2, y_0)/\partial x_2}{\partial g(x_1, x_2, y_0)/\partial x_1}. \tag{5.33}$$

Many of the curves used in intermediate economic theory are level curves; indifference curves and isoquants are probably the most familiar. Though it may not be as obvious, the IS and LM curves encountered in intermediate macroeconomics are also level curves. We will use isoquants as our example and leave a discussion of the others to the next chapter.

Isoquants are level curves of production functions. If a firm is producing q_0 units of output using labor (L) and capital (K) according to the production function $f(L, K)$, the corresponding isoquant, shown in Figure 5.4, is the graph showing all the combinations

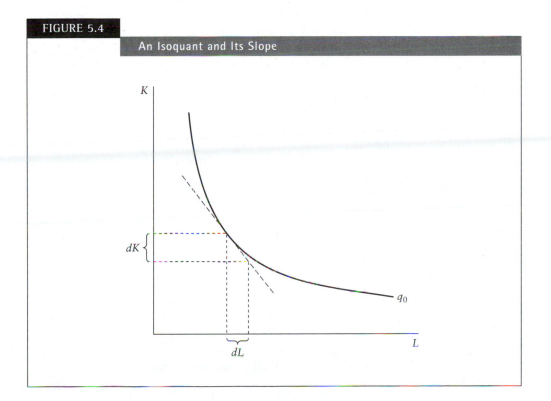

FIGURE 5.4

An Isoquant and Its Slope

of L and K that, when used together as inputs into the production function F, yield q_0 units of output. The implicit function rule says that the slope of the isoquant (if K is on the vertical axis and L on the horizontal) is

$$\frac{dK^*(L, q_0)}{dL} = -\frac{\partial F / \partial L}{\partial F / \partial K}. \tag{5.34}$$

As an example, if the firm has a Cobb-Douglas production function, $q = AL^{\alpha}K^{\beta}$, then the isoquant slope equals

$$\frac{dK^*(L, q_0)}{dL} = -\frac{\alpha AL^{\alpha-1}K^{\beta}}{\beta AL^{\alpha}K^{\beta-1}} = -\frac{\alpha}{\beta}\frac{K}{L}. \tag{5.35}$$

The absolute value of the slope of an isoquant is called the **marginal rate of technical substitution.** Since each of the partial derivatives on the right-hand side of equation (5.34) equals the change in output that results from a change in one factor of production, holding the other factor of production constant, they represent the marginal products of labor and capital. We have just shown that the marginal rate of technical substitution equals the ratio of marginal products of the two inputs.

This is typical of the way level curves are used in economics: the slope of a level curve represents the *tradeoff* of one thing for another. These tradeoffs are measured as the ratios of partial derivatives. Tradeoffs as measured by the slopes of level curves play a crucial role in constrained optimization, as discussed in Chapter 9.

5.6   HOMOGENEITY

Homogeneous functions play an important role in economic theory. Let $\mathbf{x}$ be a vector of variables and let $\lambda > 0$ be any positive constant. A function $f(\mathbf{x})$ is said to be **homogeneous of degree *k*** if $f(\lambda\mathbf{x}) = \lambda^k f(\mathbf{x})$ for all values of $\mathbf{x}$ and for any positive λ. That is, if all the arguments of the function are multiplied by the same positive constant λ and the value of the function ends up being λ^k times its old value, then the function is homogeneous of degree k. If a function is homogeneous of degree one, then doubling all its arguments doubles the value of the function, reducing every argument by a third reduces the value of the function by a third, and so on. If a function is homogeneous of degree 0, then as long as all arguments change by the same percentage the value of the function will remain constant.

Homogeneity is useful in many applications in economics, but one of the most useful is in production theory because of the relationship between homogeneity and returns to scale. A production function that is homogeneous of degree one exhibits constant returns to scale: doubling all inputs doubles output. A production function that is homogeneous of degree <1 exhibits decreasing returns to scale because doubling all inputs results in something less than double the output. A production function that is homogeneous of degree >1 exhibits increasing returns to scale.

For example, a Cobb-Douglas production function $q = AL^{\alpha}K^{\beta}$ is homogeneous of degree $\alpha + \beta$:

$$A(\lambda L)^{\alpha}(\lambda K)^{\beta} = A\lambda^{\alpha}L^{\alpha}\lambda^{\beta}K^{\beta} = \lambda^{\alpha+\beta}(AL^{\alpha}K^{\beta}). \tag{5.36}$$

Thus the production function has increasing, constant, or decreasing returns to scale depending on whether $\alpha + \beta$ is greater than, equal to, or less than 1.

Two theorems (really one theorem and a corollary) about homogeneous functions turn out to be useful in economics:

If $f(\mathbf{x})$ is a function of n variables that is homogeneous of degree k, then

$$kf(\mathbf{x}) = \sum_{i=1}^{n} \frac{\partial f}{\partial x_i} x_i.$$

As an example of the use of Euler's theorem, if a firm has a constant returns to scale production function, it will exactly use up the value of its output if it pays each input the value of its marginal product. In this example, $f(\mathbf{x})$ is the production function and $k = 1$ since constant returns to scale implies that the production function is homogeneous of degree 1. The marginal product of each input is the amount by which production increases when the use of the input is increased marginally. That is, the marginal product of input i is the partial derivative of the production function with respect to input i. Thus the right-hand side of the equation in Euler's theorem is interpreted in this example as the sum of the marginal products of each input multiplied by the amount of that input used. Each input's **value of marginal product** is defined to be its marginal product times the product price. So multiplying the right-hand side of the equation in Euler's theorem by the product price yields the total amount the firm would spend if it paid each input its value of marginal product. Since $k = 1$, the left-hand side of the equation in Euler's theorem is simply $f(\mathbf{x})$, the amount of output the firm produces given its use of inputs. Multiplying this by the product price yields the total amount of revenue the firm obtains by selling its output. So Euler's theorem implies in this context that the firm's revenue exactly equals the amount it spends on inputs if each input is paid the value of its marginal product. This is the basis for standard income distribution theory based on factor shares.

COROLLARY TO EULER'S THEOREM

If $f(\mathbf{x})$ is a function of n variables that is homogeneous of degree k, then each of the n partial derivatives of f is a function that is homogeneous of degree k − 1.

One implication of this theorem in economics is that if a firm has a constant returns to scale production function, each marginal product is a function of the firm's inputs, which is homogeneous of degree 0. Marginal products therefore depend on ratios of factors, not on the absolute levels of factor use.

SUMMARY

In this chapter we have introduced several concepts and techniques from multivariate calculus that are extremely useful in mathematical economics. They, along with the matrix algebra techniques covered in the previous chapters, especially Cramer's rule, provide the foundation for the analysis of optimization problems, with which much of the rest of the book deals. In addition, many economics problems can be investigated using just the approaches and techniques covered so far in the book, as illustrated by several applications in the next chapter.

The most important contribution of the material in this chapter is that we can now analyze problems involving nonlinear functions or functions of unknown form. In most cases we can derive comparative static results, even though explicit solutions for endogenous variables cannot be derived. This allows us to get more general results and to evaluate whether the results of linear models are due to the assumption of linearity or whether they hold more generally.

Problems

5.1 Find all the first-, second-, and third-order partial derivatives of the following functions:

(a) $f(x, y) = ax^2 + bxy + cy^2$

(b) $f(x, y, z) = ax^2 + by^2 + cz^2 + dxy + jxz + kyz$

(c) $f(x, y) = \sqrt{xy}$

(d) $f(w, x, y, z) = (wxyz)^{1/4}$

(e) $f(x, y, z) = Ax^\alpha y^\beta z^\gamma$

(f) $f(x, y, z) = a\ln(x - x_0) + b\ln(y - y_0) + c\ln(z - z_0)$

(g) $f(x, y) = A(ax^p + (1 - a)y^p)^{1/p}$

5.2 Find the total differential of each of the functions in Problem 5.1.

5.3 Suppose that $w = t^2$, $x = 4t^{-4}$, $y = -2 + 3t - 5t^2$, and $z = t^{-3/4}$. For each of the functions in Problem 5.1, find the total derivative of f with respect to t.

5.4 For each of the functions in Problem 5.1, check to see whether the function $x^*(y)$ exists. (Sometimes there will be additional arguments to the function.) If it does, find the derivative $\partial x^*/\partial y$ by using the implicit function rule and by implicit differentiation. Confirm that your solutions are identical.

5.5 Find the Jacobian of each of the following systems of equations:

(a) $f(x, y) = ax^2 + bxy + cy^2$

$g(x, y) = A(ax^p + (1 - a)y^p)^{1/p}$

(b) $f(x, y, z) = a\sqrt{xyz}$

$g(x, y, z) = Ax^a y^b z^c$

$h(x, y, z) = a\ln(x - x_0) + b\ln(y - y_0) + c\ln(z - z_0)$

(c) $f(x, y, z) = \sqrt{axyz}$

$g(x, y, z) = ax^2 + by^2 + cz^2 + dxy + jxz + kyz$

$h(x, y, z) = (xyz)^{a/3}$

5.6 For each system of equations in Problem 5.5, check to see whether the functions $x^*(a)$ and $y^*(a)$ exist. (There will sometimes be additional arguments to the functions.) If they do, implicitly differentiate the equations with respect to a and use Cramer's rule to find the derivatives $\partial x^*/\partial a$ and $\partial y^*/\partial a$.

5.7 For each of the following functions, find the formula for the slope of the level curve (the graph of the function $x^*(y)$). In each case, find the value of the slope of the level curve when $y = 1$. [*Hint:* You will have to solve for the corresponding value of x.] Would the slope increase or decrease if y were slightly greater than 1?

(a) $2x^2 - 6xy + 3y^2 = 12$

(b) $2x^2 - 6xy + 3y^2 = 36$

(c) $\sqrt{xy} = 36$

(d) $\sqrt{xy} = 100$

(e) $5x^4 y^2 = 66$

(f) $5x^4 y^2 = 99$

5.8 Evaluate whether each of the functions in Problem 5.1 is homogeneous. If so, of what degree?

6 Multivariate Calculus: Applications

6.1 INTRODUCTION

This chapter contains several macro- and microeconomic applications of the topics presented in Chapter 5. We begin with a section deriving a balanced-budget multiplier in three successively more complex macroeconomic models: first, a simple Keynesian model of the goods market, then an IS–LM model adding a money market, and then an aggregate demand-aggregate supply model adding flexible prices. This application illustrates the use of total derivatives and Cramer's rule and is also an example of the common technique in mathematical economics of starting with a simple model and then relaxing more and more simplifying assumptions to see if the results generalize. The following section is in the same vein: we examine the effectiveness of monetary policy, first in a closed-economy IS–LM model and then in an open-economy Mundell-Fleming model, deriving results that would be very difficult to obtain by reason alone. The example also illustrates the implicit function theorem, the implicit function rule, and implicit differentiation. The macroeconomic portion of the chapter concludes with a discussion of the IS and LM curves as level curves.

The microeconomic portion of the chapter begins with two tax incidence applications: first in a simple supply-and-demand model and then in a monopoly model. Then the Cournot duopoly example from previous chapters is revisited in a more general form. Section 6.8 is a labor supply example that discusses the problem of corner solutions and also shows how to convert a constrained maximization problem with two choice variables into an unconstrained maximization problem with one choice variable. The same problem is redone in Chapter 11 using more powerful techniques. A utility-maximization example follows, discussing the important concept of invariance of utility-maximization problems to monotonic transformations of the utility function. The application illustrates the use of level curves in economics. It also shows how to convert a constrained two-variable maximization problem into an unconstrained one-variable maximization problem; the problem is revisited using more powerful techniques in Chapter 9. Section 6.10 illustrates the concept of homogeneity in consumer theory. The chapter concludes with a section of problems based on the worked-out applications in the previous sections.

6.2 BALANCED–BUDGET MULTIPLIERS

In this application we derive the balanced-budget multiplier in three different macroeconomic models: a fixed-price, closed-economy model with no money market; a fixed-price IS–LM model that adds a money market; and then a flexible-price aggregate demand-aggregate supply model. We pay particular attention to how the balanced-budget multiplier changes as the model becomes more and more general; end-of-chapter problems analyze how the effects of other parameters on the endogenous variables of the model change as the model is generalized. The application also illustrates the use of total differentials and Cramer's rule in comparative static analysis. The total differentials approach is needed because more than one exogenous variable is changing. (Both taxes and government purchases of goods and services are changing, by equal but opposite amounts, in order to keep the budget surplus [or deficit] the same before and after the changes.)

6.2.1   Simple Keynesian Model

We analyzed the simple, fixed-price, closed-economy model with no money market in Section 2.7; the only change we make here is to allow for a nonlinear consumption function. Equilibrium output is defined by $Y = C + I + G$. For the time being we will assume that investment and government purchases are exogenously determined, while consumption is some (possibly nonlinear) function of disposable income, $C(Y^d)$. Disposable income Y^d equals $Y - T$, where taxes (net of transfer payments), T, are assumed for simplicity to be exogenous. Equilibrium output is thus implicitly defined by the equation

$$Y = C(Y^d) + I + G, \qquad (6.1)$$

where $Y^d = Y - T$.

To illustrate the effects of a balanced-budget fiscal policy change, let government purchases and taxes increase by the same amount. Recalling that $Y^d = Y - T$, we can see the effect on equilibrium output by taking the total differential of equation (6.1):

$$dY = C'(Y^d)\, dY - C'(Y^d)\, dT + dI + dG. \qquad (6.2)$$

In our example, investment is not changing, so $dI = 0$. Since taxes are increasing by the same amount as purchases, $dT = dG$. Thus the total differential can be rewritten as

$$dY = C'(Y^d)\, dY + (1 - C'(Y^d))\, dG. \qquad (6.3)$$

Solving for dY,

$$dY = \frac{(1 - C'(Y^d))}{(1 - C'(Y^d))}\, dG = dG. \qquad (6.4)$$

The balanced-budget multiplier is 1: the multiplier effects of the spending and tax changes exactly offset, so in the end the change in equilibrium GDP equals only the direct impact of the change in government purchases. This is the same result as we obtained in Section 2.7 with a linear consumption function; thus we have shown that the result continues to hold with a more general consumption function.

6.2.2 IS–LM Model

The simple balanced-budget multiplier derived in the previous section is often presented in introductory economics texts but assumes that there is no crowding out (by assuming that investment is exogenous or that interest rates are fixed). When we add a money market to the simple Keynesian model, the results change. We derive those results here, illustrating the technique of total differentiation of a system of equations.

The simple Keynesian model is a model of the market for goods. Adding a model of the money market yields the IS–LM model, whose equilibrium requires both the goods market and the money market to be in equilibrium. Goods-market equilibrium is given by $Y = C + I + G$, where consumption C is some (possibly nonlinear) function of disposable income (defined as $Y - T$, where the level of [net] taxes T is assumed to be exogenous) and investment I is some (possibly nonlinear) function of the interest rate r. Government purchases G are exogenous. The graph of all combinations of Y and r that satisfy the goods-market equilibrium condition is called the IS curve (because one way of defining this equilibrium is that planned investment I equals planned saving S). The IS curve is a level curve, which is explored further in Section 6.4.

The money market will be in equilibrium when the (exogenous) money supply M equals the demand for money (or liquidity) L, which is a function of income and the interest rate. The graph of all combinations of Y and r that satisfy the money market equilibrium condition is called the LM curve (because in equilibrium the demand for liquidity L equals the supply of money M). Like the IS curve, the LM curve is a level curve and is discussed further in Section 6.4.

The equilibrium levels of Y and r are those that satisfy both the goods market and money market equilibrium conditions. Graphically the equilibrium is the intersection of the IS and LM curves, as shown in Figure 6.1. This is why the model is called the IS–LM model. The model is described mathematically by the following two equations,

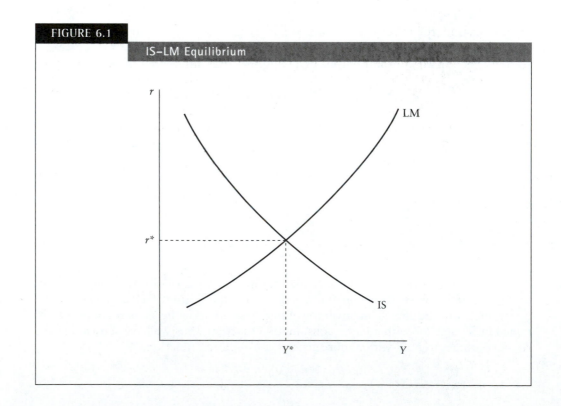

FIGURE 6.1

IS–LM Equilibrium

with the usual assumptions regarding derivatives listed to the right:

$$Y = C(Y^d) + I(r) + G, \quad 0 < C' < 1, \quad I' < 0,$$

$$M = L(Y, r), \quad L_Y \equiv \partial L/\partial Y > 0, \quad L_r \equiv \partial L/\partial r < 0$$

(6.5)

where the "prime" notation is used for the derivative of a function with only one argument, $Y^d = Y - T$, and G and M are exogenous.

Substituting $Y - T$ for Y^d, the total differential of this system of two equations is

$$dY = C'dY - C'dT + I'dr + dG$$

$$dM = L_Y \, dY + L_r \, dr.$$

(6.6)

To examine the effects of a balanced-budget fiscal policy, we will hold the money supply constant so $dM = 0$. Because it is a balanced-budget policy, $dT = dG$. We now have a system of two equations in the two endogenous differentials dY and dr:

$$dY = C'dY + (1 - C') \, dG + I'dr$$

$$0 = L_Y \, dY + L_r \, dr$$

(6.7)

which can be written in matrix notation as

$$\begin{bmatrix} 1 - C' & -I' \\ L_Y & L_r \end{bmatrix} \begin{bmatrix} dY \\ dr \end{bmatrix} = \begin{bmatrix} (1 - C')dG \\ 0 \end{bmatrix}.$$

(6.8)

Using Cramer's rule to solve for dY, we get

$$dY = \frac{\begin{vmatrix} (1 - C')dG & -I' \\ 0 & L_r \end{vmatrix}}{\begin{vmatrix} 1 - C' & -I' \\ L_Y & L_r \end{vmatrix}} = \frac{(1 - C')L_r dG}{(1 - C')L_r + L_Y I'}.$$

(6.9)

Note that the denominator of equation (6.9) is the Jacobian of the system of equations (6.5). Equation (6.9) can be rewritten as $dY = (1/(1 + \phi)) \, dG$, where $\phi = L_Y I'/(1 - C')L_r$. The numerator and denominator of ϕ are both products of a positive term and a negative term. Thus ϕ is positive and the balanced-budget multiplier is positive but less than 1.

The economics of this result is straightforward. The combination of increased spending and increased taxes is on balance stimulative, as in the simple multiplier story given in introductory textbooks and the previous section. Graphically, this stimulative policy is reflected by the rightward shift (since equilibrium output would be higher, holding the interest rate constant) of the IS curve shown in Figure 6.2.[1] But the stimulative policy leads to higher interest rates as shown in Figure 6.2 (proof of this is left as an exercise), and some private investment is crowded out.

The formula for ϕ shows that this crowding-out effect depends on four factors. The more responsive money demand is to changes in income (that is, the higher is L_Y), the more crowding out there will be, since the stimulative fiscal policy will lead to a larger excess demand for money. There will also be more crowding out if money demand is less sensitive to interest rates (that is, the less negative is L_r), since a larger increase in the interest rate will be needed to eliminate the excess demand for money. The more

[1] In fact, the magnitude of the rightward shift of the IS curve is equal to dG: this is the balanced-budget multiplier result of Section 6.2.1.

FIGURE 6.2

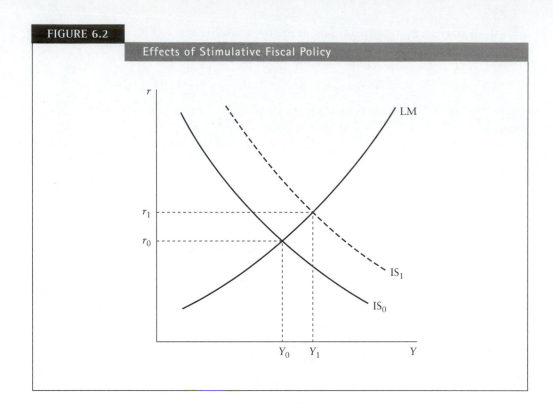

Effects of Stimulative Fiscal Policy

responsive investment is to changes in interest rates (that is, the more negative is I'), the more crowding out occurs, because investment will fall more in response to the higher interest rates. Finally, there will be more crowding out the larger is the marginal propensity to consume C', since the expenditure multiplier will be larger, leading to a larger drop in consumption and output when investment falls. (The relationships between each of these factors and the slopes of the IS or LM curves are discussed in Section 6.4.)

6.2.3 Aggregate Demand–Aggregate Supply Model

So far we have seen that the addition of a money market to the simple model reduces the balanced-budget multiplier because a stimulative balanced-budget fiscal policy crowds out some private investment. In this section we generalize the model further by making prices flexible. We continue to work with a closed-economy model and with simplistic versions of the consumption and investment functions to isolate the impact of relaxing the assumption of fixed prices. The IS–LM model focuses on the determinants of aggregate demand and has as endogenous variables real output Y and the real interest rate r. By adding a third endogenous variable, the aggregate price level P, we require a third equilibrium condition: an aggregate supply equation. This is why the model is called an aggregate demand-aggregate supply, or AD–AS, model.

With no international sector, the goods market equilibrium is defined by the familiar condition $Y = C + I + G$. Consumption C is assumed to be some (possibly nonlinear) function of disposable income $Y^d = Y - T$ where taxes are assumed to be exogenous.[2]

---

[2]Most aggregate demand-aggregate supply models would add real wealth as an argument of the consumption function. We do not do so here in order to facilitate comparisons with the simpler models in this application.

$C'(Y^d)$ is the marginal propensity to consume and therefore is bounded between 0 and 1. Investment I is assumed to be some (possibly nonlinear) function of the interest rate, with $I'(r) < 0$. Government purchases of goods and services G are exogenous. The goods market equilibrium condition is therefore

$$Y = C(Y - T) + I(r) + G, \quad \text{where } 0 < C' < 1, \quad I' < 0. \tag{6.10}$$

The money market equilibrium condition is the same as in the IS–LM model except that, when prices are flexible, we must be careful to distinguish between the nominal money supply M and the supply of real balances M/P. In equilibrium the supply of money (in real terms) equals the demand for real balances, or liquidity, which is an increasing function of real income and a decreasing function of the interest rate. In symbols, the money market equilibrium condition is

$$\frac{M}{P} = L(Y, r) \quad \text{where } L_Y > 0, \quad L_r < 0. \tag{6.11}$$

The model is closed by adding a simple aggregate supply curve in which the aggregate price level equals the (exogenous) expected price level P^E plus another term related to the difference between output and the potential (full-employment) level of output, Y^F:

$$P = P^E + g(Y - Y^F). \tag{6.12}$$

The derivative of the function $g(Y - Y^F)$ is the slope of the aggregate supply curve, which is nonnegative. Three possible aggregate supply (AS) curves are shown in Figure 6.3. If $g' = 0$, we are in a fixed-price model; that is, the model will reduce to the IS–LM model. This case is illustrated by the aggregate supply curve AS_1. If g' is infinitely large, the AS curve is AS_2, and we are in a classical model with completely flexible prices and continual full employment. If g' is positive but finite, the aggregate supply curve looks like AS_3. When output falls below potential, the aggregate price level

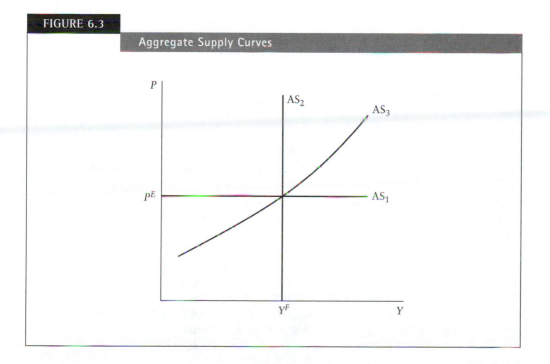

FIGURE 6.3

Aggregate Supply Curves

falls below its expected level; when output exceeds potential, the price level rises above its expected level. The magnitude of the derivative g' represents the flexibility of prices.

Together, equations (6.10), (6.11), and (6.12) comprise the AD–AS model. The endogenous variables in the system are Y, r, and P. Exogenous variables are T, G, M, P^E, and Y^F; also exogenous are the (unspecified) forms of the functions $C(Y^d)$, $I(r)$, $L(Y, r)$, and $g(Y - Y^F)$. We are interested in a balanced-budget fiscal policy action, so we take the total differentials of equations (6.10), (6.11), and (6.12), set $dM = dP^E = dY^F = 0$, and set $dT = dG$. In matrix notation, the result is

$$\begin{bmatrix} 1 - C' & -I' & 0 \\ L_Y & L_r & \dfrac{M}{P^2} \\ -g' & 0 & 1 \end{bmatrix} \begin{bmatrix} dY \\ dr \\ dP \end{bmatrix} = \begin{bmatrix} (1 - C')dG \\ 0 \\ 0 \end{bmatrix}. \tag{6.13}$$

We can now use Cramer's rule to solve for dY:

$$dY = \frac{\begin{vmatrix} (1 - C')\,dG & -I' & 0 \\ 0 & L_r & \dfrac{M}{P^2} \\ 0 & 0 & 1 \end{vmatrix}}{\begin{vmatrix} 1 - C' & -I' & 0 \\ L_Y & L_r & \dfrac{M}{P^2} \\ -g' & 0 & 1 \end{vmatrix}}. \tag{6.14}$$

Expanding both the numerator and the denominator by the third row,

$$dY = \frac{L_r(1 - C')\,dG}{g'[(M/P^2)I'] + L_r(1 - C') + L_Y I'}, \tag{6.15}$$

so the balanced-budget multiplier in the AD–AS model is

$$\frac{dY}{dG} = \frac{L_r(1 - C')}{g'[(M/P^2)I'] + L_r(1 - C') + L_Y I'}. \tag{6.16}$$

This is identical to the balanced-budget multiplier in the IS–LM model, equation (6.9), except for the first term in the denominator. Since $g' \geq 0$ and $I' < 0$, the first term is nonpositive. So the balanced-budget multiplier in the AD–AS model is positive but smaller than or equal to the balanced-budget multiplier in the IS–LM model. The two are equal only if $g' = 0$, which confirms our earlier statement that in this special case the AD–AS model reduces to the IS–LM model. If $g' > 0$ the flexibility of prices reduces the effect of a stimulative, balanced-budget fiscal policy. If prices are completely flexible, $g' \to \infty$ and the balanced-budget multiplier is zero.

Equation (6.16) indicates the economics of why the balanced-budget multiplier is smaller in the AD–AS model than in the IS–LM model. In addition to the crowding out present in the IS–LM model, when prices are flexible they will rise as output increases; prices will rise by more, the larger is g'. The higher aggregate price level reduces the supply of real balances (by the amount $-M/P^2$). This creates an excess demand for money, which leads to higher interest rates and further crowding out of private investment; the crowding out is larger the greater is the magnitude of I'.

6.2.4 Summary of Balanced-Budget Multipliers

In this application we derived balanced-budget multipliers for three models, starting with a simple model and then relaxing some of the simplifying assumptions. When simplifying assumptions were relaxed, variables that were exogenous in the simpler models (the interest rate and the aggregate price level) became endogenous. We found that the specific results of the simple model did not continue to hold in more complex models. In particular, the balanced-budget multiplier became smaller and smaller as we moved from the simple Keynesian model to the IS–LM model and then to the AD–AS model. But, except in the special case of completely flexible prices, a balanced-budget increase in both government purchases and taxes was stimulative overall in all three models.

The technique of analysis we have used in this application can be used to investigate the effects of changes in many other exogenous variables and parameters in these three models. By taking total differentials of a model's equations and making assumptions about which exogenous variables have differentials that are zero and which have nonzero differentials, Cramer's rule can be used to solve for how endogenous variables are affected by changes in any one or more exogenous variables. The object of analysis is usually to sign comparative static derivatives and to see what factors influence their magnitudes. Some examples of this can be found in the end-of-chapter problems.

6.3 MONETARY POLICY EFFECTIVENESS

In this application we illustrate the implicit function theorem and implicit differentiation by investigating the effectiveness of monetary policy as a macroeconomic tool. We begin with a traditional IS–LM model and derive the comparative static effects of a change in the money supply on the equilibrium interest rate and equilibrium output. The resulting comparative static derivatives show the influence of various parameters. Because one of our purposes in this application is to show how results are affected by various assumptions about international trade and capital flows, our IS–LM model is of a closed economy. We then generalize the IS–LM model to an open-economy macroeconomic model with a balance-of-payments constraint; this is often called the Mundell-Fleming model. We describe how to set up the cases of both flexible and fixed exchange rates and go through the flexible exchange rate case in detail. As part of our analysis we compare two extreme cases of how mobile capital is between countries: perfect capital mobility and no capital mobility.

6.3.1 IS–LM Model

The IS–LM model was described in Section 6.2.2 in the context of the balanced-budget multiplier. We give a somewhat different version of it here and use it to investigate monetary policy effectiveness. The equilibrium of the IS–LM model occurs when both the goods market and the money market are in equilibrium.

The goods market (represented graphically by the IS curve, which is investigated more in Section 6.4) requires national output Y to equal the sum of planned consumption C, planned investment I, and exogenous government purchases of goods and services G. Assuming that consumption is an increasing function of output and that investment is an increasing function of output and a decreasing function of the interest rate, the goods market equilibrium condition is

$$Y = C(Y) + I(Y, r) + G, \quad \text{where } 0 < C' < 1, \quad I_Y > 0, \quad I_r < 0. \tag{6.17}$$

We will assume that $C' + I_Y < 1$; if this were not true, then when income increases the induced increase in aggregate demand would exceed the increase in income. The money market (represented graphically by the LM curve, discussed in more detail in Section 6.4) is in equilibrium when the exogenous supply of money M equals the demand for money (or liquidity) L, which is assumed to be an increasing function of output and a decreasing function of the interest rate:

$$M = L(Y, r), \quad \text{where } L_Y > 0, \quad L_r < 0. \tag{6.18}$$

We could analyze this system of two equations in the two endogenous variables Y and r by taking their total differentials, but since in this application we are interested in the effects of changing only one exogenous variable, M, we will instead use implicit differentiation. Assuming that equilibrium solutions for Y and r exist (as functions of all the exogenous variables in the model, though we will suppress all exogenous variables other than M as arguments of functions), the model can be written as

$$Y^*(M) \equiv C(Y^*(M)) + I(Y^*(M), r^*(M)) + G$$
$$M \equiv L(Y^*(M), r^*(M)) \tag{6.19}$$

so, using implicit differentiation and suppressing arguments of functions,

$$\frac{\partial Y^*}{\partial M} = C'\frac{\partial Y^*}{\partial M} + I_Y\frac{\partial Y^*}{\partial M} + I_r\frac{\partial r^*}{\partial M}$$
$$1 = L_Y\frac{\partial Y^*}{\partial M} + L_r\frac{\partial r^*}{\partial M}. \tag{6.20}$$

In matrix notation, this system of two equations in the two unknowns $\partial Y^*/\partial M$ and $\partial r^*/\partial M$ is

$$\begin{bmatrix} 1 - C' - I_Y & -I_r \\ L_Y & L_r \end{bmatrix} \begin{bmatrix} \partial Y^*/\partial M \\ \partial r^*/\partial M \end{bmatrix} = \begin{bmatrix} 0 \\ 1 \end{bmatrix}. \tag{6.21}$$

So using Cramer's rule we get

$$\frac{\partial Y^*}{\partial M} = \frac{\begin{vmatrix} 0 & -I_r \\ 1 & L_r \end{vmatrix}}{|J|} = \frac{I_r}{|J|} \quad \text{and} \quad \frac{\partial r^*}{\partial M} = \frac{\begin{vmatrix} 1 - C' - I_Y & 0 \\ L_Y & 1 \end{vmatrix}}{|J|} = \frac{1 - C' - I_Y}{|J|}. \tag{6.22}$$

The Jacobian of this system is

$$|J| = \begin{vmatrix} 1 - C' - I_Y & -I_r \\ L_Y & L_r \end{vmatrix} = L_r(1 - C' - I_Y) + L_Y I_r. \tag{6.23}$$

Since $C' + I_Y < 1$, the Jacobian is negative. Since it is nonzero, the implicit function theorem confirms that the equilibrium functions $Y^*(M)$ and $r^*(M)$ do indeed exist. Since the Jacobian is negative, $\partial Y^*/\partial M > 0$ and $\partial r^*/\partial M < 0$. The greater supply of money creates excess supply in the money market, lowering interest rates. The lower interest rates, in turn, spur investment, which sets off a multiplier effect on output. As output increases, the demand for money increases, which tends to raise interest rates and crowd out investment. But equation (6.22) indicates that this secondary effect does not completely offset the initial impact of the monetary policy, and the equilibrium value of output rises while the equilibrium interest rate falls.

The results shown in (6.22) can also be derived using the implicit function rule. To do so, we begin by rewriting the two equations (6.17) and (6.18) as the two implicit functions

$$f(Y, r, G, M) = Y - C(Y) - I(Y, r) - G = 0$$
$$g(Y, r, G, M) = M - L(Y, r) = 0 \tag{6.24}$$

and then apply the implicit function rule:

$$\frac{\partial Y^*}{\partial M} = -\frac{\begin{vmatrix} \partial f/\partial M & \partial f/\partial r \\ \partial g/\partial M & \partial g/\partial r \end{vmatrix}}{|J|} = -\frac{\begin{vmatrix} \partial f/\partial M & \partial f/\partial r \\ \partial g/\partial M & \partial g/\partial r \end{vmatrix}}{\begin{vmatrix} \partial f/\partial Y & \partial f/\partial r \\ \partial g/\partial Y & \partial g/\partial r \end{vmatrix}} = -\frac{\begin{vmatrix} 0 & -I_r \\ 1 & -L_r \end{vmatrix}}{\begin{vmatrix} 1 - C' - I_Y & -I_r \\ -L_Y & -L_r \end{vmatrix}} \tag{6.25}$$

$$= -\frac{I_r}{-L_r(1 - C' - I_Y) - L_Y I_r} = \frac{I_r}{L_r(1 - C' - I_Y) + L_Y I_r}$$

and

$$\frac{\partial r^*}{\partial M} = -\frac{\begin{vmatrix} \partial f/\partial Y & \partial f/\partial M \\ \partial g/\partial Y & \partial g/\partial M \end{vmatrix}}{|J|} = -\frac{\begin{vmatrix} \partial f/\partial Y & \partial f/\partial M \\ \partial g/\partial Y & \partial g/\partial M \end{vmatrix}}{\begin{vmatrix} \partial f/\partial Y & \partial f/\partial r \\ \partial g/\partial Y & \partial g/\partial r \end{vmatrix}} = -\frac{\begin{vmatrix} 1 - C' - I_Y & 0 \\ -L_Y & 1 \end{vmatrix}}{\begin{vmatrix} 1 - C' - I_Y & -I_r \\ -L_Y & -L_r \end{vmatrix}} \tag{6.26}$$

$$= -\frac{1 - C' - I_Y}{-L_r(1 - C' - I_Y) - L_Y I_r} = \frac{1 - C' - I_Y}{L_r(1 - C' - I_Y) + L_Y I_r}.$$

The comparative static derivative functions in (6.22) can also be used to investigate the influence of various factors on the effectiveness of monetary policy. We will look at how the effectiveness of monetary policy on equilibrium output is affected by how sensitive investment is to output and the interest rate; other factors and the influence monetary policy has on equilibrium interest rates are investigated in the end-of-chapter problems.

If investment becomes more sensitive to changes in output, then I_Y becomes larger. From equation (6.23), this makes the Jacobian less negative since $(1 - C' - I_Y)$ becomes smaller. Since this makes the denominator of $\partial Y^*/\partial M$ smaller in absolute value, an increase in the money supply results in a larger increase in equilibrium output. The economic explanation of this is based on the multiplier effect of any increase in aggregate demand: when any component of aggregate demand increases, this sets off a multiplier effect as the increased spending becomes extra national income, which then increases consumption and, in the model we have been using, increases investment as well. This is because the higher income is seen as a signal to firms that future demand for their products will be higher, so the firms increase investment in order to have increased production capacity in the future. This increase in spending generates another increase in national income, and the multiplier process continues. As long as $C' + I_Y$ stays smaller than 1, $0 < (1 - C' - I_Y) < 1$ so this multiplier process eventually peters out, but not before increasing equilibrium output by a multiple (greater than 1) of the original change in aggregate demand. If investment is more responsive to changes in output, then the increase in demand during every round of this multiplier process becomes larger, so the eventual increase in equilibrium output is larger too. In the context of monetary policy effectiveness, monetary policy becomes more effective at changing output since, once the monetary policy action changes interest rates, investment changes, which sets off a multiplier process. So, since the multiplier effect is larger, monetary policy is more effective.

If investment is more responsive to changes in interest rates, then I_r becomes more negative. Equation (6.23) shows that the Jacobian becomes more negative; this will tend to make $\partial Y^*/\partial M$ smaller. But as can be seen from (6.22), I_r is also in the numerator of the comparative static derivative formula for $\partial Y^*/\partial M$. If I_r becomes more negative, then the numerator of $\partial Y^*/\partial M$ will be larger in absolute value, which will tend to make $\partial Y^*/\partial M$ larger. To see the overall effect, rewrite the comparative static derivative as

$$\frac{\partial Y^*}{\partial M} = \frac{I_r}{L_r(1 - C' - I_Y) + L_Y I_r} = \frac{1}{[L_r(1 - C' - I_Y)/I_r] + L_Y}. \tag{6.27}$$

Now it is clear that if I_r becomes greater in absolute value, the denominator of (6.27) becomes smaller in absolute value, so $\partial Y^*/\partial M$ becomes greater. The economic explanation is twofold, corresponding to the two places that I_r appears in the formula for $\partial Y^*/\partial M$ in (6.22). Since the Jacobian is larger in absolute value, the multiplier effect of any increase in aggregate demand is smaller; this is because more investment will be crowded out by the higher interest rates that result from increased demand for money (due to higher income). But this effect is dominated by the other effect of higher sensitivity of investment to interest rates: expansionary monetary policy works by lowering interest rates, which then increases investment. If investment is more sensitive to changes in interest rates, then this effect is larger. Equation (6.27) shows that, even though the multiplier is smaller, the fact that the initial increase in aggregate demand is higher makes monetary policy more effective when investment is more responsive to changes in interest rates. This result would be very difficult to obtain by reason alone, without the aid of the mathematical economic model.

6.3.2 Mundell–Fleming Model with Flexible Exchange Rates

The Mundell-Fleming model (named after the two economists who developed it) adds an international sector to the IS–LM model. This changes the equilibrium conditions of both the goods market and the money market, plus it adds another endogenous variable to the system. This necessitates a third equation to complete the model: a balance-of-payments equation.

The balance-of-payments equation keeps track of international exchanges of currencies. Currencies must be exchanged when there is international trade of goods and services, measured by net exports X, or when capital flows from one country to another. Said another way, the balance of payments equation combines the trade balance with the capital account balance to see whether, on net, more people want to exchange dollars for foreign currencies or foreign currencies for dollars. When exchange rates are flexible, the trade balance and the capital account balance must exactly offset each other so that there is no excess demand for or supply of dollars in the foreign exchange market. If we let f represent the net inflow[3] of capital into the United States (which requires foreign currencies to be exchanged for dollars, which is also true when net exports are positive), then the balance of payments equation with flexible exchange rates is $X + f = 0$. When exchange rates are fixed, on the other hand, the balance of payments is not generally zero. Instead, the central bank (or whatever agency maintains the fixed

---

[3]Sometimes the capital account is represented by net capital outflows rather than inflows. Mathematically, the model is the same except that the balance of payments equals $X - f$ rather than $X + f$, and when we let net capital flows depend on the difference between U.S. and foreign interest rates, the derivative f' will be negative instead of positive.

exchange rate) buys or sells dollars in the foreign exchange market to offset whatever imbalance results from private transactions. The balance of payment is now $X + f = \Delta F$; ΔF is usually defined as the change in central bank holdings of foreign currency, though it might be easier to work through the reasoning of the model if ΔF is thought of as the amount of dollars the central bank puts into circulation by buying foreign currency on the foreign exchange market. If net exports plus net capital inflows is positive, then there is an excess demand for dollars in the foreign exchange markets. The central bank offsets this by selling dollars and buying (and holding in reserve) foreign currency. If net exports plus net capital inflows is negative, there is an excess supply of dollars in the foreign exchange markets, so the central bank buys the surplus dollars, using its reserves of foreign currency; this reduces the amount of dollars in circulation.

X and f are both endogenous variables, depending on other economic variables. Net exports are usually assumed to depend on output and the exchange rate: $X = X(Y, e)$ where e is the exchange rate of domestic for foreign currency. For convenience, we will use the dollar as the domestic currency and the euro as the foreign currency, though in reality if we were modeling the U.S. economy net exports would depend on the exchange rates with respect to the currencies of all U.S. trading partners; another option would be to have net exports depend on the trade-weighted average of those exchange rates. We must also specify whether the exchange rate is dollars per euro or euros per dollar; this will determine whether $\partial X / \partial e$ is positive or negative. We will choose to measure exchange rates as dollars per euro, so if e increases the dollar is depreciating[4] (more dollars are needed to buy the same number of euros). Since when the dollar depreciates U.S. goods are less expensive compared to foreign goods, net exports will rise, so $\partial X / \partial e > 0$. Since higher domestic output means higher domestic income and more imports, $\partial X / \partial Y < 0$. We will assume that net capital inflows are a function of interest rate differentials: $f = f(r - r^F)$ where r^F represents the rate of return that can be earned on foreign assets and $f' > 0$. (Later we will consider the case of perfect capital immobility, in which case $f' = 0$.)

In summary, the balance-of-payments equilibrium condition is

$$X(Y, e) + f(r - r^F) = \Delta F. \tag{6.28}$$

When exchange rates are flexible, ΔF equals zero and the exchange rate e is endogenously determined by the demand for and supply of dollars in the foreign exchange market. When exchange rates are fixed, e is exogenous and ΔF is endogenously determined by the private demand for and supply of dollars.

Adding the international sector to the goods market equilibrium condition requires simply adding net exports to aggregate demand:

$$Y = C(Y) + I(Y, r) + G + X(Y, e). \tag{6.29}$$

In the money market, the total money supply is no longer completely exogenous in a fixed exchange rate regime, since whenever the central bank increases its stock of foreign exchange in order to keep the dollar from appreciating, this has the effect of putting more dollars in circulation. So the net change in the stock of foreign currency must be added to the otherwise exogenous money supply. The money market equilibrium condition is therefore

$$M + \Delta F = L(Y, r). \tag{6.30}$$

[4]The usual convention is to say the exchange rate depreciates when exchange rates are flexible, but that it is devalued when exchange rates are fixed. We will ignore this convention and use the term depreciate regardless of whether exchange rates are fixed or flexible.

We are now prepared to analyze the effectiveness of monetary policy in this model, which consists of equations (6.28), (6.29), and (6.30). We will go through the details for the case of flexible exchange rates; the fixed exchange rate case will be explored in the end-of-chapter problems. With flexible exchange rates, the endogenous variables are Y, r, and e, and the exogenous variables are r^F, G, and M.[5] We will begin by using the implicit function rule to derive formulas for $\partial Y^*/\partial M$ and $\partial r^*/\partial M$ and then show how to use implicit differentiation to derive the same results.

To use the implicit function rule, we need to write the three equations of our system in implicit form:

$$
\begin{aligned}
g(e, Y, r, r^F, G, M) &= X(Y, e) + f(r - r^F) = 0 \\
h(e, Y, r, r^F, G, M) &= Y - C(Y) - I(Y, r) - G - X(Y, e) = 0 \\
k(e, Y, r, r^F, G, M) &= M - L(Y, r) = 0
\end{aligned}
\tag{6.31}
$$

where we have set $\Delta F = 0$ since we are working with flexible exchange rates. We can now apply the implicit function rule:

$$
\frac{\partial Y^*}{\partial M} = -\frac{\begin{vmatrix} \partial g/\partial e & \partial g/\partial M & \partial g/\partial r \\ \partial h/\partial e & \partial h/\partial M & \partial h/\partial r \\ \partial k/\partial e & \partial k/\partial M & \partial k/\partial r \end{vmatrix}}{\begin{vmatrix} \partial g/\partial e & \partial g/\partial Y & \partial g/\partial r \\ \partial h/\partial e & \partial h/\partial Y & \partial h/\partial r \\ \partial k/\partial e & \partial k/\partial Y & \partial k/\partial r \end{vmatrix}} = -\frac{\begin{vmatrix} X_e & 0 & f' \\ -X_e & 0 & -I_r \\ 0 & 1 & -L_r \end{vmatrix}}{\begin{vmatrix} X_e & X_Y & f' \\ -X_e & 1 - C' - I_Y - X_Y & -I_r \\ 0 & -L_Y & -L_r \end{vmatrix}}
$$

$$
= -\frac{-1(-I_r X_e + f' X_e)}{X_e(-L_r(1 - C' - I_Y - X_Y) - L_Y I_r) - (-X_e)(-L_r X_Y + L_Y f')}
\tag{6.32}
$$

$$
= \frac{\overset{(-)}{\overbrace{I_r - f'}}}{\underset{(-)}{\underbrace{L_r(1 - C' - I_Y) + L_Y(I_r - f')}}} > 0
$$

and

$$
\frac{\partial r^*}{\partial M} = -\frac{\begin{vmatrix} \partial g/\partial e & \partial g/\partial Y & \partial g/\partial M \\ \partial h/\partial e & \partial h/\partial Y & \partial h/\partial M \\ \partial k/\partial e & \partial k/\partial Y & \partial k/\partial M \end{vmatrix}}{\begin{vmatrix} \partial g/\partial e & \partial g/\partial Y & \partial g/\partial r \\ \partial h/\partial e & \partial h/\partial Y & \partial h/\partial r \\ \partial k/\partial e & \partial k/\partial Y & \partial k/\partial r \end{vmatrix}} = -\frac{\begin{vmatrix} X_e & X_Y & 0 \\ -X_e & 1 - C' - I_Y - X_Y & 0 \\ 0 & -L_Y & 1 \end{vmatrix}}{\begin{vmatrix} X_e & X_Y & f' \\ -X_e & 1 - C' - I_Y - X_Y & -I_r \\ 0 & -L_Y & -L_r \end{vmatrix}}
$$

$$
= -\frac{\overset{(+)}{\overbrace{X_e(1 - C' - I_Y - X_Y) + X_e X_Y}}}{X_e(-L_r(1 - C' - I_Y - X_Y) - L_Y I_r) - (-X_e)(-L_r X_Y + L_Y f')}
\tag{6.33}
$$

$$
= \frac{\overset{(+)}{\overbrace{1 - C' - I_Y}}}{\underset{(-)}{\underbrace{L_r(1 - C' - I_Y) + L_Y(I_r - f')}}} < 0.
$$

---

[5] Many other exogenous variables could be added to a richer version of the model, for example, foreign income as an argument of the net exports function, measures of political stability as arguments in the net capital outflows function, wealth as an argument of the consumption function, domestic taxes as arguments in the consumption and/or investment equations, and trade tariffs or subsidies as arguments in the net exports function.

To derive the comparative static derivatives using implicit differentiation, we start by assuming that the model can be solved for the equilibrium values of the endogenous variables as functions of the exogenous variables. Since the only exogenous variable we will allow to change is M, we will suppress all other exogenous variables as arguments of the equilibrium functions. The system of equations can now be written as

$$X(Y^*(M), e^*(M)) + f(r^*(M) - r^F) \equiv 0$$

$$Y^*(M) \equiv C(Y^*(M)) + I(Y^*(M), r^*(M)) + G + X(Y^*(M), e^*(M)) \quad \text{(6.34)}$$

$$M \equiv L(Y^*(M), r^*(M))$$

where the three equations have been turned into identities by substituting in the equilibrium functions for the three endogenous variables.

Implicitly differentiating (6.34) with respect to M yields

$$X_Y \frac{\partial Y^*}{\partial M} + X_e \frac{\partial e^*}{\partial M} + f' \frac{\partial r^*}{\partial M} = 0$$

$$\frac{\partial Y^*}{\partial M} = C' \frac{\partial Y^*}{\partial M} + I_Y \frac{\partial Y^*}{\partial M} + I_r \frac{\partial r^*}{\partial M} + X_Y \frac{\partial Y^*}{\partial M} + X_e \frac{\partial e^*}{\partial M} \quad \text{(6.35)}$$

$$1 = L_Y \frac{\partial Y^*}{\partial M} + L_r \frac{\partial r^*}{\partial M}$$

or, in matrix notation,

$$\begin{bmatrix} X_e & X_Y & f' \\ -X_e & 1 - C' - I_Y - X_Y & -I_r \\ 0 & L_Y & L_r \end{bmatrix} \begin{bmatrix} \partial e^*/\partial M \\ \partial Y^*/\partial M \\ \partial r^*/\partial M \end{bmatrix} = \begin{bmatrix} 0 \\ 0 \\ 1 \end{bmatrix}. \quad \text{(6.36)}$$

The Jacobian of this system of equations is

$$|J| = \begin{vmatrix} X_e & X_Y & f' \\ -X_e & 1 - C' - I_Y - X_Y & -I_r \\ 0 & L_Y & L_r \end{vmatrix}$$

$$= X_e(L_r(1 - C' - I_Y - X_Y) + L_Y I_r) + X_e(L_r X_Y - L_Y f') \quad \text{(6.37)}$$

$$= X_e(L_r(1 - C' - I_Y) + L_Y(I_r - f'))$$

which, given the usual assumptions about signs of derivatives, is positive. (The fact that the Jacobian is nonzero confirms, by the implicit function theorem, that the equilibrium functions Y^*, r^*, and e^* exist.) Using Cramer's rule,

$$\frac{\partial Y^*}{\partial M} = \frac{\begin{vmatrix} X_e & 0 & f' \\ -X_e & 0 & -I_r \\ 0 & 1 & L_r \end{vmatrix}}{|J|} = \frac{-1(-I_r X_e + f' X_e)}{|J|} = \frac{X_e(I_r - f')}{|J|} > 0 \quad \text{(6.38)}$$

and

$$\frac{\partial r^*}{\partial M} = \frac{\begin{vmatrix} X_e & X_Y & 0 \\ -X_e & 1 - C' - I_Y - X_Y & 0 \\ 0 & L_Y & 1 \end{vmatrix}}{|J|}$$

$$= \frac{X_e(1 - C' - I_Y - X_Y) + X_e X_Y}{|J|} = \frac{X_e(1 - C' - I_Y)}{|J|} < 0. \quad \text{(6.39)}$$

The effect of monetary policy on exchange rates is left as an end-of-chapter problem.

To facilitate comparisons of the Mundell-Fleming results shown in equations (6.32) and (6.33) (or, equivalently, equations (6.38) and (6.39)) to the corresponding IS–LM results shown in equation (6.22), the Mundell-Fleming results can be rewritten as

$$\frac{\partial Y^*}{\partial M} = \frac{I_r - f'}{L_r(1 - C' - I_Y) + L_Y(I_r - f')} = \frac{1}{[L_r(1 - C' - I_Y)/(I_r - f')] + L_Y} \quad \textbf{(6.40)}$$

and

$$\frac{\partial r^*}{\partial M} = \frac{1 - C' - I_Y}{L_r(1 - C' - I_Y) + L_Y(I_r - f')} = \frac{1}{L_r + [L_Y(I_r - f')/(1 - C' - I_Y)]} \quad \textbf{(6.41)}$$

while the IS–LM results can be rewritten as

$$\frac{\partial Y^*}{\partial M} = \frac{I_r}{L_r(1 - C' - I_Y) + L_Y I_r} = \frac{1}{[L_r(1 - C' - I_Y)/I_r] + L_Y} \quad \textbf{(6.42)}$$

and

$$\frac{\partial r^*}{\partial M} = \frac{(1 - C' - I_Y)}{L_r(1 - C' - I_Y) + I_Y I_r} = \frac{1}{L_r + [L_Y I_r/(1 - C' - I_Y)]}. \quad \textbf{(6.43)}$$

The differences between the Mundell-Fleming results and the IS–LM results are caused by the appearance of the term f'. This term represents how much capital flows across international boundaries in response to interest rate differentials. Comparing equations (6.40) and (6.42), it is clear that as long as $f' > 0$, $\partial Y^*/\partial M$ will be greater (since the denominator will be smaller) in the open-economy, Mundell-Fleming model than in the closed-economy, IS–LM model. Comparing equations (6.41) and (6.43), if $f' > 0$ then $\partial r^*/\partial M$ will be smaller (since the denominator will be greater) in an open economy than in a closed economy.

We now turn to two special cases: perfect capital immobility and perfect capital mobility. If $f' = 0$, there are no capital flows across borders: capital is perfectly immobile. In this case, the Mundell-Fleming model results are exactly the same as the IS–LM results, even though the Mundell-Fleming model includes net exports. The reason is that, with completely flexible exchange rates and no international capital flows, the only way for the balance of payments to equal zero is for the trade balance to be zero. Exchange rates will move in whichever direction and by however much is necessary to offset any initial change in net exports. Thus even though the model includes net exports, it is set up in a way that ensures net exports will always equal zero, so in effect we are in an IS–LM model.

The other extreme case is perfect capital mobility, in which case $f' \to \infty$. This makes the Jacobian for the Mundell-Fleming model infinitely large, implying that $\partial r^*/\partial M = 0$. The economic explanation is that when capital is perfectly mobile, interest rates must be equal everywhere: as soon as the interest rate in one country exceeds the interest rates available elsewhere, capital will immediately flow into the country, chasing the higher returns. This arbitrage ensures that the interest rate differential is always zero, so the domestic interest rate always equals the exogenous foreign interest rate and monetary policy has no effect on the equilibrium domestic interest rate. In general, the more mobile is capital, which means f' is higher, the less interest rates change in response to monetary policy actions.

To ascertain the effectiveness of monetary policy on output when capital is perfectly mobile, consider equation (6.40). As $f' \to \infty$, $\partial Y^*/\partial M \to 1/L_Y$. Since domestic interest rates aren't changing, the only way the money market can be in equilibrium is if equilibrium output rises just enough to create the right increase in the demand for money. The goods market is irrelevant in terms of the comparative static outcome, though it plays a role in how the new equilibrium is achieved: when the money supply increases, interest rates start to fall; but capital flows immediately out of the country in response, which puts downward pressure on the dollar in foreign exchange markets; this in turn increases net exports, thereby increasing domestic output and income. The higher income increases the demand for money, which puts upward pressure on interest rates in the money market, offsetting the original effect of monetary policy. At the same time, higher domestic income increases the demand for imports, which offsets the initial change caused by the depreciated dollar. The mathematical model shows that, in the new equilibrium, interest rates will return to match the foreign interest rate, net exports will return to zero, and output will rise just enough to restore equilibrium in the money market.

6.3.3   Summary of Monetary Policy Effectiveness

In an open economy, the more mobile is capital, the less effective monetary policy is in changing interest rates but the more effective monetary policy is in changing output. As long as capital is at least somewhat mobile, the effectiveness of monetary policy at changing output is greater in the Mundell-Fleming model than in the IS–LM model. Monetary policy effectiveness also depends on several other parameters, as shown at the end of Section 6.3.1; similar exercises, some of which are included in the end-of-chapter problems, would show the impact of changes in those parameters on the effectiveness of monetary policy in an open economy.

6.4   SLOPES OF IS AND LM CURVES

The effects of government fiscal and monetary policy actions depend in part on the slopes of the IS and LM curves. In this example we will derive formulas for the slopes of these curves using the implicit function theorem and rule. Typical IS and LM curves are shown in Figure 6.4.

The LM curve represents all the combinations of income and the interest rate that satisfy the equilibrium condition for the money market. That is, it represents all the combinations for which money demand equals money supply. In the simple model

---

FIGURE 6.4

Typical IS and LM Curves

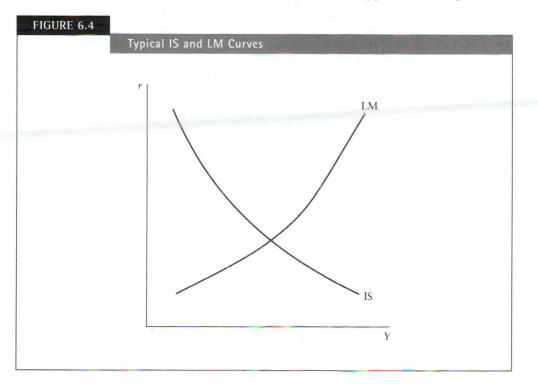

usually used in intermediate macroeconomics, money supply is assumed to be exogenous (M_0) while money demand (L) depends on income (Y) and the interest rate (r). Thus the LM curve can be viewed as a level curve of the money demand function $L(Y, r)$, with the "level" being M_0.

The LM curve is drawn with the interest rate r on the vertical axis and income Y on the horizontal axis. Thus, by the implicit function rule, the slope of the LM curve is given by

$$\frac{\partial r}{\partial Y} = -\frac{\partial L/\partial Y}{\partial L/\partial r}. \tag{6.44}$$

Since money demand depends positively on income and negatively on the interest rate, $\partial r/\partial Y$ is positive and the LM curve slopes upward.

The LM curve becomes steeper ($\partial r/\partial Y$ increases) when money demand is more responsive to changes in income and less responsive to changes in the interest rate. The economic explanation is as follows. Suppose that income increases, raising money demand and creating excess demand in the money market. Equilibrium will be restored by an increase in the interest rate, which will lower money demand and eliminate the excess demand. If money demand is highly sensitive to income, then the original excess demand will have been large, so a large increase in the interest rate is needed to restore equilibrium. Moreover, if money demand is not very sensitive to the interest rate, large increases in the interest rate will be necessary to drive down the demand for money. Thus small increases in income necessitate large increases in the interest rate to maintain money market equilibrium, and the LM curve is steep.

Although it may not be as obvious, the IS curve is also a level curve, showing the combinations of income (output) and the interest rate that correspond to equilibrium in the goods market. In a simple version of the IS–LM model, goods-market equilibrium is reached when output (Y) equals the sum of consumption (C), which for simplicity is assumed to depend only on output (income); investment (I), which is assumed to depend on the interest rate (r); and government purchases of goods and services (G), which is assumed to be exogenous. Thus we can define equilibrium in the goods market by the equation $Y = C(Y) + I(r) + G$. Equilibrium is thus defined implicitly by

$$f(Y, r, G) = Y - C(Y) - I(r) - G = 0. \tag{6.45}$$

The IS curve is drawn with the interest rate on the vertical axis and output on the horizontal axis. The implicit function rule can be used to derive its slope:

$$\frac{\partial r}{\partial Y} = -\frac{\partial f/\partial Y}{\partial f/\partial r} = -\frac{1 - C'}{-I'} = \frac{1 - C'}{I'}, \tag{6.46}$$

where the "prime" notation is used to denote derivatives of functions with only one argument. The numerator of the right-hand side of equation (6.46) is positive. The denominator is negative because investment demand is negatively related to the interest rate, so the IS curve slopes downward.

The IS curve becomes steeper ($\partial r/\partial Y$ becomes larger in absolute value) if the marginal propensity to consume decreases or if the responsiveness of investment demand to the interest rate decreases. The economic explanation is the usual multiplier story: suppose the interest rate increases, lowering investment. This lowers output, which lowers consumer income, which lowers consumption and output again, and so forth until equilibrium is restored at a lower output level. If investment is not very sensitive to interest rates, then interest rates can be considerably higher without reducing investment (and thus equilibrium output) much. Moreover, if the marginal propensity to consume is low, the reduction in consumption (and the resulting reduction in output) will be smaller on each round of the multiplier process. Once again, equilibrium will be restored without a large decrease in output.

The relative effectiveness of fiscal and monetary policies (at least in this fixed-price model) depends a great deal on the slopes of the IS and LM curves. This dependence was one of the issues (though probably not the most important) that divided monetarists from Keynesians in the vigorous debates of the 1960s. For example, expansionary fiscal policy will shift the IS curve to the right. If, as the monetarists believed, the LM curve is steep while the IS curve is flat, equilibrium output will not increase very much. This case is shown in Figure 6.5. If the Keynesian belief in a relatively flat LM curve and steep IS curve is correct, expansionary fiscal policy will be effective at increasing equilibrium output, as shown in Figure 6.6. Expansionary monetary policy, on

FIGURE 6.5

Expansionary Fiscal Policy: Monetarist Assumptions

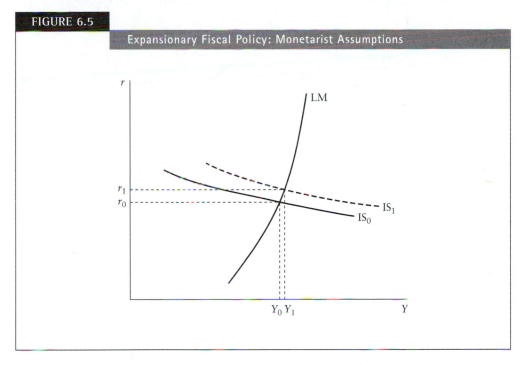

FIGURE 6.6

Expansionary Fiscal Policy: Keynesian Assumptions

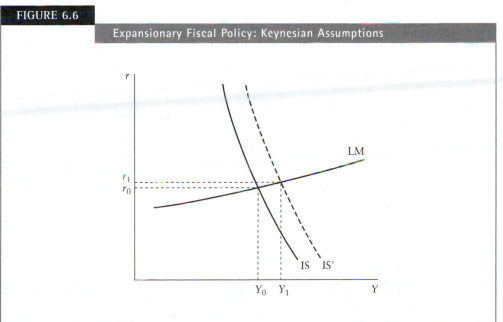

FIGURE 6.7

Incidence of a Per–Unit Tax

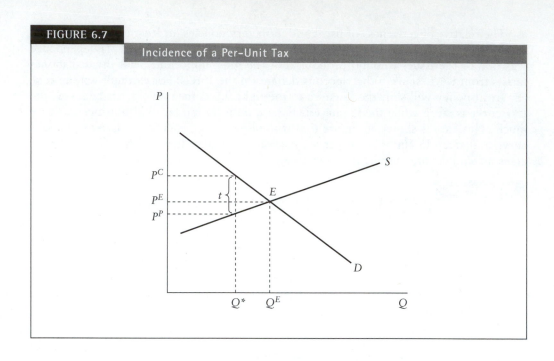

the other hand, shifts the LM curve to the right. This yields a much larger increase in equilibrium output under monetarist beliefs than under Keynesian.

6.5   TAX INCIDENCE IN A SUPPLY–AND–DEMAND MODEL

When tax incidence is discussed in introductory economics courses, it is usually in the context of supply and demand curves, using a graph similar to Figure 6.7. With no tax, the market equilibrium is at E, with price P^E and quantity Q^E. When a constant per-unit tax at rate t is imposed, it drives a wedge between consumer and producer prices. The new equilibrium is at a lower quantity, Q^*; consumers pay the after-tax price $P^C > P^E$ while producers get the after-tax price $P^P < P^E$. There are three important results of this tax incidence analysis:[6]

1. The after-tax equilibrium is the same whether the tax is imposed on consumers or producers.

2. Consumers and producers generally share the burden[7] of the tax: the tax makes the consumer price go up and the producer price go down, but each by less than t.

3. Consumers pay a higher share of the tax, the smaller is the elasticity of demand and the larger is the elasticity of supply.

In this section we will analyze the incidence of a constant per-unit tax in a supply-and-demand mathematical economics model. We will see whether the three results just

---

[6]This is a partial-equilibrium tax incidence analysis; when interactions with other markets are taken into consideration, the results change.

[7]We use this term the way it is often used in the incidence literature, to indicate what happens to consumer and producer prices. A better way to measure burden would be to find the changes in consumer and producer surpluses.

listed are generally true or whether they are special results dependent on how the supply and demand curves are drawn.

To develop a mathematical economics model that can be used to analyze the incidence of a tax, it is important to recognize that when there is a tax there will be two after-tax prices: a price that consumers pay and a price that firms get when they sell their product. The demand for the good will depend on what price consumers have to pay, including the tax; the supply of the good will depend on what price firms receive, after the tax, for selling the good. But there will be a single equilibrium output, where the amount consumers want to buy just equals the amount producers want to sell.

If we use P^C to represent the consumers' after-tax price and P^P to represent the firms' after-tax price, the inverse demand function is

$$P^C = D(Q) \tag{6.47}$$

while the (inverse) supply function is

$$P^P = S(Q). \tag{6.48}$$

If a constant, per-unit tax at rate t is placed on consumers, then

$$P^C = P^P + t \tag{6.49}$$

whereas if a constant, per-unit tax at rate τ is placed on producers, then

$$P^P = P^C - \tau. \tag{6.50}$$

Clearly, if $t = \tau$ then equations (6.49) and (6.50) are equivalent, so the mathematical solution of the supply-and-demand model will be the same regardless of whether the tax is imposed on consumers or producers. This establishes tax incidence result (1) in our list: the after-tax equilibrium will be the same regardless of who is legally responsible for the tax. Since it doesn't matter whether we use a consumer tax or a producer tax, we will arbitrarily choose to use equation (6.49) to close our mathematical model. We therefore have three equations [(6.47), (6.48), and (6.49)] in three endogenous variables (Q, P^C, and P^P); the only exogenous variable in our system is t, though the model might also include parameters of the demand or supply functions.

To analyze the incidence of the per-unit tax, we will imagine starting from a no-tax equilibrium and then imposing a small tax. We will derive the comparative static derivatives dP^{*C}/dt and dP^{*P}/dt and investigate whether consumers and producers share the tax, and if so, whether their shares depend on elasticities of demand and supply. We will use the total differentials approach, though we could alternatively derive the results using the implicit function rule or implicit differentiation. The total differentials of the three equations in our system are

$$\begin{aligned}
dP^C &= D'(Q)\, dQ \\
dP^P &= S'(Q)\, dQ \\
dP^C &= dP^P + dt
\end{aligned} \tag{6.51}$$

or, in matrix notation,

$$\begin{bmatrix} 1 & 0 & -D' \\ 0 & 1 & -S' \\ 1 & -1 & 0 \end{bmatrix} \begin{bmatrix} dP^C \\ dP^P \\ dQ \end{bmatrix} = \begin{bmatrix} 0 \\ 0 \\ dt \end{bmatrix}. \tag{6.52}$$

Using Cramer's rule to solve for dP^{*C}/dt and dP^{*P}/dt,

$$dP^{*C} = \frac{\begin{vmatrix} 0 & 0 & -D' \\ 0 & 1 & -S' \\ dt & -1 & 0 \end{vmatrix}}{\begin{vmatrix} 1 & 0 & -D' \\ 0 & 1 & -S' \\ 1 & -1 & 0 \end{vmatrix}} = \frac{D'\,dt}{-S' + D'} \quad \text{so} \quad \frac{dP^{*C}}{dt} = \frac{D'}{D' - S'} = \frac{1}{1 - (S'/D')}, \quad \textbf{(6.53)}$$

and

$$dP^{*P} = \frac{\begin{vmatrix} 1 & 0 & -D' \\ 0 & 0 & -S' \\ 1 & dt & 0 \end{vmatrix}}{\begin{vmatrix} 1 & 0 & -D' \\ 0 & 1 & -S' \\ 1 & -1 & 0 \end{vmatrix}} = \frac{-dt(-S')}{-S' + D'} \quad \text{so} \quad \frac{dP^{*P}}{dt} = \frac{S'}{D' - S'} = \frac{-1}{1 - (D'/S')}. \quad \textbf{(6.54)}$$

If the demand curve slopes downward, so that $D' < 0$, and the supply curve slopes upward, so that $S' > 0$, then $0 < dP^{*C}/dt < 1$ and $0 > dP^{*P}/dt > -1$, which establishes the second in our list of incidence results: consumers and producers typically share the tax burden. But while market demand curves can be safely assumed to slope downward (perfectly horizontal or vertical demand curves are not realistic), it is not always the case that market supply curves slope upward, especially if they are long-run market supply curves. So we should consider three cases: $S' > 0$, $S' = 0$, and $S' < 0$.

We have already analyzed the case of $S' > 0$. If $S' = 0$ then $dP^{*C}/dt = 1$ and $dP^{*P}/dt = 0$: consumers bear the entire burden of the tax and producers bear none of it (even if the producers are legally responsible for the tax). If $S' < 0$ then the signs of dP^{*C}/dt and dP^{*P}/dt depend on whether the demand curve or supply curve has a steeper slope. If $D' < S'$, which, since both are negative, implies that the demand curve has a steeper slope than the supply curve, then, from equation (6.53), consumers end up paying *more* than 100% of the tax: S'/D' is positive but less than one,[8] so $1 - (S'/D')$ is positive but less than one. Since $dP^{*C}/dt > 1$, dP^{*P}/dt must be greater than zero, which can be confirmed using equation (6.54). So if the demand and supply curves both slope downward, but the demand curve is steeper, then when a tax is imposed producers end up getting a higher price than they did before, and the consumer price goes up by more than the tax. This is illustrated in Figure 6.8. If the supply curve is steeper, then when a tax is imposed the equilibrium quantity will *rise*, and equations (6.53) and (6.54) show that *both* consumer and producer prices will fall, with producers bearing more than 100% of the tax burden since $dP^{*P}/dt < -1$. This case is shown in Figure 6.9.

To show that the burden of the tax depends on demand and supply elasticities, we will convert the formulas in equations (6.53) and (6.54) into equivalent formulas using elasticities. To make the conversions, recall that the elasticity of demand is defined as[9]

$$\eta^D = -\frac{dQ^D}{dP}\frac{P}{Q} \quad \textbf{(6.55)}$$

or, since we are using the inverse demand function,

$$\eta^D = -\frac{1}{D'}\frac{P}{Q}. \quad \textbf{(6.56)}$$

---

[8]Remember that both S' and D' are negative. So if $D' < S'$, then $1 > (S'/D')$.

[9]The negative sign is to make the elasticity positive for downward-sloping demand curves.

FIGURE 6.8

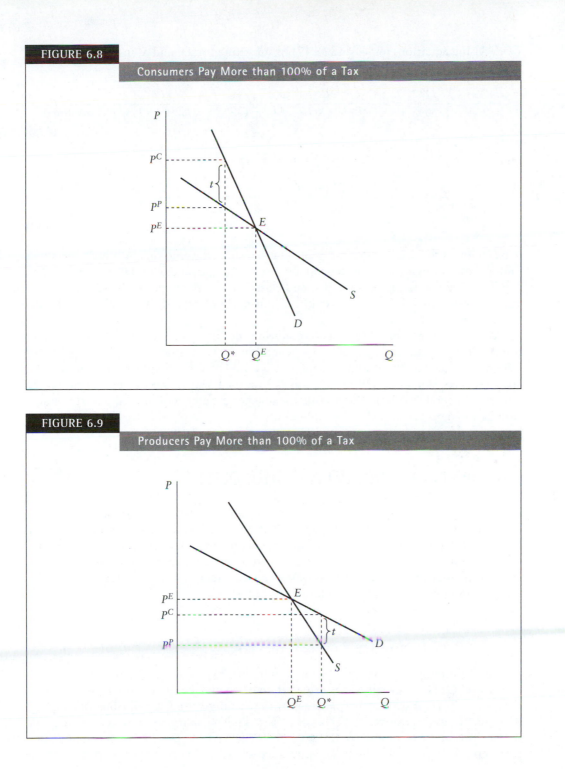

FIGURE 6.9

Producers Pay More than 100% of a Tax

The elasticity of supply is

$$\eta^S = \frac{dQ^S}{dP}\frac{P}{Q} = \frac{1}{S'}\frac{P}{Q}. \qquad (6.57)$$

Since we are considering the imposition of a tax starting from a tax rate of zero, we can evaluate these elasticities at the original, no-tax equilibrium price and quantity P^E and

Q^E.[10] Using equations (6.56) and (6.57) to substitute for D' and S' in equations (6.53) and (6.54), we get

$$\frac{dP^{*C}}{dt} = \frac{D'}{D' - S'} = \frac{-(P^E/\eta^D Q^E)}{-(P^E/\eta^D Q^E) - (P^E/\eta^S Q^E)} = \frac{(1/\eta^D)}{(1/\eta^D) + (1/\eta^S)} = \frac{\eta^S}{\eta^S + \eta^D}$$

(6.58)

and

$$\frac{dP^{*P}}{dt} = \frac{S'}{D' - S'} = \frac{(P^E/\eta^S Q^E)}{-(P^E/\eta^D Q^E) - (P^E/\eta^S Q^E)} = -\frac{(1/\eta^S)}{(1/\eta^D) + (1/\eta^S)} = -\frac{\eta^D}{\eta^S + \eta^D}.$$

(6.59)

This confirms the third on our list of incidence results: the amounts by which consumer and producer prices change depend on both demand and supply elasticities. The smaller (in absolute value) is the demand elasticity and the larger is the supply elasticity, the greater will be the increase in the consumer price and the smaller will be the decrease in the producer price.

We have shown that the partial-equilibrium tax incidence results usually stated in introductory classes and textbooks are general in the sense that they hold regardless of whether the demand and supply curves are linear. The usual results do not hold when supply curves slope downward, but this is true for linear as well as nonlinear supply curves. The models used in this section assume perfect competition;[11] the next application investigates tax incidence in a monopoly.

6.6 AN EXCISE TAX ON A MONOPOLIST

In Chapter 2 we determined the effects of an excise tax on a monopolist's profit-maximizing price and quantity for the case of linear demand and cost functions. Here we extend that analysis to the case of more general demand and cost functions and investigate what portion of the tax can be passed on by the monopolist to its consumers. We also include an application of the envelope theorem to find the effect of an excise tax on a monopolist's equilibrium level of profit.

6.6.1 Tax Incidence

Let the monopolist's (inverse) demand function be given by $P(Q)$ and its cost function be $C(Q)$. It is usually assumed that the demand curve is downward sloping, $P'(Q) \equiv dP/dQ < 0$, and that, at least in the vicinity of the equilibrium value of Q, marginal cost is positive, $C'(Q) \equiv dC/dQ > 0$. The marginal cost curve may be upward-sloping, horizontal, or downward-sloping, so $C''(Q) \equiv d^2C/dQ^2$ can be positive, zero, or negative. The monopolist is taxed at a rate of t per unit of output. Its profit

---

[10]It is not necessary to evaluate the elasticities at the no-tax equilibrium, but it simplifies the analysis some since the demand and supply elasticities can be evaluated at the same price.

[11]This is because market supply curves only exist in perfect competition; when firms have market power they control price as a choice variable, so it doesn't make sense to think about the quantity they are willing to supply at different prices.

is therefore given by[12]

$$\Pi(Q, t) = P(Q)Q - C(Q) - tQ \tag{6.60}$$

and the first-order condition (FOC) and second-order condition (SOC) for its profit-maximization problem are given by

$$\text{FOC:} \quad \Pi'(Q) = QP'(Q) + P(Q) - C'(Q) - t = 0$$
$$\text{SOC:} \quad \Pi''(Q) = P' + QP'' + P' - C'' < 0. \tag{6.61}$$

The first two terms on the left-hand side of the first-order condition equal the firm's marginal revenue, so the first-order condition can be interpreted to say that the firm will choose its output level so that marginal revenue equals the sum of marginal production cost C' and the tax rate. The second-order condition, in which arguments of the functions have been suppressed since they are not needed in what follows, has the interpretation that the slope of the marginal cost curve, C'', must be algebraically greater than the slope of the marginal revenue curve, which is equal to the sum of the first three terms on the left-hand side of the second-order condition. Ordinarily we draw downward-sloping marginal revenue curves and upward-sloping marginal cost curves, so this condition is automatically satisfied. But the second-order condition can be satisfied by many marginal revenue and marginal cost curves, for example, a downward-sloping marginal cost curve, as long as the marginal cost curve MC cuts the marginal revenue curve MR from below at their intersection, as illustrated in Figure 6.10.

The first-order condition implicitly defines quantity, the endogenous variable in the equation, as a function of the tax rate, which is the exogenous variable in the equation, and whatever parameters there might be in the demand and cost functions. (Equilibrium price is determined completely by the [inverse] demand function once equilibrium quantity is determined.) The implicit function theorem says that the function $Q^*(t)$, which gives equilibrium quantity Q^* as a function of the tax rate t, exists as long as the derivative of the left-hand side of the first-order condition in (6.61) with respect to Q does not equal zero. This derivative equals the left-hand side of the second-order condition in (6.61), so if the second-order condition is satisfied, then $Q^*(t)$ does exist. So $Q^*(t)$ can be substituted into the first-order condition to get an identity,

$$Q^*(t)P'(Q^*(t)) + P(Q^*(t)) - C'(Q^*(t)) - t \equiv 0 \tag{6.62}$$

which can be implicitly differentiated term by term:

$$P'\frac{dQ^*}{dt} + QP''\frac{dQ^*}{dt} + P'\frac{dQ^*}{dt} - C''\frac{dQ^*}{dt} - 1 = 0. \tag{6.63}$$

Solving for $\partial Q^*/\partial t$,

$$\frac{dQ^*}{dt} = \frac{1}{P' + QP'' + P' - C''}. \tag{6.64}$$

---

[12]Note that if the same per-unit tax is placed on consumers instead of the firm, the firm will face the demand curve $\tilde{P}(Q) = P(Q) - t$ since consumers will be willing to buy the same quantity only if their pre-tax price falls by t so that the consumer price remains at its former level. So the firm's profit function will be $\Pi = \tilde{P}(Q)Q - C(Q) = (P(Q) - t)Q - C(Q) = P(Q)Q - C(Q) - tQ$. Since this is identical to equation (6.60), the results of the model will be identical regardless of whether consumers or the monopolist are legally responsible for paying the tax. This is the same tax incidence result as in the perfectly competitive models of Section 6.5.

FIGURE 6.10

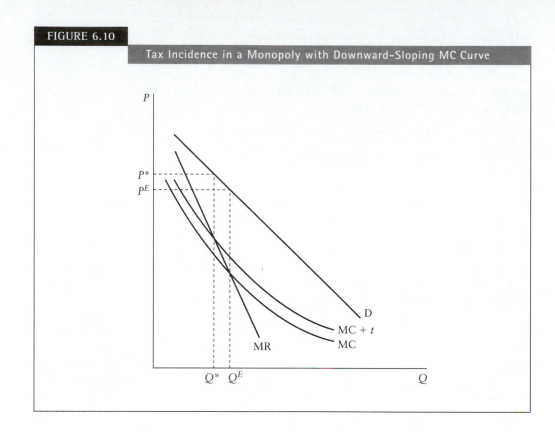

The second-order condition says that the denominator of this fraction is negative, so $\partial Q^*/\partial t < 0$. As long as there is an interior solution to the firm's maximization problem, a higher tax rate will lead to lower output. It is also worth remarking that in this example, as in many other optimization problems, the second-order condition is what enables us to sign the comparative static derivative.

The effect on equilibrium price can be found from the inverse demand function:

$$\frac{dP^*}{dt} = \frac{dP(Q^*(t))}{dt} = P'\frac{dQ^*}{dt}. \tag{6.65}$$

We showed above that $\partial Q^*/\partial t < 0$; as long as the demand curve is downward sloping, $P' < 0$ as well, so $dP^*/dt > 0$. To analyze tax incidence we are interested in whether $dP^*/dt < 1$; if $dP^*/dt < 1$ then the monopolist, who after paying the excise tax gets $P^* - t$ for each unit of its output, cannot pass the entire increase in the tax rate on to its consumers: after paying the tax, it will get less per unit of output. Substituting equation (6.64) into equation (6.65) yields

$$\frac{dP^*}{dt} = \frac{P'}{P' + QP'' + P' - C''}. \tag{6.66}$$

Without knowing more about the demand function, we cannot say whether $dP^*/dt < 1$. In order for $dP^*/dt \geq 1$ while the second-order condition holds we must have $P' \leq 2P' + QP'' - C'' < 0$, which implies $P' + QP'' - C'' \geq 0$. Unfortunately, this has no obvious economic interpretation. But for some special cases we can show that the monopolist will have to bear some of the burden of the tax. If the demand curve is linear and the marginal cost curve is horizontal or slopes upward, then $P'' = 0$, $C'' \geq 0$, and

$$\frac{dP^*}{dt} = \frac{P'}{2P' - C''} = \frac{1}{2 - (C''/P')} < 1 \tag{6.67}$$

where the last step follows from the fact that $C''/P' \leq 0$. If the marginal cost curve slopes downward, then $dP^*/dt < 1$ as long as the marginal cost curve slopes downward less steeply than does the demand curve, so that $C''/P' < 1$.

We have shown that a monopolist will always be able to pass at least some of an excise tax on to its consumers. For the case of linear demand curves, the monopolist will always have to share some of the burden of the tax unless the marginal cost curve is downward sloping and steeper than the demand curve. But for nonlinear demand curves, the monopolist may end up raising price by more than the tax, thereby shifting more than 100% of the burden of the tax onto consumers.

6.6.2 Envelope Theorem Results

Before leaving this application, we will use the envelope theorem, described in Section 1.4, to derive the impact on the monopolist's equilibrium profit level of an increase in an excise tax. We begin by noting that, in equilibrium, the monopolist's quantity, and therefore price and cost as well, are functions of the exogenous parameter t. Substituting these functions into the monopolist's profit function (6.60) yields the monopolist's value function:

$$\Pi^*(t) = P(Q^*(t))Q^*(t) - C(Q^*(t)) - tQ^*(t). \tag{6.68}$$

It might appear at first that discovering the impact of a higher excise tax on a monopolist's profit would be complicated. After all, as equation (6.68) shows, profit depends on price, quantity, and costs, all of which depend on the tax rate, as well as on the tax rate itself. But the power of the envelope theorem is that, when calculating the effect of an exogenous variable on the variable being optimized, the effects of changes in the choice variable can be ignored.

In our example, price, quantity, and costs all change only because the firm's choice variable, Q, changes. So the envelope theorem says that the effect of the higher tax on the equilibrium value of profit is just the partial derivative of the profit function (6.60) with respect to t, holding constant Q. (Since they depend on Q but not t, P and C are held constant also.) So the comparative static effect on equilibrium profit is

$$\frac{\partial \Pi^*(t)}{\partial t} = \frac{\partial \Pi(Q, t)}{\partial t} = -Q < 0. \tag{6.69}$$

(We have used the partial derivative sign even though t is the only explicit argument of the optimal value function $\Pi^*(t)$, since in general there will be other exogenous variables in the model, for example parameters governing the demand or cost functions.) The envelope theorem result can be confirmed by finding the total derivative of the value function (6.68) with respect to t:

$$\frac{\partial \Pi^*(t)}{\partial t} = Q\frac{\partial P}{\partial Q}\frac{\partial Q^*(t)}{\partial t} + P\frac{\partial Q^*(t)}{\partial t} - \frac{\partial C}{\partial Q}\frac{\partial Q^*(t)}{\partial t} - t\frac{\partial Q^*(t)}{\partial t} - Q$$

$$= \left(Q\frac{\partial P}{\partial Q} + P - \frac{\partial C}{\partial Q} - t\right)\frac{\partial Q^*(t)}{\partial t} - Q. \tag{6.70}$$

But the first-order condition shown in (6.61) implies that the term in parentheses is equal to zero, which confirms the envelope theorem result.

Equation (6.69) should be evaluated at the profit-maximizing output level Q^*, so when the excise tax rate increases, profit falls by the amount of extra tax paid on the profit-maximizing output level. The envelope theorem says we can ignore the fact that this output level will change, since the firm's first-order condition ensures that small changes in output have no effect on profit.

We can also confirm the envelope theorem by taking the total differential of the profit function (6.60) (ignoring whatever parameters other than t there may be in the model):

$$d\Pi = (QP' + P - C' - t)dQ - Qdt. \qquad (6.71)$$

But, according to the first-order condition in (6.61), the term in parentheses on the right-hand side of (6.71) equals zero. So the change in the profit-maximizing output level, dQ, does not affect profit in equilibrium, and the effect on profit is $d\Pi = -Qdt$, which confirms the comparative static derivative (6.69).

In Chapter 2 we illustrated the application of the envelope theorem to a case in which the monopolist's demand function is linear and her marginal cost is constant. In that example the envelope theorem is superfluous for comparative static analyses because we could easily obtain the reduced form of the value function. In the present chapter the demand and cost functions are general, rather than linear or constant, so we cannot obtain the value function in an explicit form. But the envelope theorem enables us to obtain the comparative static result (6.69) without having the explicit value function.

6.7   COURNOT DUOPOLY MODEL WITH NONLINEAR COSTS

In Chapter 4 we analyzed the duopoly model with linear demand and linear costs. In this example we analyze the Cournot duopoly model with a linear demand curve but a cubic total cost function for each firm. While we could find comparative static derivatives by solving explicitly for each firm's supply function, we would need to use the quadratic formula as well as a symmetry condition. It is easier to obtain the comparative static derivatives by implicitly differentiating the first-order conditions for both firms' profit-maximization problems. This technique is also more powerful in the sense that if the firms' costs are not identical, implicit differentiation (or, equivalently, the implicit function rule or total differentials) can still be used to derive comparative static derivatives whereas solving explicitly for the firms' supply functions would be difficult since we would not be able to use symmetry to solve the model.

The two firms have identical cost functions, $TC_i = cq_i^3$, where $c > 0$. Market price is a linear function of market quantity, which in turn is the sum of the two firms' quantities: $P = a - q_1 - q_2$, where $a > 0$. Thus the two firms' profit functions are given by

$$\pi_1 = (a - q_1 - q_2)q_1 - cq_1^3$$

and

$$\pi_2 = (a - q_1 - q_2)q_2 - cq_2^3. \qquad (6.72)$$

Consistent with the Cournot assumption, each firm maximizes its own profit, assuming that the other firm's quantity is fixed. The first-order conditions for these two maximization problems are

$$\frac{\partial \pi_1}{\partial q_1} = a - 2q_1 - q_2 - 3cq_1^2 = 0$$

and

$$\frac{\partial \pi_2}{\partial q_2} = a - q_1 - 2q_2 - 3cq_2^2 = 0.$$

(6.73)

The second-order conditions for the two maximization problems are satisfied since each is of the form $-2 - 6cq_i < 0$.

The two first-order conditions make up a system of two equations in the two unknowns q_1 and q_2. The equilibrium for this problem is described by the values of q_1 and q_2 that solve this system of equations. We can use symmetry to solve these equations, but we need the quadratic formula to do it; there is an easier way to get comparative static derivatives. The two equations implicitly define the two outputs as functions of the constants and parameters of the equations; that is, $q_1^* = q_1^*(a, c)$ and $q_2^* = q_2^*(a, c)$. Thus we can implicitly differentiate the two first-order conditions with respect to one of the parameters. For example, we can find the effect of a shift in the demand curve by implicitly differentiating the first-order conditions (6.73) with respect to a:

$$1 - (2 + 6cq_1)\frac{\partial q_1^*}{\partial a} - \frac{\partial q_2^*}{\partial a} = 0$$

$$1 - \frac{\partial q_1^*}{\partial a} - (2 + 6cq_2)\frac{\partial q_2^*}{\partial a} = 0$$

(6.74)

or, written in matrix notation,

$$\begin{bmatrix} -2 - 6q_1 & -1 \\ -1 & -2 - 6q_2 \end{bmatrix} \begin{bmatrix} \partial q_1^*/\partial a \\ \partial q_2^*/\partial a \end{bmatrix} = \begin{bmatrix} -1 \\ -1 \end{bmatrix}.$$

(6.75)

Using Cramer's rule to solve for $\partial q_1^*/\partial a$,

$$\frac{\partial q_1^*}{\partial a} = \frac{\begin{vmatrix} -1 & -1 \\ -1 & -2 - 6q_2 \end{vmatrix}}{\begin{vmatrix} -2 - 6q_1 & -1 \\ -1 & -2 - 6q_2 \end{vmatrix}} = \frac{2 + 6q_2 - 1}{(2 + 6q_1)(2 + 6q_2) - 1} > 0.$$

(6.76)

Either by invoking symmetry or by using Cramer's rule, we can see that firm 2 will also increase output in response to an upward shift in the market demand curve.

The solution for the comparative static derivative includes the choice variables q_1 and q_2, which is common when using the technique of implicit differentiation. While the magnitudes of the derivatives cannot be determined without explicit solutions for the choice variables, we do not usually need the values of the choice variables to sign the comparative static derivatives because the signs of the values of the choice variables themselves are almost always known.

6.8 LABOR SUPPLY WITH A STONE–GEARY UTILITY FUNCTION

This exercise investigates the labor-leisure choice when utility is given by the Stone-Geary utility function. It illustrates the possibility of getting corner solutions to optimization problems and shows how to do a constrained maximization problem with two choice variables by substituting the constraint into the objective function. We also explore the comparative statics of a change in the wage, both by differentiating the explicit labor supply function and by implicitly differentiating the first-order condition.

The Stone-Geary utility function for this problem is

$$U(C, T - L) = a \ln(C - C_0) + (1 - a) \ln(T - L) \tag{6.77}$$

where a is an exogenous parameter between 0 and 1, C is consumption, C_0 is the (exogenous) subsistence level of consumption, T is the (exogenous) total time available, and L is the amount of time spent working. Thus $T - L$ is leisure time, where T is defined in such a way that the subsistence level of leisure is 0. The budget constraint for this problem is that consumption must equal wage income wL plus (exogenous) non-wage income I: $C = wL + I$. Individuals choose consumption C and labor supply L to maximize utility subject to this budget constraint.

Although constrained optimization problems are really the province of Chapter 9, we can solve this example by turning it into a one-variable unconstrained maximization problem. To do this, we substitute the budget constraint into the utility function:

$$U(C, T - L) = U(wL + I, T - L) = a \ln(wL + I - C_0) + (1 - a) \ln(T - L). \tag{6.78}$$

The first-order condition for the problem of choosing L to maximize this "constrained" utility function is

$$\frac{dU}{dL} = \frac{a}{wL + I - C_0} w + \frac{1 - a}{T - L} (-1) = 0. \tag{6.79}$$

Solving for L, we have

$$\frac{a}{wL + I - C_0} w = \frac{1 - a}{T - L}$$

$$aw(T - L) = (1 - a)(wL + I - C_0)$$

$$awT - awL = (1 - a)wL + (1 - a)(I - C_0) \tag{6.80}$$

$$awT - (1 - a)(I - C_0) = (1 - a)wL + awL = wL$$

$$aT - \frac{(1 - a)(I - C_0)}{w} = L^*.$$

Assuming that the second-order condition is satisfied, this is the labor supply function *if* there is an interior solution to the problem. This may not be so in this case because labor supply cannot be negative. For example, let $a = \frac{1}{2}$ so that consumption and leisure are valued equally, let T equal 10 hours and the wage rate w be 3 dollars per hour, and let the subsistence level of consumption C_0 be 10 units. Then, according to equation (6.80), labor supply is given by

$$L^* = \frac{1}{2}(10) - \frac{\frac{1}{2}(I - 10)}{3} = 5 - \frac{I - 10}{6}. \tag{6.81}$$

If nonlabor income I is greater than 40, this equation gives a negative value for labor supply, which is impossible. In this case, the maximization problem yields a corner solution where $L^* = 0$ and $C^* = I$. (A corner solution might also be obtained if the consumption implied by the optimal labor supply were less than the subsistence level C_0. But that is impossible in our numerical example.)

FIGURE 6.11

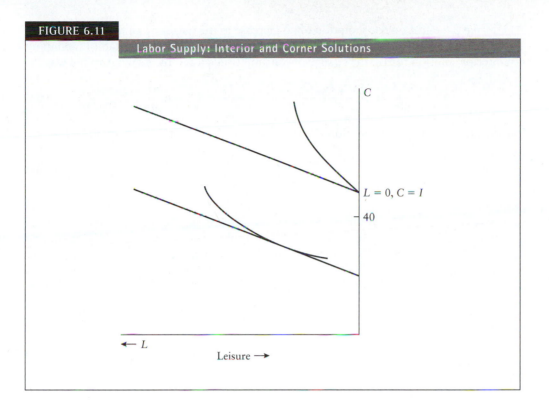

Labor Supply: Interior and Corner Solutions

Figure 6.11 illustrates the possibilities of interior and corner solutions, depending on the size of nonlabor income. As is conventional, we draw the graph with leisure on the horizontal axis while placing the vertical axis at the maximum possible value of leisure, which is $T = 10$ in our example. Thus leftward movement in the graph corresponds to an increase in labor supply L. With the values of a, w, and C_0 used in our example, a tangency exists between the budget constraint and indifference curves if $I < 40$. But if $I > 40$, the indifference curve that goes through the point $(L = 0, C = I)$ is steeper than the budget constraint. Thus there is no interior solution since utility is higher when $L = 0$ than for any positive value of L consistent with the budget constraint.

Whenever constraints exist on the permissable values of the choice variables, the possibility of corner solutions must be investigated. We will examine a more formal and powerful way of incorporating inequality constraints (such as $L \geq 0$) in Chapter 11.

To find the comparative statics of a change in the wage, differentiate the labor supply function in (6.80) to get

$$\frac{\partial L^*}{\partial w} = \frac{(1-a)(I - C_0)}{w^2}. \tag{6.82}$$

As long as nonlabor income is greater than the subsistence level of consumption, the labor supply curve will be upward-sloping. If nonlabor income is less than C_0, though, labor supply will decrease when the wage increases: the substitution effect of an increase in the wage, which would result in more labor as the individual substitutes away from leisure, is dominated by the income effect, which results in more leisure since the higher wage increases the individual's income. So the labor supply curve will be backward-bending.

Multivariate Calculus: Applications

To see the effect on consumption of an increase in the wage, recall that $C = wL + I$. Since in equilibrium L is a function of w and the other exogenous variables (assuming an interior solution for L), the change in consumption caused by a change in the wage is

$$\frac{\partial C^*}{\partial w} = \frac{\partial (wL^*(w, C_0, T, I, a) + I)}{\partial w} = L^* + w\frac{\partial L^*}{\partial w}. \tag{6.83}$$

Substituting for L^* from equation (6.80) and for $\partial L^*/\partial w$ from equation (6.82), we get

$$\frac{\partial C^*}{\partial w} = aT - \frac{(1-a)(I - C_0)}{w} + w\frac{(1-a)(I - C_0)}{w^2} = aT, \tag{6.84}$$

which is unambiguously positive. An increase in the wage increases consumption regardless of whether the labor supply curve is upward-sloping or backward-bending.

We could have found the effects on labor supply of a change in w without solving explicitly for L by implicitly differentiating the first-order condition. The first-order condition implicitly defines the choice variable L as a function of all the parameters (exogenous variables) of the model. Implicitly differentiating with respect to w the first-order condition (6.79), thinking of L as a function of w (and the other exogenous variables), we get

$$\frac{a}{wL + I - C_0} - \frac{aw}{(wL + I - C_0)^2}\left(L + w\frac{\partial L^*}{\partial w}\right) + \frac{1-a}{(T - L)^2}\left(-\frac{\partial L^*}{\partial w}\right) = 0. \tag{6.85}$$

Solving for $\partial L/\partial w$,

$$\frac{a}{wL + I - C_0} = \frac{awL}{(wL + I - C_0)^2} + \frac{\partial L^*}{\partial w}\left(\frac{aw^2}{(wL + I - C_0)^2} + \frac{1-a}{(T - L)^2}\right)$$

$$\frac{awL + a(I - C_0)}{(wL + I - C_0)^2} = \frac{awL}{(wL + I - C_0)^2} + \frac{\partial L^*}{\partial w}\left(\frac{aw^2}{(wL + I - C_0)^2} + \frac{1-a}{(T - L)^2}\right)$$

$$\frac{a(I - C_0)}{(wL + I - C_0)^2} = \frac{\partial L^*}{\partial w}\left(\frac{aw^2}{(wL + I - C_0)^2} + \frac{1-a}{(T - L)^2}\right) \tag{6.86}$$

$$\frac{\partial L^*}{\partial w} = \left(\frac{a(I - C_0)}{(wL + I - C_0)^2}\right)\left(\frac{aw^2}{(wL + I - C_0)^2} + \frac{1-a}{(T - L)^2}\right)^{-1}.$$

As long as consumption exceeds its subsistence level and some leisure is taken, all terms on the right-hand side of the last equation in (6.86) are positive with the possible exception of $(I - C_0)$. Again we get the result that if income is greater than the subsistence level of consumption, an increase in the wage will increase labor supply.

The technique of implicit differentiation yields a formula for the comparative static derivative that contains the level of the endogenous variable itself. This is often the case, but does not usually cause difficulty in signing the comparative static derivative. It may appear that we have derived two entirely different formulas for $\partial L^*/\partial w$, but this is not true. We obtain the explicit formula (6.82) if we substitute the explicit choice function for L^* from (6.80) into the comparative static derivative from (6.86).

6.9 UTILITY MAXIMIZATION AND THE ORDINALITY OF UTILITY FUNCTIONS

In this example we look at the consumer's utility-maximization problem in a two-good model. Using the concept of level curves, we examine the familiar condition that the consumer's budget constraint must be tangent to an indifference curve and show that the solution is invariant to monotonic transformations of the utility function—that is, the same solution will result for all utility functions that maintain the same ranking of preferences. This is an important result because economists do not believe that a cardinal measure of utility is meaningful. Since any monotonic transformation of a utility function will yield the same solution to the consumer's utility-maximization problem, it is necessary only to believe that consumers can rank their preferences consistently.

We use a particular functional form of the utility function to illustrate this result. The derivation also shows how to do a two-variable constrained optimization problem by substituting in the constraint and turning it into a one-variable unconstrained optimization problem. (The same utility-maximization problem is redone in Chapter 10 using a more powerful technique.)

Utility is derived from the consumption of two goods, x and y, according to the utility function $U(x, y)$. From intermediate microeconomic theory we know that a consumer with the linear budget constraint $p_x x + p_y y = I$ will maximize utility by choosing x and y so that her indifference curve is tangent to her budget constraint. (This result is derived mathematically in Chapter 9.) Indifference curves are level curves of the utility function, so the slope of the indifference curve that passes through a particular combination of x and y equals

$$\frac{\partial y^*}{\partial x} = -\frac{\partial U/\partial x}{\partial U/\partial y}. \tag{6.87}$$

That is, the marginal rate of substitution, which is defined as the absolute value of the slope of the indifference curve, is equal to the ratio of the marginal utilities of the goods.[13] The graph of the budget constraint can also be viewed as a level curve, since it shows all the combinations of x and y that will cost the same total amount (I). That is, $p_x x + p_y y - I = g(x, y) = 0$. Thus the slope of the budget constraint is given by

$$\frac{\partial y^*}{\partial x} = -\frac{\partial g/\partial x}{\partial g/\partial y} = -\frac{p_x}{p_y}. \tag{6.88}$$

Setting the slope of the indifference curve equal to the slope of the budget constraint yields the familiar result that, in equilibrium, the consumer's marginal rate of substitution must equal the price ratio.

Now suppose that the utility function is $W(x, y)$ where $W(x, y) = G(U(x, y))$ and $dG/dU > 0$. G is said to be a **monotonic transformation** of U since the value of G will increase if and only if the value of U increases. Thus the utility functions W and U maintain the same rankings of preferences: any changes in x and y that make a consumer with utility function U better off will also make a consumer with utility function W better off. Similarly, whenever U decreases, W will also decrease; and if x and y change in such a way that U is unchanged, W will not change either. Therefore the level curves of the function U must be the same as the level curves of the function W (except for the label that is attached: the same curve might represent $U = 100$ and $W = 5$). Consider the slopes of the level curves of U and W. As shown in equation (6.87), the slope of the level curve of U passing through a particular x and y combination is $-[(\partial U/\partial x)/(\partial U/\partial y)]$. The slope of the level curve of W passing through the same x and y equals

$$\frac{\partial y^*}{\partial x} = -\frac{\partial W/\partial x}{\partial W/\partial y} = -\frac{(dW/dU)(\partial U/\partial x)}{(dW/dU)(\partial U/\partial y)} = -\frac{\partial U/\partial x}{\partial U/\partial y}. \tag{6.89}$$

Since this is true for any combination of x and y, the level curves $y = y^*(x)$ of the functions W and U must coincide. Moreover, the utility-maximizing choices of x and y must

---

[13]Notice the complete symmetry between this and the slope of a firm's isoquant, discussed in Chapter 5.

be the same for the two utility functions, since utility is maximized when the slope of an indifference curve is equated to the slope of the budget constraint.

We now turn to a particular utility function to illustrate these results further. Let $U(x, y) = \ln x + \ln y$. Since the consumer will spend all of her income I, we can solve the budget constraint for y and substitute it into her utility function:

$$U(x, y(x)) = U\left(x, \frac{I - p_x x}{p_y}\right) = \ln x + \ln\left(\frac{I - p_x x}{p_y}\right), \tag{6.90}$$

which turns the two-variable constrained maximization problem into a one-variable unconstrained maximization problem. The first-order condition for this problem is

$$\frac{\partial U(x, y(x))}{\partial x} = \frac{1}{x} + \left(\frac{I - p_x x}{p_y}\right)^{-1}\left(\frac{-p_x}{p_y}\right) = 0. \tag{6.91}$$

This is the same equation that we would have obtained by setting the marginal rate of substitution equal to the price ratio,

$$-\frac{1/x}{1/y} = -\frac{p_x}{p_y}, \tag{6.92}$$

and then substituting from the budget constraint for y. Solving the first-order condition for x, we get

$$\begin{aligned} \frac{p_x}{I - p_x x} &= \frac{1}{x} \\ p_x x &= I - p_x x \\ 2p_x x &= I \\ x^* &= \frac{I}{2p_x}. \end{aligned} \tag{6.93}$$

Using the utility function $W(x, y) = xy$ would yield the same result since W is a monotonic transformation of U:

$$W(x, y) = xy = e^{\ln x + \ln y} = e^{U(x,y)} = W(U), \qquad \frac{dW}{dU} = e^U > 0. \tag{6.94}$$

The verification of this result is left as a problem.

6.10   HOMOGENEITY OF CONSUMER DEMAND FUNCTIONS

The demand for a good is a function of its price, prices of other goods, and income. (Other things can affect demand, such as tastes. Economists usually model these as parameters of the demand function rather than as arguments of the function.) In this example we will use differentials to show that the demand for a good does not change when prices and income change by an equal percentage. That is, the demand function is homogeneous of degree 0 in prices and income. While there are easier ways to show homogeneity, this approach provides good practice with the use of differentials. For our example we will use the demand function coming from a two-good linear expenditure system.[14]

In the linear expenditure system there are some amounts of the two goods that are absolute minima in the sense that under no conditions will the consumer ever choose to consume less than these amounts. Let α be the minimum amount of good 1 and β be

---

[14]This is the system of demand functions that comes from the Stone-Geary utility function, used in Section 6.8.

the minimum amount of good 2. The demand for good 1 is given by

$$x_1(p_1, p_2, I) = \alpha + \frac{\gamma}{p_1}(I - \alpha p_1 - \beta p_2) \quad \text{where } \alpha, \beta > 0 \quad \text{and} \quad 0 < \gamma < 1. \quad \textbf{(6.95)}$$

Thus the consumer buys the "necessary" amounts of goods 1 and 2 and spends some fraction γ of her remaining income on purchasing good 1. (The rest of her income goes to buying "excess" amount of good 2.)

In this problem, x_1 and x_2 are the endogenous variables; α, β, and γ are unchanging parameters; and p_1, p_2, and I are exogenous variables whose values can change. We are interested in showing that the amount consumed of each good does not change if p_1, p_2, and I all change by the same percentage. Let this percentage be $100 \times \theta$. For any variable z the percentage change in z equals $100 \times (dz/z)$ since dz is the change in the variable and z is its level. So if income and both prices all change by $(100 \times \theta)$ percent, $\theta = (dI/I) = (dp_1/p_1) = (dp_2/p_2)$. Since our choice of which good is listed as good 1 and which as good 2 is arbitrary, it suffices to show that the demand for one of them is homogeneous of degree 0 in prices and income. One way to show this is by taking the total differential of the demand function for good 1:

$$\begin{aligned}
dx_1 &= \frac{\gamma}{p_1}(dI - \alpha dp_1 - \beta dp_2) - \frac{\gamma}{p_1^2}(I - \alpha p_1 - \beta p_2)dp_1 \\
&= \frac{\gamma}{p_1}\left(I\frac{dI}{I} - \alpha p_1\frac{dp_1}{p_1} - \beta p_2\frac{dp_2}{p_2}\right) - \frac{\gamma}{p_1}\frac{dp_1}{p_1}(I - \alpha p_1 - \beta p_2) \\
&= \frac{\gamma}{p_1}(I\theta - \alpha p_1\theta - \beta p_2\theta) - \frac{\gamma}{p_1}\theta(I - \alpha p_1 - \beta p_2) \\
&= \theta\left(\frac{\gamma}{p_1}(I - \alpha p_1 - \beta p_2) - \frac{\gamma}{p_1}(I - \alpha p_1 - \beta p_2)\right) = 0.
\end{aligned} \quad \textbf{(6.96)}$$

Since x_1 does not change when prices and income change by the same percentage, the demand function is homogeneous of degree 0. By symmetry, the same is true for x_2.

When the demand function is known explicitly, as it is here, it is usually easier to show homogeneity directly. For example, the demand function used in this example is homogeneous of degree 0 since for any positive constant λ,

$$\begin{aligned}
x_1(\lambda p_1, \lambda p_2, \lambda I) &= \alpha + \frac{\gamma}{\lambda p_1}(\lambda I - \alpha\lambda p_1 - \beta\lambda p_2) = \alpha + \frac{\lambda\gamma}{\lambda p_1}(I - \alpha p_1 - \beta p_2) \\
&= \lambda^0 x_1(p_1, p_2, I).
\end{aligned} \quad \textbf{(6.97)}$$

But the technique of using differentials to show homogeneity is useful in cases when explicit demand functions are either impossible or too inconvenient to derive. The next example is of such a case.

6.11   HOMOGENEITY OF COBB–DOUGLAS INPUT DEMANDS

In this example we will show that the input demand functions resulting from a two-good Cobb-Douglas production function are homogeneous of degree 0 in the input and output prices. Since the production function is highly nonlinear, it will be inconvenient to solve explicitly for the input demand functions. Nonetheless we can show their homogeneity using differentials.

A Cobb-Douglas production function with two inputs, labor (L) and capital (K), takes the form

$$F(L, K) = AL^\alpha K^\beta \quad \textbf{(6.98)}$$

where A, α, and β are all positive constants. We know from microeconomic theory that perfectly competitive firms will hire each input up to the point that the value of marginal product (VMP) of the input equals the input's price. (This result can be derived using the techniques shown in Chapter 7.) The VMP of an input is the marginal product (MP) of the input times the price (P) of the firm's output. So a firm with the production function (6.98) will hire labor and capital in order to satisfy the two equations[15]

$$w = \text{VMP}_L = P \times \text{MP}_L$$

and

$$r = \text{VMP}_K = P \times \text{MP}_K.$$

(6.99)

Since the marginal product of an input is the partial derivative of the production function with respect to that input, for the production function (6.98) the marginal products of labor and capital are $\text{MP}_L = \partial F / \partial L = \alpha A L^{\alpha-1} K^\beta$ and $\text{MP}_K = \partial F / \partial K = \beta A L^\alpha K^{\beta-1}$. Substituting these into (6.99), the firm will be choosing L and K to solve the two-equation system

$$w = P\alpha A L^{\alpha-1} K^\beta$$
$$r = P\beta A L^\alpha K^{\beta-1}.$$

(6.100)

Since these are nonlinear equations in L and K, it may not be obvious how to solve them explicitly to get the input demand functions. But we can show that the input demand functions are homogeneous of degree 0 by taking the differentials of the two equations:

$$dw = \alpha A L^{\alpha-1} K^\beta \, dP + P\alpha(\alpha-1) A L^{\alpha-2} K^\beta \, dL + P\alpha\beta A L^{\alpha-1} K^{\beta-1} dK$$
$$dr = \beta A L^\alpha K^{\beta-1} \, dP + P\alpha\beta A L^{\alpha-1} K^{\beta-1} \, dL + P\beta(\beta-1) A L^\alpha K^{\beta-2} \, dK$$

(6.101)

or, in matrix notation,

$$\begin{bmatrix} P\alpha(\alpha-1) A L^{\alpha-2} K^\beta & P\alpha\beta A L^{\alpha-1} K^{\beta-1} \\ P\alpha\beta A L^{\alpha-1} K^{\beta-1} & P\beta(\beta-1) A L^\alpha K^{\beta-2} \end{bmatrix} \begin{bmatrix} dL \\ dK \end{bmatrix} = \begin{bmatrix} dw - \alpha A L^{\alpha-1} K^\beta \, dP \\ dr - \beta A L^\alpha K^{\beta-1} \, dP \end{bmatrix}.$$

(6.102)

The determinant of the first matrix in this equation is the Jacobian of the system, which can be shown to be positive as long as $\alpha + \beta < 1$.[16] Using Cramer's rule to solve for dL,

$$dL = \frac{\begin{vmatrix} dw - \alpha A L^{\alpha-1} K^\beta \, dP & P\alpha\beta A L^{\alpha-1} K^{\beta-1} \\ dr - \beta A L^\alpha K^{\beta-1} \, dP & P\beta(\beta-1) A L^\alpha K^{\beta-2} \end{vmatrix}}{|J|}$$

$$= \frac{(dw - \alpha A L^{\alpha-1} K^\beta \, dP)(P\beta(\beta-1) A L^\alpha K^{\beta-2})}{|J|}$$

$$- \frac{(dr - \beta A L^\alpha K^{\beta-1} \, dP)(P\alpha\beta A L^{\alpha-1} K^{\beta-1})}{|J|}.$$

(6.103)

For any variable z, the percentage change in z equals $100 \times (dz/z)$ since dz is the change in the variable and z is its level. So if w, r, and P all change by the same percentage, say

---

[15]These turn out to be the first-order conditions for the firm's profit-maximizing choices of L and K.

[16]This comes from the second-order condition for the firm's profit-maximizing choices of L and K.

$100 \times \theta$, then $(dw/w) = (dr/r) = (dP/P) = \theta$, so $dw = w\theta$, $dr = r\theta$, and $dP = P\theta$. Making these substitutions, we can write the numerator of equation (6.103) as

$$
\begin{aligned}
(w\theta &- \alpha AL^{\alpha-1}K^{\beta}P\theta)(P\beta(\beta-1)AL^{\alpha}K^{\beta-2}) \\
&- (r\theta - \beta AL^{\alpha}K^{\beta-1}P\theta)(P\alpha\beta AL^{\alpha-1}K^{\beta-1}) \\
&= \theta(w - \alpha AL^{\alpha-1}K^{\beta}P)(P\beta(\beta-1)AL^{\alpha}K^{\beta-2}) \qquad\text{(6.104)} \\
&\quad - \theta(r - \beta AL^{\alpha}K^{\beta-1}P)(P\alpha\beta AL^{\alpha-1}K^{\beta-1}) \\
&= 0
\end{aligned}
$$

where the last step comes from applying the two equations in (6.100). In words, when the input and output prices all change by the same percentage, the firm's demand for labor does not change: it is homogeneous of degree 0 in input and output prices. The proof that the demand for capital is homogeneous of degree 0 is analogous. So differentials can be used to show homogeneity without solving explicitly for the input demand functions.

Problems

6.1 Suppose that a new president successfully enacts a program that would cut government purchases of goods and services *and* cut taxes by *twice* the decrease in spending. Using the model of Section 6.2.1, derive the impact on equilibrium output.

6.2 The government wants to increase government purchases but have equilibrium output stay constant. Using the model of Section 6.2.1, by how much and in which direction should taxes change, relative to the increase in government purchases?

6.3 Show that $dr > 0$ in the IS–LM model of Section 6.2.2 when there is a balanced-budget increase in government purchases and taxes.

6.4 Suppose taxes are a function of income, $T = t(Y)Y$, instead of being exogenous. If the tax system is progressive, then $dt/dY > 0$. Let G increase and let t increase such that, in equilibrium, $dT = dG$. What is the value of the balanced-budget multiplier in the IS–LM model of Section 6.2.2?

6.5 In an IS–LM model, suppose that government purchases of goods and services increase while taxes remain constant. The Fed changes the money supply in such a way that interest rates stay constant. (Does the Fed have to increase or decrease the money supply?) Find and sign (if possible) a formula for dY. How is dY affected by various parameters? Explain the economics of your results.

6.6 For the aggregate demand-aggregate supply model of Section 6.2.3, suppose that there is an increase in government spending and the Fed changes the money supply in such a way that interest rates do not change. Find and sign (if possible) expressions for dY and dP and explain the economics of your results.

6.7 Add real wealth as an argument of the consumption function in the aggregate demand-aggregate supply model of Section 6.2.3: $C = C(Y - T, W/P)$. Find and sign (if possible) the comparative static effects on output, interest rate, and price of an increase in nominal wealth W. Explain the economics of your result.

6.8 Find the comparative static effects on output, interest rate, and price of a decrease in the nominal money supply in the aggregate demand-aggregate supply model of Section 6.2.3. In what ways do the results depend on the sensitivities of investment and money demand to the interest rate?

6.9 In a closed-economy aggregate demand-aggregate supply model, consumption is a (possibly nonlinear) function of disposable income $Y - T$ and real wealth W/P. There is a progressive tax system so $T = T(Y)$, $0 < T' < 1$. Investment is a (possibly nonlinear) function of the interest rate. Contrary to the usual assumption, government purchases of goods and services are not exogenous. Instead, the government automatically adjusts spending on public works projects (such as highways, bridges, and water treatment plants) depending on how much unemployment there is, so G is a function of the difference between GDP and full-employment GDP: $G(Y - Y^F)$, $G' < 0$. The nominal money supply M is exogenous; in equilibrium the real money supply M/P equals the demand for real balances $L(Y, r)$. The (inverse) aggregate supply function is $P = P^E + g(Y - Y^F)$. For simplicity, some of the usual arguments of some functions have been omitted, but otherwise the usual assumptions about signs of the functions' derivatives apply. Write out the total differentials of the three equations in the AD–AS model and use them to find and sign (if possible) the comparative static effect on the aggregate price level of an increase in full-employment GDP. Explain the economics.

6.10 For the model of Problem 6.9, find and sign (if possible) the comparative static effect on output of a change in wealth. How is the magnitude of your answer affected if G' becomes more negative? Explain the economics.

6.11 For the model of Problem 6.9, if the Fed uses monetary policy to ensure that an increase in wealth does not lead to a change in the aggregate price level, how large will dM have to be compared to dW?

6.12 Based on equation (6.26), what factors influence the impact of monetary policy on interest rates? Explain the economics.

6.13 Based on equation (6.25), how is the effectiveness of monetary policy on equilibrium output affected by the marginal propensity to consume and by the sensitivities of money demand to income and the interest rate? Explain the economics.

6.14 Add to the IS–LM model of Section 6.3.1 an equation relating the health of the natural environment, N, to the pollution caused by the production of output: $N = e(Y)$, where $e' < 0$. This results in a system of three equations in the three endogenous variables Y, r, and N. Find the total differentials of the three equations and use them to find and sign, if possible, $\partial Y^*/\partial G$. Explain why your answer is either the same or different than the usual IS–LM result.

6.15 For the model of Problem 6.14, find and sign, if possible, $\partial N^*/\partial G$ and $\partial N^*/\partial M$. Explain the economics.

6.16 For the model of Problem 6.14, suppose the government commits to environmental sustainability, and assume this means that the health of the natural environment must stay at its current level: $dN = 0$. What monetary policy must accompany an expansionary fiscal policy in order to ensure sustainability? (That is, find a relationship between dM and dG when $dN = 0$.)

6.17 For the Mundell-Fleming model of Section 6.3.2, find and sign (if possible) the effect of monetary policy on the exchange rate (when exchange rates are flexible). If the sensitivity of investment to the interest rate increases, does that make your answer larger or smaller? Explain the economics.

6.18 For the flexible exchange rate version of the Mundell-Fleming model of Section 6.3.2, explain how each of the following parameters affects the change in output caused by an increase in the money supply: the sensitivity of investment to interest rates, the sensitivity of money demand to output, and the marginal propensity to consume.

6.19 Find $\partial Y^*/\partial M$ and $\partial r^*/\partial M$ in a fixed exchange rate Mundell-Fleming model. Evaluate for the cases of perfectly capital mobility and perfect capital immobility, and explain the economics of your results.

6.20 Find the comparative static effect on output of expansionary fiscal policy in a fixed exchange rate Mundell-Fleming model. What factors influence the effectiveness of fiscal policy in this model? Explain the economics.

6.21 Build a Mundell-Fleming model that shows the effect of political instability in a country on the effectiveness of fiscal policy in that country (see Footnote 5 on page 150). Does more political instability make fiscal policy more or less effective at changing output?

6.22 Consider the following changes that make the IS–LM model used in Section 6.9 more general:

- Consumption is a function of disposable income, $Y - T$

- Taxes T are a (possibly nonlinear) function of income

- Investment depends on output as well as the interest rate

- Money supply is an increasing function of the interest rate

Find the slopes of the IS and LM curves for this more general case. Explain what factors influence the slopes of the two curves and give economic explanations.

6.23 For a monopolist like the one in Section 6.6, suppose there is an *ad valorem* tax (where the tax is applied as a percentage of sales) so that the firm's after-tax profit is given by $\Pi(Q) = P(Q)Q(1 - t) - C(Q)$. Find the effects on equilibrium price and quantity of an increase in the tax.

6.24 For the model of Problem 6.23, use the envelope theorem to find the effect of an increase in the *ad valorem* tax on the monopolist's after-tax profit.

6.25 For the Cournot duopoly model as set up in Section 6.7, find the comparative static effects on each firm's output, their combined output, and price of an increase in the firms' costs (that is, an increase in the parameter b).

6.26 In the Cournot duopoly model, let each firm's costs be cubic but not equal: $TC_i = b_i q_i^3$. Find out how each firm changes its output in response to an upward shift of the market demand curve. Explain the economics of this result.

6.27 Redo the analysis of Section 6.7 with the (inverse) market demand curve

$$P = \frac{a}{q_1 + q_2}.$$

6.28 In the duopoly model, let the (inverse) market demand curve be $P = P(a, Q)$, where $\partial P/\partial a > 0$, $\partial P/\partial Q < 0$, and $Q = q_1 + q_2$. Let each firm's costs be $TC_i = C(q_i)$ where $C'(q_i) > 0$ and $C''(q_i) > 0$. Derive the formula for the effect on firm 1's quantity of an upward shift of the market demand curve (that is, an increase in the parameter a) for each of the following cases:

(a) The Cournot model, where $\partial q_i/\partial q_j = 0$

(b) Each firm has the conjecture $\partial q_i/\partial q_j = \frac{1}{2}$

(c) The firms' conjectures are $\partial q_1/\partial q_2 = 1$ and $\partial q_2/\partial q_1 = 0$

6.29 Two firms compete with each other in the output market but have geographically separate, competitive labor markets, and labor is immobile. The profit functions for the two firms are therefore $R^1(q^1, q^2) - w^1 L^1$ and $R^2(q^1, q^2) - w^2 L^2$, where q stands for output, L stands for labor input, w stands for the wage rate, and superscripts identify firms. The revenue function is left general, but each firm has some market power, so, using subscripts to signify derivatives, $R_i^i > 0$, $R_j^i < 0$: revenue for firm i will increase when its own output increases, but its revenue will decrease when its competitor increases output. Each firm produces output using labor as the only variable input: $q^1 = F^1(L^1)$ and $q^2 = F^2(L^2, \alpha)$ where α is a parameter measuring technological change for firm 2's production function: an increase in α increases output for a given amount of labor and it also increases the marginal product of a given amount of labor. Each firm chooses its labor input in order to maximize profits, treating the other firm's quantity as fixed. Find the first- and second-order conditions for both firms' profit-maximization problems and use them to find and sign (if possible) the comparative static derivative $\partial L^{1*}/\partial w^1$. Explain the economics of your answer.

6.30 For the model of Section 6.8, confirm by substituting in the explicit formula for L that the formula for dL/dw obtained by implicit differentiation of the first-order condition is identical to that obtained by differentiating the explicit choice function $L^*(w)$.

6.31 Show that the utility-maximizing choice of x when utility is $W(x, y) = xy$ is given by $x^* = I/2P_x$.

6.32 Show that the utility-maximizing choices of x and y are identical for the following two utility functions: $U(x, y) = a \ln(x - x_0) + b \ln(y - y_0)$ and $V(x, y) = (x - x_0)^a (y - y_0)^b$.

6.33 Find the marginal rates of substitution between x and y for the following utility functions:

(a) (Cobb-Douglas) $U(x, y) = Ax^\alpha y^\beta$

(b) (Stone-Geary) $U(x, y) = a \ln(x - x_0) + b \ln(y - y_0)$

(c) (CES) $U(x, y) = A(\alpha x^\rho + (1 - \alpha)y^\rho)^{1/\rho}$

(d) (linear) $U(x, y) = ax + by$

6.34 For each of the cases in problem 6.33 sketch the indifference curves. [Hint: Since the indifference curve is the graph of the function $y(x)$, its curvature depends on $d^2 y/dx^2$.]

6.35. Find the marginal rate of technical substitution (that is, the absolute value of the slope of the isoquant) for the constant elasticity of substitution (CES) production function $f(L, K) = A(\alpha L^\rho + (1 - \alpha)K^\rho)^{1/\rho}$.

6.36 Several demand functions frequently used in applied work are listed below. For each, show whether or under what conditions demand is homogeneous of degree zero in prices and income.

(a) (Linear) $\qquad\qquad\qquad\qquad\quad x_1 = \alpha + \beta_1 p_1 + \beta_2 p_2 + \beta_3 I$

(b) (Log-linear) $\qquad\qquad\qquad\quad\; \ln x_1 = \alpha + \beta_1 p_1 + \beta_2 p_2 + \beta_3 I$

(c) (Log-log) $\qquad\qquad\qquad\qquad\; \ln x_1 = \alpha + \beta_1 \ln p_1 + \beta_2 \ln p_2 + \beta_3 \ln I$

(d) (Cobb-Douglas) $\qquad\qquad\qquad x_1 = \alpha p_1^{\beta_1} p_2^{\beta_2} I^{\beta_3}$

(e) (Almost ideal demand system) $\quad x_1 = (I/p_1)(\alpha + \beta \ln(p_1/p_2) + \gamma \delta (p_1/p_2)^{\delta})$

6.37 Show that the demand for capital is homogeneous of degree 0 in the example of Section 6.11.

6.38 Show that the Cobb-Douglas production function is homogeneous of degree $(\alpha + \beta)$ in labor and capital and that the marginal products of labor and capital are each homogeneous of degree $(\alpha + \beta - 1)$ in labor and capital.

6.39 $D(x)$ gives the damage done by pollution x. The costs of pollution control θ are $C(\theta, t)$ where $\theta = \bar{x} - x$, $\bar{x}$ is the amount of pollution there would be if none were controlled, and t is a parameter related to technology: when t increases, pollution control technology has become better. Find the first- and second-order conditions for the problem of choosing the pollution level that minimizes the total costs associated with pollution, $TC(x) = D(x) + C(\bar{x} - x, t)$. Make reasonable assumptions about the signs of the derivatives D', C_θ, C_t, $C_{\theta\theta}$, C_{tt}, and $C_{\theta t}$ and use them to find and sign, if possible, the comparative static effect on the optimal pollution level of an increase in t. Explain the economics of your result

Multivariable Optimization without Constraints: Theory

7.1 INTRODUCTION

A central mathematical tool of microeconomics is calculus-based optimization. We began this text by examining one variable optimization and its application to economics. Simple optimization proved to be a useful tool with many economic applications, but many economic models include agents who have more than one choice variable. The theory of the firm, in particular, makes extensive use of multivariable optimization. Multivariable choice models for the firm include choosing the right mix of several inputs, choosing production levels when a firm operates more than one factory, and choosing prices or quantities when a firm sells in more than one market (all of these will be covered as applications in Chapter 8). In these models, and others, we need the appropriate mathematical tools for maximizing functions of several variables.

This chapter extends our previous study of one variable optimization: first to functions of two variables and then to functions of n variables. We will make extensive use of the mathematics from Chapters 3 and 5. Matrix algebra will be useful in characterizing the second-order conditions for multivariable optimization problems. In addition, we will use both matrix algebra and the implicit function theorem to derive comparative static results in multivariable optimization problems.

To introduce the unconstrained multivariable optimization we start this chapter by setting up an economic example in which consumer demand for a product depends on both price and advertising. A firm then controls two choice variables. How should the firm choose the optimum levels of price and advertising? Since both a lower price and higher advertising could increase demand for a product, what is the optimal mix of the two strategies? How does this mix depend on parameters such as production cost? To answer these questions we need to develop the mathematics of multivariable optimization.

We begin the mathematical theory by reexamining single variable optimization models. In Chapter 1 we characterized the first- and second-order conditions for these types of problems in terms of derivatives. Here, we offer an alternative interpretation that focuses on differentials. The differential approach is useful because we can easily extend it to the case of two or more choice variables. We will also show how the differential approach allows us to check second-order conditions by examining the properties of a special matrix called a Hessian matrix. In addition to the differential approach to optimization, we offer a geometric interpretation using the concepts of concavity (for maximization) and convexity (for minimization). We conclude our theoretical discussion by combining optimization and the implicit function theorem to derive comparative static results in a multivariable optimization model. Section 7.9 presents a solvable example that illustrates how the mathematical theory can be applied to an actual problem.

The techniques in this chapter are quite powerful tools for economic analysis. Nevertheless, this chapter only covers **unconstrained** optimization—situations in which agents are free to pick any value of the economic choice variables. In many other cases agents face **constrained** optimization problems—problems in which economic choice variables are constrained to satisfy certain relationships (for example, a consumer's expenditures are constrained by the consumer's level of income). Still, the study of unconstrained optimization theory is quite useful: it not only allows us to model many interesting economic problems, it also provides an excellent background for our later coverage of constrained optimization.

7.2 AN ECONOMIC EXAMPLE

For some products consumer demand depends on both the product's price and consumers' exposure to advertising. Advertising might be informative or persuasive, but what will matter in our model is that in either case higher advertising leads to a higher quantity demanded. We will assume that the firm in our model is a monopolist (this simplifies matters) and that it faces a demand function that gives quantity demanded as a function of price and advertising.

Let the quantity demanded of the product be

$$Q = Q(P, A), \tag{7.1}$$

where Q is quantity demanded, P is price, and A is advertising (measured in dollars). We will keep this initial example simple by assuming that the demand function takes the form

$$Q = (\alpha - P)A^\beta, \tag{7.2}$$

where α and β are positive parameters. Like some of the demand functions that we have used earlier, this one is linear in price. In addition, however, quantity demanded is an increasing, but nonlinear, function of advertising.

We will assume that the firm operates with a constant marginal cost of \$c/unit. The profit function for the firm can be written as

$$\Pi = PQ - cQ - A,$$
$$\Pi = (P - c)Q - A, \quad \text{or} \tag{7.3}$$
$$\Pi = (P - c)(\alpha - P)A^\beta - A.$$

The middle line of equation (7.3) has an easy economic interpretation. Price minus production cost is the firm's per-unit profit margin, so that profits are net revenue (the per-unit profit margin times the volume of output sold) minus advertising costs.

In the following sections of this chapter we will learn how to solve for the optimal levels of P^* and A^*, how to check that these values generate a profit maximum, and how to find the comparative static effects of changes in the parameters. After learning the relevant mathematical theory we will return to this problem (in Section 7.10) and solve for the firm's price and advertising choices.

7.3 ONE VARIABLE OPTIMIZATION REVISITED

In Chapter 1 we reviewed the calculus of maximizing or minimizing a function of one variable. Given such a function, $y = f(x)$, the first-order necessary condition for a local extreme point is

$$\frac{dy}{dx} = f'(x) = 0. \tag{7.4}$$

The respective second order sufficient conditions are

$$\frac{d^2y}{dx^2} = f''(x) < 0 \quad \text{(maximum)} \quad \text{or}$$

$$\frac{d^2y}{dx^2} = f''(x) > 0 \quad \text{(minimum)}. \tag{7.5}$$

Each of these equations is in the form of a derivative equals zero or has a particular sign. There is, however, another way of writing and interpreting these equations. This alternative will be very helpful in explaining multivariable optimization.

Suppose we write the first-order condition as

$$dy = f'(x)\,dx = 0. \tag{7.6}$$

Note first that we have stated the first-order condition as $dy = 0$, i.e., that a necessary condition for a maximum or minimum is that for any (small) change in the choice variable, x, there be no change in the variable, y, that is being maximized or minimized. Second, the equation implies that, because changes in x are arbitrary, the only way to guarantee that the value of dy will equal zero is by finding a value of x such that $f'(x) = 0$. In the next section we will see that the necessary condition $dy = 0$ generalizes to multivariable optimization problems. The condition $f'(x) = 0$ will need to be extended, however, to include zero partial derivatives for all choice variables for the first-order conditions that ensure $dy = 0$.

The second-order condition can be rewritten in an analogous fashion as

$$d^2y = f''(x)dx^2 < 0 \quad \text{(maximum)} \quad \text{or}$$
$$d^2y = f''(x)dx^2 > 0 \quad \text{(minimum)}. \tag{7.7}$$

The term d^2y is the change in dy, i.e., the change in the change in y. The term dx^2 is dx times dx, i.e., the square of the change in x. The second-order condition can then be interpreted as a sign restriction on d^2y, which since dx^2 is always positive, is equivalent to a sign restriction on $f''(x)$. For multivariable problems the same sign conditions on d^2y will be sufficient for a maximum or minimum, but the relationship between the sign of d^2y and the second partial derivatives of the function f will be more complex.

7.4 TWO–VARIABLE OPTIMIZATION

We now turn to two-variable optimization. Consider maximizing or minimizing a function

$$y = f(x_1, x_2). \tag{7.8}$$

The differential of this equation is

$$dy = f_1\,dx_1 + f_2\,dx_2. \tag{7.9}$$

where $f_i = \partial f(x_1, x_2)/\partial x_i$ is the partial derivative of the function f with respect to x_i. For a maximum (minimum) it must be the case that there is no way to increase (decrease) y. In either case, an extreme point requires that for any arbitrary changes in the x variables, there must be no change in y. The necessary condition that $dy = 0$ for all possible small combinations of the dx terms will hold only if

$$f_1(x_1, x_2) = 0 \quad \text{and} \quad f_2(x_1, x_2) = 0 \quad \text{or}$$
$$f_i(x_1, x_2) = 0 \quad \text{for all } i.$$

(7.10)

The arguments, x_1 and x_2, are included as a reminder that partial derivatives are generally functions of all the original x variables. These first-order conditions are necessary for any local extreme point: maximum or minimum. For a maximum, one way of interpreting the first-order conditions is that for each x variable a zero partial derivative ensures that changes in that x variable alone cannot increase the value of y.

The second-order condition places a sign restriction on d^2y; therefore, we need to derive the equation for d^2y. To do this we find the total differential of the right-hand side of equation (7.9) with respect to the x variables (while treating the dx variables as constants). This gives

$$d^2y = (f_{11}\, dx_1 + f_{21}\, dx_2)\, dx_1 + (f_{12}\, dx_1 + f_{22}\, dx_2)\, dx_2,$$
$$d^2y = f_{11}\, dx_1^2 + f_{21}\, dx_2\, dx_1 + f_{12}\, dx_1\, dx_2 + f_{22}\, dx_2^2, \quad \text{or}$$
$$d^2y = f_{11}\, dx_1^2 + 2f_{12}\, dx_1\, dx_2 + f_{22}\, dx_2^2.$$

(7.11)

where the last step uses Young's Theorem that cross-partial derivatives are identical regardless of the order of differentiation. Now the dx squared terms are clearly positive, but, since the x variables can change in either direction, there is no way to specify a sign for the $dx_1\, dx_2$ term and no immediate way to determine the sign of d^2y.

Before trying to sign d^2y let us first write equation (7.11) in matrix form:

$$d^2y = (dx_1 \; dx_2) \begin{pmatrix} f_{11} & f_{12} \\ f_{21} & f_{22} \end{pmatrix} \begin{pmatrix} dx_1 \\ dx_2 \end{pmatrix}.$$

(7.12)

The reader can verify that multiplying out equation (7.12) yields equation (7.11). This equation for d^2y is called a **quadratic form**.[1] The matrix of second partial derivatives in equation (7.9) is called a **Hessian** matrix, usually abbreviated as H. After some derivations we will find that the sufficient conditions for signing d^2y can be written as sign restrictions on the subdeterminants of H.

The connection between H and equation (7.11) can be drawn using a trick called completing the square. Adding and subtracting the term $(f_{12})^2 dx_2^2 / f_{11}$ in equation (7.11) yields

$$d^2y = f_{11}\, dx_1^2 + 2f_{12}\, dx_1\, dx_2 + \frac{(f_{12})^2\, dx_2^2}{f_{11}} + f_{22}\, dx_2^2 - \frac{(f_{12})^2\, dx_2^2}{f_{11}}.$$

(7.13)

Combining the first three terms and rearranging the last two terms allow us to write the equation as

$$d^2y = f_{11}\left(dx_1 + \frac{f_{12}\, dx_2}{f_{11}}\right)^2 + \left(\frac{f_{11} f_{22} - (f_{12})^2}{f_{11}}\right) dx_2^2.$$

(7.14)

---

[1] A quadratic form in two variables, x and y, is defined as function $f(x, y) = ax^2 + 2bxy + cy^2$, where a, b, and c are constants.

Since the second and fourth terms are positive squares, the sign of d^2y depends on the signs of the first and third terms. We can now write the sufficient conditions for a maximum ($d^2y < 0$) and a minimum ($d^2y > 0$). These are

$$d^2y < 0 \quad \text{if } f_{11} < 0 \quad \text{and} \quad f_{11}f_{22} - (f_{12})^2 > 0 \quad \text{(maximum)} \quad \text{or}$$
$$d^2y > 0 \quad \text{if } f_{11} > 0 \quad \text{and} \quad f_{11}f_{22} - (f_{12})^2 > 0 \quad \text{(minimum)}. \tag{7.15}$$

Note that the second term on the right-hand side of (7.15) can only be positive if f_{11} and f_{22} have the same sign. Also observe that the second-order conditions for a maximum are opposite (or alternate) in sign, but for a minimum both second-order conditions must be positive. We will see in Section 7.5 that this pattern carries over to n variable optimization problems.

7.5 HESSIAN MATRICES AND LEADING PRINCIPAL MINORS

The second-order conditions derived in Section 7.4 can be represented using properties of the Hessian matrix, H. Let us define a submatrix H_k to be the matrix that is left when $n - k$ rows and columns are deleted from the matrix H. If we delete the same rows and columns (e.g., second row and second column), then each H_k is referred to as a **principal submatrix of order k** and the determinant, $|H_k|$, of this submatrix is called a **principal minor of order k**. For optimization problems we focus on **leading principal minors**: the kth leading principal minor is the determinant of the submatrix H_k composed of the a_{ij} remaining when all elements with subscripts (either i or j) greater than k have been eliminated from a matrix H. Thus, for a two variable maximization problem, we have

$$H_1 = (f_{11}) \quad \text{and} \quad |H_1| = f_{11}$$
$$H_2 = \begin{pmatrix} f_{11} & f_{12} \\ f_{21} & f_{22} \end{pmatrix} = H \quad \text{and} \quad |H_2| = f_{11}f_{22} - (f_{12})^2 = |H|. \tag{7.16}$$

From this we see that we can write the second-order conditions for the sign of d^2y in terms of leading principal minors as

$$d^2y < 0 \quad \text{if } |H_1| = f_{11} < 0 \quad \text{and} \quad |H_2| = f_{11}f_{22} - (f_{12})^2 > 0 \quad \text{(maximum)} \quad \text{or}$$
$$d^2y > 0 \quad \text{if } |H_1| = f_{11} > 0 \quad \text{and} \quad |H_2| = f_{11}f_{22} - (f_{12})^2 > 0 \quad \text{(minimum)}.$$
$$\tag{7.17}$$

Leading principal minors are defined in the same way for larger Hessian matrices. For example, the 3×3 Hessian

$$H = \begin{pmatrix} f_{11} & f_{12} & f_{13} \\ f_{21} & f_{22} & f_{23} \\ f_{31} & f_{32} & f_{33} \end{pmatrix} \tag{7.18}$$

has the leading principal minors

$$|H_1| = |f_{11}|,$$
$$|H_2| = \begin{vmatrix} f_{11} & f_{12} \\ f_{21} & f_{22} \end{vmatrix}, \quad \text{and}$$
$$|H_3| = \begin{vmatrix} f_{11} & f_{12} & f_{13} \\ f_{21} & f_{22} & f_{23} \\ f_{31} & f_{32} & f_{33} \end{vmatrix} = |H|. \tag{7.19}$$

The same rules can be used to generate the leading principal minors of larger Hessian matrices. Note that since the ordering of variables is arbitrary all principal minors will have the same signs as the leading principal minors. This fact is often useful in signing comparative statics results.

7.6 MULTIVARIABLE OPTIMIZATION

Multivariable optimization is a straightforward extension of the two variable case. Suppose an economic agent faces a problem of optimizing a function with n choice variables. We write the function being optimized as

$$y = f(x_1, x_2, \ldots, x_n). \tag{7.20}$$

The total differential of this equation is

$$dy = \sum_{i=1}^{n} f_i \, dx_i. \tag{7.21}$$

Since an agent can vary the x variables in any manner (i.e., the values of the dx terms are arbitrary) the only way to ensure a local extreme point, $dy = 0$, is for the all of partial derivatives to equal zero. This gives n first-order necessary conditions:

$$
\begin{aligned}
f_1(x_1, x_2, \ldots, x_n) &= 0, \\
f_2(x_1, x_2, \ldots, x_n) &= 0, \\
&\cdot \\
&\cdot \\
&\cdot \\
f_n(x_1, x_2, \ldots, x_n) &= 0.
\end{aligned}
\tag{7.22}
$$

The arguments are included as yet another reminder that partial derivatives of a function are themselves generally functions of all of the variables of the original function.

To derive the equation for d^2y we totally differentiate equation (7.21). The reader should verify that the result, derived in the same manner as for the two variable case, is a quadratic form

$$d^2y = \mathbf{dx}' \, H \, \mathbf{dx}, \tag{7.23}$$

where $\mathbf{dx}'$ is an $1 \times n$ row vector of the dx terms, $\mathbf{dx}$ is an $n \times 1$ column vector of the dx terms, and H is the $n \times n$ Hessian matrix:

$$
H = \begin{pmatrix}
f_{11} & f_{12} & \cdot & \cdot & \cdot & f_{1n} \\
f_{21} & f_{22} & \cdot & \cdot & \cdot & f_{2n} \\
\cdot & \cdot & \cdot & \cdot & \cdot & \cdot \\
\cdot & \cdot & \cdot & \cdot & \cdot & \cdot \\
\cdot & \cdot & \cdot & \cdot & \cdot & \cdot \\
f_{n1} & f_{n2} & \cdot & \cdot & \cdot & f_{nn}
\end{pmatrix}.
\tag{7.24}
$$

The sufficient conditions for a maximum ($d^2y < 0$) and a minimum ($d^2y > 0$) can now be written in terms of the leading principal minors of H. The n second-order conditions are:[2]

$$d^2y < 0 \quad \text{if } (-1)^i |H_i| > 0 \quad \text{for } i = 1, \ldots, n \quad \text{(maximum)} \quad \text{or}$$

$$d^2y > 0 \quad \text{if } |H_i| > 0 \qquad \text{for } i = 1, \ldots, n \quad \text{(minimum).}$$

(7.25)

When the conditions for $d^2y < 0$ hold the quadratic form is said to be **negative definite** (remember this as d^2y is definitely negative) and when the conditions for $d^2y > 0$ hold the quadratic form is **positive definite** (d^2y is definitely positive). Note that the multivariable case follows the same pattern as the two-variable case: a maximum requires that the first leading principal minor be negative and that the rest of the higher-order leading principal minors alternate in sign, while a minimum requires that all leading principal minors be positive.

7.7 CONCAVITY, CONVEXITY, AND OPTIMIZATION PROBLEMS

In this section we present an alternative method of characterizing extreme points of a function of several variables. This alternative is based on the mathematical concepts of concavity and convexity. A formal definition of these terms will be given shortly, but the essence is that a stationary point, where the first-order conditions hold, will be a global (or local) maximum if a function is globally (or locally) concave and a stationary point will be a global (or local) minimum if a function is globally (or locally) convex.

The advantage of the concavity/convexity approach is that it is more general than the approach based on differentials. First, it holds even when the function being examined is not differentiable. Second, there may be cases where the second-order conditions are ambiguous, i.e., $d^2y = 0$, but the concavity or convexity conditions are sufficient to establish whether a point is a maximum or minimum. The main disadvantage of the concavity/convexity approach is that it is often difficult to verify directly: the differential requirement is usually the easier method for checking whether an extreme point is a maximum or a minimum.

We will begin with definitions of concavity and convexity that apply to any continuous function (a second set of definitions for differentiable functions will be given below). Consider a function $y = f(\mathbf{x})$, where $\mathbf{x} = (x_1, x_2, \ldots, x_n)$.

DEFINITION **7.1**

Concavity

The function $y = f(\mathbf{x})$ is globally concave if and only if for any two distinct points $\bar{\mathbf{x}} = (\bar{x}_1, \bar{x}_2, \ldots, \bar{x}_n)$ and $\hat{\mathbf{x}} = (\hat{x}_1, \hat{x}_2, \ldots, \hat{x}_n)$ and for $0 < \lambda < 1$:

$$\lambda f(\bar{\mathbf{x}}) + (1 - \lambda) f(\hat{\mathbf{x}}) \leq f(\lambda \bar{\mathbf{x}} + (1 - \lambda)\hat{\mathbf{x}}).$$

---

[2]It is worth noting that, since the ordering of the x variables is arbitrary, if $|H_1|$, composed of partial derivatives f_{jk}, has a specific sign (and all of the other second-order conditions hold), then the same sign holds for permutations of j and k. For example, in an n variable problem if $|H_2| = f_{11} f_{22} - (f_{12})^2 > 0$ and all of the other $|H_i|$ have the proper sign then all terms of the form $f_{jj} f_{kk} - (f_{jk})^2$ will also be positive. This correspondence between leading principal minors of order k and nonleading principal minors of order k will often be helpful in signing comparative static derivatives.

Convexity

The function $y = f(\mathbf{x})$ is globally convex if and only if for any two distinct points $\bar{\mathbf{x}} = (\bar{x}_1, \bar{x}_2, \ldots, \bar{x}_n)$ and $\hat{\mathbf{x}} = (\hat{x}_1, \hat{x}_2, \ldots, \hat{x}_n)$ and for $0 < \lambda < 1$:

$$\lambda f(\bar{\mathbf{x}}) + (1 - \lambda) f(\hat{\mathbf{x}}) \geq f(\lambda \bar{\mathbf{x}} + (1 - \lambda)\hat{\mathbf{x}}).$$

If the last lines of Definitions 7.1 and 7.2 hold with strict inequality then the function is *strictly concave* or *strictly convex*. Also, if the requirements only hold over some range of x values then the function is *locally concave* or *locally convex* in that neighborhood.

For a function of two variables, a concave function is a hill and a convex function a valley. Consider, for example, setting an eggshell on end and cutting it in half. The top half would constitute a concave function and the bottom half a convex function. The concave case is illustrated in Figure 7.1. In Figure 7.1 the line segment $\lambda f(\bar{\mathbf{x}}) + (1 - \lambda) f(\hat{\mathbf{x}})$ lies strictly below the surface of the function $f(\mathbf{x})$.

The importance of these definitions is given in the following theorem:

THEOREM 7.1

If the function $y = f(\mathbf{x})$ is strictly and globally concave (convex) and $\tilde{\mathbf{x}} = (\tilde{x}_1, \tilde{x}_2, \ldots, \tilde{x}_n)$ yields an extreme point of $f(\mathbf{x})$, then this extreme point is a global maximum (minimum) of $f(\mathbf{x})$.

FIGURE 7.1

A Strictly Concave Function

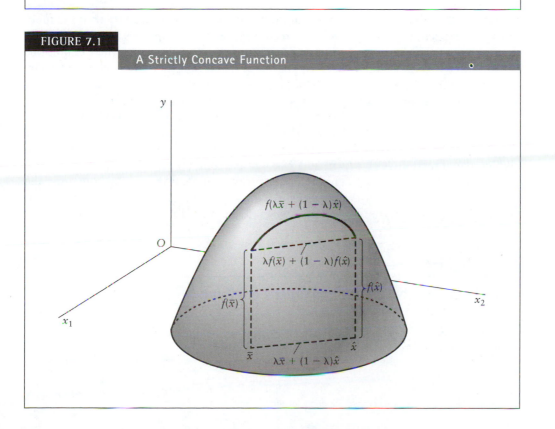

In other words, strict global concavity (convexity) is sufficient to ensure that any local extreme point is a global maximum (minimum).

The descriptions above hold for all functions, whether or not the functions are differentiable. For differentiable functions $y = f(\mathbf{x})$ there are alternative definitions of concavity and convexity, which are equivalent to those given previously:

DEFINITION **7.3**

Concavity of Differentiable Functions

A differentiable function $y = f(\mathbf{x})$ is globally concave if and only if for any two distinct points $\bar{\mathbf{x}} = (\bar{x}_1, \bar{x}_2, \ldots, \bar{x}_n)$ and $\hat{\mathbf{x}} = (\hat{x}_1, \hat{x}_2, \ldots, \hat{x}_n)$:

$$f(\bar{\mathbf{x}}) \leq f(\hat{\mathbf{x}}) + \sum_{i=1}^{n} \frac{\partial f(\hat{\mathbf{x}})}{\partial x_i}(\bar{x}_i - \hat{x}_i),$$

DEFINITION **7.4**

Convexity of Differentiable Functions

A differentiable function $y = f(\mathbf{x})$ is globally convex if and only if for any two distinct points $\bar{\mathbf{x}} = (\bar{x}_1, \bar{x}_2, \ldots, \bar{x}_n)$ and $\hat{\mathbf{x}} = (\hat{x}_1, \hat{x}_2, \ldots, \hat{x}_n)$:

$$f(\bar{\mathbf{x}}) \geq f(\hat{\mathbf{x}}) + \sum_{i=1}^{n} \frac{\partial f(\hat{\mathbf{x}})}{\partial x_i}(\bar{x}_i - \hat{x}_i).$$

As before when the last lines of the definitions hold with strict inequality then the functions are, respectively, *strictly concave* and *convex;* or, if the properties are local then the functions are *locally concave* or *convex*. The geometric interpretation of these definitions is that, for an n variable functions, the hyperplane that is tangent to the function at a particular point lies above the function (concavity) or below the function (convexity).

The implications of these definitions for maxima and minima are straightforward. Suppose that $\hat{\mathbf{x}}$ yields a stationary point of the function $y = f(\mathbf{x})$, i.e., at $\hat{\mathbf{x}}$, $\partial f(\hat{\mathbf{x}})/\partial x_i = 0$ for all i. Then for a strictly (and globally) concave function we have $f(\bar{\mathbf{x}}) < f(\hat{\mathbf{x}})$, i.e., the point $\hat{\mathbf{x}}$ is a global maximum. Similarly, for a strictly (and globally) convex function $f(\bar{\mathbf{x}}) > f(\hat{\mathbf{x}})$ and $\hat{\mathbf{x}}$ is a global minimum.

So far we have given two definitions of concavity and convexity. We now draw the connection between concavity/convexity and the second-order conditions for an n variable optimization problem. The connection is given in the following theorem:

THEOREM **7.2**

The Relation between Second-Order Conditions and Concavity/Convexity

A function $y = f(\mathbf{x})$ is globally strictly concave (convex) if d^2y is everywhere negative (positive).[3]

---

[3]As before the theorem is easily adapted to apply locally or non-strictly.

Note that the sign restriction on d^2y is sufficient but not necessary. By this we mean that a global negative or positive sign for d^2y establishes whether a function is strictly concave or convex, but a function might still be strictly concave or convex even if $d^2y = 0$.[4]

Concavity and convexity are very general tools for characterizing maxima and minima. In practice, however, economists (at least for most applied theory) rely on the weaker test of second-order conditions. When second-order conditions give a clear answer we can be sure that a function has the appropriate properties. Only when the second-order conditions are ambiguous do we need to investigate further. On the other hand, if a function is known, or assumed to be, strictly and globally concave (or convex) it is not necessary to check the second-order conditions.[5]

7.8 COMPARATIVE STATICS AND MULTIVARIABLE OPTIMIZATION

In this section we explore two methods for deriving comparative static results. Both methods use the fact that the optimal solution for the x variables is defined by the first-order conditions. In some cases this solution can be explicitly computed and we can then take partial derivatives of the solutions for the x variables. If this method fails, or is inconvenient, we can instead turn to the implicit function theorem to derive the implicitly defined comparative static results.

Consider a three-variable maximization example that includes a single exogenous parameter c. The problem can be written as, maximize

$$y = f(x_1, x_2, x_3; c). \tag{7.26}$$

The first-order conditions are

$$
\begin{aligned}
f_1(x_1, x_2, x_3; c) &= 0, \\
f_2(x_1, x_2, x_3; c) &= 0, \quad \text{and} \\
f_3(x_1, x_2, x_3; c) &= 0.
\end{aligned}
\tag{7.27}
$$

The second-order conditions for a maximum are

$$
\begin{aligned}
|H_1| &= |f_{11}| < 0, \\
|H_2| &= \begin{vmatrix} f_{11} & f_{12} \\ f_{21} & f_{22} \end{vmatrix} > 0, \quad \text{and} \\
|H_3| &= \begin{vmatrix} f_{11} & f_{12} & f_{13} \\ f_{21} & f_{22} & f_{23} \\ f_{31} & f_{32} & f_{33} \end{vmatrix} = |H| < 0.
\end{aligned}
\tag{7.28}
$$

Since there are three first-order equations and three endogenous variables, it will, in some cases, be possible to solve for each of the x variables in terms of the parameter c.

---

[4]As a simple example, consider the function $y = x^4$. This function is clearly convex, yet, at the minimum point of $x = 0$ we have $d^2y = f''(x)\,dx^2 = 12x^2\,dx^2 = 0$.

[5]Many papers in economic theory use this. A model will assume that a functional form is concave without placing any further restrictions on the function. The concavity assumption is then sufficient to guarantee than an extreme point is a global maximum.

For example, if the equations were linear we could use Cramer's rule or matrix inversion to solve the equation system. Solutions would be of the form

$$x_i^* = g^i(c). \tag{7.29}$$

Comparative static effects of changes in the parameter (or parameters in the more general case) can be found by taking the partial derivatives of the solution functions.

There are, however, cases in which solving nonlinear first-order conditions is difficult or impossible. Furthermore, the most general attempts at deriving economic predictions will leave the function f unspecified (although economic theory generally places some restrictions on f, such as specifying the signs of its partial and second partial derivatives). Comparative static results are still possible even when the first-order conditions are intractable. For this we need to use the implicit function theorem.

The equations in (7.27) constitute a system with three equations, three endogenous variables, and one parameter. The reader should check that taking the total differentials of the three equations yields the system:

$$J \begin{pmatrix} dx_1 \\ dx_2 \\ dx_3 \end{pmatrix} = \begin{pmatrix} -f_{1c}\, dc \\ -f_{2c}\, dc \\ -f_{3c}\, dc \end{pmatrix}. \tag{7.30}$$

where J, the Jacobian matrix for the three (first-order condition) equations, is

$$J = \begin{pmatrix} f_{11} & f_{12} & f_{13} \\ f_{21} & f_{22} & f_{23} \\ f_{31} & f_{32} & f_{33} \end{pmatrix}. \tag{7.31}$$

Note that J (the Jacobian matrix) is the same as H (the Hessian matrix), that $|J| = |H|$, and that the second-order conditions ensure that $|J| \neq 0$ so that the implicit function theorem can be used. This correspondence between the two matrices will be very helpful in signing comparative static derivatives.

To solve for changes in the x variables we can apply Cramer's rule to equation (7.30). For example, the solution for dx_3 is

$$dx_3 = \frac{\begin{vmatrix} f_{11} & f_{12} & -f_{1c}\, dc \\ f_{21} & f_{22} & -f_{2c}\, dc \\ f_{31} & f_{32} & -f_{3c}\, dc \end{vmatrix}}{|H|}, \quad \text{or}$$

$$\frac{dx_3}{dc} = \frac{\begin{vmatrix} f_{11} & f_{12} & -f_{1c} \\ f_{21} & f_{22} & -f_{2c} \\ f_{31} & f_{32} & -f_{3c} \end{vmatrix}}{|H|}. \tag{7.32}$$

By the implicit function theorem, the value of the derivative in equation (7.32) is equal to that of the value of the derivative of the explicit solution function in equation (7.29).

At this level of generality it is often difficult to sign dx_3, but in many economic applications economic theory will place restrictions on the partial derivatives, and those restrictions will help us to sign the comparative static results. Suppose, for example, that the parameter c measures the cost of activity x_3 so that c does not appear in the first-order conditions for x_1 and x_2, i.e., economic theory tells us that $f_{1c} = f_{2c} = 0$.

The solution for dx_3/dc becomes

$$dx_3 = \frac{\begin{vmatrix} f_{11} & f_{12} & 0 \\ f_{21} & f_{22} & 0 \\ f_{31} & f_{32} & -f_{3c}dc \end{vmatrix}}{|H|} = \frac{-f_{3c}dc \begin{vmatrix} f_{11} & f_{12} \\ f_{21} & f_{22} \end{vmatrix}}{|H|} \tag{7.33}$$

or

$$\frac{dx_3}{dc} = \frac{-f_{3c}|H_2|}{|H|}$$

Since the second-order conditions for a maximum require that principal minors alternate in sign, we have

$$\frac{-|H_2|}{|H|} > 0. \tag{7.34}$$

This gives us the following result for the sign of the derivative:

$$sign\left(\frac{dx_3}{dc}\right) = sign(f_{3c}), \tag{7.35}$$

and if c measures the cost of economic activity x_3, f_{3c} and thus the derivative will be negative.[6] In many economic applications such *a priori* sign restrictions from economic theory, combined with sign restrictions from the second-order conditions, will allow us to derive this type of result.

7.9   A MATHEMATICAL EXAMPLE

In this section we will work through an algebraic example of a two-variable maximization problem. We will first solve explicitly for the x variables and find the partial derivatives of the solution functions with respect to parameters. Then we will use total differentiation and the implicit function theorem to derive the same comparative static results. Finally, we will verify that the two methods are equivalent.

Suppose we have a function to be maximized that takes the form:

$$y = f(x_1, x_2; c_1, c_2) = x_1 x_2 - \frac{1}{3}c_1 x_1^3 - \frac{1}{3}c_2 x_2^3, \tag{7.36}$$

where the x variables are endogenous choice variables and the c variables are positive parameters. The first-order conditions for an extreme point are

$$f_1(x_1, x_2, c_1, c_2) = x_2 - c_1 x_1^2 = 0 \quad \text{and}$$
$$f_2(x_1, x_2, c_1, c_2) = x_1 - c_2 x_2^2 = 0. \tag{7.37}$$

Since these first-order conditions are nonlinear we cannot use matrix algebra. We can, however, solve by substitution. Solving for x_2 in the $f_1 = 0$ equation gives

$$x_2 = c_1 x_1^2. \tag{7.38}$$

---

[6]Compare this to the result derived in the benefit and cost example in Section 1.3.2. Both results are fairly general: higher costs lead to lower levels of economic activity.

Substituting this into the $f_2 = 0$ equation yields

$$x_1 - c_2 \left(c_1 x_1^2\right)^2 = 0. \tag{7.39}$$

Rearranging and manipulating this equation provides the solution for x_1^*

$$
\begin{aligned}
x_1 &= c_2 c_1^2 x_1^4, \quad \text{or} \\
x_1^{-3} &= c_2 c_1^2, \qquad \text{so that} \\
x_1^* &= c_2^{-1/3} c_1^{-2/3}.
\end{aligned}
\tag{7.40}
$$

We could find the solution for x_2 by an analogous method of substituting and re-arranging. There is, however, an easier method. An examination of equation (7.36) shows that there is a symmetry between the two x variables: if we interchanged all subscripts the equation would still be the same. Because of this symmetry the solution for x_2 will be symmetric to the solution for x_1: we can take the solution for x_1 and interchange the subscripts to get the solution for x_2.[7] This is

$$x_2^* = c_1^{-1/3} c_2^{-2/3}. \tag{7.41}$$

The next step is to make sure that these solutions correspond to a maximum. The Hessian matrix is

$$H = \begin{pmatrix} f_{11} & f_{12} \\ f_{21} & f_{22} \end{pmatrix} = \begin{pmatrix} -2c_1 x_1 & 1 \\ 1 & -2c_2 x_2 \end{pmatrix}. \tag{7.42}$$

The two second-order conditions are

$$
\begin{aligned}
|H_1| &= f_{11} = -2c_1 x_1 < 0 \quad \text{and} \\
|H_2| &= |H| = \begin{vmatrix} f_{11} & f_{12} \\ f_{21} & f_{22} \end{vmatrix} = 4c_1 c_2 x_1 x_2 - 1 > 0.
\end{aligned}
\tag{7.43}
$$

Since c_1 and x_1 are positive the $|H_1| < 0$ condition is definitely satisfied. The condition for $|H_2| > 0$, however, depends on the magnitudes of the variables, so that the function is not globally concave. What matters, however, is whether the function is locally concave so that the $|H_2| > 0$ second-order condition holds at the values of x that solve the first-order conditions. To check this we substitute the x solutions into the expression for $|H_2|$, finding that

$$|H_2| = |H| = 4c_1 c_2 x_1 x_2 - 1 = 4c_1 c_2 c_2^{-1/3} c_1^{-2/3} c_1^{-1/3} c_2^{-2/3} - 1 = 3 > 0. \tag{7.44}$$

Now that we have verified that the solutions correspond to a maximum we can derive comparative static results. Since the solutions for x_1 and x_2 are symmetric their partial derivatives will be symmetric as well, so that we need consider only one of the variables. For x_1 we get:

$$
\begin{aligned}
\frac{\partial x_1^*}{\partial c_1} &= \left(\frac{-2}{3}\right) c_2^{-1/3} c_1^{-5/3} < 0 \quad \text{and} \\
\frac{\partial x_1^*}{\partial c_2} &= \left(\frac{-1}{3}\right) c_2^{-4/3} c_1^{-2/3} < 0.
\end{aligned}
\tag{7.45}
$$

---

[7]Symmetry is a nice time-saving trick. Yet, if you are unsure of whether symmetry applies, or of how it should be used, a problem can be solved by the longer method of explicitly working through the steps that yield a solution.

Next we consider the implicit function approach to finding these derivatives and verify that both approaches yield the same solutions. To start we take the total differentials of the first-order conditions in equation (7.37).[8] In matrix form this gives

$$\begin{pmatrix} f_{11} & f_{12} \\ f_{21} & f_{22} \end{pmatrix} \begin{pmatrix} dx_1 \\ dx_2 \end{pmatrix} = \begin{pmatrix} -f_{1c_1} dc_1 \\ -f_{2c_2} dc_2 \end{pmatrix} \quad \text{or}$$

$$\begin{pmatrix} -2c_1 x_1 & 1 \\ 1 & -2c_2 x_2 \end{pmatrix} \begin{pmatrix} dx_1 \\ dx_2 \end{pmatrix} = \begin{pmatrix} x_1^2 dc_1 \\ x_2^2 dc_2 \end{pmatrix}. \tag{7.46}$$

The Jacobian matrix on the left is, of course, the Hessian matrix for the second-order conditions.

Using Cramer's rule to solve equation (7.46) for dx_1 gives

$$dx_1 = \frac{\begin{vmatrix} x_1^2 dc_1 & 1 \\ x_2^2 dc_2 & -2c_2 x_2 \end{vmatrix}}{|H|} = \frac{-2c_2 x_2 x_1^2 dc_1 - x_2^2 dc_2}{|H|}. \tag{7.47}$$

This yields the two derivatives:

$$\frac{dx_1}{dc_1}\bigg|_{dc_2=0} = \frac{-2c_2 x_2 x_1^2}{|H|} < 0 \quad \text{and}$$

$$\frac{dx_1}{dc_2}\bigg|_{dc_1=0} = \frac{-x_2^2}{|H|} < 0. \tag{7.48}$$

where the vertical line notation (followed by additional information) indicates the conditions under which a derivative is being evaluated. This is interpreted as the derivative *given* the restriction $dc_i = 0$, i.e., $dx_1/dc_1|_{dc_2=0} \equiv \partial x_1/\partial c_1$.

Although the comparative static results in equations (7.45) and (7.48) have the same signs, they do not, at first glance, appear to be identical. The implicit derivatives in (7.48) include endogenous variables on the right-hand side, while the explicit derivatives in (7.45) are solely in terms of exogenous parameters. To show that the solutions are equivalent we substitute the solutions for x_1, x_2 and $|H|$ into (7.48):

$$\frac{dx_1}{dc_1}\bigg|_{dc_2=0} = \frac{-2c_2 \left(c_1^{-1/3} c_2^{-2/3}\right) \left(c_2^{-1/3} c_1^{-2/3}\right)^2}{3} = \left(\frac{-2}{3}\right) c_2^{-1/3} c_1^{-5/3} = \frac{\partial x_1^*}{\partial c_1} \quad \text{and}$$

$$\frac{dx_1}{dc_2}\bigg|_{dc_1=0} = \frac{-\left(c_1^{-1/3} c_2^{-2/3}\right)^2}{3} = \left(\frac{-1}{3}\right) c_2^{-4/3} c_1^{-2/3} = \frac{\partial x_1^*}{\partial c_2}. \tag{7.49}$$

The explicit and implicit comparative static results are identical, so that we are free to use either method in a particular problem. Since we are generally only interested in the qualitative results (i.e., signs of the derivatives), the form in equation (7.48) is fine. Only in the rare cases where quantitative magnitudes matter do we need to transform comparative static results into the form of equation (7.49).

---

[8]We could also implicitly differentiate with respect to one parameter at a time. The total differential approach, although a bit more work to set up initially, has the advantage of yielding all comparative static derivatives.

7.10   A SOLVED ECONOMIC EXAMPLE

We now have all the mathematical theory to solve the economic example that we began in Section 7.2. The final version of our profit function from equation (7.3) was

$$\Pi = (P - c)(\alpha - P) A^{\beta} - A. \tag{7.50}$$

The firm's profit maximizing choices are determined by the two first-order conditions for price and advertising. Taking the partial derivatives and setting them equal to zero gives

$$\frac{\partial \Pi}{\partial P} = (\alpha - P) A^{\beta} - (P - c) A^{\beta} = 0 \quad \text{and}$$

$$\frac{\partial \Pi}{\partial A} = \beta(P - c)(\alpha - P) A^{\beta-1} - 1 = 0. \tag{7.51}$$

Each first-order condition requires that, at a profit maximum, the marginal impact on profits of the relevant choice variable must be zero.

Note that the advertising term will drop out of the first-order condition for price, so that we get an immediate solution

$$P^* = \frac{\alpha + c}{2}. \tag{7.52}$$

To find the solution for A^* we substitute the price solution into the first-order condition for advertising. This yields

$$\frac{\partial \Pi}{\partial A} = \beta \left(\frac{\alpha + c}{2} - c \right) \left(\alpha - \frac{\alpha + c}{2} \right) A^{\beta-1} - 1 = 0 \quad \text{or}$$

$$\beta \left(\frac{\alpha - c}{2} \right) \left(\frac{\alpha - c}{2} \right) A^{\beta-1} = 1. \tag{7.53}$$

Rearranging and then solving gives A^* as

$$A^* = \left(\frac{\beta}{4} (\alpha - c)^2 \right)^{\frac{1}{1-\beta}}. \tag{7.54}$$

The next step is to verify that these solutions do in fact yield a profit maximum. The Hessian matrix for this problem is

$$H = \begin{pmatrix} \dfrac{\partial^2 \Pi}{\partial P^2} & \dfrac{\partial^2 \Pi}{\partial P \, \partial A} \\ \dfrac{\partial^2 \Pi}{\partial A \partial P} & \dfrac{\partial^2 \Pi}{\partial A^2} \end{pmatrix} = \begin{pmatrix} -2 A^{\beta} & \beta(\alpha - 2P + c) A^{\beta-1} \\ \beta(\alpha - 2P + c) A^{\beta-1} & \beta(\beta - 1)(P - c)(\alpha - P) A^{\beta-2} \end{pmatrix}. \tag{7.55}$$

Since we are interested in the behavior of the profit function when evaluated at P^* and A^*, we need only check the local concavity of the profit function at the solutions to the first-order conditions. This simplifies the Hessian because at P^* both cross-partial derivatives are zero. Thus,

$$H = \begin{pmatrix} -2 A^{\beta} & 0 \\ 0 & \beta(\beta - 1)(P - c)(\alpha - P) A^{\beta-2} \end{pmatrix}. \tag{7.56}$$

The second-order conditions for a maximum are

$$|H_1| = -2A^\beta < 0 \quad \text{and}$$

$$|H_2| = |H| = -2A^\beta \beta(\beta - 1)(P - c)(\alpha - P)A^{\beta-2} > 0. \tag{7.57}$$

The latter condition will hold only if $\beta < 1$. In economic terms this means that there must be diminishing marginal returns to advertising, i.e., that an increase in advertising leads to a less than proportional increase in quantity demanded.

The main comparative static results in this model are how price and advertising decisions respond to changes in the demand parameter α and the marginal cost parameter c. These derivatives are left as an end-of-chapter problem.

SUMMARY

This chapter has developed the mathematical theory for multivariable optimization problems. The new elements in the chapter were not so much the mathematical tools, but were, instead, how the tools from previous theory chapters—first- and second-order conditions, matrix algebra, and the implicit function theorem—could be combined to solve more complex mathematical models. Since optimization is at the center of microeconomic analysis, we are now ready to widen the set of economic situations that we can model and analyze. The next chapter presents a number of economic applications that use multivariable optimization. The reader should be aware, however, that even with the theory in this chapter there are still many examples of optimization that we are not yet prepared to cover; the mathematical theory developed here is directly applicable only to unconstrained optimization. The mathematics needed for situations where agents face constraints will be presented later in the text. Fortunately, we will see that many of the basic ideas and techniques from this chapter carry over quite directly to constrained optimization.

Problems

7.1 For the following functions find numerical values of x and y that satisfy the first-order conditions for a local extreme points. Check whether these extreme points are minima or maxima:

(a) $f(x, y) = 2x^2 + y^2 + 2xy - 2x - y$

(b) $f(x, y) = xy - 0.5x^2 - y^2 - 2y$

(c) $f(x, y) = 3\sqrt[3]{xy} - 2x - y$

(d) $f(x, y) = \ln(xy) - (x + y)^2$

(e) $f(x, y, z) = (2 - x)x + (1 - y)y + (3 - z)z - (x + y + z)^2$

7.2 For the following functions find values of x and y that satisfy the first-order conditions for local extreme points. Assume that a and b are positive parameters. Check whether the second-order conditions are sufficient to determine whether these extreme points are minima or maxima:

(a) $f(x, y) = ax - x^2 + by - y^2$

(b) $f(x, y) = ax - x^2 + by - y^2 - (x + y)^2$

(c) $f(x, y) = ax^2 + by^2 - xy - x - y$

(d) $f(x, y) = x^{0.5} y^{0.25} - ax - by$

(e) $f(x, y) = 3\ln(x - a) + \ln(y) - x - by$

7.3 For each of the solutions in Problem 7.2 that correspond to a local extreme point:

(a) find the partial derivatives of the solution for x with respect to the parameters a and b.

(b) find the partial derivatives of the solution for y with respect to the parameters a and b.

7.4 For the following functions find values of x, y, and z that satisfy the first-order conditions for local extreme points. Assume that a, b, and c are positive parameters. Check whether the second-order conditions are sufficient to determine whether these extreme points are minima or maxima:

(a) $f(x, y, z) = x + y + z - ax^2 - by^2 - c^2 z$

(b) $f(x, y, z) = x^{0.25} y^{0.25} + \ln(z - c) - ax - by - z$

(c) $f(x, y, z) = ax + by + cz - \ln(xy) - z^2$

(d) $f(x, y, z) = 4x^{0.25} y^{0.25} z^{0.25} - ax - by - cz$

(e) $f(x, y, z) = (a - x)x + (b - y)y + (c - z)z - (x + y + z)^2$

7.5 For each of the solutions in Problem 7.4 that correspond to a local extreme point:

(a) find the partial derivatives of the solution for x with respect to the parameters a, b, and c.

(b) find the partial derivatives of the solution for y with respect to the parameters a, b, and c.

(c) find the partial derivatives of the solution for z with respect to the parameters a, b, and c.

7.6 Show that if $f(\mathbf{x})$ is a linear function in n variables, then $f(\mathbf{x})$ is both concave *and* convex, but is neither strictly concave nor strictly convex.

7.7 Show that if a function of n variables, $f(\mathbf{x})$, is concave, then the negative of the function, i.e., $-f(\mathbf{x})$, is convex.

7.8 Show that if two functions of n variables, $f(\mathbf{x})$ and $g(\mathbf{x})$, are both strictly concave, then the function $h(\mathbf{x}) = f(\mathbf{x}) + g(\mathbf{x})$ is also strictly concave.

7.9 For the following functions find the first- and second-order conditions for a maximum and find the change in x^* for a change in the parameter a. [Hint: Use the implicit function theorem.]

(a) $f(x, y) = x^2 y^2 - a(x + y)^3$

(b) $f(x, y) = \ln(x + y) - x^2 - y^2 + x + ay$

(c) $f(x, y) = \sqrt[3]{xy} - x^2 - y^2$

(d) $f(x, y) = xy(a - x - y)$

(e) $f(x, y) = 2x - x^2 + ay - y^2 - (x + y)^3$

7.10 For the economic model in Section 7.10:

    (a)   Find the comparative static effects of a change in the marginal cost parameter, c, by implicitly differentiating the first-order conditions (i.e., do not use the explicit solutions for P^* and A^*).

    (b)   Find the comparative static effects of a change in the demand parameter, α, by implicitly differentiating the first-order conditions (i.e., do not use the explicit solutions for P^* and A^*).

    (c)   Verify your answers above by taking the partial derivatives of the explicit solution functions P^* and A^*.

8

Multivariable Optimization without Constraints: Applications

8.1   INTRODUCTION

Each of the earlier applications chapters has included examples of single variable economic optimization problems. Few economic agents, however, face such a simple environment that they need to make only one choice. In reality most agents must make many choices and indeed must consider the tradeoffs between multiple choice variables. This chapter provides several economic examples in which agents optimize over more than one choice variable.

Most of the applications in this chapter are drawn from the theory of the firm. There is a reason for this: consumer choices are generally constrained (by the consumer's income), whereas many, but certainly not all, choices by firms are unconstrained. In later chapters we will see examples of constrained optimization by both consumers and firms.

In Section 2.2 we examined the profit maximizing labor input choice of a perfectly competitive firm. At that point we were forced to limit the model to the firm's short-run choice because we did not yet have the tools to consider the multiple input choices that a firm faces in the long run. The first two applications in our chapter extend the earlier model to the long run. We present one model with a specific production technology: the Cobb-Douglas production function. We then consider the general case in which the production function is left unspecified.

The next application covers the efficiency wage model of labor demand. The efficiency wage theory is a theory that explains why firms may pay above-market wages in order to encourage workers to put forth more effort and be more productive. The model has important macroeconomic implications: it can help to explain why firms fail to lower wages in the face of excess labor supply and unemployment.

The next two applications cover firm decision making with multiple output choices. We cover both multiplant firms, where production levels at different factories are chosen simultaneously, and multimarket firms, where firms must choose what levels of output to sell in different markets.

We then turn to an example from environmental economics and model the case of a firm subject to a tax on pollution emissions. We examine the effect of the tax on both pollution control activities and the firm's output.

The final application in this chapter is quite different from the rest. Rather than illustrating another example of choices by economic agents, we instead show how economists use mathematical optimization in applied statistical work. Most statistical work in economics uses some form of regression equations—equations that use data to estimate the relationships between economic variables. Our application shows how optimization analysis is used in deriving the best formulas for statistical estimators in regression equations.

8.2   COMPETITIVE FIRM INPUT CHOICES: COBB–DOUGLAS TECHNOLOGY

One way of writing a profit function is revenue from output minus the costs of hiring inputs,

$$\Pi = TR - TC = PQ - wL - rK. \tag{8.1}$$

where P is price, Q is output, L is labor, K is capital, and w and r are the prices of L and K. For a competitive price-taking firm, L, K, and Q are endogenous choice variables, while P, w, and r are exogenous variables. The formulation in equation (8.1) is not very useful by itself. The three endogenous variables are not independent: output depends on labor and capital. The technical relationship between inputs and output is captured by a production function of the form

$$Q = F(L, K). \tag{8.2}$$

One often-used example of a production technology is the Cobb-Douglas production function

$$Q = L^\alpha K^\beta, \tag{8.3}$$

where α and β are positive parameters. The Cobb-Douglas function is widely used because of its analytical tractability. In this section we simplify the presentation by working with the special case of $\alpha = \beta$,

$$Q = L^\alpha K^\alpha. \tag{8.4}$$

The more general (and algebraically messier) case of $\alpha \neq \beta$ is covered in a problem at the end of the chapter.

Substituting the Cobb-Douglas production technology into equation (8.1) we can write the firm's profits as a function of two endogenous variables, L and K,

$$\Pi = PL^\alpha K^\alpha - wL - rK. \tag{8.5}$$

The first-order conditions for a profit maximum are

$$\Pi_L = P\alpha L^{\alpha-1} K^\alpha - w = 0 \quad \text{and} \quad \Pi_K = P\alpha L^\alpha K^{\alpha-1} - r = 0. \tag{8.6}$$

The first term in the each equation is the change in revenue for a change in the relevant input. This change in revenue contains two terms: the output price is multiplied by the marginal product of the input. This combination is often called the value of the marginal product or VMP, and the economic interpretation of the first-order condition for labor is

$$VMP_L = w. \tag{8.7}$$

This equation is yet another example of the microeconomic rule that marginal benefit (the revenue generated from hiring an extra unit of labor) must equal marginal cost (the wage rate paid for an extra unit of labor).

For the second-order conditions we need to find the Hessian matrix, which was defined in Chapter 7. For this profit maximization problem the Hessian matrix is

$$H = \begin{pmatrix} \Pi_{LL} & \Pi_{LK} \\ \Pi_{KL} & \Pi_{KK} \end{pmatrix} = \begin{pmatrix} P F_{LL} & P F_{LK} \\ P F_{KL} & P F_{KK} \end{pmatrix} \quad \text{or}$$

$$H = \begin{pmatrix} P\alpha(\alpha-1)L^{\alpha-2}K^\alpha & P\alpha^2 L^{\alpha-1}K^{\alpha-1} \\ P\alpha^2 L^{\alpha-1}K^{\alpha-1} & P\alpha(\alpha-1)L^\alpha K^{\alpha-2} \end{pmatrix}.$$

(8.8)

The two second-order conditions are

$$|H_1| = P F_{LL} = P\alpha(\alpha-1)L^{\alpha-2}K^\alpha < 0 \quad \text{and}$$

$$|H_2| = |H| = P^2 F_{LL}F_{KK} - P^2 F_{LK}F_{KL}$$

$$= P^2\alpha^2(\alpha-1)^2 L^{2\alpha-2}K^{2\alpha-2} - P^2\alpha^4 L^{2\alpha-2}K^{2\alpha-2} > 0.$$

(8.9)

Both second-order conditions have economic interpretations. $|H_1|$ is price times the change in the marginal product of labor for a change in labor. $|H_1|$ will be negative when $\alpha < 1$, or, in economic terms, when increases in labor cause the marginal product of labor, F_L, to decline. A declining marginal product of labor is referred to (in what should be a familiar term) as diminishing marginal returns to labor.

The interpretation of $|H|$ requires a bit more work. To begin, note that we can rewrite $|H|$ in successive steps as:

$$|H| = P^2\alpha^2 L^{2\alpha-2}K^{2\alpha-2}((\alpha-1)^2 - \alpha^2) > 0$$

$$|H| = P^2\alpha^2 L^{2\alpha-2}K^{2\alpha-2}(\alpha^2 - 2\alpha + 1 - \alpha^2) > 0$$

(8.10)

$$|H| = P^2\alpha^2 L^{2\alpha-2}K^{2\alpha-2}(1 - 2\alpha) > 0.$$

Thus, the second-order condition for $|H|$ holds when $2\alpha < 1$, or $\alpha < 1/2$, which is more restrictive than the second-order condition, $\alpha < 1$, for $|H_1|$. The key to interpreting this second-order condition lies in examining returns to scale.[1] Returns to scale are determined by the percentage change in output when *all* inputs are simultaneously increased by $x\%$. The three cases are increasing (output rises by more than $x\%$), constant (output rises by exactly $x\%$), and decreasing (output rises by less than $x\%$).

Mathematically, returns to scale depend on the degree of homogeneity of the production function. Recall from Chapter 5 that a production function is homogenous of degree k when

$$\lambda^k Q = F(\lambda L, \lambda K),$$

(8.11)

for all $\lambda > 0$. Returns to scale are increasing for $k > 1$, constant for $k = 1$, and decreasing for $k < 1$. For our version of the Cobb-Douglas production function we have

$$\lambda^k Q = (\lambda L)^\alpha(\lambda K)^\alpha = \lambda^{2\alpha}L^\alpha K^\alpha = \lambda^{2\alpha}Q \quad \text{or} \quad k = 2\alpha.$$

(8.12)

The second-order condition of $2\alpha < 1$ therefore corresponds to decreasing returns to scale.

What do the second-order conditions requiring diminishing marginal returns to labor and decreasing returns to scale mean?[2] The easiest way to answer this is to use

---

[1] The material below is a brief review of Section 5.6. See Section 5.6 of Chapter 5 for more detail on homogeneity and returns to scale.

[2] Note that since the ordering of the input variables is arbitrary the two second-order conditions also imply that there will be diminishing marginal returns to capital.

FIGURE 8.1

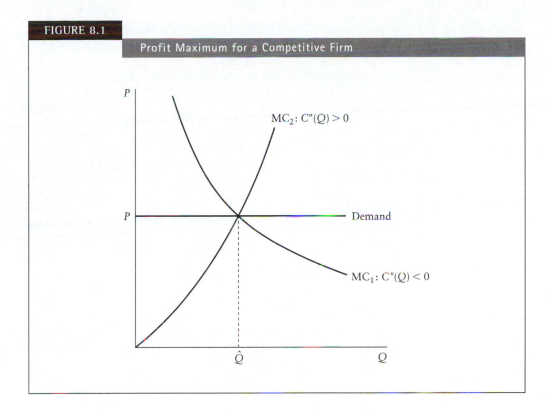

Profit Maximum for a Competitive Firm

the microeconomic theory relating production and cost. Diminishing marginal returns to labor imply an upward-sloping short-run marginal cost curve: each additional unit of output requires an increasing increment of labor input and therefore creates a higher marginal cost. In the long run, decreasing returns to scale imply increasing marginal capital and labor input requirements for additional output, and hence an upward-sloping long-run marginal cost curve. For a given market price, the profit maximizing condition for a competitive firm (price equal to marginal cost) requires that the marginal cost curve cuts the marginal revenue (i.e., price) curve from below.[3] Figure 8.1 shows two marginal cost curves. Each curve has price equal to marginal cost at an output level $\hat{Q}$. For the downward-sloping marginal cost curve $\hat{Q}$ fails the second-order condition: the firm could improve profits by expanding output beyond $\hat{Q}$. For the upward-sloping marginal cost curve $\hat{Q}$ satisfies the second-order conditions: any change in output would lead to a lower profit.

Proceeding under the assumption that the second-order conditions hold, we are now ready to solve the first-order conditions to find the optimal choices of labor and capital. Since the equations are nonlinear we solve by substitution. We first solve, in steps, for the value of K implied by the $\Pi_L = 0$ equation:

$$P\alpha L^{\alpha-1}K^{\alpha} = w$$

$$K^{\alpha} = \frac{w}{P\alpha}L^{1-\alpha} \tag{8.13}$$

$$K = \left(\frac{w}{P\alpha}L^{1-\alpha}\right)^{1/\alpha}.$$

---

[3]The equivalent conditions for a monopoly firm's second-order condition were covered in Section 6.6.

Note carefully that the solution for K in the last line is *not* the reduced form solution (K^*) for our model since the last line includes L on the right-hand side, whereas the solution K^* should be solely a function of the exogenous variables.

To find L^* we next substitute for K in the $\Pi_K = 0$ equation. This yields

$$P\alpha L^\alpha \left(\left(\frac{w}{P\alpha} L^{1-\alpha} \right)^{1/\alpha} \right)^{\alpha-1} = r. \tag{8.14}$$

Rearranging and then combining exponents gives

$$P\, P^{(1-\alpha)/\alpha}\alpha\alpha^{(1-\alpha)/\alpha}w^{(\alpha-1)/\alpha}L^\alpha L^{(1-\alpha)(\alpha-1)/\alpha} = r$$

$$P^{\alpha/\alpha}P^{(1-\alpha)/\alpha}\alpha^{\alpha/\alpha}\alpha^{(1-\alpha)/\alpha}w^{(\alpha-1)/\alpha}L^{\alpha^2/\alpha}L^{(-1+2\alpha-\alpha^2)/\alpha} = r \tag{8.15}$$

$$P^{1/\alpha}\alpha^{1/\alpha}w^{(\alpha-1)/\alpha}L^{(-1+2\alpha)/\alpha} = r$$

$$L^{(-1+2\alpha)/\alpha} = P^{-1/\alpha}\alpha^{-1/\alpha}w^{(1-\alpha)/\alpha}r.$$

With some algebraic manipulation, the solution for L^* is[4]

$$L^* = (\alpha P w^{\alpha-1} r^{-\alpha})^{1/(1-2\alpha)}. \tag{8.16}$$

We can find K^* by substituting L^* into equation (8.13). Alternatively, we could use the symmetry between the L and K variables: the profit function has L and K raised to the same power and then the cost of each input is subtracted. Using symmetry means that we take the solution for L^* and interchange the input prices to find K^*. The reader can verify that either method gives

$$K^* = (\alpha P r^{\alpha-1} w^{-\alpha})^{1/(1-2\alpha)}. \tag{8.17}$$

These solutions for L^* and K^* also allow us to find the firm's supply function, i.e., quantity produced as a function of the output price and the input prices. To find this solution we substitute L^* and K^* into the production function and simplify:

$$Q^* = L^{*\alpha} K^{*\alpha}$$

$$Q^* = (\alpha P w^{\alpha-1} r^{-\alpha})^{\alpha/(1-2\alpha)} (\alpha P r^{\alpha-1} w^{-\alpha})^{\alpha/(1-2\alpha)} \tag{8.18}$$

$$Q^* = (\alpha^2 P^2 w^{-1} r^{-1})^{\alpha/(1-2\alpha)}$$

From these solutions we can find the comparative static effects of changes in P, w, and r. There are several ways to take the comparative static partial derivatives. The easiest method is, for each particular exogenous variable under consideration, to think of the equations for L^* and Q^* as a group of constant terms times the particular exogenous variable raised to a power. The derivative can then be taken using the power rule. For L^*, from equation (8.16), the power rule gives the following partial derivatives:

$$\frac{\partial L^*}{\partial P} = (\alpha w^{\alpha-1} r^{-\alpha})^{1/(1-2\alpha)} \left(\frac{1}{1-2\alpha} \right) P^{[1/(1-2\alpha)]-1} > 0,$$

$$\frac{\partial L^*}{\partial w} = (\alpha P r^{-\alpha})^{1/(1-2\alpha)} \left(\frac{\alpha-1}{1-2\alpha} \right) w^{[(\alpha-1)/(1-2\alpha)]-1} < 0, \quad \text{and} \tag{8.19}$$

$$\frac{\partial L^*}{\partial r} = (\alpha P w^{\alpha-1})^{1/(1-2\alpha)} \left(\frac{-\alpha}{1-2\alpha} \right) r^{[-\alpha/(1-2\alpha)]-1} < 0.$$

---

[4]Students will discover that there are often a variety of (mathematically equivalent) ways to express reduced form solutions. In general it is best to simplify right-hand side expressions as much as possible. Try to avoid writing $(1/(3+2)^2)^{-1/2}$ when the answer is 5.

Note that it is the second-order condition for $|H|$, which requires $(1 - 2\alpha) > 0$, that allows us to sign each of these derivatives. The results indicate that L^* is increasing in the output price, but decreasing in both input prices. Since L^* and K^* are symmetric, similar comparative static outcomes can also be derived for K^*.

For output, Q^* in equation (8.18), the comparative static derivatives are

$$\frac{\partial Q^*}{\partial P} = (\alpha^{2\alpha} w^{-\alpha} r^{-\alpha})^{1/(1-2\alpha)} \left(\frac{2\alpha}{1 - 2\alpha} \right) P^{2\alpha/(1-2\alpha)-1} > 0,$$

$$\frac{\partial Q^*}{\partial w} = (\alpha^{2\alpha} r^{-\alpha} P^{2\alpha})^{1/(1-2\alpha)} \left(\frac{-\alpha}{1 - 2\alpha} \right) w^{-\alpha/(1-2\alpha)-1} < 0, \qquad \text{(8.20)}$$

$$\frac{\partial Q^*}{\partial r} = (\alpha^{2\alpha} w^{-\alpha} P^{2\alpha})^{1/(1-2\alpha)} \left(\frac{-\alpha}{1 - 2\alpha} \right) r^{-\alpha/(1-2\alpha)-1} < 0.$$

Thus, increases in the product's price lead to increases in output, while increases in input prices lead to decreases in output. The derivations of input demand and output supply elasticities are covered as an end-of-chapter problem.

These results (the signs of the derivatives) are unambiguous. Yet, we must keep in mind that the results were derived using a specific example of a production function. We cannot jump from these conclusions to blanket generalizations about how a competitive firm's input and output decisions respond to changes in prices. The next section examines the extent to which our specific results for the Cobb-Douglas case hold for unrestricted production functions.

8.3 COMPETITIVE FIRM INPUT CHOICES: GENERAL PRODUCTION TECHNOLOGY

This section examines a model of a profit-maximizing competitive firm, but instead of using a specific production function we leave the production function as

$$Q = F(L, K). \qquad \text{(8.21)}$$

The profit function is

$$\Pi = PF(L, K) - wL - rK, \qquad \text{(8.22)}$$

where w and r are the prices of labor and capital. The advantage of using this general form is that any results will apply to all possible specific examples of production functions. The disadvantage is that some results that apply in many reasonable cases may not be derivable in a general model that subsumes all possibilities.

The first-order conditions for a profit maximum are

$$\Pi_L = PF_L(L, K) - w = 0$$

and   (8.23)

$$\Pi_K = PF_K(L, K) - r = 0.$$

In economic terms both first-order conditions specify that for a profit maximum the value of the marginal product (i.e., price times marginal product) for each input must equal the price of the input.

The Hessian matrix for this problem is

$$H = \begin{pmatrix} \Pi_{LL} & \Pi_{LK} \\ \Pi_{KL} & \Pi_{KK} \end{pmatrix} = \begin{pmatrix} PF_{LL} & PF_{LK} \\ PF_{KL} & PF_{KK} \end{pmatrix}, \tag{8.24}$$

and the two second-order conditions are

$$|H_1| = PF_{LL} < 0$$

and $\hspace{8cm}$ (8.25)

$$|H_2| = |H| = P^2\left(F_{LL}F_{KK} - F_{LK}^2\right) > 0.$$

The first condition requires a diminishing marginal product of labor (F_L is the marginal product of labor and F_{LL} is the slope of that marginal product curve). A diminishing marginal product of labor is equivalent to an upward-sloping short-run marginal cost curve. The second condition ensures that the firm's long-run marginal cost curve is upward sloping.[5] Also, note that the two conditions, when viewed in combination, imply diminishing marginal returns to capital, i.e., $F_{KK} < 0$. Since the second-order conditions are sufficient to ensure that the production function is strictly concave (see Section 7.6), we have a direct link between concavity and the economic properties of the production function.

It is not possible to solve for L^* and K^*; therefore, the only avenue for comparative static results is through the implicit function theorem. Taking the total differentials of the two first-order conditions in equation (8.23) and writing the result in matrix form yields

$$\begin{pmatrix} PF_{LL} & PF_{LK} \\ PF_{KL} & PF_{KK} \end{pmatrix} \begin{pmatrix} dL \\ dK \end{pmatrix} = \begin{pmatrix} dw - F_L\,dP \\ dr - F_K\,dP \end{pmatrix}. \tag{8.26}$$

The solutions for dL and dK are given by Cramer's rule as

$$dL = \frac{\begin{vmatrix} dw - F_L\,dP & PF_{LK} \\ dr - F_K\,dP & PF_{KK} \end{vmatrix}}{|H|} = \frac{PF_{KK}\,dw - PF_{LK}\,dr + (F_K PF_{LK} - F_L PF_{KK})\,dP}{|H|}$$

and $\hspace{8cm}$ (8.27)

$$dK = \frac{\begin{vmatrix} PF_{LL} & dw - F_L\,dP \\ PF_{LK} & dr - F_K\,dP \end{vmatrix}}{|H|} = \frac{-PF_{LK}\,dw + PF_{LL}\,dr + (F_L PF_{LK} - F_K PF_{LL})\,dP}{|H|}.$$

From these solutions we can solve for the partial effects of changes in w, r, and P. Since L and K have symmetric solutions we focus just on the derivatives of K. These

---

[5]The proof of this requires using constrained optimization techniques that will be introduced in the next chapter.

derivatives are:

$$\frac{\partial K^*}{\partial r} = \frac{dK}{dr}\bigg|_{dw=dP=0} = \frac{P F_{LL}}{|H|} = \frac{|H_1|}{|H|} < 0,$$

$$\frac{\partial K^*}{\partial w} = \frac{dK}{dw}\bigg|_{dr=dP=0} = \frac{-P F_{LK}}{|H|}, \quad \text{and} \tag{8.28}$$

$$\frac{\partial K^*}{\partial P} = \frac{dK}{dP}\bigg|_{dr=dw=0} = \frac{(F_L P F_{LK} - F_K P F_{LL})}{|H|}.$$

Since the second-order conditions place no sign restriction on F_{LK}, only the first of our derivatives has a definite sign: input demand will decline with an increase in the input's price. We are unable to provide a definite prediction on the effects of an increase in the price of the other input or of an increase in output price. We might think it is reasonable to expect that labor and capital are complements in production so that an increase in capital increases the marginal productivity of labor, i.e., $F_{LK} > 0$.[6] In this case we would have

$$\frac{dK}{dw}\bigg|_{dr=dP=0} < 0$$

and $\hspace{10cm}$ (8.29)

$$\frac{dK}{dP}\bigg|_{dr=dw=0} > 0.$$

Nevertheless, this assumption is by no means necessary (especially in a model with more than two inputs) and the results will not hold for all production functions.

We next turn to the comparative static effects on output. From the production function, equation (8.21), we get the total differential for output as a function of changes in L and K

$$dQ = F_L \, dL + F_K \, dK. \tag{8.30}$$

Substituting the solutions for dL and dK and then rearranging yields

$$dQ = F_L \left[\frac{P F_{KK} \, dw - P F_{LK} \, dr + (F_K P F_{LK} - F_L P F_{KK}) \, dP}{|H|} \right]$$
$$+ F_K \left[\frac{-P F_{LK} \, dw + P F_{LL} \, dr + (F_L P F_{LK} - F_K P F_{LL}) \, dP}{|H|} \right]$$

or $\hspace{10cm}$ (8.31)

$$dQ = \frac{1}{|H|} \big[(F_L P F_{KK} - F_K P F_{LK}) \, dw + (-F_L P F_{LK} + F_K P F_{LL}) \, dr$$
$$+ (2 F_L F_K P F_{LK} - F_K^2 P F_{LL} - F_L^2 P F_{KK}) \, dP \big].$$

Here we find that, because the sign of F_{LK} is indeterminate, the effects of changes in w or r on output cannot be signed. As shown, however, complementary inputs, $F_{LK} > 0$, would be a sufficient condition for increases in input prices to cause decreases

---

[6]Other inputs may be substitutes in production, $F_{LK} < 0$, which would yield opposite results for dk/dw.

in output. For the effect of changes in output price we get:

$$\frac{\partial Q^*}{\partial P} = \left.\frac{dQ}{dP}\right|_{dr=dw=0} = \frac{P F_L F_K (2 F_L F_K F_{LK} - F_K^2 F_{LL} - F_L^2 F_{KK})}{|H|}. \qquad \textbf{(8.32)}$$

The second-order condition of $|H| > 0$ implies that the term in parentheses in the numerator must be positive as well, so that increases in price definitely lead to increases in output. The algebra for proving this is left as an exercise for the reader.[7]

8.4 EFFICIENCY WAGES

Efficiency wage theory is an attempt to explain the macroeconomic phenomenon of sticky wages in the face of unemployment. Why, for example, do firms turn away qualified job applicants who are willing to work for less than the firm's current wage offer? Wouldn't lowering wages increase the firm's profits? The basic insight of efficiency wage theory is that worker on-the-job effort and productivity are positively associated with the wages paid by the firm. Thus, wage reductions, which would reduce unemployment and allow the labor market to clear, may result in profit-reducing productivity declines.

Before we turn to a firm's wage choice it is useful to consider why wages and effort levels might be positively correlated. Consider a very simple model where workers either put forth a positive effort level, $e > 0$, or shirk, $e = 0$. Firms can detect shirking imperfectly. Let x be the probability that a shirking worker is caught and fired. Finally, suppose that an employed worker earns the wage, w, offered by the firm, and a fired worker gains employment elsewhere at an alternative wage $\bar{w}$.

Assume that a worker's net utility is her wage minus the level of effort expended. The utility for a worker who puts forth effort is

$$U^e = w - e. \qquad \textbf{(8.33)}$$

The expected utility for a worker who shirks is[8]

$$U^s = (1 - x)w + x\bar{w}. \qquad \textbf{(8.34)}$$

In order for the firm to induce effort it must offer a wage rate, w, such that $U^e \geq U^s$. This equation is often referred to as the "no shirking condition." Mathematically, the no shirking condition requires

$$w - e \geq (1 - x)w + x\bar{w}, \quad \text{or}$$
$$w \geq \bar{w} + \frac{e}{x}, \quad \text{or} \qquad \textbf{(8.35)}$$
$$w \geq \bar{w} + e + \frac{(1 - x)e}{x}.$$

The last line in equation (8.35) shows that to induce effort the firm must pay more than the alternative wage plus the cost of effort. If the firm only paid $\bar{w} + e$ then the worker would not put forth effort, both because shirking is imperfectly detected and because the opportunity cost to getting caught shirking would be zero. The solution for w also

---

[7]Hint: $|H| > 0$ requires that either F_{LL} or F_{KK} or both be greater in absolute value than F_{LK} and that their product be greater than $(F_{LK})^2$.

[8]Expected values, used when outcomes are probabilistic, are found by weighting the values of each particular outcome by the probability that the particular outcome occurs and summing over all possible outcomes.

implies that increases in x, the probability that shirking is detected, lead to decreases in the wage premium, $w - \bar{w}$, required to induce effort. In essence, a firm can induce effort through either a carrot approach (higher wages) or a stick approach (more careful monitoring).

Having explained how higher wages induce effort, we now turn to a more general model in which effort choices are continuous rather than dichotomous. We will simply assume that the level of worker effort is an increasing function of the wage rate:

$$e = e(w). \tag{8.36}$$

We will also assume that the effort function is concave, i.e., $(de/dw) > 0$ and $(d^2e/dw^2) < 0$.

For the firm we will consider a short-run production function with fixed capital. Output as a function of labor is:

$$Q = F(L), \tag{8.37}$$

where L, the effective labor input, is the number of workers, N, times the effort level per worker. The production function can then be written as

$$Q = F(Ne(w)). \tag{8.38}$$

If we let $R(Q)$ be the firm's revenue function, profits are:

$$\Pi = R(F(Ne(w))) - wN \tag{8.39}$$

Since the firm's profits depend on the number of workers and on the wage rate (which determines effort), we must take first-order conditions with respect to both choice variables. These first-order conditions are

$$\Pi_w = R'F'N\frac{de}{dw} - N = 0 \quad \text{or} \quad R'F'\frac{de}{dw} = 1$$

and $\tag{8.40}$

$$\Pi_N = R'F'e - w = 0 \quad \text{or} \quad R'F'e = w.$$

The second-order conditions are left as an end-of-chapter problem.

By combining the two first-order conditions we find that

$$\frac{de}{dw}\frac{w}{e} = 1. \tag{8.41}$$

The left-hand side of this equation is in the form of an elasticity, i.e., it gives the percentage change in effort for a percentage change in the wage rate. The firm's profit maximizing wage rate must therefore be chosen so that the effort function is unit elastic with respect to changes in the wage rate.

Although there is no easy economic intuition for the unit elasticity result, there are important economic implications of the result that are worth noting. First, the result is quite general: because we left the production function and revenue function unspecified, the result applies across any production technology or market structure. Second, there is no particular reason why the firm's wage choice should be identical to the market-clearing wage in the labor market. Thus, profit-maximizing microeconomic choices do not guarantee macroeconomic full employment.

Several aspects of the efficiency wage model are explored further in the end-of-chapter problems.

8.5 A MULTIPLANT FIRM

In this section we examine the output choices of a firm that operates a number, n, of separate factories. Each factory is characterized by a cost function that depends on the output level at that factory. We will write the cost function for each of the firm's n factories as

$$TC_i = C_i(q_i) \quad \text{for } i = 1, \ldots, n \tag{8.42}$$

where q_i denotes output at factory i and the subscripts on the cost functions indicate that those functions need not be identical.

We will assume that the firm sells its output in a single market. This means that total revenue depends on the total level of output, or

$$TR = R(Q), \tag{8.43}$$

where $R(Q) = P(Q)Q$ and $Q = \sum_{i=1}^{n} q_i$. For a competitive firm price, or $P(Q)$, would simply be a constant function, i.e., the (exogenous) market price would not depend on the firm's output. For a monopoly $P(Q)$ would be the downward-sloping market demand function. Our analysis can therefore encompass both types of market structure.

The firm's profit function can be written as

$$\Pi = R(Q) - \sum_{i=1}^{n} C_i(q_i) = P(Q)Q - \sum_{i=1}^{n} C_i(q_i). \tag{8.44}$$

For profit maximization there will be n first-order conditions, one for each factory's output. Letting Π_i be the partial derivative of profits with respect to output from factory i, the first-order conditions are

$$\frac{\partial \Pi}{\partial q_i} = R'(Q) - C_i'(q_i) = P'(Q)Q + P(Q) - C_i'(q_i) = 0 \quad \text{for } i = 1, \ldots, n. \tag{8.45}$$

The derivative of revenue uses the chain rule: a change in output from factory i changes Q by one unit and the change in Q changes revenue. All first-order conditions take the form of marginal revenue (in the single market) and must equal marginal cost at each individual factory.

An immediate result from the first-order conditions is

$$C_1'(q_1) = C_2'(q_2) = \cdots = C_n'(q_n) = R'(Q). \tag{8.46}$$

This means that profit maximization requires that the levels of output be chosen so that all factories operate at the same level of marginal cost and that this common marginal cost should equal the marginal revenue from the factories' combined output. Note in particular that there is no presumption (except in the case where cost functions are identical) that output levels will be the same.

For the second-order conditions we find the Hessian matrix:

$$H = \begin{pmatrix} \Pi_{11} & \Pi_{12} & . & . & . & \Pi_{1n} \\ \Pi_{21} & \Pi_{22} & . & . & . & \Pi_{2n} \\ . & . & . & . & . & . \\ . & . & . & . & . & . \\ . & . & . & . & . & . \\ \Pi_{n1} & \Pi_{n2} & . & . & . & \Pi_{nn} \end{pmatrix} = \begin{pmatrix} R'' - C_1'' & R'' & . & . & . & R'' \\ R'' & R'' - C_2'' & . & . & . & R'' \\ . & . & . & . & . & . \\ . & . & . & . & . & . \\ . & . & . & . & . & . \\ R'' & R'' & . & . & . & R'' - C_n'' \end{pmatrix},$$

$$\tag{8.47}$$

where all elements on the main diagonal take the form $R'' - C_i''$, all elements off the diagonal take the form R'' (since costs at factory i depend on output at factory i only), and $R'' = P''(Q)Q + 2P'(Q)$.

The second-order conditions require that the n leading principal minors of H must alternate in sign. We write out the first and second leading principal minors and then give the rule for higher-order leading principal minors:

$$|H_1| = R'' - C_1'' < 0,$$
$$|H_2| = (R'' - C_1'')(R'' - C_2'') - R''R'' = C_1''C_2'' - R''(C_1'' + C_2'') > 0, \quad \text{and} \quad \textbf{(8.48)}$$
$$(-1)^i|H_i| > 0 \quad \text{for } i = 3, \ldots, n.$$

For perfect competition the interpretation of these conditions for a profit maximum is straightforward. In perfect competition price is exogenous for an individual firm. This means that $P'(Q)$, $P''(Q)$ and R'' are all zero. Thus the second-order conditions require $C_i'' > 0$, upward-sloping marginal cost, for every factory.

For monopoly, the $|H_1| < 0$ second-order condition only requires that the downward-sloping marginal revenue curve be steeper than the marginal cost curve for factory 1 at a profit maximum. In particular, $|H_1| < 0$ does not rule out the possibility of downward-sloping marginal cost. Nevertheless, the second-order conditions will not hold if all of the marginal cost curves slope downward. As an exercise the reader should prove that if marginal cost is downward-sloping at both factory 1 and factory 2 then the $|H_2| > 0$ condition must fail.[9] Why? Because if there are economies of scale then the cheapest production method is to operate a single factory at a high volume of output (and low level of cost) rather than operating two factories at lower volumes and higher costs. In other words, with economies of scale the monopolist should concentrate production at a single factory and close the rest.

8.6   MULTIMARKET MONOPOLY

In this section we examine a firm that produces in a single factory, but sells in two separate markets. We assume that the firm is a monopolist in each market.

In our model we will assume that the two markets have revenue functions that are similar, but that the first market is "more important" to the firm in the sense that, if equal quantities were sold in both markets, total revenue will be higher in the first market. Formally we will write the revenue functions for markets 1 and 2 as

$$TR_1 = \alpha R_1(q_1) = \alpha P(q_1)q_1$$

and

$$\textbf{(8.49)}$$

$$TR_2 = R_2(q_2) = P(q_2)q_2,$$

where $\alpha > 1$. We have written the demand equations as if, for any quantity that yields a positive price, consumers in the first market are willing to pay a higher price ($\alpha > 1$) than the price that consumers in the second market are willing to pay. Thus, α is a parameter that rotates D_1, the market 1 demand curve, upward around a fixed horizontal intercept. This is shown in Figure 8.2.

Before we proceed we need to understand the notation in equation (8.49). R_1 and R_2 are both functions of one variable, q_1 and q_2 respectively, but have identical functional forms. We have chosen to place the subscripts on the functions as a notational convenience to indicate which market is being referred to. In the analysis below when

[9]Hint: Use the expression for $|H_1| < 0$.

FIGURE 8.2

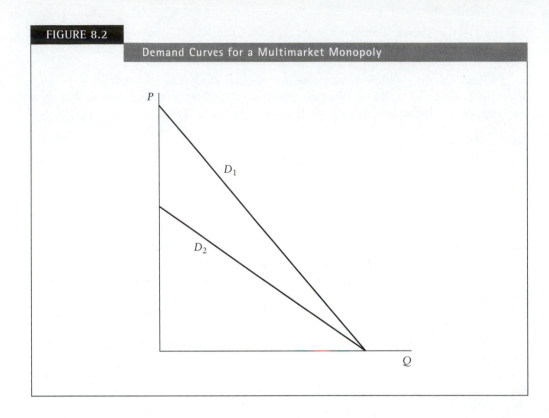

we write R_1' this will be the derivative of the revenue function, R_1, with respect to q_1. Similarly, R_2' will be the derivative of the revenue function, R_2, with respect to q_2.

On the cost side we will assume that total production costs depend on total output, Q, which is the sum of the outputs for the individual markets. We will also assume that the firm incurs an additional cost of $\$t$/unit when selling in the second market, but that there are no extra costs for selling in the first market. The parameter t might represent a transport cost, but it could also represent a tax or tariff that applies only in the second market. Total costs can be written as

$$TC = C(Q) + tq_2, \tag{8.50}$$

where $Q = q_1 + q_2$. We will not put any restrictions on the slope of the marginal cost curve, i.e., $C''(Q)$ may be positive, negative, or zero. Later in the model we will see that some comparative static results depend on whether marginal cost slopes up or down.

Given these assumptions about the revenue and cost functions, the profit function for the firm is

$$\Pi = \alpha R_1(q_1) + R_2(q_2) - C(Q) - tq_2 \quad \text{or}$$
$$\Pi = \alpha P(q_1)q_1 + P(q_2)q_2 - C(q_1 + q_2) - tq_2. \tag{8.51}$$

Let Π_i be the partial derivative of profits with respect to q_i. The first-order conditions for profit maximization are

$$\Pi_1 = \alpha R_1'(q_1) - C'(Q) = \alpha P'(q_1)q_1 + \alpha P(q_1) - C'(Q) = 0$$

and $\tag{8.52}$

$$\Pi_2 = R_2'(q_2) - C'(Q) - t = P'(q_2)q_2 + P(q_2) - C'(Q) - t = 0.$$

Each first-order condition states that the marginal revenue from selling in a market must equal the marginal cost of serving that market. Since marginal revenue is higher and marginal cost lower in market 1, the quantity sold will be higher in market 1.

The Hessian matrix for this problem is

$$H = \begin{pmatrix} \alpha R_1'' - C'' & -C'' \\ -C'' & R_2'' - C'' \end{pmatrix}, \tag{8.53}$$

where $\alpha R_1'' = \alpha P''(q_1)q_1 + 2\alpha P'(q_1)$ and $R_2'' = P''(q_2)q_2 + 2P'(q_2)$. This yields the two second-order conditions:

$$|H_1| = (\alpha R_1'' - C'') < 0 \quad \text{and}$$
$$|H_2| = |H| = (\alpha R_1'' - C'')(R_2'' - C'') - C''^2 \tag{8.54}$$
$$= \alpha R_1'' R_2'' - \alpha R_1'' C'' - R_2'' C'' > 0.$$

Since marginal revenue is, by assumption, downward-sloping ($R'' < 0$) the condition on $|H_1|$ is automatically satisfied for constant ($C'' = 0$) or upward-sloping ($C'' > 0$) marginal cost curves. For downward-sloping marginal cost curves the requirement is that marginal revenue be steeper than marginal cost. A similar economic interpretation applies to $|H|$.

We now turn to the comparative static effects of changes in the parameters. To find the effects of a change in demand in market 1, or of a change in t, we can take the total differentials of the first-order conditions. Note first that we can write the first-order conditions in equation (8.52) as

$$\Pi_1(q_1, q_2; \alpha) = 0 \tag{8.55}$$

and

$$\Pi_2(q_1, q_2; t) = 0. \tag{8.56}$$

The differentials are

$$d\Pi_1 = \Pi_{11}dq_1 + \Pi_{12}dq_2 + \Pi_{1\alpha}\, d\alpha = 0 \tag{8.57}$$

and

$$d\Pi_2 = \Pi_{21}dq_1 + \Pi_{22}dq_2 + \Pi_{2t}\, dt = 0. \tag{8.58}$$

The next step is to substitute the specific functions for the partial derivatives and write the two equations in matrix form:

$$\begin{pmatrix} \alpha R_1'' - C'' & -C'' \\ -C'' & R_2'' - C'' \end{pmatrix} \begin{pmatrix} dq_1 \\ dq_2 \end{pmatrix} = \begin{pmatrix} -R_1'\, d\alpha \\ dt \end{pmatrix}, \tag{8.59}$$

where the leftmost matrix is the Hessian matrix. We can solve for the changes in output using Cramer's rule. These are

$$dq_1 = \frac{\begin{vmatrix} -R_1'\, d\alpha & -C'' \\ dt & R_2'' - C'' \end{vmatrix}}{|H|} = \frac{-R_1'\,(R_2'' - C'')\, d\alpha + C''\, dt}{|H|},$$

$$dq_2 = \frac{\begin{vmatrix} \alpha R_1'' - C'' & -R_1'\, d\alpha \\ -C'' & dt \end{vmatrix}}{|H|} = \frac{-R_1'C''\, d\alpha + (\alpha R_1'' - C'')\, dt}{|H|}, \quad \text{and} \tag{8.60}$$

$$dQ = dq_1 + dq_2 = \frac{-R_1'R_2''\, d\alpha + \alpha R_1''\, dt}{|H|}.$$

For changes in the level of demand in market 1 we find that

$$\frac{\partial q_1^*}{\partial \alpha} = \frac{dq_1}{d\alpha}\bigg|_{dt=0} = \frac{-R_1'(R_2'' - C'')}{|H|} > 0,$$

$$\frac{\partial q_2^*}{\partial \alpha} = \frac{dq_2}{d\alpha}\bigg|_{dt=0} = \frac{-R_1' C''}{|H|}, \; sign\left(\frac{dq_2}{d\alpha}\bigg|_{dt=0}\right) = sign\left(-C''\right), \quad \text{and} \qquad \textbf{(8.61)}$$

$$\frac{\partial Q^*}{\partial \alpha} = \frac{dQ}{d\alpha}\bigg|_{dt=0} = \frac{-R_1' R_2''}{|H|} > 0.$$

For all three derivatives we use $R' > 0$ (marginal revenue is positive) and $|H| > 0$ (by the second-order conditions). The sign of $dq_1/d\alpha$ is derived from the fact that the second-order conditions require both terms of the form $R_i'' - C''$ to be negative, while the sign of $dQ/d\alpha$ is derived from the fact that marginal revenue slopes down, i.e., $R_i'' < 0$. Thus, we use a combination of assumptions from economic theory and results from the second-order conditions to derive our results.

To interpret these results, note that the changes in q_1 and Q are quite intuitive. An increase in demand in market 1 leads to an increase in output for that market and to an increase in total output. The result for q_2 may be less obvious since it depends on the slope of the marginal cost curve. The three possibilities are shown in Figure 8.3. Note that the relative cost of any two total outputs, $\hat{Q}$ and $\bar{Q}$, depends on the slope of the marginal cost curve.

Consider first diseconomies of scale, where marginal cost slopes up, $C'' > 0$. Here, an increase in demand in market 1 leads to an increase in output for market 1. Because the firm's marginal cost slopes up, the increase in q_1 raises the marginal cost of serving market 2 and leads to a decline in q_2. For the case of economies of scale, $C'' < 0$, increases in q_1 decrease the marginal cost of output, so q_2 increases. Finally, with

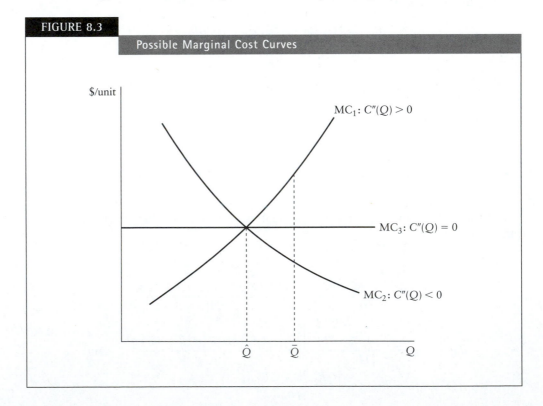

FIGURE 8.3

Possible Marginal Cost Curves

constant returns to scale, $C'' = 0$, the marginal cost of serving the second market is unaffected, so q_2 does not change.

For changes in t, the extra cost of serving the second market, we get

$$\frac{\partial q_1^*}{\partial t} = \frac{dq_1}{dt}\bigg|_{d\alpha=0} = \frac{C''}{|H|}, \; sign\left(\frac{dq_1}{dt}\bigg|_{d\alpha=0}\right) = sign(C''),$$

$$\frac{\partial q_2^*}{\partial t} = \frac{dq_2}{dt}\bigg|_{d\alpha=0} = \frac{(\alpha R_1'' - C'')}{|H|} < 0, \quad \text{and} \tag{8.62}$$

$$\frac{\partial Q^*}{\partial t} = \frac{dQ}{dt}\bigg|_{d\alpha=0} = \frac{\alpha R_1''}{|H|} < 0.$$

Here the changes in q_2 and Q are as expected—higher costs lead to reduced outputs—and it is the change in q_1 that depends on the slope of the marginal cost curve. The interpretation is much the same as for the case of a demand change. For example, when marginal cost slopes up, $C'' > 0$, the decline in q_2 causes the firm to move back down its marginal cost curve. This lowers the marginal cost of serving market 1 and leads to an increase in q_1.

Our basic result here is that for a multimarket firm the way that markets are connected is through the marginal cost curve. The effect of changes in one market on output and price in the other market will depend on the nature of returns to scale, i.e., on whether marginal cost slopes up, down, or is constant.

8.7   POLLUTION TAXES AND POLLUTION EMISSIONS

Economics has played a key role in shaping environmental policies. One key insight is that direct regulation, such as requiring firms to implement specific pollution control technologies or requiring all firms to reduce pollution emissions by a specified percentage, is generally inefficient. Instead, economists generally advocate market-based approaches that give incentives to firms to internalize the external costs imposed by pollution. One way of accomplishing this is to tax firms directly on pollution emissions, with the tax rate set equal to the external costs imposed by the emissions. In this application we model the effects of such a tax.

Let us assume that a firm derives revenue from output according to the function $R(Q)$ and incurs direct production costs $C(Q)$; this is general enough to cover both monopoly and perfect competition. Let us also assume that the firm can engage in pollution control, X, and that X costs α per unit. We will write the firm's pollution emissions as

$$E = E(Q, X). \tag{8.63}$$

We won't use a specific function for E but will place reasonable restrictions on the signs of its derivatives. First, we assume $E_Q > 0$ and $E_{QQ} \geq 0$; higher output increases pollution, and as output increases, the marginal pollution from a unit of output is also increasing. Second, we assume $E_X < 0$ and $E_{XX} > 0$; pollution control expenditures are effective in reducing emissions but the marginal effectiveness of X is declining. Finally, we will assume that $E_{QX} < 0$. This cross-partial derivative measures the effect of an increase in pollution control activities on the marginal pollution from an extra unit of output. The derivative will normally be negative since increased pollution control activities should reduce the pollution generated by an extra unit of output.

The firm's profit function is

$$\Pi = R(Q) - C(Q) - \alpha X - tE(Q, X), \tag{8.64}$$

where t is the per-unit tax on pollution emissions. The first-order conditions for a profit maximum are

$$\Pi_Q = R_Q - C_Q - tE_Q = 0$$

and

$$\Pi_X = -\alpha - tE_X = 0.$$

(8.65)

The first-order condition for Q is simply that marginal revenue must equal marginal cost inclusive of the emissions tax that results from an extra unit of output. The first-order condition for X is that the marginal reduction in the emissions tax bill must equal α, the marginal cost of pollution abatement activity.

The Hessian for the second-order conditions is

$$H = \begin{pmatrix} \Pi_{QQ} & \Pi_{QX} \\ \Pi_{QX} & \Pi_{XX} \end{pmatrix} = \begin{pmatrix} R_{QQ} - C_{QQ} - tE_{QQ} & -tE_{QX} \\ -tE_{QX} & -tE_{XX} \end{pmatrix}.$$

(8.66)

The second-order conditions are

$$|H_1| = \Pi_{QQ} < 0 \quad \text{and} \quad |H| = \Pi_{QQ}\Pi_{XX} - \Pi_{QX}^2 > 0.$$

(8.67)

To find the effects of changes in parameters on the levels of output and pollution control, we totally differentiate the first-order conditions. This gives

$$\begin{pmatrix} \Pi_{QQ} & \Pi_{QX} \\ \Pi_{QX} & \Pi_{XX} \end{pmatrix} \begin{pmatrix} dQ \\ dX \end{pmatrix} = \begin{pmatrix} E_Q \, dt \\ E_X \, dt + d\alpha \end{pmatrix}.$$

(8.68)

Solving for dQ and dX yields

$$dQ = \frac{\left(\overset{(-)}{\Pi_{XX}} \overset{(+)}{E_Q} - \overset{(+)}{\Pi_{QX}} \overset{(-)}{E_x} \right) dt - \overset{(+)}{\Pi_{QX}} d\alpha}{|H|}$$

and

(8.69)

$$dX = \frac{\left(\overset{(-)}{\Pi_{QQ}} \overset{(-)}{E_X} - \overset{(+)}{\Pi_{QX}} \overset{(+)}{E_Q} \right) dt + \overset{(-)}{\Pi_{QQ}} d\alpha}{|H|}.$$

The effects of a change in α are unambiguous. An improvement in pollution control technology (i.e., a decline in α) will lead to more pollution abatement activities and higher output. The effects of an increase in the pollution tax are ambiguous. One might expect that an increase in the tax will raise X (which lowers emissions and helps avoid the tax) and decrease Q. This intuition would indeed be unambiguously true if changes in X did not affect marginal pollution from an extra unit of output so that $E_{QX} = 0$, and hence $\Pi_{QX} = 0$. However, when changes in X lead to changes in the marginal pollution from an extra unit of output, the assumptions we have made so far cannot rule out other outcomes, such as an increase in the tax causing output to decline so that pollution control activities become less profitable and fall as well.[10]

---

[10]One outcome is not possible: if X falls then Q cannot increase. This is left as an end-of-chapter problem.

8.8 STATISTICAL ESTIMATION

Economists are not just theorists. Empirical work is needed to test whether theories are supported by observations and data. Econometrics is the branch of economics that applies statistical techniques in an attempt to uncover regularities and relationships in economic data. Yet, even empirical work with data requires mathematical theory to develop the best methods for estimating relationships between economic variables.

A typical economic data relationship is shown in Figure 8.4. The individual points on the diagram represent observations of data. The x variable is the independent variable, while the y variable is the dependent variable. As examples, x might be years of education and y weekly earnings, or x might be the nominal money supply and y the level of GDP.

The issue for an econometrician is to find a mathematical function that best describes the relationship between the data on the economic variables. In this application we will assume that the true relationship is linear, but with random "noise," so that the data points do not simply lie directly on a straight line. We write the relationship between the dependent variable, y, and the independent variable, x, as

$$y_i = \alpha + \beta x_i + \varepsilon_i. \tag{8.70}$$

where y_i and x_i are the ith (out of n total) observations of the variables, ε_i is a random error term, and α and β are the true (but unknown) values of the intercept and slope parameters.

If there were no randomness then the observations would all lie exactly on the straight line described by the equation $y_i = \alpha + \beta x_i$. Because of randomness, however, the data are "scattered" about this line and the task is to estimate the true values of α and β based on the available data. Let $\hat{\alpha}$ and $\hat{\beta}$ denote the estimated values of the true

---

FIGURE 8.4

A Scatter Plot of Statistical Data

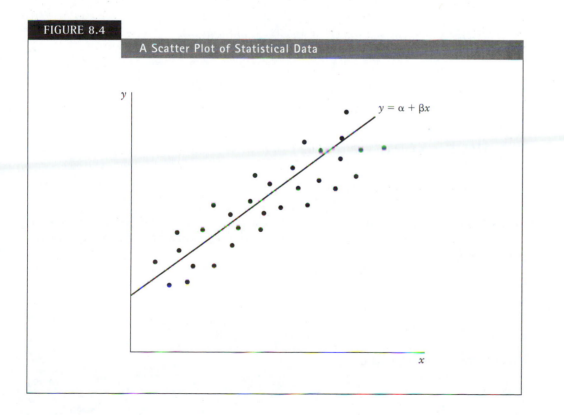

parameters α and β. Then the predicted values, $\hat{y}_i$, of the dependent variable are described by the equation

$$\hat{y}_i = \hat{\alpha} + \hat{\beta}x_i. \tag{8.71}$$

The standard criterion for the best estimates of α and β is to find estimators, $\hat{\alpha}$ and $\hat{\beta}$, that minimize the sum of squared residuals, i.e., the sum of the squared deviations between the mathematically predicted value of the dependent variable, $\hat{y}_i$, and the actual value.

Formally, the sum of squared residuals, SSR, is

$$SSR = \sum_{i=1}^{n} (y_i - \hat{y}_i)^2, \tag{8.72}$$

where $\hat{y}_i$ is the predicted value of y_i. Since the predicted value of y_i is $\hat{y}_i = \hat{\alpha} + \hat{\beta}x_i$, we have

$$SSR = \sum_{i=1}^{n} (y_i - \hat{\alpha} - \hat{\beta}x_i)^2. \tag{8.73}$$

Minimizing this with respect to the two coefficients, $\hat{\alpha}$ and $\hat{\beta}$, yields the two first-order conditions

$$\frac{\partial SSR}{\partial \hat{\alpha}} = \sum_{i=1}^{n} -2(y_i - \hat{\alpha} - \hat{\beta}x_i) = 0$$

and

$$\frac{\partial SSR}{\partial \hat{\beta}} = \sum_{i=1}^{n} -2x_i(y_i - \hat{\alpha} - \hat{\beta}x_i) = 0. \tag{8.74}$$

The Hessian matrix and the second-order conditions are

$$H = \begin{pmatrix} 2n & \sum_{i=1}^{n} 2x_i \\ \sum_{i=1}^{n} 2x_i & \sum_{i=1}^{n} 2x_i^2 \end{pmatrix},$$

$$|H_1| = 2n > 0, \quad \text{and} \tag{8.75}$$

$$|H_2| = |H| = 4\left(\sum_{i=1}^{n} nx_i^2 - \left(\sum_{i=1}^{n} x_i\right)\left(\sum_{i=1}^{n} x_i\right)\right) = 4n\left(\sum_{i=1}^{n} (x_i - \bar{x})^2\right) > 0,$$

where $\bar{x} = 1/n \sum_{i=1}^{n} x_i$ is the mean of the observations of the independent variables. The reader can verify the last step in the derivation by expanding the sum of squares and working backward.

Since the second-order conditions for a minimum are satisfied, we now turn to the solutions for $\hat{\alpha}$ and $\hat{\beta}$. The first order equation for $\hat{\alpha}$ can be rewritten as

$$\sum_{i=1}^{n} (y_i - \hat{\alpha} - \hat{\beta}x_i) = n\bar{y} - n\hat{\alpha} - n\hat{\beta}\bar{x} = 0. \tag{8.76}$$

Solving for $\hat{\alpha}$ in terms of $\hat{\beta}$ gives

$$\hat{\alpha} = \bar{y} - \hat{\beta}\bar{x}. \tag{8.77}$$

In words, the estimate for $\hat{\alpha}$ depends on the observed means of the dependent and independent variables and on the estimate for $\hat{\beta}$.

Substituting the equation for $\hat{\alpha}$ into the first-order condition for $\hat{\beta}$, rearranging, and solving gives us the solution

$$-2\sum_{i=1}^{n} x_i y_i + 2\hat{\alpha}\sum_{i=1}^{n} x_i + 2\hat{\beta}\sum_{i=1}^{n} x_i^2 = 0, \quad \text{or}$$

$$-\sum_{i=1}^{n} x_i y_i + (\bar{y} - \hat{\beta}\bar{x})\sum_{i=1}^{n} x_i + \hat{\beta}\sum_{i=1}^{n} x_i^2 = 0, \quad \text{or}$$

$$\left(\sum_{i=1}^{n} x_i^2 - \bar{x}\sum_{i=1}^{n} x_i\right)\hat{\beta} = \sum_{i=1}^{n} x_i y_i - \bar{y}\sum_{i=1}^{n} x_i, \quad \text{or} \tag{8.78}$$

$$\hat{\beta} = \frac{\left(\sum_{i=1}^{n} x_i y_i\right) - n\bar{x}\bar{y}}{\left(\sum_{i=1}^{n} x_i^2\right) - n\bar{x}^2}.$$

We now have a solution for the slope estimator as a function of the observed data points. (For those with knowledge of statistics: $\hat{\beta}$ is the ratio of the covariance of x and y to the variance of x.) This in turn allows us to find the estimated intercept. All of the statistical calculations can of course be done by a computer. The same basic mathematical methods can also be used to derive estimators for more complex cases where there are multiple independent variables.[11]

Problems

8.1 Using the solutions for L and Q in Section 8.2, find the elasticities of labor demand and output supply with respect to w, r, and P.

8.2 Suppose that a competitive firm receives a price of P for its output, pays prices of w and r for its labor and capital inputs, and operates with the production function $Q = L^a K^b$.

(a) Write profits as a function of L and K (with P, w, and r as parameters). Derive the first-order conditions. Provide an economic interpretation of the first-order conditions.

(b) Solve for the optimal levels of L^* and K^* (these solutions should be expressed as functions of only exogenous parameters).

(c) Check the second-order conditions. What restrictions on the values of a and b are necessary for a profit maximum? Provide an economic interpretation of these restrictions.

(d) Find the signs of the partial derivatives of L^* with respect to P, w, and r.

(e) Derive the firm's long-run supply curve, i.e., Q^* as a function of the exogenous parameters. Find the elasticities of supply with respect to w, r,

---

[11]Multiple regression, with many independent variables, is most easily represented using matrix algebra.

and P. Do these elasticities sum to zero? Provide an economic explanation for this fact.

(f) Find the firm's marginal cost curve. (Hint: Remember that for a competitive firm output is chosen so that marginal cost equals price at a profit maximum.) Is marginal cost upward or downward sloping?

8.3 Suppose that a perfectly competitive firm operates with the production function $Q = F(L, K, R)$ where R is raw materials. Let P be the output price and w, r, and v be the respective input prices. What are the first- and second-order conditions for profit maximization? Find the sign of dL^*/dw given $dP = dr = dv = 0$. (Hint: For the last result you may want to re-read Footnote 2 in Chapter 7.)

8.4 Suppose that a perfectly competitive firm uses three inputs—L, K, and R—pays input prices of w, r, and v, sells its output at a price of P, and operates with a production function of: $Q = 3(LK)^{1/3} + \ln R$.

(a) Write the expression for the firm's profits. What are the first-order conditions? Give an economic interpretation of the first-order conditions.

(b) Check the second-order conditions.

(c) Without explicitly solving for L^*:

(i) Find the change in L for a change in r when all other parameters are constant.

(ii) Find the change in L for a change in v when all other parameters are constant.

(d) Solve for L^*. Take partial derivatives of L^* to confirm the results derived in part c.

8.5 Suppose that a monopolist faces the demand curve $P = 3Q^{-1/2}$, operates with the production function $Q = (LK)^{2/3}$, and pays input prices of w and r.

(a) Find the first-order conditions for profit maximization. Solve for L^* and K^* as a functions of parameters.

(b) Check the second-order conditions.

(c) Find the change in L when r changes, but w is held constant.

8.6 Derive the second-order conditions for the efficiency wage model presented in Section 8.4. For the case of a perfectly competitive firm, where revenue equals an exogenous price times output, explain when the second-order conditions will hold.

8.7 Suppose that in the efficiency wage model of Section 8.4, worker effort depends on the difference between the wage, w, offered by the firm and the alternative wage, $\bar{w}$. Specifically, assume that $e(w) = (w - \bar{w})^{\beta}$ where β is a parameter between zero and one. Find the equilibrium level of w. How does this solution depend on the alternative wage $\bar{w}$?

8.8 Consider a possible extension of the efficiency wage model presented in Section 8.4. Suppose that the amount of effort expended by workers is $e = e(w, x)$ where x, the probability that shirking is detected, is determined by the firm. Assume that $e_x > 0$ and $e_{xx} < 0$. Suppose that the cost of a given level of detection is $c(x)$, where $c'(x) > 0$ and $c''(x) > 0$. Find the first-order conditions for a profit maximum. Write out the second-order conditions [Note: These are fairly messy].

Does the unit elasticity result derived in the text still hold in this more complex model?

8.9 Suppose that a monopolist faces a demand curve of $P = 10 - Q$ and operates two factories. Total costs at the respective factories are:

$$TC_1 = aq_1^2 \quad \text{and} \quad TC_2 = bq_2^2 + 2cq_2.$$

(a) Find the first-order conditions and solve for the optimal output levels. Check the second-order conditions.

(b) Solve for q_1^*. Find $\partial q_1^*/\partial c$.

(c) Find (by implicitly differentiating the first-order conditions) dq_1/dc when the parameters a and b are held constant.

8.10 Suppose that a monopolist sells its output in two separate markets. The demand equations for the two markets are given by

$$P_1 = a - q_1 \quad \text{and} \quad P_2 = b - q_2.$$

Total production costs are $TC = Q^3$ where $Q = q_1 + q_2$ [Hint: Do not multiply out the cube]

(a) What are the first-order conditions?

(b) Do the second-order conditions hold?

(c) Suppose that the parameter b increases while the parameter a remains constant. Does q_1 increase or decrease?

8.11 Consider a Cournot duopoly market where demand is $P = a - Q$ and Q is total output. Suppose that firm 1 operates two factories. Total costs at factory 1 are $TC_{11} = cq_{11}^3$ and total costs at factory two are $TC_{12} = eq_{12}$, where the first subscript identifies firm 1 and the second subscript identifies the factory. Suppose that firm 2 operates a single factory with total costs of $TC_2 = fq_2$.

(a) Find the first- and second-order condition(s) for each firm's profit maximization problem.

(b) Find the change in q_{11} when the parameters e and f increase by equal amounts while the other parameters do not change.

8.12 Consider a two-country, X and Y, model with one firm located in each country. Suppose that demand in each country is $P = a - Q$, where Q is the total quantity being sold in the relevant country. Each firm operates with total costs of $TC_i = c_i q_i^3$, where the subscript $i = x$ or y, and q_i is firm i's total output. Suppose that firm X sells in both markets, its home market X and the foreign market Y, while firm Y sells only in its own market. Let q_{xj} be the output that firm X sells in market j so that $q_x = q_{xx} + q_{xy}$. Assume that the firms both operate under the Cournot assumption when they compete in country Y.

(a) Write the expression for each firm's profit. Find the first- and second-order condition(s) for each firm's profit maximization problem.

(b) Find the change in firm X's quantity sold in its home market when c_y increases, but other parameters are held constant.

8.13 Using the pollution emissions tax model from Section 8.7, show that if $dX/dt < 0$ then dQ/dt must be negative. [Hint: Show that the conditions for

$dX/dt < 0$ and $dQ/dt > 0$ taken together would imply that the second-order condition would fail to hold.]

8.14 Using the model from Section 8.7, suppose that a polluting firm is a monopolist with linear demand and constant marginal cost. Assume that pollution emissions are given by the function $E = Q^2/X$.

(a) Find the first- and second-order conditions.

(b) Solve for Q^* and X^*.

(c) Find (and sign if possible) the effects of changes in t and α on Q^* and X^*.

Constrained Optimization: Theory

9.1 INTRODUCTION

Economics is the study of the allocation of scarce resources. Scarcity implies constraints, so much of economic theory has to do with agents who are optimizing some objective function subject to a constraint, as is the case, for example, when a consumer maximizes utility subject to a budget constraint. In Figure 9.1 we see an illustration of a one-variable constrained maximization problem. The objective function is $f(x)$, which is maximized when $x = x^U$. If the choices of x are constrained to those that satisfy the inequality $g(x) \leq g_0$, however, the highest value the function $f(x)$ can take is $f(x^C)$. That is, the value of x that leads to the highest value of the function $f(x)$ while maintaining consistency with the constraint $g(x) \leq g_0$ is x^C. This is a **binding constraint** because the constraint changes the optimal value of the choice variable x (and the maximal value of the objective function as well) and the constraint holds as an equality at the optimal value of x. If the constraint were relaxed to $g(x) \leq g_1$, where g_1 is the higher value shown in Figure 9.1, we would have an example of a **nonbinding** (or **slack**) **constraint** because the objective function would be maximized when $x = x^U$ with or without the imposition of the constraint. That is, the solution to the constrained maximization problem is identical to the solution to the unconstrained maximization problem. In addition, at the optimal value of x the constraint holds as a strict inequality.

Figure 9.2 illustrates a constrained maximization problem with two choice variables. The constraint $g(x, y) = g_0$ is represented by its level curve. That is, the graph of the constraint shows all of the combinations of x and y for which $g(x, y) = g_0$. The objective function $f(x, y)$ is also represented by level curves, each showing the combinations of x and y that yield a particular value of the function. We assume that higher values of the function $f(x, y)$ correspond to level curves that are farther up and to the right in Figure 9.2, and that the level curves of $f(x, y)$ are convex to the origin. These represent the most common situation in maximization problems in economics. Graphically, we find the solution to the constrained maximization problem by examining all points on the level curve of the constraint and picking the one that yields the highest value of the objective function. As illustrated in Figure 9.2, if there is an interior solution, it will be at a point of tangency between the level curve representing the constraint and a level curve of the objective function. This means that, at the margin, the tradeoff between x and y allowed by the constraint must just equal the tradeoff

FIGURE 9.1

Constrained Maximization: One Variable

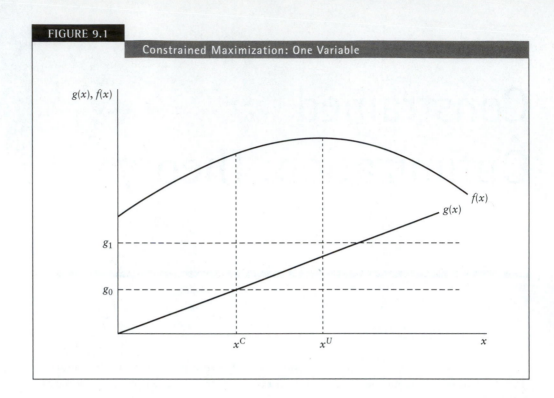

FIGURE 9.2

Constrained Maximization: Two Variables

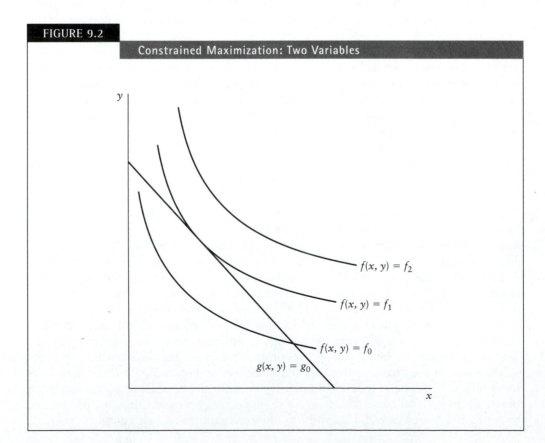

between x and y that will leave the value of the objective function unchanged. Both the tangency condition and its interpretation in words should be familiar from intermediate microeconomic theory. Later in this chapter we will obtain similar tangency results from a general optimization problem with many choice variables and an equality constraint.

Almost all constrained optimization problems in economics really have inequality constraints, so we will set up the formal modeling in this chapter accordingly. But it is also true that most (though not all) constraints in economics problems are binding, in which case treating the constraint as an equality will yield the correct solution. We will assume in this chapter that all constraints are binding. In Chapter 11 we introduce a technique for analyzing problems in which constraints may or may not be binding. This technique also addresses the possibility of corner solutions by explicitly including the constraints that the choice variables are nonnegative. These nonnegativity constraints are often ignored in the formal analysis (as we have usually done earlier in this book and will continue to do in this chapter) by assuming interior solutions.

Although we have already analyzed a few constrained optimization problems, we have so far dealt with the constraints by transforming the problems into unconstrained optimization problems as in the labor supply example of Section 6.8 and the utility-maximization problem of Section 6.9. In this chapter we introduce and develop a powerful technique, called the **Lagrangian method,** for analyzing constrained optimization problems. We will begin by discussing maximization problems, then explain the differences if the problem is to minimize an objective function. Our focus in this chapter is comparative static analysis of the optimal values of the choice variables, thereby laying the foundation for the following several chapters. In Chapter 10 we provide several economic applications of the Lagrangian method, again focusing on the choice variables. In Chapters 11 and 12 we discuss the analysis of optimization problems with inequality constraints. Then in Chapters 13 and 14 we study the value taken by the objective function when it is optimized, and the comparative statics of how that value changes in response to changes in parameters. For example, we will analyze how a firm's profits are affected when an input price changes and how a consumer's utility level is affected when income changes. These chapters will be previewed by the discussion in Section 9.4.

9.2   THE LAGRANGIAN METHOD

Consider the problem of maximizing the function $f(\mathbf{x}, \boldsymbol{\alpha})$ subject to the constraint $g(\mathbf{x}, \boldsymbol{\alpha}) \leq g_0$, where $\mathbf{x}$ is a vector of choice variables, $\boldsymbol{\alpha}$ is a vector of parameters, and g_0 is another parameter. (Not all elements of $\mathbf{x}$ need to appear as arguments in the function g; elements of $\boldsymbol{\alpha}$ can appear in the function f, the function g, or both.) The technique we have used on occasion earlier in the book (for example, in Section 6.8) to solve problems of this sort is the **substitution method.** First, we assume the constraint is binding (and therefore is an equality) and use it to solve for one of the choice variables as a function of the others. Then we substitute this function into the objective function $f(\mathbf{x}, \boldsymbol{\alpha})$ and maximize the resulting function with respect to the remaining choice variables. This method often works, provided the function g has a simple enough form. But even when the substitution method works, an alternative technique called the Lagrangian method gives more information about the solution and is usually easier to implement.

The **Lagrangian method** combines the objective function and the constraint into a single function called the **Lagrange function,** or simply the **Lagrangian.** The Lagrangian is usually denoted with the symbol $\mathscr{L}$ and is formed by rewriting the constraint so

FIGURE 9.3

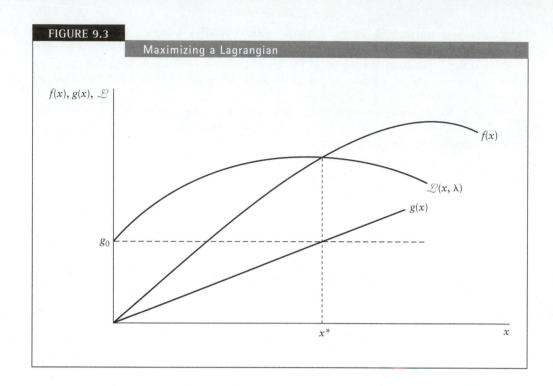

FIGURE 9.3 Maximizing a Lagrangian

that it is nonnegative,[1] multiplying it by a new variable λ called the **Lagrange** (or **Lagrangian**) **multiplier,** and adding the product to the objective function:

$$\mathscr{L} = \mathscr{L}(\mathbf{x}, \boldsymbol{\alpha}, \lambda, g_0) = f(\mathbf{x}, \boldsymbol{\alpha}) + \lambda(g_0 - g(\mathbf{x}, \boldsymbol{\alpha})). \tag{9.1}$$

It turns out that maximizing the Lagrangian with respect to both $\mathbf{x}$ and λ yields the values of $\mathbf{x}$ that maximize $f(\mathbf{x}, \boldsymbol{\alpha})$ subject to the (binding) constraint $g(\mathbf{x}, \boldsymbol{\alpha}) \leq g_0$ and gives a value of λ that has a useful interpretation.[2]

Before examining the first- and second-order conditions for maximizing the function (9.1), we will begin with a more intuitive description of the Lagrangian method. As was the case with the substitution method, the Lagrangian transforms the constrained optimization problem into an equivalent unconstrained optimization problem. Instead of transforming the problem into one with one less choice variable, however, the Langrangian method adds one extra choice variable, the Lagrange multiplier. Assuming that the constraint is binding,[3] that is, that the optimal values of $\mathbf{x}$ are such that $g(\mathbf{x}, \boldsymbol{\alpha}) \leq g_0$, the value of the Lagrangian will equal the value of the function f, since the Lagrangian will be adding zero to the value of f. Figure 9.3 illustrates this case for a single choice variable. It shows the function f, the function g, and the Lagrangian $\mathscr{L}$. At x^* the constraint is binding (any value of x higher than x^* would make $g(x) > g_0$), the Lagrangian is maximized, and the value of the Lagrangian equals the value of $f(x)$.

---

[1]Writing the constraint as nonpositive would change nothing except the sign and interpretation of the Lagrange multiplier.

[2]For a proof that the Lagrangian method yields values of $\mathbf{x}$ that maximize the objective function subject to the constraint, see Carl P. Simon and Lawrence Blume, *Mathematics for Economists,* New York: Norton, 1994, pp. 478–480. The interpretation of λ is discussed briefly in this chapter and in more detail in Chapter 13.

[3]If the constraint is not binding, the solution to the problem will be identical to that of an unconstrained maximization of the function *f.* There is a simple indicator of whether the constraint turns out to bind: if the value of λ that satisfies the first-order conditions of the maximization of the Lagrangian turns out to be non-positive, then the constraint is not binding. We will discuss this issue further in Chapter 11.

By choosing $\mathbf{x}$ and λ to maximize[4] equation (9.1), we ensure that the constraint is satisfied as an equality and that the function f reaches its largest value consistent with the constraint. Adding λ as a choice variable is what guarantees that the solution to the first-order conditions will make the constraint binding. (Recall that we are *assuming* that the constraint is binding; in Chapter 11 we will discuss the possibility of nonbinding constraints, which will allow us to investigate whether, for particular problems, constraints do or do not bind.)

9.2.1 First–Order Conditions

Since the constrained maximization problem has been converted into an equivalent unconstrained maximization problem, the structure of the first-order conditions is already familiar. To derive the first-order conditions, we set equal to zero the partial derivatives of the function (9.1) with respect to the choice variables ($\mathbf{x}$ and λ). If there are n elements in the vector $\mathbf{x}$ the result is the $n+1$ equations

$$\frac{\partial \mathcal{L}}{\partial x_i} = \frac{\partial f}{\partial x_i}(\mathbf{x}, \boldsymbol{\alpha}) - \lambda \frac{\partial g}{\partial x_i}(\mathbf{x}, \boldsymbol{\alpha}) = 0 \quad i = 1, \dots, n$$

$$\frac{\partial \mathcal{L}}{\partial \lambda} = (g_0 - g(\mathbf{x}, \boldsymbol{\alpha})) = 0.$$

(9.2)

The last of these equations states that the constraint is binding. If the functional forms of f and g are known and are simple enough, these $n+1$ equations can be solved simultaneously for the $n+1$ variables $x_1, \dots, x_n$ and λ. Even if the functional forms of f and g are unknown, we can derive comparative static derivatives by implicit differentiation or by finding total differentials, as shown in Section 9.3.

The first-order conditions can be rewritten in a way that is very useful for interpretation. Consider two of the first-order conditions (9.2):

$$\frac{\partial f}{\partial x_i} - \lambda \frac{\partial g}{\partial x_i} = 0$$

$$\frac{\partial f}{\partial x_j} - \lambda \frac{\partial g}{\partial x_j} = 0.$$

(9.3)

If we eliminate λ from these equations (for example, by using one equation to solve for λ and then substituting into the other equation), we can write the result as

$$\frac{\partial f / \partial x_i}{\partial f / \partial x_j} = \frac{\partial g / \partial x_i}{\partial g / \partial x_j}$$

(9.4)

The left-hand side of equation (9.4) is the negative of the slope of a level curve[5] of the function $f(\mathbf{x}, \boldsymbol{\alpha})$; the right-hand side is the negative of the slope of a level curve of the function $g(\mathbf{x}, \boldsymbol{\alpha})$. Thus the first-order conditions imply that the optimal values $\mathbf{x}^*$ of the choice variables must be such that the slopes of the level curves of the objective function and the constraint function must be equal, and (by the last of the first-order conditions (9.2)) the constraint is satisfied as an equality. That is, the level curves of the objective function and the constraint must be tangent at $\mathbf{x}^*$.

As a simple example, let $f(x_1, x_2) = 2x_1^2 + x_2^2 + 10x_1x_2$ and $g(x_1, x_2) = x_1 + 2x_2$. The Lagrangian for the problem of maximizing $f(x_1, x_2)$ subject to the constraint that

---

[4]The Lagrangian is maximized with respect to the choice variables x but is actually minimized with respect to the Lagrange multiplier λ.

[5]Level curves and their slopes are discussed in Chapter 5.

$g(x_1, x_2) \leq g_0$ is

$$\mathcal{L}(x_1, x_2, \lambda, g_0) = 2x_1^2 + x_2^2 + 10x_1x_2 + \lambda(g_0 - x_1 - 2x_2). \qquad \textbf{(9.5)}$$

The first-order conditions are

$$\frac{\partial \mathcal{L}}{\partial x_1} = 4x_1 + 10x_2 - \lambda = 0$$

$$\frac{\partial \mathcal{L}}{\partial x_2} = 2x_2 + 10x_1 - 2\lambda = 0 \qquad \textbf{(9.6)}$$

$$\frac{\partial \mathcal{L}}{\partial \lambda} = g_0 - x_1 - 2x_2 = 0.$$

Eliminating the Lagrange multiplier using the first two equations yields $4x_1 + 10x_2 = (2x_2 + 10x_1)/2$, which we can rewrite as

$$\frac{4x_1 + 10x_2}{2x_2 + 10x_1} = \frac{1}{2}. \qquad \textbf{(9.7)}$$

This is the formulation that shows the tangency of level curves of the objective function and constraint. We can then solve equation (9.7) simultaneously with the last of the first-order conditions (9.6) to find that $x_1^* = 9g_0/11$ and $x_2^* = g_0/11$. Substituting x_1^* and x_2^* into either of the first two of the first-order conditions (9.6), we find that $\lambda^* = 46g_0/11$. In the next section we will see that the second-order conditions for this problem are satisfied.

As an economic example of the Lagrangian method, let us look at the consumer's utility-maximization problem. We will use the conventional notation: x_i is the quantity consumed of good i and p_i is the price of good i, so p_ix_i is the consumer's expenditure on good i; I is the consumer's income. The objective function is the utility function $U(x_1, \ldots, x_n)$ while the constraint is $p_1x_1 + \cdots + p_nx_n \leq I$. The Lagrangian for this problem is

$$\mathcal{L}(x_1, \ldots, x_n, \lambda, I, p_1, \ldots, p_n) = U(x_1, \ldots, x_n) + \lambda(I - p_1x_1 - \cdots - p_nx_n) \quad \textbf{(9.8)}$$

and the first-order conditions are

$$\frac{\partial \mathcal{L}}{\partial x_i} = \frac{\partial U}{\partial x_i} - \lambda p_i = 0 \quad i = 1, \ldots, n$$

$$\frac{\partial \mathcal{L}}{\partial \lambda} = I - p_1x_1 - \cdots - p_nx_n = 0. \qquad \textbf{(9.9)}$$

The last of the first-order conditions (9.9) ensures that the budget constraint is binding. Solving the first-order conditions will yield equilibrium functions for the choice variables of the form $x_i^*(I, p_1, \ldots, p_n)$. Called the **ordinary** or **Marshallian demand functions,** these functions show how the demand for each good is affected by income and market prices. (In Section 9.2.3 we will show how to derive the *compensated* or *Hicksian demand functions,* which have as arguments the market prices and the consumer's utility level.)

Taking any two of the first n of the first-order conditions (9.9) and eliminating the Lagrange multiplier yields

$$\frac{\partial U/\partial x_i}{\partial U/\partial x_j} = \frac{p_i}{p_j}. \qquad \textbf{(9.10)}$$

The left-hand side of equation (9.10) is the marginal rate of substitution between goods i and j (see Section 6.9), so equation (9.10) says that the marginal rate of substitution

FIGURE 9.4

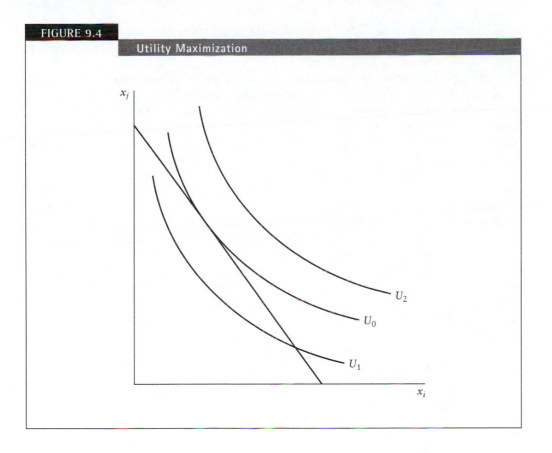

between any two goods must equal their price ratio. Graphically, this (combined with the first-order condition that says that the budget constraint is binding) gives the familiar tangency between the consumer's indifference curve (the level curve of the objective function) and budget constraint (the level curve of the constraint), as shown in Figure 9.4.

A different manipulation of the first-order conditions (9.9) yields a different insight into the utility-maximization problem and illustrates the useful interpretation of Lagrange multipliers.[6] The first-order condition for good i can be rewritten as

$$\frac{\partial U/\partial x_i}{p_i} = \lambda. \tag{9.11}$$

Since $1/p_i$ is the quantity of good i that can be purchased for one dollar and $\partial U/\partial x_i$ is the marginal utility of consuming good i, the left-hand side of equation (9.11) is the extra utility the consumer can get by spending one more dollar on good i. Since equation (9.11) holds for each good, the first-order conditions imply that the marginal utility of spending a dollar must be the same regardless of how it is spent. Furthermore, since this marginal utility equals λ, the Lagrange multiplier in this problem measures the marginal utility of income.

9.2.2 Second-Order Conditions: The Bordered Hessian

So far we have considered only the first-order conditions for the problem of maximizing the Lagrangian. As always, the first-order conditions give only the necessary

---

[6]We will discuss briefly the interpretation of Lagrange multipliers in general in Section 9.4. Chapter 13 includes a detailed discussion.

conditions for a maximum. To be sure that the solutions to the first-order conditions yield a maximum of the Lagrangian, the second-order conditions must be satisfied as well.[7]

Although the first-order conditions for a constrained maximization problem are essentially the same as those for an unconstrained maximization problem (once the problem has been transformed by constructing the Lagrangian), the same is not true for the second-order conditions. This is because the second-order conditions for an unconstrained maximization problem ensure that all small changes in any choice variable (or set of choice variables) lead to lower values of the objective function. For a constrained maximization problem, the second-order conditions play a similar role; but we do not need to consider all changes in the choice variables because some changes will result in a violation of the constraint. So the second-order conditions ensure that there is no combination of changes in the choice variables that both satisfies the constraint and leads to a higher value of the objective function.

As is the case with unconstrained optimization problems, the second-order conditions for a constrained maximization problem are conditions on the matrix of second derivatives of the function being maximized. The matrix of second derivatives of the Lagrangian with respect to $\mathbf{x}$ and λ is called a **bordered Hessian** because it contains the second derivatives of the Lagrangian with respect to the choice variables $\mathbf{x}$ (that is, the *Hessian*) *bordered* by the second derivatives of the Lagrangian with respect to λ. These second derivatives of the Lagrangian with respect to λ turn out to be the same as the first derivatives of the constraint function $g(\mathbf{x}, \boldsymbol{\alpha})$ with respect to the choice variables and the Lagrange multiplier. In symbols, the bordered Hessian of the Lagrangian (9.1) is denoted $\bar{H}$ and equals[8]

$$\bar{H} = \begin{bmatrix} \dfrac{\partial^2 \mathcal{L}}{\partial x_1^2} & \cdots & \dfrac{\partial^2 \mathcal{L}}{\partial x_1 \partial x_n} & \dfrac{\partial^2 \mathcal{L}}{\partial x_1 \partial \lambda} \\ \vdots & \ddots & \vdots & \vdots \\ \dfrac{\partial^2 \mathcal{L}}{\partial x_n \partial x_1} & \cdots & \dfrac{\partial^2 \mathcal{L}}{\partial x_n^2} & \dfrac{\partial^2 \mathcal{L}}{\partial x_n \partial \lambda} \\ \dfrac{\partial^2 \mathcal{L}}{\partial \lambda \partial x_1} & \cdots & \dfrac{\partial^2 \mathcal{L}}{\partial \lambda \partial x_n} & \dfrac{\partial^2 \mathcal{L}}{\partial \lambda^2} \end{bmatrix}$$

$$= \begin{bmatrix} \dfrac{\partial^2 f}{\partial x_1^2} - \lambda \dfrac{\partial^2 g}{\partial x_1^2} & \cdots & \dfrac{\partial^2 f}{\partial x_1 \partial x_n} - \lambda \dfrac{\partial^2 g}{\partial x_1 \partial x_n} & -\dfrac{\partial g}{\partial x_1} \\ \vdots & \ddots & \vdots & \vdots \\ \dfrac{\partial^2 f}{\partial x_n \partial x_1} - \lambda \dfrac{\partial^2 g}{\partial x_n \partial x_1} & \cdots & \dfrac{\partial^2 f}{\partial x_n^2} - \lambda \dfrac{\partial^2 g}{\partial x_n^2} & -\dfrac{\partial g}{\partial x_n} \\ -\dfrac{\partial g}{\partial x_1} & \cdots & -\dfrac{\partial g}{\partial x_n} & 0 \end{bmatrix}.$$

(9.12)

---

[7]As always, the second-order conditions only ensure a local optimum. There may be other local optima that must also be considered, and there may be corner solutions. If corner solutions are ruled out and the second-order conditions hold for all values of the choice variables, then the local optimum will also be a global optimum. In unconstrained optimization problems, there is a relationship between second-order conditions and concavity of the objective function. There is a similar relationship for constrained optimization problems, as explained in Section 9.2.5.

[8]Some authors place the border in the first row and first column. This has no effect on the second-order conditions.

As is the case with unconstrained optimization, the second-order conditions have to do with principal minors of this matrix. However, for a constrained optimization problem only border-preserving principal minors matter. A **border-preserving principal minor** of order r of the bordered Hessian (9.12) is the determinant of the matrix obtained by deleting $n - r$ rows and the corresponding columns from (9.12), with the proviso that the border row and border column cannot be among those deleted.[9] In this way, the border is preserved. Note that a border-preserving principal minor of order r has $r + 1$ rows and columns: r rows and columns of the Hessian plus the border.

If $f(x_1, x_2) = 2x_1^2 + x_2^2 + 10x_1x_2$ and $g(x_1, x_2) = x_1 + 2x_2$, the bordered Hessian would be

$$\bar{H} = \begin{bmatrix} 4 & 10 & -1 \\ 10 & 2 & -2 \\ -1 & -2 & 0 \end{bmatrix} \tag{9.13}$$

with first-order border-preserving principal minors

$$\begin{vmatrix} 2 & -2 \\ -2 & 0 \end{vmatrix} \quad \text{and} \quad \begin{vmatrix} 4 & -1 \\ -1 & 0 \end{vmatrix} \tag{9.14}$$

and a second-order border-preserving principal minor that is the determinant of the bordered Hessian (9.13) itself. (The second-order border-preserving principal minor is the determinant of the matrix formed by deleting $n - 2$ rows and columns. But in this example $n = 2$, so no rows or columns are deleted.)

We are now ready for a statement of the second-order conditions for a constrained maximization problem:

Second-Order Conditions (Maximization)

When the first-order conditions (9.2) are satisfied, a sufficient condition for the Lagrangian (9.1) to be maximized is for the border-preserving principal minors of order r of the bordered Hessian (9.12) to be of sign $(-1)^r$ for $r = 2, \ldots, n$. That is, the border-preserving principal minors alternate in sign, starting with positive second-order minors.

First-order principal minors do not appear in the second-order conditions because they are always negative. This is because, as illustrated by the determinants (9.14), each is the determinant of a (2×2) symmetric matrix with a zero in the lower right-hand corner. Also, it is unnecessary to check all of the principal minors of any order to see that they all have the correct sign. Because (by Young's Theorem) the bordered Hessian is symmetric, it turns out that all minors of the same order will have the same sign. Therefore we sometimes give the second-order conditions only in terms of the **leading principal minors**, the principal minors formed by deleting the *last* $n - r$ rows and columns of the Hessian (leaving the border of the bordered Hessian). This formulation of the second-order conditions is convenient when checking whether the second-order conditions are satisfied for particular cases. But when trying to sign comparative static derivatives, it is useful to remember that the second-order conditions give the signs of *all* border-preserving principal minors.

For the bordered Hessian (9.13), the second-order border-preserving principal minor is $-(-20 + 2) + 2(-8 + 10) = 22$, which is the determinant of $\bar{H}$ itself, obtained by expanding the third row. Since it is positive, the second-order conditions are

---

[9]The elements of the border corresponding to the deleted rows and columns are deleted, however.

satisfied and the solution to the first-order conditions (9.6) yields a (constrained) maximization of the objective function.

As an example of the second-order conditions for a constrained maximization problem, consider a three-good utility-maximization problem with the utility function $U(x_1, x_2, x_3) = x_1 x_2 x_3$. The consumer wants to maximize this function subject to the constraint that $p_1 x_1 + p_2 x_2 + p_3 x_3 \leq I$. The Lagrangian for the problem is

$$\mathcal{L}(x_1, x_2, x_3) = x_1 x_2 x_3 + \lambda(I - p_1 x_1 - p_2 x_2 - p_3 x_3) \tag{9.15}$$

and the first-order conditions are

$$\frac{\partial \mathcal{L}}{\partial x_1} = x_2 x_3 - \lambda p_1 = 0$$

$$\frac{\partial \mathcal{L}}{\partial x_2} = x_1 x_3 - \lambda p_2 = 0$$

$$\frac{\partial \mathcal{L}}{\partial x_3} = x_1 x_2 - \lambda p_3 = 0 \tag{9.16}$$

$$\frac{\partial \mathcal{L}}{\partial \lambda} = I - p_1 x_1 - p_2 x_2 - p_3 x_3 = 0.$$

The bordered Hessian of the Lagrangian (9.15) is

$$\bar{H} = \begin{bmatrix} 0 & x_3 & x_2 & -p_1 \\ x_3 & 0 & x_1 & -p_2 \\ x_2 & x_1 & 0 & -p_3 \\ -p_1 & -p_2 & -p_3 & 0 \end{bmatrix}. \tag{9.17}$$

Assuming that the first-order conditions are satisfied, the second-order conditions require the second-order border-preserving principal minors of the bordered Hessian (9.17) to be positive and the third-order border-preserving principal minor to be negative.

The leading second-order border-preserving principal minor is the determinant

$$\begin{vmatrix} 0 & x_3 & -p_1 \\ x_3 & 0 & -p_2 \\ -p_1 & -p_2 & 0 \end{vmatrix} = 0 - x_3 \begin{vmatrix} x_3 & -p_1 \\ -p_2 & 0 \end{vmatrix} - p_1 \begin{vmatrix} x_3 & -p_1 \\ 0 & -p_2 \end{vmatrix} = 2 p_1 p_2 x_3 > 0. \tag{9.18}$$

The third-order border-preserving principal minor is simply the determinant of the whole matrix in (9.17), since $(n - r) = (3 - 3) = 0$ rows and columns are deleted. Expanding by the first column, we find that

$$|\bar{H}| = 0 - x_3 \begin{vmatrix} x_3 & x_2 & -p_1 \\ x_1 & 0 & -p_3 \\ -p_2 & -p_3 & 0 \end{vmatrix} + x_2 \begin{vmatrix} x_3 & x_2 & -p_1 \\ 0 & x_1 & -p_2 \\ -p_2 & -p_3 & 0 \end{vmatrix} + p_1 \begin{vmatrix} x_3 & x_2 & -p_1 \\ 0 & x_1 & -p_2 \\ x_1 & 0 & -p_3 \end{vmatrix}$$

$$= -x_3(-p_2(-p_3 x_2) + p_3(-p_3 x_3 + p_1 x_1)) + x_2(-p_2(-p_2 x_2 + p_1 x_1) + p_3(-p_2 x_3))$$

$$\quad + p_1(x_1(-p_2 x_2 + p_1 x_1) - p_3(x_1 x_3))$$

$$= -x_2 p_2 x_3 p_3 + (p_3 x_3)^2 - x_1 p_1 x_3 p_3 + (p_2 x_2)^2 - x_1 p_1 x_2 p_2 - x_2 p_2 x_3 p_3$$

$$\quad - x_1 p_1 x_2 p_2 + (p_1 x_1)^2 - x_1 p_1 x_3 p_3, \tag{9.19}$$

which can be shown to be negative since the first-order conditions (9.16) indicate that $x_1 p_1 = x_2 p_2 = x_3 p_3$.[10] The determinant (9.19) then reduces to the sum of three negative terms.

---

[10]This can be shown by multiplying through the first first-order condition by x_1, the second by x_2, and the third by x_3. For each good, $x_i p_i = x_1 x_2 x_3$.

9.2.3 Minimization Problems

Some optimization problems in economics are constrained minimization problems rather than constrained maximization problems. For example, a firm's cost-minimization problem is to choose the level of inputs to minimize the cost of production, subject to the constraint that the amount produced is at least equal to some predetermined level. We can use the Lagrangian function to solve such problems. The Lagrangian for the problem of minimizing $f(\mathbf{x}, \boldsymbol{\alpha})$ subject to the constraint that $g(\mathbf{x}, \boldsymbol{\alpha}) \geq g_0$ is

$$\mathscr{L}(\mathbf{x}, \boldsymbol{\alpha}, \lambda, g_0) = f(\mathbf{x}, \boldsymbol{\alpha}) + \lambda(g_0 - g(\mathbf{x}, \boldsymbol{\alpha})). \tag{9.20}$$

The only difference between this Lagrangian and the one for a maximization problem is that we write the constraint so it is nonpositive. (As is the case with maximization problems, the only reason this matters is for the interpretation of the Lagrange multiplier.)

We derive the first-order conditions for the minimization of (9.20) exactly as in the maximization case, by setting equal to zero the derivatives of the Lagrangian with respect to the Lagrange multiplier and each of the choice variables. The second-order conditions are slightly different:

Second–Order Conditions (Minimization)

Assuming that the first-order conditions (9.2) are satisfied, a sufficient condition for the Lagrangian (9.20) to be minimized is that the border-preserving principal minors of all orders (greater than 1) of the bordered Hessian of Lagrangian (9.20) are negative.

As an example we will return to the utility-maximization example used in Section 9.2.1 and examine the **dual problem** of expenditure minimization. The utility-maximization and expenditure-minimization problems are *dual to each other* in the sense that for every utility-maximization problem, there is an equivalent expenditure-minimization problem and *vice versa*. (We will discuss expenditure-minimization problems and duality in consumer theory in more detail in Chapters 13 and 14.) The expenditure-minimization problem for the consumer is to choose the consumption level of each good to minimize the total expenditures necessary to achieve a given level of utility. The expenditure-minimization problem is mathematically identical to a firm's costs-minimization problem, which is one of the applications discussed in Chapter 10.

The Lagrangian for the problem of minimizing total expenditures $p_1 x_1 + \cdots + p_n x_n$ subject to the constraint that utility $U(x_1, \ldots, x_n)$ equals at least the level U_0 is

$$\mathscr{L}(x_1, \ldots, x_n, \lambda, p_1, \ldots, p_n, U_0) = p_1 x_1 + \cdots + p_n x_n + \lambda(U_0 - U(x_1, \ldots, x_n)). \tag{9.21}$$

The first-order conditions are

$$\frac{\partial \mathscr{L}}{\partial x_i} = p_i - \lambda \frac{\partial U}{\partial x_i} = 0 \quad i = 1, \ldots, n$$

$$\frac{\partial \mathscr{L}}{\partial \lambda} = U_0 - U(x_1, \ldots, x_n) = 0. \tag{9.22}$$

The last of these first-order conditions ensures that the consumer gets utility of exactly U_0. The solutions to the first-order conditions will give the **compensated** or **Hicksian demand functions**, which are the expenditure-minimizing consumptions of each good as a function of prices and the consumer's utility level:

$$x_i^*(p_1, \ldots, p_n, U_0), \quad i = 1, \ldots, n. \tag{9.23}$$

By eliminating the Lagrange multiplier λ from the first two of the first-order conditions (9.22), we obtain

$$\frac{p_1}{p_2} = \frac{\partial U / \partial x_1}{\partial U / \partial x_2}. \tag{9.24}$$

The right-hand side of equation (9.24) is the ratio of marginal utilities of goods 1 and 2. This ratio, called the marginal rate of substitution between those goods (see Section 6.9), is the absolute value of the slope of a level curve of the utility function, which is the constraint function in this minimization problem. The left-hand side of equation (9.24) is the ratio of prices, which is the absolute value of the slope of a level curve of the objective function (total expenditures). Thus the first-order conditions state that when minimizing the cost of achieving a particular level of utility U_0, a consumer must find the combination of goods that results in a tangency between a budget line and the indifference curve corresponding to the utility level U_0, as illustrated in Figure 9.5. Comparing this figure with Figure 9.4 clearly shows the duality of the expenditure-minimization and

FIGURE 9.5

Expenditure Minimization

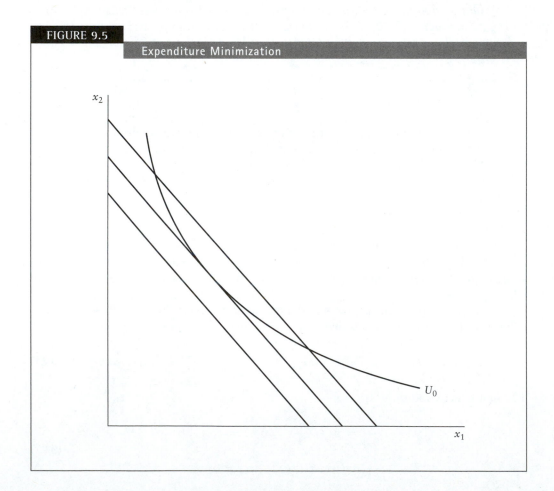

utility-maximization problems, as does a comparison of equation (9.24) with the corresponding equation (9.10) from the utility-maximization problem.

Assuming that the first-order conditions are satisfied, the second-order conditions are that the border-preserving principal minors of order 2 though n of the bordered Hessian of the problem are negative. (Border-preserving principal minors of order 1 do not appear in the second-order conditions, because they are always negative.) The bordered Hessian for this problem is

$$\bar{H} = \begin{bmatrix} -\lambda \dfrac{\partial^2 U}{\partial x_1^2} & \cdots & -\lambda \dfrac{\partial^2 U}{\partial x_1 \partial x_n} & -\dfrac{\partial U}{\partial x_1} \\ \vdots & \ddots & \vdots & \vdots \\ -\lambda \dfrac{\partial^2 U}{\partial x_n \partial x_1} & \cdots & -\lambda \dfrac{\partial^2 U}{\partial x_n^2} & -\dfrac{\partial U}{\partial x_n} \\ -\dfrac{\partial U}{\partial x_1} & \cdots & -\dfrac{\partial U}{\partial x_n} & 0 \end{bmatrix}. \tag{9.25}$$

The second-order conditions place restrictions on the properties of the utility function, as we will explain in Section 9.2.5. If we are using a particular utility function, we can check the second-order conditions. Frequently, however, we leave the utility function in general form and use the second-order conditions to help sign comparative static derivatives, as shown in Section 9.3.

9.2.4   Multiple Constraints

Although most problems in economics have a single constraint (except for nonnegativity constraints, which we consider in Chapter 13), some have more than one constraint. The Lagrangian technique can be extended easily to deal with multiple constraints. To maximize the function $f(\mathbf{x}, \boldsymbol{\alpha})$ subject to m constraints $g^1(\mathbf{x}, \boldsymbol{\alpha}) \leq g_1, \ldots, g^m(\mathbf{x}, \boldsymbol{\alpha}) \leq g_m$, we form the Lagrangian using m Lagrange multipliers:[11]

$$\mathscr{L}(\mathbf{x}, \boldsymbol{\alpha}, \lambda_1, \ldots, \lambda_m, g_1, \ldots, g_m) = f(\mathbf{x}, \boldsymbol{\alpha}) + \lambda_1(g_1 - g^1(\mathbf{x}, \boldsymbol{\alpha})) + \cdots$$
$$+ \lambda_m(g_m - g^m(\mathbf{x}, \boldsymbol{\alpha})). \tag{9.26}$$

The first-order conditions are that the derivatives of $\mathscr{L}$ with respect to the choice variables $\mathbf{x}$ and all the Lagrange multipliers must equal zero. The second-order conditions are that the border-preserving principal minors of order r of the bordered Hessian of (9.26) must be of sign $(-1)^r$, for $r = m + 1, \ldots, n$. The bordered Hessian has m rows and columns in the border. For a minimization problem, the border-preserving principal minors of order $r = m + 1, \ldots, n$ must all have sign $(-1)^m$. If there are two constraints, for example, all of the relevant principal minors must be positive; if there are three constraints, all of the relevant principal minors must be negative. Economics problems that have multiple constraints include the utility-maximizing choice of consumption goods when both income and time constraints are binding and the analysis of the Pareto-optimal allocation of goods among consumers. Both of these examples are discussed in Chapter 10.

---

[11]The number of choice variables in x must be greater than the number of constraints, for reasons we discuss in Section 10.6.

9.2.5 Quasiconcavity, Quasiconvexity, and Constrained Optimization Problems

In unconstrained optimization problems a relationship exists between the second-order conditions and the concavity/convexity of the objective function, as discussed in Section 7.7. In this section we explain a similar relationship between the second-order conditions of a constrained optimization problem and the properties of the objective and constraint functions. Instead of concavity and convexity, the relevant properties turn out to be quasiconcavity and quasiconvexity, which we now define:

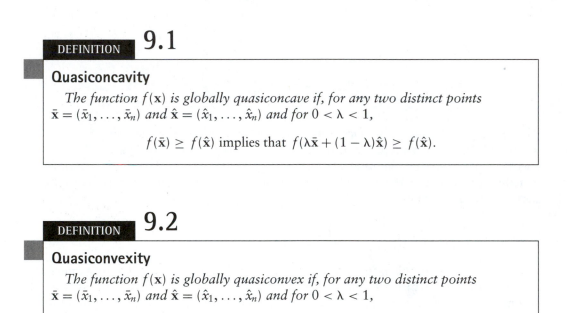

DEFINITION **9.1**

Quasiconcavity

The function $f(\mathbf{x})$ is globally quasiconcave if, for any two distinct points $\bar{\mathbf{x}} = (\bar{x}_1, \ldots, \bar{x}_n)$ *and* $\hat{\mathbf{x}} = (\hat{x}_1, \ldots, \hat{x}_n)$ *and for* $0 < \lambda < 1$,

$$f(\bar{\mathbf{x}}) \geq f(\hat{\mathbf{x}}) \text{ implies that } f(\lambda \bar{\mathbf{x}} + (1 - \lambda)\hat{\mathbf{x}}) \geq f(\hat{\mathbf{x}}).$$

DEFINITION **9.2**

Quasiconvexity

The function $f(\mathbf{x})$ is globally quasiconvex if, for any two distinct points $\bar{\mathbf{x}} = (\bar{x}_1, \ldots, \bar{x}_n)$ *and* $\hat{\mathbf{x}} = (\hat{x}_1, \ldots, \hat{x}_n)$ *and for* $0 < \lambda < 1$,

$$f(\bar{\mathbf{x}}) \leq f(\hat{\mathbf{x}}) \text{ implies that } f(\lambda \bar{\mathbf{x}} + (1 - \lambda)\hat{\mathbf{x}}) \leq f(\hat{\mathbf{x}}).$$

If the last lines of Definitions 9.1 and 9.2 hold with strict inequality, the function is **strictly quasiconcave** or **strictly quasiconvex**. If the last lines of Definitions 9.1 and 9.2 hold only in the vicinity of particular values of **x**, the function is **locally quasiconcave** or **locally quasiconvex**. All concave (convex) functions are quasiconcave (quasiconvex), and all monotonic transformations[12] of concave (convex) functions are quasiconcave (quasiconvex); but there are quasiconcave (quasiconvex) functions that are neither concave (convex) nor monotonic transformations of concave (convex) functions. Linear functions, being both concave and convex (though neither strictly), are both quasiconcave and quasiconvex.

The characterization of quasiconcave and quasiconvex functions that is most useful for our purposes has to do with their level curves. A quasiconcave function has level curves that are convex to the origin, while a quasiconvex function has level curves that are concave to the origin. Figure 9.6 shows level curves of three functions: Figure 9.6a shows a level curve of a quasiconcave function, Figure 9.6b shows a level curve of a

---

[12]Monotonic transformations were defined in Section 6.9. A function $g(f(\mathbf{x}))$ is a monotonic transformation of $f(\mathbf{x})$ if $g' > 0$; the value of g will increase if and only if the value of f increases.

FIGURE 9.6

Level Curves of Functions

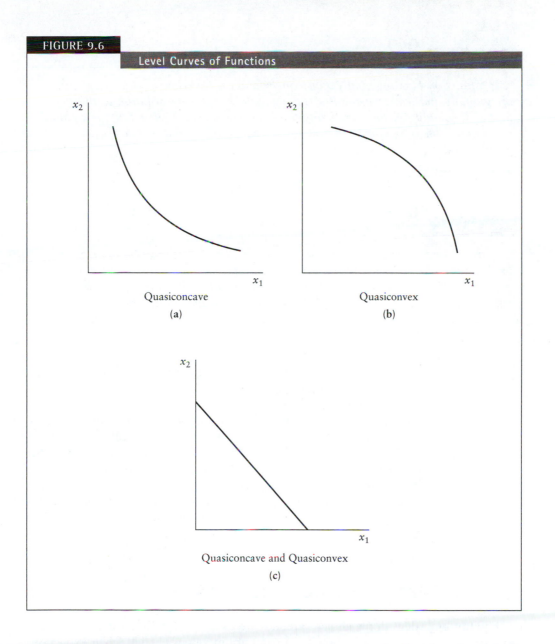

Quasiconcave

(a)

Quasiconvex

(b)

Quasiconcave and Quasiconvex

(c)

quasiconvex function, and Figure 9.6c shows a level curve of a function that is both quasiconcave and quasiconvex.

The following set of theorems describes the relationship between, on the one hand, quasiconcavity and quasiconvexity of objective and constraint functions, and, on the other hand, the second-order conditions of a contrained optimization problem:

THEOREM   **9.1**

If a function $f(\mathbf{x})$ is globally quasiconcave and the functions $g^1(\mathbf{x}), \ldots, g^m(\mathbf{x})$ are globally quasiconvex and if $\mathbf{x}^*$ (along with $\boldsymbol{\lambda}^* > 0$) is the solution to the first-order conditions for maximizing the Lagrangian (9.26), then $f(\mathbf{x}^*)$ is a global maximum value of the function $f(\mathbf{x})$ subject to the constraints $g^1(\mathbf{x}) \le g_1, \ldots, g^m(\mathbf{x}) \le g_m$.

> *If a function $f(\mathbf{x})$ is globally quasiconvex and the functions $g^1(\mathbf{x}), \ldots, g^m(\mathbf{x})$ are globally quasiconcave and if $\mathbf{x}^*$ (along with $\boldsymbol{\lambda}^* > 0$) is the solution to the first-order conditions for minimizing the Lagrangian (9.26), then $f(\mathbf{x}^*)$ is a global minimum value of the function $f(\mathbf{x})$ subject to the constraints $g^1(\mathbf{x}) \geq g_1, \ldots, g^m(\mathbf{x}) \geq g_m$.*

> *If the second-order conditions for a constrained maximization (minimization) problem are satisfied, then the objective function is (locally) quasiconcave (quasiconvex). If the second-order conditions are satisfied for all values of the choice variables, the objective function is globally quasiconcave (quasiconvex).*

The implications of this set of theorems are similar to the results discussed in Section 7.7 for unconstrained optimization problems. With quasiconvex constraints (which include linear constraints), the knowledge (or assumption) of (global) quasiconcavity of the objective function implies that the first-order conditions are both necessary and sufficient conditions for a constrained maximization problem. With quasiconcave constraints, the knowledge (or assumption) of (global) quasiconvexity of the objective function implies that the first-order conditions are both necessary and sufficient for constrained minimization problems. Since showing quasiconcavity (quasiconvexity) is usually quite difficult, analysts typically rely on the second-order conditions; if the second-order conditions hold, we can be sure that the objective function has the appropriate property.

9.3 COMPARATIVE STATIC DERIVATIVES

Equilibrium values of the choice variables are usually of less interest than comparative static derivatives, which show how equilibrium values change when exogenous variables or parameters change. We derive comparative static derivatives in constrained optimization problems in essentially the same way that we used in earlier chapters. If the first-order conditions have explicit solutions for the equilibrium values of the choice variables, we can take the derivatives directly. Usually, however, explicit solutions are difficult or impossible to obtain; in that case we can obtain comparative static derivatives by implicit differentiation or by finding total differentials.

9.3.1 The Implicit Differentiation Approach

The first-order conditions (9.2) implicitly define the equilibrium values of $\mathbf{x}$ and λ as functions of the exogenous variables and parameters $\boldsymbol{\alpha}$ and g_0. To find out how the equilibrium changes when one of the parameters, say α_0, changes, we first implicitly differentiate the first-order conditions with respect to α_0. That is, we differentiate through the first-order conditions with respect to α_0, treating λ and the elements of $\mathbf{x}$ as functions of α_0. This gives a system of $n+1$ linear equations in the $n+1$ unknowns $\partial x_1^* / \partial \alpha_0, \ldots, \partial x_n^* / \partial \alpha_0$, and $\partial \lambda^* / \partial \alpha_0$, the first of which is

$$
\left(\frac{\partial^2 f}{\partial x_1^2} - \lambda \frac{\partial^2 g}{\partial x_1^2} \right) \frac{\partial x_1^*}{\partial \alpha_0} + \cdots + \left(\frac{\partial^2 f}{\partial x_1 \partial x_n} - \lambda \frac{\partial^2 g}{\partial x_1 \partial x_n} \right) \frac{\partial x_n^*}{\partial \alpha_0}
$$
$$
- \frac{\partial g}{\partial x_1} \frac{\partial \lambda^*}{\partial \alpha_0} + \frac{\partial^2 f}{\partial x_1 \partial \alpha_0} - \lambda \frac{\partial^2 g}{\partial x_1 \partial \alpha_0} = 0. \tag{9.27}
$$

Written in matrix notation, these equations are

$$
\begin{bmatrix}
\dfrac{\partial^2 f}{\partial x_1^2} - \lambda \dfrac{\partial^2 g}{\partial x_1^2} & \cdots & \dfrac{\partial^2 f}{\partial x_1 \partial x_n} - \lambda \dfrac{\partial^2 g}{\partial x_1 \partial x_n} & -\dfrac{\partial g}{\partial x_1} \\
\vdots & \ddots & \vdots & \vdots \\
\dfrac{\partial^2 f}{\partial x_n \partial x_1} - \lambda \dfrac{\partial^2 g}{\partial x_n \partial x_1} & \cdots & \dfrac{\partial^2 f}{\partial x_n^2} - \lambda \dfrac{\partial^2 g}{\partial x_n^2} & -\dfrac{\partial g}{\partial x_n} \\
-\dfrac{\partial g}{\partial x_1} & \cdots & -\dfrac{\partial g}{\partial x_n} & 0
\end{bmatrix}
\begin{bmatrix}
\dfrac{\partial x_1^*}{\partial \alpha_0} \\
\vdots \\
\dfrac{\partial x_n^*}{\partial \alpha_0} \\
\dfrac{\partial \lambda^*}{\partial \alpha_0}
\end{bmatrix}
$$
$$
\tag{9.28}
$$
$$
= \begin{bmatrix}
-\dfrac{\partial^2 f}{\partial x_1 \partial \alpha_0} + \lambda \dfrac{\partial^2 g}{\partial x_1 \partial \alpha_0} \\
\vdots \\
-\dfrac{\partial^2 f}{\partial x_n \partial \alpha_0} + \lambda \dfrac{\partial^2 g}{\partial x_n \partial \alpha_0} \\
\dfrac{\partial g}{\partial \alpha_0}
\end{bmatrix}.
$$

The determinant of the first matrix in equation (9.28) is the Jacobian of the system of first-order equations (9.2). But this first matrix is also the bordered Hessian of the Lagrangian (9.1). Thus the Jacobian of the first-order conditions is the determinant of the bordered Hessian, which is in turn a border-preserving principal minor (of order n) of itself. Thus, as is the case with unconstrained optimization, the second-order conditions give the sign of the Jacobian: it will be negative for a minimization problem and have sign $(-1)^n$ for a maximization problem.

If we can sign the numerator when we use Cramer's rule to solve the system of equations (9.28) for a comparative static derivative, we can sign the comparative static derivative. Signing the numerator requires one of two things. We must either know enough about the functional forms of the objective function and the constraint, or else we must be able to write the numerator in terms of border-preserving principal minors of the bordered Hessian. Otherwise, the sign of the comparative static derivative will be indeterminate.

As a simple example of the procedure, we look again at the consumer's expenditure-minimization problem, whose first-order conditions are given by (9.22). For simplicity we will consider the case of just two goods. (The general case is explored as an end-of-chapter problem.) To find out how the consumer's expenditure-minimizing demand for good 1 changes when there is an increase in its price, we implicitly differentiate (9.22) with respect to p_1 (treating x_1, x_2, and λ as functions of p_1) to get

$$
\begin{bmatrix}
-\lambda \dfrac{\partial^2 U}{\partial x_1^2} & -\lambda \dfrac{\partial^2 U}{\partial x_1 \partial x_2} & -\dfrac{\partial U}{\partial x_1} \\[2ex]
-\lambda \dfrac{\partial^2 U}{\partial x_2 \partial x_1} & -\lambda \dfrac{\partial^2 U}{\partial x_2^2} & -\dfrac{\partial U}{\partial x_2} \\[2ex]
-\dfrac{\partial U}{\partial x_1} & -\dfrac{\partial U}{\partial x_2} & 0
\end{bmatrix}
\begin{bmatrix}
\dfrac{\partial x_1^*}{\partial p_1} \\[2ex]
\dfrac{\partial x_2^*}{\partial p_1} \\[2ex]
\dfrac{\partial \lambda^*}{\partial p_1}
\end{bmatrix}
=
\begin{bmatrix}
-1 \\ 0 \\ 0
\end{bmatrix}.
\tag{9.29}
$$

Note that the first matrix in equation (9.29) is the bordered Hessian (9.25) of the minimization problem. Using Cramer's rule to solve for $\partial x_1^*/\partial p_1$, we have

$$
\frac{\partial x_1^*}{\partial p_1} =
\frac{
\begin{vmatrix}
-1 & -\lambda(\partial^2 U/\partial x_1 \partial x_2) & -\partial U/\partial x_1 \\
0 & -\lambda(\partial^2 U/\partial x_2^2) & -\partial U/\partial x_2 \\
0 & -\partial U/\partial x_2 & 0
\end{vmatrix}
}{|J|}.
\tag{9.30}
$$

Since the Jacobian is the determinant of the bordered Hessian for the expenditure-minimization problem, and since this is a constrained minimization problem, the Jacobian is negative. Expanding the numerator by the first column, the numerator equals $-1(0 - (-\partial U/\partial x_2)^2) > 0$. Thus if the price of good 1 increases, the consumer's *compensated* demand (recall that the expenditure-minimization problem yields compensated, not ordinary, demand functions) for good 1 decreases.

9.3.2 The Total Differential Approach

Sometimes it is more convenient to use the total differential approach to deriving comparative statics. This is especially true when we want the derivatives with respect to several different parameters or when more than one parameter is changing. The total differential approach starts with taking the total differentials of the first-order conditions (9.2). Then we solve for the differentials of all endogenous variables as functions of the differentials of the parameters (exogenous variables). These functions show how the equilibrium values of the choice variables change when the environment changes.

The total differentials of the first-order conditions (9.2) are given by

$$
\left(\frac{\partial^2 f}{\partial x_i \partial x_1} - \lambda \frac{\partial^2 g}{\partial x_i \partial x_1} \right) dx_1 + \cdots + \left(\frac{\partial^2 f}{\partial x_i \partial x_n} - \lambda \frac{\partial^2 g}{\partial x_i \partial x_n} \right) dx_n - \frac{\partial g}{\partial x_i}\, d\lambda
$$

$$
+ \left(\frac{\partial^2 f}{\partial x_1 \partial \alpha_1} - \lambda \frac{\partial^2 g}{\partial x_1 \partial \alpha_1} \right) d\alpha_1 + \cdots
$$

$$
\tag{9.31}
$$

$$
+ \left(\frac{\partial^2 f}{\partial x_1 \partial \alpha_k} - \lambda \frac{\partial^2 g}{\partial x_1 \partial \alpha_k} \right) d\alpha_k = 0, \quad i = 1, \ldots, n
$$

and $\quad -\dfrac{\partial g}{\partial x_1} dx_1 - \cdots - \dfrac{\partial g}{\partial x_n} dx_n - \dfrac{\partial g}{\partial \alpha_1} d\alpha_1 - \cdots - \dfrac{\partial g}{\partial \alpha_k} d\alpha_k = 0$

where k is the number of parameters. Writing equations (9.31) in matrix notation, we have

$$\bar{H}\begin{bmatrix} dx_1 \\ \vdots \\ dx_n \\ d\lambda \end{bmatrix} = -\begin{bmatrix} \left(\dfrac{\partial^2 f}{\partial x_1 \partial \alpha_1} - \lambda\dfrac{\partial^2 g}{\partial x_1 \partial \alpha_1}\right)d\alpha_1 + \cdots + \left(\dfrac{\partial^2 f}{\partial x_1 \partial \alpha_k} - \lambda\dfrac{\partial^2 g}{\partial x_1 \partial \alpha_k}\right)d\alpha_k \\ \vdots \\ \left(\dfrac{\partial^2 f}{\partial x_n \partial \alpha_1} - \lambda\dfrac{\partial^2 g}{\partial x_n \partial \alpha_1}\right)d\alpha_1 + \cdots + \left(\dfrac{\partial^2 f}{\partial x_n \partial \alpha_k} - \lambda\dfrac{\partial^2 g}{\partial x_n \partial \alpha_k}\right)d\alpha_k \\ -\dfrac{\partial g}{\partial \alpha_1}d\alpha_1 - \cdots - \dfrac{\partial g}{\partial \alpha_k}d\alpha_k \end{bmatrix}$$

$$(9.32)$$

where

$$\bar{H} = \begin{bmatrix} \dfrac{\partial^2 f}{\partial x_1^2} - \lambda\dfrac{\partial^2 g}{\partial x_1^2} & \cdots & \dfrac{\partial^2 f}{\partial x_1 \partial x_n} - \lambda\dfrac{\partial^2 g}{\partial x_1 \partial x_n} & -\dfrac{\partial g}{\partial x_1} \\ \vdots & \ddots & \vdots & \vdots \\ \dfrac{\partial^2 f}{\partial x_n \partial x_1} - \lambda\dfrac{\partial^2 g}{\partial x_n \partial x_1} & \cdots & \dfrac{\partial^2 f}{\partial x_n^2} - \lambda\dfrac{\partial^2 g}{\partial x_n^2} & -\dfrac{\partial g}{\partial x_n} \\ -\dfrac{\partial g}{\partial x_1} & \cdots & -\dfrac{\partial g}{\partial x_n} & 0 \end{bmatrix} \qquad (9.33)$$

is the bordered Hessian of the Lagrangian (9.1). The matrices in (9.32) and (9.33) look formidable, but are often much simpler than suggested by the general case shown. Frequently either the objective function or the constraint is linear in the choice variables, making second derivatives zero. Also, we are usually interested in changes in just one or a few of the parameters, in which case many of the terms in the matrix on the right-hand side of equation (9.32) will equal zero.

The determinant of matrix (9.33) is the Jacobian of the first-order conditions (9.2). Since the matrix is a bordered Hessian, its determinant is a border-preserving principal minor (of order n) of itself. Thus the second-order conditions give the sign of the Jacobian: it will be negative for a minimization problem and it will have sign $(-1)^n$ for a maximization problem.

As is the case with the implicit differentiation approach to comparative statics, we can use Cramer's rule to do comparative static analysis; we will be able to find the sign of the comparative static differentials $dx_1^*, \ldots, dx_n^*$ and $d\lambda^*$ if we can sign the numerator. This is possible if we know enough about the functional forms of the objective and constraint functions or if we can write the numerator in terms of border-preserving principal minors of the bordered Hessian.

To illustrate, we will return to the utility-maximization problem with a specific utility function discussed at the end of Section 9.2.2. The Lagrangian for this problem is given by equation (9.15). The first-order conditions (9.16) are repeated here for convenience:

$$\frac{\partial \mathscr{L}}{\partial x_1} = x_2 x_3 - \lambda p_1 = 0$$

$$\frac{\partial \mathscr{L}}{\partial x_2} = x_1 x_3 - \lambda p_2 = 0$$

$$\frac{\partial \mathscr{L}}{\partial x_3} = x_1 x_2 - \lambda p_3 = 0 \qquad (9.34)$$

$$\frac{\partial \mathscr{L}}{\partial \lambda} = I - p_1 x_1 - p_2 x_2 - p_3 x_3 = 0.$$

The total differentials of these equations are

$$x_3\, dx_2 + x_2\, dx_3 - p_1\, d\lambda - \lambda\, dp_1 = 0$$
$$x_3\, dx_1 + x_1\, dx_3 - p_2\, d\lambda - \lambda\, dp_2 = 0$$
$$x_2\, dx_1 + x_1\, dx_2 - p_3\, d\lambda - \lambda\, dp_3 = 0$$
$$dI - x_1\, dp_1 - p_1\, dx_1 - x_2\, dp_2 - p_2\, dx_2 - x_3\, dp_3 - p_3\, dx_3 = 0. \qquad (9.35)$$

Written in matrix notation, this system of equations is

$$
\begin{bmatrix}
0 & x_3 & x_2 & -p_1 \\
x_3 & 0 & x_1 & -p_2 \\
x_2 & x_1 & 0 & -p_3 \\
-p_1 & -p_2 & -p_3 & 0
\end{bmatrix}
\begin{bmatrix}
dx_1 \\ dx_2 \\ dx_3 \\ d\lambda
\end{bmatrix}
=
\begin{bmatrix}
\lambda\, dp_1 \\
\lambda\, dp_2 \\
\lambda\, dp_3 \\
-dI + x_1\, dp_1 + x_2\, dp_2 + x_3\, dp_3
\end{bmatrix}. \qquad (9.36)
$$

Comparing the first matrix of (9.36) with matrix (9.17), it is evident that the matrix of coefficients on the left-hand side of equation (9.36) is the bordered Hessian of the Lagrangian (9.15). When we use Cramer's rule to solve for, say, dx_1^*, the Jacobian in the denominator is the determinant of this bordered Hessian. Since this determinant is a border-preserving principal minor of order 3 of the bordered Hessian, the second-order conditions of the utility-maximization problem say that this determinant is negative. The determinant in the numerator of the solution for dx_1^* is

$$
\begin{vmatrix}
\lambda\, dp_1 & x_3 & x_2 & -p_1 \\
\lambda\, dp_2 & 0 & x_1 & -p_2 \\
\lambda\, dp_3 & x_1 & 0 & -p_3 \\
-dI + x_1\, dp_1 + x_2\, dp_2 + x_3\, dp_3 & -p_2 & -p_3 & 0
\end{vmatrix}. \qquad (9.37)
$$

Suppose that we are interested in finding the comparative static effect on the utility-maximizing consumption of good 1 of an increase in its price. That is, we are trying to show whether the ordinary demand curve slopes downward. For this task, we will set dI, dp_2, and dp_3 equal to zero. Now we can write the determinant (9.37) as

$$
\begin{vmatrix}
\lambda\, dp_1 & x_3 & x_2 & -p_1 \\
0 & 0 & x_1 & -p_2 \\
0 & x_1 & 0 & -p_3 \\
x_1\, dp_1 & -p_2 & -p_3 & 0
\end{vmatrix}
= \lambda\, dp_1
\begin{vmatrix}
0 & x_1 & -p_2 \\
x_1 & 0 & -p_3 \\
-p_2 & -p_3 & 0
\end{vmatrix}
- x_1\, dp_1
\begin{vmatrix}
x_3 & x_2 & -p_1 \\
0 & x_1 & -p_2 \\
x_1 & 0 & -p_3
\end{vmatrix}. \qquad (9.38)
$$

The determinant being multiplied by $\lambda\, dp_1$ on the right-hand side of (9.38) is the border-preserving principal minor of order 2 of the bordered Hessian obtained by deleting the first row and first column. The second-order conditions thus say that it is positive. The determinant being multiplied by $x_1\, dp_1$, however, is not a border-preserving principal minor because it is obtained from the bordered Hessian by deleting the fourth row and first column. The second-order conditions do not therefore give us the sign of this determinant.[13] In fact the second-order conditions can never give us the sign

---

[13]Thus, even though the second-order conditions of the expenditure-minimization problem of Section 9.3.1 were sufficient to ensure that the compensated demand curves slope downward, the second-order conditions of utility-maximization problems are not sufficient to rule out the possibility of **Giffen goods** (inferior goods for which the income effect is stronger than the substitution effect), which have upward-sloping demand curves.

of comparative static effects when there is more than one nonzero element in the vector of exogenous differentials on the right-hand side of equations like (9.36).[14]

When the second-order conditions do not yield enough information to allow us to sign comparative static derivatives, knowledge of the functional forms of the objective and constraint functions can sometimes enable us to sign the derivatives. When we expand the second determinant on the right-hand side of (9.38) by the third row, we get $x_1(-x_2 p_2 + x_1 p_1) - p_3(x_3 x_1)$, the sign of which at first glance appears to be indeterminate. For this example, though, the first-order conditions (9.34) imply that $x_1 p_1 = x_2 p_2$,[15] so the determinant is negative, with the result that the right-hand side of (9.38) has the same sign as dp_1. Thus $\partial x_1^*/\partial p_1$ (which is dx_1^*/dp_1 conditional on $dp_2 = dp_3 = dI = 0$) is a positive divided by a negative, and so the demand curve for x_1 slopes downward.

9.4 VALUE FUNCTIONS AND THE LAGRANGE MULTIPLIER

In this chapter we have focused on the comparative static effects on the choice variables of constrained maximization and minimization problems. In Chapters 13 and 14 we will investigate the envelope theorem, mentioned briefly in several earlier chapters. The envelope theorem gives the comparative static effects of changes in exogenous variables on the value obtained by the objective function when it is maximized or minimized. In this section we give a brief discussion of the envelope theorem in the context of constrained optimization problems so that we may provide and explain an interpretation of Lagrange multipliers.

9.4.1 Value Functions

In either a constrained or unconstrained optimization problem, the implicit function theorem says that as long as the Jacobian of the problem is nonzero, the first-order conditions implicitly define a set of functions giving equilibrium values for the choice variables as functions of the system's exogenous variables and parameters. We obtain the **(optimal) value function,** also called the **indirect objective function,** by substituting into the objective function the functions giving equilibrium values of the choice variables. Thus the value function gives, for any values of the exogenous variables and parameters, the value taken by the objective function when it is optimized. For example, in a one-variable constrained optimization problem with objective function $f(x, \alpha)$, the first-order conditions (which include the constraint as an equality) implicitly define the equilibrium value of x as a function of α (and, from the constraint, g_0), $x^*(\alpha, g_0)$. Substituting $x^*(\alpha, g_0)$ into the objective function gives us the value function $f^*(\alpha, g_0) \equiv f(x^*(\alpha, g_0), \alpha)$. For the many-variable constrained maximization problem with Lagrangian (9.1), the first-order conditions (9.2) implicitly define the equilibrium values of the choice variables as functions of the exogenous variables and parameters. Substituting these back into the objective function yields the value function

$$f^*(\boldsymbol{\alpha}, g_0) \equiv f(\mathbf{x}^*(\boldsymbol{\alpha}, g_0), \boldsymbol{\alpha}). \tag{9.39}$$

[14]This is so because the solution will involve determinants that are not border-preserving principal minors of the bordered Hessian. The same is true for the implicit differentiation approach: if more than one element is nonzero in the vector of constants on the right-hand side of equations like (9.28), the second-order conditions will not be sufficient for signing comparative static derivatives.

[15]The very astute reader will have seen that this implication follows from the symmetry of the problem. It can be shown by eliminating λ from the first two first-order conditions (9.34).

There is a value function associated with every optimization problem. Some of the most important for economic theory are the **profit function**, which gives a firm's maximal profits as a function of the exogenous variables of the profit-maximization problem; the **cost function**, which gives a firm's minimal costs as a function of input prices (and sometimes quantity); the **indirect utility function**, which gives a consumer's maximal utility level as a function of prices and income, and the **expenditure function**, which gives as a function of prices and utility the minimum expenditure necessary for a consumer to achieve a given utility level.

9.4.2   Interpretation of Lagrange Multipliers

One of the advantages of the Lagrangian technique in economics is that the Lagrange multiplier almost always has an interesting interpretation. When the Lagrangian is set up as described in this chapter, the value of the Lagrange multiplier is the marginal effect of a relaxation of its associated constraint on the value of the value function.[16] We will demonstrate this result mathematically, followed by a further discussion of the economic interpretation of the Lagrange multiplier.

Consider the constrained maximization problem with Lagrangian (9.1), first-order conditions (9.2), and value function (9.39). An increase in the value of g_0 relaxes the constraint and changes the equilibrium values of the choice variables $\mathbf{x}$ and λ. To see how these changes affect the value function, differentiate (9.39) with respect to g_0:

$$\frac{\partial f^*}{\partial g_0} = \frac{\partial f}{\partial x_1}\frac{\partial x_1^*}{\partial g_0} + \cdots + \frac{\partial f}{\partial x_n}\frac{\partial x_n^*}{\partial g_0}. \tag{9.40}$$

But, from the first-order conditions (9.2),

$$\frac{\partial f}{\partial x_i}(\mathbf{x}, \boldsymbol{\alpha}) = \lambda\frac{\partial g}{\partial x_i}(\mathbf{x}, \boldsymbol{\alpha}) \quad i = 1, \ldots, n \tag{9.41}$$

so

$$\frac{\partial f^*}{\partial g_0} = \lambda\frac{\partial g}{\partial x_1}\frac{\partial x_1^*}{\partial g_0} + \cdots + \lambda\frac{\partial g}{\partial x_n}\frac{\partial x_n^*}{\partial g_0} = \lambda\left(\frac{\partial g}{\partial x_1}\frac{\partial x_1^*}{\partial g_0} + \cdots + \frac{\partial g}{\partial x_n}\frac{\partial x_n^*}{\partial g_0}\right). \tag{9.42}$$

Equation (9.42) can be further simplified by implicitly differentiating the first-order condition for λ (that is, the first-order condition that ensures that the constraint is binding) with respect to g_0:

$$1 - \frac{\partial g}{\partial x_1}\frac{\partial x_1^*}{\partial g_0} - \cdots - \frac{\partial g}{\partial x_n}\frac{\partial x_n^*}{\partial g_0} = 0. \tag{9.43}$$

This, when substituted into (9.42), implies

$$\frac{\partial f^*}{\partial g_0} = \lambda \tag{9.44}$$

which is what we set out to prove. This mathematical result is an example of the envelope theorem, which we discuss in detail in Chapter 13.

In the context of the utility-maximization problem we have used in this chapter, the value function shows the relationship between the maximum possible utility level and

---

[16]If, in a constrained maximization problem, the constraint is written so that it is nonpositive instead of nonnegative, the only difference is that the equilibrium value of the Lagrange multiplier will change sign. It therefore measures the effect on the optimal value function of making the constraint more binding.

the parameters of the model: prices and income.[17] Since the Lagrange multiplier measures how this optimal value function changes when the constraint is relaxed, the Lagrange multiplier in this case measures the marginal utility of income, as we mentioned at the end of Section 9.2.1.

In general, the economic interpretation of the Lagrange multiplier is that it measures the **imputed value** of the constraint—the amount by which the value of the objective function increases when the constraint is relaxed. Looked at another way, it represents the maximum amount that the economic agent would be willing to "pay" (in whatever units the objective function is measured in) for a relaxation of the constraint. For this reason, the Lagrange multiplier is often called the **shadow price** of the constraint, an interpretation we discuss more fully in Chapter 13.

SUMMARY

The Lagrangian technique is a powerful tool for analyzing economics problems because it allows us to solve constrained optimization problems and yields results with important economic interpretations. First-order conditions are generally simple to derive and, assuming that the second-order conditions hold, we can obtain comparative static results by implicit differentiation or by using total differentials. Since so many problems in microeconomics are constrained optimization problems, the Lagrangian technique is widely used in economic theory. The next chapter contains several worked-out examples to illustrate both the Lagrangian technique and the way economists set up, solve, and interpret constrained optimization problems. Chapters 11 and 12 then discuss inequality constraints, including nonnegativity constraints. Chapters 13 and 14 discuss comparative static effects on the optimal values of the objective functions in constrained optimization problems of changes in exogenous variables and develop some important results in consumer theory and the theory of the firm.

Problems

9.1 Write the Lagrangian and the first-order conditions for the following problems. Assume in each case that the constraint will be binding and that α, β, γ, a, and g_0 are positive constants.

(a) maximize $f(x, y) = (x + 2)(y + 1)$ subject to $ax + 5y \leq 20$

(b) maximize $f(x, y, z) = xyz$ subject to $ax + y + z \leq 5$

(c) maximize $f(x, y) = \sqrt{xy}$ subject to $ax + y \leq g_0$

(d) maximize $f(x, y, z) = A\ln(x^\alpha y^\beta z^\gamma)$ subject to $ax^2 + y^2 + z^2 \leq g_0$

(e) minimize $f(x, y, z) = x^2 + y^2 + z^2$ subject to $ax + y + z \geq g_0$

(f) minimize $f(x, y) = x - y$ subject to $a\sqrt{xy} \geq g_0$

(g) minimize $f(x, y, z) = ax + y + z$ subject to $A\ln(x^\alpha y^\beta z^\gamma) \geq g_0$

Constrained Optimization: Theory

---

[17]This is the indirect utility function, discussed at length in Chapters 13 and 14.

9.2 For each part of Problem 9.1, show whether, or under what conditions, the second-order conditions are satisfied.

9.3 For each part of Problem 9.1, find and sign (if possible) $\partial x^*/\partial a$ and $\partial y^*/\partial a$.

9.4 For parts (d) and (e) of Problem 9.1, find and sign (if possible) $\partial x^*/\partial g_0$ and $\partial y^*/\partial g_0$.

9.5 Consider the problem of a consumer who is minimizing total expenditures on n goods subject to the constraint that utility must equal at least U_0. Is it possible, without knowing the form of the utility function, to sign the derivative of the compensated demand for good 1 with respect to its own price? Is it possible, without knowing the form of the utility function, to sign the derivative of the compensated demand for good 1 with respect to the price of good 2?

Constrained Optimization: Applications

10.1 INTRODUCTION

In this chapter we examine several examples of applications of constrained optimization. These examples represent just a small sample of applications of the Lagrangian technique so commonly used in economic analysis. We assume throughout this chapter that constraints are binding and that internal solutions exist; then in Chapters 11 and 12 we formally address the possibilities of nonbinding constraints and corner solutions. The focus of the applications in this chapter is on comparative static analysis of the choice variables, while Chapters 13 and 14 explore the effects of changes in parameters on the values of the objective functions.

In Sections 10.2 and 10.3 we look at applications from the theory of the firm, emphasizing the derivation of conditional and unconditional input demand functions from the firm's cost-minimization and profit-maximization problems, respectively. Sections 10.4 through 10.6 are applications from consumer theory, starting with a simple utility-maximization problem using a logarithmic utility function. Section 10.5 discusses an individual's labor supply decision, while the following section deals with utility maximization subject to both budget and time constraints. Section 10.7 is an example from welfare economics of dealing with multiple constraints; it derives the conditions necessary for Pareto-efficient allocations. The final three sections of the chapter relate to macroeconomics. Section 10.8 explores the problem of intertemporal consumption choice, in which the consumer must decide whether to save some current income to finance additional future consumption or to borrow against future income to finance additional current consumption. Although this is a traditional consumer choice problem in microeconomics, it also forms the foundation for important macroeconomic insights and models. Similarly, Section 10.9 presents the Baumol-Tobin microfoundations of a typical macroeconomic money demand function. Section 10.10 discusses a model of a macroeconomic policymaker minimizing a loss function subject to the Phillips curve as a constraint. Some of this chapter's applications revisit problems analyzed earlier in the text to show the usefulness and power of the Lagrangian technique even when other techniques are possible. Others are revisited later in the text when nonbinding constraints and corner solutions are analyzed formally.

10.2　COST MINIMIZATION AND CONDITIONAL INPUT DEMAND

The example of constrained minimization we used in Chapter 9 was that of a consumer minimizing the expenditures necessary to achieve a given level of utility. A more familiar example is that of a firm minimizing the cost of producing a given amount of output. The mathematics of these two constrained minimization problems are identical. In this application we explore the cost-minimization problem for a firm that is perfectly competitive in the input markets. First we will analyze the case of two inputs, and then we will discuss the more general case of many inputs.

10.2.1　Two Inputs

For the two-input case, a perfectly competitive firm's total cost is $wL + rK$, where w is the wage rate, r is the rental rate of capital, and L and K are the amounts of labor and capital hired, respectively. The firm wants to produce at least Q_0 units of output using the production function $Q = F(L, K)$. Recalling that for minimization problems the Lagrangian is formed with the constraint written so that it is nonpositive, we write the Lagrangian for the cost-minimization problem

$$\mathcal{L}(L, K, w, r, \lambda, Q_0) = wL + rK + \lambda(Q_0 - F(L, K)). \tag{10.1}$$

The first-order conditions are

$$\frac{\partial \mathcal{L}}{\partial L} = w - \lambda F_L = 0$$

$$\frac{\partial \mathcal{L}}{\partial K} = r - \lambda F_K = 0 \tag{10.2}$$

$$\frac{\partial \mathcal{L}}{\partial \lambda} = Q_0 - F(L, K) = 0$$

where $F_L = \partial F/\partial L$ and $F_K = \partial F/\partial K$. The last of these first-order conditions ensures that the firm produces exactly Q_0. By eliminating the Lagrange multiplier from the first two of the first-order conditions (10.2), we obtain

$$\frac{w}{r} = \frac{F_L}{F_K}. \tag{10.3}$$

The right-hand side of equation (10.3) is the ratio of marginal products of labor and capital. This ratio, called the *marginal rate of technical substitution* between those inputs (see Section 5.5), is the absolute value of the slope of the level curve of the production function, which is the constraint in this minimization problem. The left-hand side of equation (10.3) is the ratio of input prices, which is the absolute value of the slope of a level curve of the objective function. Thus the first-order conditions state the familiar result that if a perfectly competitive firm is minimizing the costs of producing a particular level of output Q_0, it must find the combination of inputs that results in a tangency between an isocost line and the isoquant corresponding to Q_0 units of output, as illustrated in Figure 10.1.

Assuming that the first-order conditions are satisfied, the second-order condition is that the determinant of the bordered Hessian of the problem be negative. (The only border-preserving principal minor of order greater than 1 is that of second order, which

FIGURE 10.1

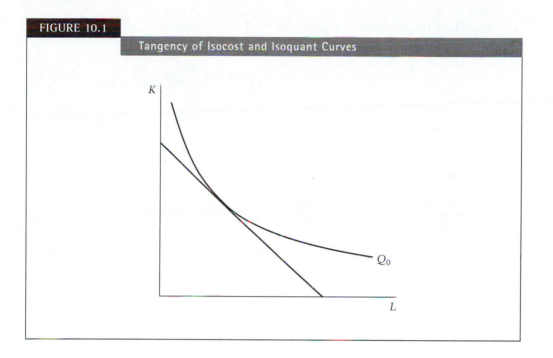

is the determinant of the whole 3×3 bordered Hessian.) The bordered Hessian for this problem is

$$\bar{H} = \begin{bmatrix} -\lambda F_{LL} & -\lambda F_{LK} & -F_L \\ -\lambda F_{KL} & -\lambda F_{KK} & -F_K \\ -F_L & -F_K & 0 \end{bmatrix} \tag{10.4}$$

the determinant of which is (expanding by the third row)

$$\begin{aligned} |\bar{H}| &= -F_L(\lambda F_{LK} F_K - \lambda F_{KK} F_L) + F_K(\lambda F_{LL} F_K - \lambda F_{KL} F_L) \\ &= \lambda(F_{LL}(F_K)^2 - 2F_{LK} F_L F_K + F_{KK}(F_L)^2). \end{aligned} \tag{10.5}$$

The second-order condition, then, is that the determinant (10.5) be negative. If the second-order condition is satisfied, the first-order conditions (10.2) implicitly define the choice variables L, K, and λ as functions of the parameters of the problem. In this case, the parameters include w, r, and Q_0, though in principle other parameters may influence the form of the production function $F(L, K)$, which we have left in general form. The functions $L^*(w, r, Q_0)$ and $K^*(w, r, Q_0)$ that are implicitly defined by the first-order conditions give the firm's cost-minimizing choices of labor and capital, conditional on producing Q_0 units of output. Thus they are called **conditional input demand functions**.

To see how these conditional input demands are affected when parameters change, we totally differentiate the first-order conditions (10.2) and get

$$\begin{aligned} dw - \lambda F_{LL}\, dL - \lambda F_{LK}\, dK - F_L\, d\lambda &= 0 \\ dr - \lambda F_{KL}\, dL - \lambda F_{KK}\, dK - F_K\, d\lambda &= 0 \\ dQ_0 - F_L\, dL - F_K\, dK &= 0 \end{aligned} \tag{10.6}$$

which can be written in matrix notation as

$$
\begin{bmatrix}
-\lambda F_{LL} & -\lambda F_{LK} & -F_L \\
-\lambda F_{KL} & -\lambda F_{KK} & -F_K \\
-F_L & -F_K & 0
\end{bmatrix}
\begin{bmatrix}
dL \\
dK \\
d\lambda
\end{bmatrix}
=
\begin{bmatrix}
-dw \\
-dr \\
-dQ_0
\end{bmatrix}. \tag{10.7}
$$

The first matrix in equation (10.7) is the bordered Hessian matrix (10.4). We can now use Cramer's rule to derive comparative static results. For example,

$$
dL = \frac{\begin{vmatrix} -dw & -\lambda F_{LK} & -F_L \\ -dr & -\lambda F_{KK} & -F_K \\ -dQ_0 & -F_K & 0 \end{vmatrix}}{|\bar{H}|} = \frac{dw(F_K)^2 - dr\,F_K F_L - dQ_0(\lambda F_{LK} F_K - \lambda F_{KK} F_L)}{|\bar{H}|} \tag{10.8}
$$

which implies that the conditional demand for labor curve slopes downward:

$$
\left.\frac{dL}{dw}\right|_{dr=dQ_0=0} = \frac{(F_K)^2}{|\bar{H}|} < 0 \tag{10.9}
$$

since the second-order condition says that the denominator is negative. This result is general since we did not specify the production function: conditional input demand functions slope downward for all two-input production functions that satisfy the second-order conditions of the firm's cost-minimization problem. Shortly, we will see that this result continues to hold when there are many inputs.

We can also use equation (10.8) to show that, in the two-input case, an increase in the rental rate of capital leads to an increase in the conditional demand for labor:[1]

$$
\left.\frac{dL}{dr}\right|_{dw=dQ_0=0} = \frac{-F_K F_L}{|\bar{H}|} > 0. \tag{10.10}
$$

Because output is held constant, this derivative represents a pure *factor-substitution effect*—the change in the use of labor as the firm moves along an isoquant in response to a change in the relative prices of labor and capital. Figure 10.2 illustrates this factor-substitution effect. The increase in the rental rate of capital changes the slopes of the firm's isocost curves. Thus the firm will move along the isoquant corresponding to output level Q_0 until if finds the point where the marginal rate of technical substitution equals the new input price ratio. The second-order condition for this problem ensures that the firm's production function is quasiconcave, which implies that the level curves of the production function (that is, the isoquants) are convex to the origin. This convexity in turn implies that the new tangency point must lie below and to the right of the old tangency point. Thus an increase in the rental rate of capital will reduce the conditional demand for capital and increase the conditional demand for labor.

In the following section we will see that this result is not general: when there are many inputs, an increase in one input price will increase the conditional demands for some inputs but decrease the conditional demands for other inputs. In Section 10.3 we

---

[1]To sign this derivative, we must use the facts that the marginal products of labor and capital are both positive. This is ensured by the first-order conditions (10.2) combined with the fact that $\lambda > 0$ as long as the constraint is binding.

FIGURE 10.2

Factor–Substitution Effect of Higher Rental Rate of Capital

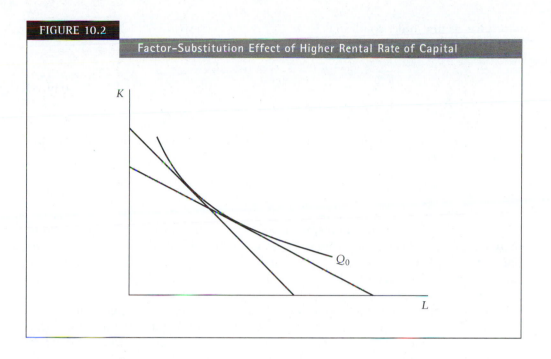

will also see that even in the two-input case, the effects of input price changes on unconditional input demand functions are ambiguous.

10.2.2 Many Inputs

In this section we extend the analysis of the previous section to the case of many inputs and show which of the results derived previously continue to hold. In particular, the first-order conditions imply that the isocost lines and isoquants are tangent and that the conditional input demand curves slope downward. We shall see, however, that the effect of an increase in the price of one input on the conditional demand for another input is now ambiguous. We also use the Lagrange multiplier to show the effect on long-run marginal cost[2] of an increase in one input price.

A firm that is perfectly competitive in the input markets takes input prices as given. Thus, if x_i is quantity used of the ith input, which has input price w_i, $\mathbf{x}$ and $\mathbf{w}$ denote the vectors of input quantities and prices, and there are n inputs, the firm's total costs are given by

$$C(\mathbf{x}, \mathbf{w}) = \sum_{i=1}^{n} w_i x_i. \qquad (10.11)$$

The firm is interested in producing Q_0 units of output using the production function $Q = F(\mathbf{x})$, which is assumed to be twice differentiable. In this problem, the input quantities are the choice variables (along with the Lagrange multiplier), input prices and the output level Q_0 are exogenous variables, and other parameters may govern the shape of the production function, which we have left in a general form. The Lagrangian for this cost-minimization problem is

$$\mathscr{L}(\mathbf{x}, \mathbf{w}, \lambda, Q_0) = w_1 x_1 + \cdots + w_n x_n + \lambda(Q_0 - F(\mathbf{x})) \qquad (10.12)$$

[2]Since all inputs are assumed to be variable, we are in the long run.

which has as first-order conditions

$$\frac{\partial \mathcal{L}}{\partial x_i} = w_i + \lambda \left(-\frac{\partial F}{\partial x_i} \right) = 0, \quad i = 1, \ldots, n$$
$$\frac{\partial \mathcal{L}}{\partial \lambda} = Q_0 - F(\mathbf{x}) = 0.$$

(10.13)

By taking any two of the first n first-order conditions and eliminating λ from them, we get

$$\frac{w_i}{w_j} = \frac{\partial F / \partial x_i}{\partial F / \partial x_j},$$

(10.14)

which says that for any two inputs, the ratio of input prices has to equal the ratio of marginal products. The ratio of input prices is the negative of the slope of an isocost line showing all combinations of inputs i and j that would lead to the same total cost (holding constant the use of other inputs). The ratio of marginal products is the marginal rate of technical substitution between that pair of inputs and equals the absolute value of the slope of an isoquant showing (again, holding all other inputs constant) all the combinations of inputs i and j that would yield the same output (Q_0 in this case). As always, the first-order conditions imply that the slope of a level curve of the objective function equals the slope of a level curve of the constraint. With many inputs, this condition holds for any pair.

The first-order conditions, used a different way, indicate that the Lagrange multiplier in this example measures marginal cost. Rearranging the first-order condition for input i, we have

$$\lambda = \frac{w_i}{\partial F / \partial x_i}.$$

(10.15)

Since the denominator of the right-hand side is the marginal product of input i, the economic interpretation of the right-hand side is that it equals the price of input i times the amount of input i needed to produce one more unit of the final good Q. (Since one more unit of input i produces $\partial F / \partial x_i$ units of good Q, one more unit of good Q requires $1/(\partial F / \partial x_i)$ more units of input i.) Thus, the right-hand side equals the cost to the firm of hiring enough extra of input i to produce one more unit of good Q. That is, λ equals the marginal cost of producing more output, using input i. But since we used an arbitrary first-order condition and since *all* the first-order conditions hold when the firm is minimizing costs, marginal cost must be the same no matter which input is being increased (or, indeed, if all inputs are being increased simlutaneously) to produce the extra output.

If the second-order conditions for this minimization problem hold, the conditions of the implicit function theorem are met. The first-order conditions then implicitly define the solutions for the choice variables as functions of the exogenous variables:

$$x_i^* = x_i^*(w_1, \ldots, w_n, Q_0), \quad i = 1, \ldots, n$$
$$\lambda^* = \lambda^*(w_1, \ldots, w_n, Q_0).$$

(10.16)

The first n of these functions are the conditional input demand functions. Given our interpretation of the Lagrange multiplier, the last of the functions shows the long-run marginal cost function of the firm.

Comparative static results can be obtained by implicitly differentiating the first-order conditions (10.13) with respect to one or more exogenous variables. We can, for example, differentiate the first-order conditions (10.13) with respect to w_j:

$$-\lambda \sum_{i=1}^{n} \frac{\partial^2 F}{\partial x_1 \partial x_i} \frac{\partial x_i^*}{\partial w_j} - \frac{\partial F}{\partial x_1} \frac{\partial \lambda^*}{\partial w_j} = 0$$

$$\vdots$$

$$1 - \lambda \sum_{i=1}^{n} \frac{\partial^2 F}{\partial x_j \partial x_i} \frac{\partial x_i^*}{\partial w_j} - \frac{\partial F}{\partial x_j} \frac{\partial \lambda^*}{\partial w_j} = 0 \qquad \textbf{(10.17)}$$

$$\vdots$$

$$-\lambda \sum_{i=1}^{n} \frac{\partial^2 F}{\partial x_n \partial x_i} \frac{\partial x_i^*}{\partial w_j} - \frac{\partial F}{\partial x_n} \frac{\partial \lambda^*}{\partial w_j} = 0$$

$$-\frac{\partial F}{\partial x_1} \frac{\partial x_1^*}{\partial w_j} - \cdots - \frac{\partial F}{\partial x_n} \frac{\partial x_n^*}{\partial w_j} = 0.$$

Written in matrix notation, these equations are

$$\begin{bmatrix} -\lambda \dfrac{\partial^2 F}{\partial x_1^2} & \cdots & -\lambda \dfrac{\partial^2 F}{\partial x_1 \partial x_j} & \cdots & -\lambda \dfrac{\partial^2 F}{\partial x_1 \partial x_n} & -\dfrac{\partial F}{\partial x_1} \\ \vdots & \ddots & \vdots & \ddots & \vdots & \vdots \\ -\lambda \dfrac{\partial^2 F}{\partial x_j \partial x_1} & \cdots & -\lambda \dfrac{\partial^2 F}{\partial x_j^2} & \ddots & -\lambda \dfrac{\partial^2 F}{\partial x_j \partial x_n} & -\dfrac{\partial F}{\partial x_j} \\ \vdots & \ddots & \vdots & \ddots & \vdots & \vdots \\ -\lambda \dfrac{\partial^2 F}{\partial x_n \partial x_1} & \cdots & -\lambda \dfrac{\partial^2 F}{\partial x_n \partial x_j} & \cdots & -\lambda \dfrac{\partial^2 F}{\partial x_n^2} & -\dfrac{\partial F}{\partial x_n} \\ -\dfrac{\partial F}{\partial x_1} & \cdots & -\dfrac{\partial F}{\partial x_j} & \cdots & -\dfrac{\partial F}{\partial x_n} & 0 \end{bmatrix} \begin{bmatrix} \dfrac{\partial x_1^*}{\partial w_j} \\ \vdots \\ \dfrac{\partial x_j^*}{\partial w_j} \\ \vdots \\ \dfrac{\partial x_n^*}{\partial w_j} \\ \dfrac{\partial \lambda^*}{\partial w_j} \end{bmatrix} = \begin{bmatrix} 0 \\ \vdots \\ -1 \\ \vdots \\ 0 \\ 0 \end{bmatrix}. \qquad \textbf{(10.18)}$$

The first matrix on the left-hand side is the bordered Hessian of the Lagrangian, which we will denote as $\bar{H}$. To see that the conditional input demand function for input j is downward-sloping, we will use Cramer's rule to solve for $\partial x_j^*/\partial w_j$. In so doing, we will replace the jth column of $\bar{H}$ with the vector on the right-hand side of equation (10.18):

$$\frac{\partial x_j^*}{\partial w_j} = \frac{\begin{vmatrix} -\lambda \dfrac{\partial^2 F}{\partial x_1^2} & \cdots & 0 & \cdots & -\lambda \dfrac{\partial^2 F}{\partial x_1 \partial x_n} & -\dfrac{\partial F}{\partial x_1} \\ \vdots & \ddots & \vdots & \ddots & \vdots & \vdots \\ -\lambda \dfrac{\partial^2 F}{\partial x_j \partial x_1} & \cdots & -1 & \ddots & -\lambda \dfrac{\partial^2 F}{\partial x_j \partial x_n} & -\dfrac{\partial F}{\partial x_j} \\ \vdots & \ddots & \vdots & \ddots & \vdots & \vdots \\ -\lambda \dfrac{\partial^2 F}{\partial x_n \partial x_1} & \cdots & 0 & \cdots & -\lambda \dfrac{\partial^2 F}{\partial x_n^2} & -\dfrac{\partial F}{\partial x_n} \\ -\dfrac{\partial F}{\partial x_1} & \cdots & 0 & \cdots & -\dfrac{\partial F}{\partial x_n} & 0 \end{vmatrix}}{|\bar{H}|}$$

$$= \frac{(-1)^{j+j}(-1)|\bar{H}_{jj}|}{|\bar{H}|} = \frac{-|\bar{H}_{jj}|}{|\bar{H}|} \qquad \textbf{(10.19)}$$

where $|\bar{H}_{jj}|$ is the minor of $\bar{H}$ obtained by deleting the jth row and the jth column (that is, $(-1)^{j+j}|\bar{H}_{jj}|$ is the cofactor of the element in the jth row and j^{th} column). Since the same row as column has been deleted, $|\bar{H}_{jj}|$ is a *principal* minor, as is $|\bar{H}|$ itself. Both of these principal minors will be negative if the second-order conditions hold, since all border-preserving principal minors (of order 2 and higher) are negative for a con-strained minimization problem with one constraint. Thus we have shown that the con-ditional input demand function is downward-sloping because

$$\frac{\partial x_j^*}{\partial w_j} = \frac{-|\bar{H}_{jj}|}{|\bar{H}|} = \frac{(-) \cdot (-)}{(-)} < 0. \tag{10.20}$$

In general we cannot say whether the conditional demand for input k will increase or decrease when the price of input j increases. Using Cramer's rule to solve equa-tion (10.18) for $\partial x_k^* / \partial w_j$, we obtain

$$\frac{\partial x_k^*}{\partial w_j} = \frac{(-1)^{k+j}(-1)|\bar{H}_{jk}|}{|\bar{H}|}. \tag{10.21}$$

We now have a nonprincipal minor in the numerator, and $k + j$ may be either odd or even. So, without further assumptions about the production function $F(\mathbf{x})$, we cannot say whether the factor substitution effect in this case is to increase or decrease the use of input k.

This result differs from that of the two-input case because the firm may now substi-tute away from input j in many possible ways when the price of input j increases. De-pending on which factors are complements in production and which are substitutes, the conditional demands for some inputs may increase while the demands for others may fall.

Since the Lagrange multiplier represents marginal cost, we can also use equa-tion (10.18) to find out whether the marginal cost curve shifts up or down when there is an increase in the price of input j. Solving for $\partial \lambda^* / \partial w_j$, we get

$$\frac{\partial \lambda^*}{\partial w_j} = \frac{(-1)^{(n+1)+j}(-1)|\bar{H}_{j\lambda}|}{|\bar{H}|} \tag{10.22}$$

where $|\bar{H}_{j\lambda}|$ is the minor obtained by deleting the jth row and the last column from $\bar{H}$. This is clearly not a principal minor, much less a border-preserving principal minor, so we cannot say from the second-order conditions alone that marginal costs increase when an input price rises.

Although this result is not intuitive, it can be explained by considering the effect on the (conditional) demand for input j of an increase in the output level Q_0. Since Q_0 appears explicitly only in the last first-order condition, implicitly differentiating the first-order conditions (10.13) with respect to Q_0 yields

$$\bar{H} \begin{bmatrix} \dfrac{\partial x_1^*}{\partial Q_0} \\ \vdots \\ \dfrac{\partial x_j^*}{\partial Q_0} \\ \vdots \\ \dfrac{\partial x_n^*}{\partial Q_0} \\ \dfrac{\partial \lambda^*}{\partial Q_0} \end{bmatrix} = \begin{bmatrix} 0 \\ \vdots \\ 0 \\ \vdots \\ 0 \\ -1 \end{bmatrix}. \tag{10.23}$$

Thus, using Cramer's rule,

$$\frac{\partial x_j^*}{\partial Q_0} = \frac{(-1)^{j+(n+1)}(-1)|\bar{H}_{\lambda j}|}{|\bar{H}|} \tag{10.24}$$

$|\bar{H}_{\lambda j}|$ is obtained by deleting the last row and the jth column of $\bar{H}$. This is not a border-preserving principal minor, so the second-order conditions do not give us its sign. The mathematical structure of the cost-minimization problem, then, is not sufficient to rule out **inferior inputs**, the demand for which actually declines when output increases. But notice the similarity of the equations for $\partial x_j^*/\partial Q_0$ and $\partial \lambda^*/\partial w_j$ (equations (10.24) and (10.22), respectively). In fact, since the bordered Hessian is symmetric, $|\bar{H}_{\lambda j}| = |\bar{H}_{j\lambda}|$. So as long as the input is *not* inferior, marginal cost will increase when the input price rises.

It is also worth noting that by using Cramer's rule to solve for $\partial \lambda^*/\partial Q_0$ from equation (10.23), we have

$$\frac{\partial \lambda^*}{\partial Q_0} = \frac{(-1)^{(n+1)+(n+1)}(-1)|\bar{H}_{\lambda\lambda}|}{|\bar{H}|} \tag{10.25}$$

which cannot be signed since $|\bar{H}_{\lambda\lambda}|$ is not a *border-preserving* principal minor. This result is not surprising, since economic theory allows long-run marginal cost curves to be downward-sloping, upward-sloping, or horizontal.

10.3 PROFIT MAXIMIZATION AND UNCONDITIONAL INPUT DEMAND

In the previous section we analyzed a firm's cost-minimization problem, which resulted in demand functions for inputs conditional on a particular output level. In Sections 8.2 and 8.3, we derived a competitive firm's demand for inputs by analyzing the firm's profit-maximization decision, using the production function to relate inputs to the output sold by the firm. We also investigated the firm's profit-maximization decision in Chapter 2, when the focus was on the output level and the cost-minimization problem was assumed to have been done already, resulting in a cost function relating total costs to output.

In this section we analyze the profit-maximizing choices of inputs by replicating the analysis of Section 8.3; but instead of substituting the production function into the objective function, as we did there, in this section we treat the production function as a constraint. The result of the firm's profit-maximization problem is a set of functions relating the demand for inputs to prices of all inputs and to the output price. These functions are called **unconditional input demand functions** because the firm's output is no longer treated as exogenous.

The profit function for a perfectly competitive firm using two inputs is $\Pi = PQ - wL - rK$, as described in equation (8.1); that is, revenues (price P times quantity sold Q) minus costs (payments to factors of production). For a firm using the two inputs, labor L and capital K, with wage rate w and rental rate of capital r, costs equal $wL + rK$. The quantity of output sold, however, can be no greater than the amount of output produced from the inputs used; this level of output is given by the production function $Q = F(L, K)$. Thus the Lagrangian for the firm's profit-maximization problem is

$$\mathscr{L}(Q, L, K, P, w, r, \lambda) = PQ - wL - rK + \lambda(F(L, K) - Q). \qquad \textbf{(10.26)}$$

There are four choice variables, Q, L, K, and λ, and three exogenous variables, P, w, and r. The first-order conditions for this constrained maximization problem are

$$\frac{\partial \mathscr{L}}{\partial Q} = P - \lambda = 0$$

$$\frac{\partial \mathscr{L}}{\partial L} = -w + \lambda \frac{\partial F}{\partial L} = 0$$

$$\frac{\partial \mathscr{L}}{\partial K} = -r + \lambda \frac{\partial F}{\partial K} = 0 \qquad \textbf{(10.27)}$$

$$\frac{\partial \mathscr{L}}{\partial \lambda} = F(L, K) - Q = 0.$$

Note that the last three of these are identical to the first-order conditions (10.2) coming from the firm's cost-minimization problem. It is the addition of the first-order condition with respect to Q that ensures that not only has the cost-minimizing combination of inputs been used, but that the profit-maximizing level of quantity has been produced as well.

In addition, when λ is eliminated by substitution from the first-order conditions, the first-order conditions for L and K are the same as those derived in Section 8.3 where the production function was substituted into the objective function (see equations (8.23)). The only difference is that we now have the relationship between inputs and output explicitly given as part of the first-order conditions. This enables us to get comparative static derivatives of output supply as well as input demands from the first-order conditions. For example, in Section 8.3 the comparative static effects on labor and capital were derived directly from the first-order conditions (see equations (8.27)), whereas the effects on output were derived by substituting the effects on input demands into the total differential of the production function (see equations (8.30) through (8.32)). Here we will be able to obtain all the derivatives directly from the first-order conditions.

The second-order conditions for the profit-maximization problem are conditions on the bordered Hessian matrix

$$\bar{H} = \begin{bmatrix} 0 & 0 & 0 & -1 \\ 0 & \lambda\dfrac{\partial^2 F}{\partial L^2} & \lambda\dfrac{\partial^2 F}{\partial L \partial K} & \dfrac{\partial F}{\partial L} \\ 0 & \lambda\dfrac{\partial^2 F}{\partial K \partial L} & \lambda\dfrac{\partial^2 F}{\partial K^2} & \dfrac{\partial F}{\partial K} \\ -1 & \dfrac{\partial F}{\partial L} & \dfrac{\partial F}{\partial K} & 0 \end{bmatrix}. \qquad \textbf{(10.28)}$$

Since this is a maximization problem with one constraint, the border-preserving principal minors of order 2 and above of $\bar{H}$ alternate in sign, starting with positive.

To get comparative statics result, we totally differentiate the first-order conditions. Writing the results in matrix notation yields

$$\bar{H} \begin{bmatrix} dQ \\ dL \\ dK \\ d\lambda \end{bmatrix} = \begin{bmatrix} -dP \\ dw \\ dr \\ 0 \end{bmatrix}. \tag{10.29}$$

Using Cramer's rule to solve for dK, we get[3]

$$dK = \frac{\begin{vmatrix} 0 & 0 & -dP & -1 \\ 0 & \lambda\dfrac{\partial^2 F}{\partial L^2} & dw & \dfrac{\partial F}{\partial L} \\ 0 & \lambda\dfrac{\partial^2 F}{\partial K \partial L} & dr & \dfrac{\partial F}{\partial K} \\ -1 & \dfrac{\partial F}{\partial K} & 0 & 0 \end{vmatrix}}{|\bar{H}|} = \frac{\begin{vmatrix} 0 & -dP & -1 \\ \lambda\dfrac{\partial^2 F}{\partial L^2} & dw & \dfrac{\partial F}{\partial L} \\ \lambda\dfrac{\partial^2 F}{\partial K \partial L} & dr & \dfrac{\partial F}{\partial K} \end{vmatrix}}{|\bar{H}|}$$

$$= \frac{dP\left(\lambda\dfrac{\partial^2 F}{\partial L^2}\dfrac{\partial F}{\partial K} - \lambda\dfrac{\partial^2 F}{\partial K \partial L}\dfrac{\partial F}{\partial L}\right) - \left(\lambda\dfrac{\partial^2 F}{\partial L^2} dr - \lambda\dfrac{\partial^2 F}{\partial K \partial L} dw\right)}{|\bar{H}|}. \tag{10.30}$$

Thus

$$\left.\frac{dK}{dr}\right|_{dw=dP=0} = \frac{-\lambda(\partial^2 F/\partial L^2)}{|\bar{H}|}. \tag{10.31}$$

Since the determinant of the bordered Hessian is a third-order border-preserving principal minor of itself, the second-order conditions imply that it must be negative.[4] In addition, the sign of the numerator is determined by the second-order conditions. This can be seen be examining the border-preserving principal minor of order 2 obtained by deleting the third row and third column of $\bar{H}$:

$$\begin{vmatrix} 0 & 0 & -1 \\ 0 & \lambda\dfrac{\partial^2 F}{\partial L^2} & \dfrac{\partial F}{\partial L} \\ -1 & \dfrac{\partial F}{\partial L} & 0 \end{vmatrix} = -1\left(0 + \lambda\dfrac{\partial^2 F}{\partial L^2}\right) = -\lambda\dfrac{\partial^2 F}{\partial L^2}. \tag{10.32}$$

Since this is a second-order border-preserving principal minor, it must be positive. Thus

$$\left.\frac{dK}{dr}\right|_{dw=dP=0} = \frac{(+)}{(-)} < 0. \tag{10.33}$$

Like the conditional input demands derived in the previous section, unconditional input demand curves slope downward. This result holds for all two-input production functions that satisfy the second-order conditions of the profit-maximization problem. We leave the extension to many inputs for you to do in Problem 10.9.

---

[3]The solution for dL can be derived in the same way. We focus on dK to facilitate comparisons with Section 8.3.

[4]In fact, this determinant is exactly equal to the negative of the determinant of the Hessian from Section 8.3. This equality, combined with the first-order condition that $P = \lambda$, confirms that the solutions for dL and dK are exactly the same as those from Section 8.3 (see equations (8.27)).

Constrained Optimization: Applications

We can also find the comparative static effects on output by using Cramer's rule and the first-order conditions:

$$dQ = \frac{\begin{vmatrix} -dP & 0 & 0 & -1 \\ dw & \lambda\dfrac{\partial^2 F}{\partial L^2} & \lambda\dfrac{\partial^2 F}{\partial L\,\partial K} & \dfrac{\partial F}{\partial L} \\ dr & \lambda\dfrac{\partial^2 F}{\partial K\,\partial L} & \lambda\dfrac{\partial^2 F}{\partial K^2} & \dfrac{\partial F}{\partial K} \\ 0 & \dfrac{\partial F}{\partial L} & \dfrac{\partial F}{\partial K} & 0 \end{vmatrix}}{|\bar{H}|}$$

$$= \frac{-dP\begin{vmatrix} \lambda\dfrac{\partial^2 F}{\partial L^2} & \lambda\dfrac{\partial^2 F}{\partial L\,\partial K} & \dfrac{\partial F}{\partial L} \\ \lambda\dfrac{\partial^2 F}{\partial K\,\partial L} & \lambda\dfrac{\partial^2 F}{\partial K^2} & \dfrac{\partial F}{\partial K} \\ \dfrac{\partial F}{\partial L} & \dfrac{\partial F}{\partial K} & 0 \end{vmatrix} + \begin{vmatrix} dw & \lambda\dfrac{\partial^2 F}{\partial L^2} & \lambda\dfrac{\partial^2 F}{\partial L\,\partial K} \\ dr & \lambda\dfrac{\partial^2 F}{\partial K\,\partial L} & \lambda\dfrac{\partial^2 F}{\partial K^2} \\ 0 & \dfrac{\partial F}{\partial L} & \dfrac{\partial F}{\partial K} \end{vmatrix}}{|\bar{H}|} \tag{10.34}$$

$$= \frac{-dP|\bar{H}_{11}| + dw\left(\lambda\dfrac{\partial^2 F}{\partial K\,\partial L}\dfrac{\partial F}{\partial K} - \lambda\dfrac{\partial^2 F}{\partial K^2}\dfrac{\partial F}{\partial L}\right) - dr\left(\lambda\dfrac{\partial^2 F}{\partial L^2}\dfrac{\partial F}{\partial K} - \lambda\dfrac{\partial^2 F}{\partial K\,\partial L}\dfrac{\partial F}{\partial L}\right)}{|\bar{H}|}.$$

As is the case in Section 8.3, the second-order conditions give us the sign of $\partial Q/\partial P$ but not the signs of either $\partial Q/\partial w$ or $\partial Q/\partial r$.[5] This is because, in equation (10.34), dP is multiplied by a border-preserving principal minor of $\bar{H}$ while dw and dr are each multiplied by a minor of $\bar{H}$, but not a border-preserving principal minor. So, while we can show in general that an increase in output price increases the profit-maximizing output level, we need to put more structure on the problem before we can say unambiguously whether increases in input prices increase or decrease the firm's output. Problem 10.2 investigates this, and other, issues for some specific production functions.

10.4   UTILITY MAXIMIZATION: LOGARITHMIC UTILITY

In Section 6.9 we analyzed utility maximization by substituting the income constraint into the utility function. Here we revisit the problem as a simple, straightforward example of using the Lagrangian technique. Suppose that utility is derived from two goods, x and y, according to the utility function $U(x, y) = \ln x + \ln y$. The consumer has income I to spend; the prices of the two goods are P_x and P_y. The Lagrangian for the utility-maximization problem is thus

$$\mathcal{L}(x, y, \lambda, I, P_x, P_y) = \ln x + \ln y + \lambda(I - P_x x - P_y y). \tag{10.35}$$

The choice variables are x, y, and λ, and the exogenous variables are I, P_x, and P_y. The first-order conditions for this problem are

$$\frac{\partial \mathcal{L}}{\partial x} = \frac{1}{x} - \lambda P_x = 0$$

$$\frac{\partial \mathcal{L}}{\partial y} = \frac{1}{y} - \lambda P_y = 0 \tag{10.36}$$

$$\frac{\partial \mathcal{L}}{\partial \lambda} = I - P_x x - P_y y = 0.$$

---

[5]In fact, equation (10.34) is identical to equation (8.31) because the denominators of the two equations are negatives of each other and $P = \lambda$.

Eliminating λ from the first two first-order conditions yields

$$-\frac{1/x}{1/y} = -\frac{P_x}{P_y}.$$ (10.37)

As described in Section 6.9, the interpretation of equation (10.37) is that the marginal rate of substitution between x and y equals their price ratio.

Since there are only two choice variables (not counting λ), the second-order condition for this problem is that the determinant of the bordered Hessian be positive:

$$|\bar{H}| = \begin{vmatrix} -\dfrac{1}{x^2} & 0 & -P_x \\ 0 & -\dfrac{1}{y^2} & -P_y \\ -P_x & -P_y & 0 \end{vmatrix} = -\frac{1}{x^2}(-P_y^2) - P_x\left(-\frac{P_x}{y^2}\right) = \frac{P_y^2}{x^2} + \frac{P_x^2}{y^2} > 0.$$ (10.38)

The first-order conditions (10.36) are three equations in the three unknowns x, y, and λ, but they are nonlinear equations. The solutions can be derived by successive substitution, however. For example, to derive the demand for x, we first use the last two first-order conditions to solve for λ as a function of x and then substitute that function into the first of the first-order conditions:

$$\lambda = \frac{1}{P_y y} = \frac{1}{I - P_x x}$$

$$\frac{1}{x} = \left(\frac{1}{I - P_x x}\right) P_x$$

$$x = \frac{I}{P_x} - x$$ (10.39)

$$x^* = \frac{I}{2P_x}$$

which is the result derived in Section 6.9 (see equation (6.93)).

10.5 LABOR SUPPLY

We can also use the utility-maximization model to investigate the labor/leisure choice, as we did in Section 6.8. It is conventional to assume that utility depends on two arguments: consumption of a composite good and leisure. There are two constraints in this problem. One is the budget constraint $C \leq wL + I$, where C is the amount of the composite good, L is the amount of time spent supplying labor (working), w is the wage rate, and I is nonlabor income. The budget constraint shows that the amount of income available for buying the composite good depends on how much labor is supplied. The other constraint is a time constraint showing that time spent working reduces the amount of leisure time. If T is the total time available, $T - L$ is the amount of leisure time. It is also conventional to use labor rather than leisure as a choice variable and to substitute the time constraint directly into the utility function while using the Lagrangian technique to account for the budget constraint.[6]

---

[6]In Section 6.9 we showed how to do a labor supply example by substituting the budget constraint into the utility function as well.

The Lagrangian for this utility-maximization problem is

$$\mathscr{L}(C, L, T, \lambda, I, w) = U(C, T - L) + \lambda(I + wL - C) \tag{10.40}$$

and the first-order conditions are

$$\frac{\partial \mathscr{L}}{\partial C} = \frac{\partial U}{\partial C} - \lambda = 0$$

$$\frac{\partial \mathscr{L}}{\partial L} = -\frac{\partial U}{\partial (T - L)} + \lambda w = 0 \tag{10.41}$$

$$\frac{\partial \mathscr{L}}{\partial \lambda} = I + wL - C = 0.$$

Since there are just two choice variables (not counting λ), the second-order condition is that the determinant of the bordered Hessian must be positive:

$$|\bar{H}| = \begin{vmatrix} \dfrac{\partial^2 U}{\partial C^2} & -\dfrac{\partial^2 U}{\partial C \, \partial(T - L)} & -1 \\ -\dfrac{\partial^2 U}{\partial(T - L)\partial C} & \dfrac{\partial^2 U}{\partial(T - L)^2} & w \\ -1 & w & 0 \end{vmatrix} > 0. \tag{10.42}$$

To obtain comparative static results, we will find the total differentials of the first-order conditions (10.41). Putting the results in matrix notation, we have

$$\bar{H} \begin{bmatrix} dC \\ dL \\ d\lambda \end{bmatrix} = \begin{bmatrix} 0 \\ -\lambda \, dw \\ -L \, dw - dI \end{bmatrix}. \tag{10.43}$$

Using Cramer's rule to solve for dL, we obtain

$$dL = \frac{\begin{vmatrix} \dfrac{\partial^2 U}{\partial C^2} & 0 & -1 \\ -\dfrac{\partial^2 U}{\partial(T - L)\,\partial C} & -\lambda \, dw & w \\ -1 & -L \, dw - dI & 0 \end{vmatrix}}{|\bar{H}|}$$

$$= \frac{\dfrac{\partial^2 U}{\partial C^2}\,(w(L \, dw + dI)) - \left(\dfrac{\partial^2 U}{\partial(T - L)\,\partial C}(L \, dw + dI) - \lambda \, dw \right)}{|\bar{H}|} \tag{10.44}$$

$$= \frac{\left(w\dfrac{\partial^2 U}{\partial C^2} - \dfrac{\partial^2 U}{\partial(T - L)\,\partial C} \right) L \, dw + \lambda \, dw + \left(w\dfrac{\partial^2 U}{\partial C^2} - \dfrac{\partial^2 U}{\partial(T - L)\,\partial C} \right) dI}{|\bar{H}|}.$$

Thus, the effect on labor supply of an increase in the wage rate is

$$\frac{\partial L^*}{\partial w} = \frac{dL}{dw}\bigg|_{dI=0} = \frac{\left(w\dfrac{\partial^2 U}{\partial C^2} - \dfrac{\partial^2 U}{\partial(T - L)\,\partial C} \right) L + \lambda}{|\bar{H}|}. \tag{10.45}$$

The second-order condition is that the denominator of equation (10.45) must be positive, but we cannot sign the numerator without more information about the form of the utility function. This should not be surprising to students who have had an intermediate

FIGURE 10.3

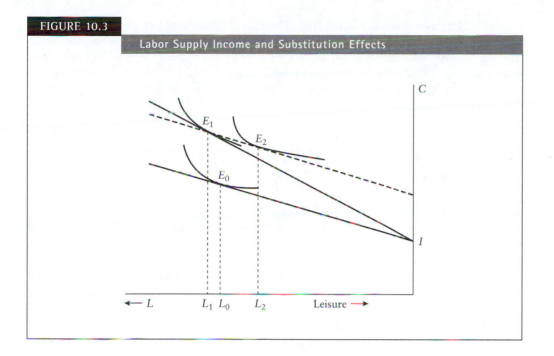

Labor Supply Income and Substitution Effects

microeconomics course: with the usual assumption that leisure is a normal good, the income and substitution effects of an increased wage work in opposite directions. As equation (10.44) makes clear, if an increase in income leads to a decrease in labor supply, as should be the case if leisure is a normal good, the expression in parentheses in the numerator of the right-hand side of equation (10.45) must be negative. Multiplying this expression by L and dividing by $|\bar{H}|$ gives the income effect of the higher wage: when the wage goes up by dw, income goes up by $L\,dw$; and this increased income reduces labor supply by an amount equal to the change in income times the derivative of labor supply with respect to income. The remaining term $\lambda/|\bar{H}|$ in equation (10.45) represents the substitution effect of the higher wage. Since Lagrange multipliers are positive and $|\bar{H}| > 0$, the substitution effect of a higher wage increases labor supply.

The substitution and income effects of an increased wage rate are illustrated in Figure 10.3, which will be familiar to readers who have studied intermediate microeconomics. In the diagram, labor supply increases as we move to the left and a movement to the right indicates that more leisure is taken. The diagram is drawn this way so that indifference curves will have the usual shape. Consumption is measured on the vertical axis. Point E_0 shows the original equilibrium, which is a point of tangency between the budget line and an indifference curve, and L_0 is the amount of labor supplied at that equilibrium. When the wage rate increases, the budget line becomes steeper and a new tangency point is reached at point E_1; the new equilibrium amount of labor supply is L_1.

The change from L_0 to L_1 can be divided into substitution and income effects by considering the effects of keeping the wage at its original level but increasing income enough that point E_1 can be chosen.[7] That is, we consider the effects of having a budget line that is parallel to the original but goes through point E_1; this budget line is

---

[7]This method is one of several ways to identify income and substitution effects, all of which are equivalent for small changes in the wage rate. Perhaps the most common is to find the point on the old indifference curve that has the same slope as the new budget line, and identify substitution and income effects with reference to that point. We use the method that corresponds to the interpretation of the terms in equation (10.45).

shown as the dashed line in Figure 10.3. This budget line would be tangent to an indifference curve at a point like E_2, with associated labor supply L_2. Now we can identify the income and substitution effects of the higher wage rate: the income effect equals $L_2 - L_0$, while the substitution effect equals $L_1 - L_2$. If the form of the utility function were known, equation (10.45) could be used to find the magnitudes of these two effects.

10.6 UTILITY MAXIMIZATION SUBJECT TO BUDGET AND TIME CONSTRAINTS

Suppose you were going to New York City to spend a weekend. You would have to choose among many activities: going shopping, seeing a show, walking in Central Park, visiting a museum, and attending a sporting event, among many other choices. Some of these activities would have a monetary cost, such as the admission fee to the sporting event. Other activities might be "free" in the sense that they require no admission fees or other related monetary expenses. Each activity would take some time, however, thus preventing you from engaging in some other activity during that time. In making your choice of how to spend your time, you would take into consideration not only the monetary cost of each activity but also the opportunity cost of the time each activity requires.

This is an example of a utility-maximization problem subject to two constraints: an income constraint and a time constraint. Because these are the two most important constraints consumers face in their daily lives, understanding this two-constraint model allows us to explain a great deal of human behavior.

Consider first the problem of maximizing the utility function $U(x_1, x_2, \ldots, x_n)$ subject to constraints on income and time. Each good i has a monetary price[8] (in dollars) p_i, so the consumer with total money income of I must satisfy the budget constraint $\sum_{i=1}^{n} p_i x_i \leq I$. In addition, the consumption of each good i requires a certain amount of time (in minutes) t_i. If T total minutes are available, then the consumer must also satisfy a time constraint $\sum_{i=1}^{n} t_i x_i \leq T$. Defining $\mathbf{x}$, $\mathbf{p}$, and $\mathbf{t}$ as the vectors of consumption, monetary prices, and time requirements, respectively, the Lagrangian for the constrained maximization problem is

$$\mathscr{L}(\mathbf{x}, \lambda_1, \lambda_2, I, \mathbf{p}, \mathbf{t}) = U(x_1, \ldots, x_n) + \lambda_1(I - p_1 x_1 - \cdots - p_n x_n)$$
$$+ \lambda_2(T - t_1 x_1 - \cdots - t_n x_n). \tag{10.46}$$

The first-order conditions for this constrained maximization problem are

$$\frac{\partial \mathscr{L}}{\partial x_i} = U_i - \lambda_1 p_i - \lambda_2 t_i = 0, \quad i = 1, \ldots, n$$

$$\frac{\partial \mathscr{L}}{\partial \lambda_1} = I - p_1 x_1 - \cdots - p_n x_n = 0 \tag{10.47}$$

$$\frac{\partial \mathscr{L}}{\partial \lambda_2} = T - t_1 x_1 - \cdots - t_n x_n = 0$$

---

[8]In our example of vacationing in New York City, most of the activities have an admission fee, although here we have modeled it more like a price per minute of activity. The economic effects of admission fees are often analyzed using as a model a two-part tariff. We will analyze two-part tariffs in Chapter 14.

FIGURE 10.4

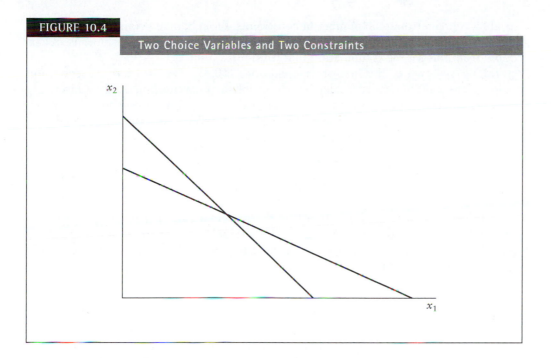

where U_i stands for $\partial U / \partial x_i$. If the second-order conditions are satisfied, these first-order conditions give the solution, assuming that an interior solution exists[9] and that both the income and time constraints are binding. The second-order conditions (see Section 9.2.4) are that the border-preserving principal minors of order $r = 3, \ldots, n$ of the bordered Hessian

$$\bar{H} = \begin{bmatrix} U_{11} & \cdots & U_{1n} & -p_1 & -t_1 \\ \vdots & \ddots & \vdots & \vdots & \vdots \\ U_{n1} & \cdots & U_{nn} & -p_n & -t_n \\ -p_1 & \cdots & -p_n & 0 & 0 \\ -t_1 & \cdots & -t_n & 0 & 0 \end{bmatrix}$$ (10.48)

have sign $(-1)^r$.

Note that the problem must have at least three choice variables for the first- and second-order conditions to make sense. If the number of choice variables equals the number of constraints, the consumer really has no choice: as long as all constraints are binding (and all are independent), the constraints will form a system of n equations in n unknowns. Typically, only one combination of values of the choice variables will satisfy the constraints. The two-variable, two-constraint case is illustrated in Figure 10.4. Only one combination of x_1 and x_2 satisfies both the budget and time constraints as graphed: the combination indicated by the intersection of the two constraints. So if both constraints are binding, this must be the solution. When the number of choice

---

[9]In our New York City vacation example, we would probably have corner solutions: no vacationer engages in *every* possible activity! Chapters 11 and 12 address corner solutions formally.

variables is greater than the number of constraints, though, many combinations of values of the choice variables typically satisfy the constraints, so the consumer is free to choose from this set the combination that maximizes utility.

Taking any two of the first-order conditions (10.47), we can derive a condition analogous to condition (9.10) which, for the problem of maximizing utility subject only to an income constraint, showed that the marginal rate of substitution between any two goods should equal the ratio of the two goods' prices. When both the income and time constraints are binding, the relevant condition is

$$\frac{U_i}{U_j} = \frac{\lambda_1 p_i + \lambda_2 t_i}{\lambda_1 p_j + \lambda_2 t_j}. \tag{10.49}$$

The left-hand side of equation (10.49) is the marginal rate of substitution between goods i and j (see Section 6.9 or 9.2.1). The interpretation of the right-hand side of equation (10.49) will be clearer if we define a new variable $\tilde{\lambda} = \lambda_2/\lambda_1$. We can now rewrite equation (10.49) as

$$\frac{U_i}{U_j} = \frac{p_i + \tilde{\lambda} t_i}{p_j + \tilde{\lambda} t_j}. \tag{10.50}$$

Since Lagrange multipliers measure the effect on the objective function of relaxing the associated constraint, λ_1 is the marginal utility of income, and λ_2 is the marginal utility of time. Thus λ_2 measures the amount by which utility will go up if we increase the consumer's endowment of time by a small amount. Since λ_1 is the marginal utility of income, $1/\lambda_1$ is the amount of income necessary to raise utility by one unit. Thus $\tilde{\lambda} = \lambda_2/\lambda_1$ converts time into money: in our New York City vacation example, $\tilde{\lambda}$ is the amount of extra spending money that would be equivalent, in terms of its ability to provide utility, to a slight increase in the length of the vacation. In our mathematical example, $\tilde{\lambda} t_i$ is the amount of income that would be equivalent to the amount of time needed for consuming a little bit more of good i.

The interpretation of equation (10.50) is now easy: the marginal rate of substitution between goods i and j must equal the ratio of effective prices of the two goods, where the effective price of good i is the monetary price p_i plus the opportunity cost $\tilde{\lambda} t_i$ (measured in dollars) of the time needed to consume good i.

A very important implication of equation (10.50) is that even if the monetary price of a good is zero, consumers will not want to "purchase" it in unlimited quantities. Even a "free" good (for example, a walk in the park) has a positive price if it takes some time to consume (assuming that the time constraint is binding). This insight is the basis for two well-known and important maxims: "time is money" and "there's no such thing as a free lunch."

10.7 PARETO EFFICIENCY (MULTIPLE CONSTRAINTS)

The concept of **Pareto efficiency** (sometimes called **Pareto optimality**) is central to welfare economics. A Pareto-efficient allocation is one in which no consumer can be made better off without making another consumer worse off. Our example will consider exchange efficiency: the Pareto-efficient allocation among consumers of given amounts of final goods. The concept of Pareto efficiency can also be applied to a production economy; but as an illustration of both the economic interpretation of efficiency and the mathematics of dealing with more than one constraint, the exchange efficiency example suffices. We begin with a two-person, two-good example, which is familiar from the Edgeworth-Bowley box diagrams of intermediate microeconomics, and then extend the analysis to a three-person, two-good model to illustrate how the results from the simpler model generalize.

Consider first an example with two consumers, indicated by superscripts 1 and 2, and two goods, X and Y. The two consumers' utility functions, not necessarily identical, show the utility derived by the consumption of the two goods. The first consumer's utility is given by $U^1(X^1, Y^1)$, while the second consumer's utility is given by $U^2(X^2, Y^2)$. There are fixed quantities of the two goods, $\bar{X}$ and $\bar{Y}$, available to be allocated. An allocation is efficient, or Pareto optimal, if neither consumer can be made better off without making the other worse off.

Mathematically we can find efficient allocations by considering the problem of choosing X^1, Y^1, X^2, and Y^2 to maximize the utility of one of the consumers, say consumer 1, subject to three constraints: that the other consumer's utility be at least equal to some arbitrary level $\overline{U^2}$ and that the total amounts of goods X and Y allocated not be greater than the amount available. The Lagrangian for this problem is thus

$$\mathcal{L} = U^1(X^1, Y^1) + \lambda_1(U^2(X^2, Y^2) - \overline{U^2}) + \lambda_2(\bar{X} - X^1 - X^2) + \lambda_3(\bar{Y} - Y^1 - Y^2)$$

$$(10.51)$$

which has as first-order conditions

$$\frac{\partial \mathcal{L}}{\partial X^1} = \frac{\partial U^1}{\partial X^1} - \lambda_2 = 0 \qquad \frac{\partial \mathcal{L}}{\partial X^2} = \lambda_1 \frac{\partial U^2}{\partial X^2} - \lambda_2 = 0$$

$$\frac{\partial \mathcal{L}}{\partial Y^1} = \frac{\partial U^1}{\partial Y^1} - \lambda_3 = 0 \qquad \frac{\partial \mathcal{L}}{\partial Y^2} = \lambda_1 \frac{\partial U^2}{\partial Y^2} - \lambda_3 = 0$$

$$\frac{\partial \mathcal{L}}{\partial \lambda_1} = U^2(X^2, Y^2) - \overline{U^2} = 0 \qquad (10.52)$$

$$\frac{\partial \mathcal{L}}{\partial \lambda_2} = \bar{X} - X_1 - X_2 = 0 \qquad \frac{\partial \mathcal{L}}{\partial \lambda_3} = \bar{Y} - Y_1 - Y_2 = 0.$$

This gives us seven (nonlinear) equations in the seven unknowns X^1, Y^1, X^2, Y^2, λ_1, λ_2, and λ_3. We can gain insight into the efficient allocation by using the first four of the first-order conditions (10.52) and eliminating the Lagrange multipliers from them:

$$\frac{\partial U^1}{\partial X^1} = \lambda_2 = \lambda_1 \frac{\partial U^2}{\partial X^2} \quad \text{and} \quad \frac{\partial U^1}{\partial Y^1} = \lambda_3 = \lambda_1 \frac{\partial U^2}{\partial Y^2},$$

so

$$\frac{\partial U^1/\partial X^1}{\partial U^1/\partial Y^1} = \frac{\partial U^2/\partial X^2}{\partial U^2/\partial Y^2}. \qquad (10.53)$$

Thus the marginal rate of substitution between the two goods for consumer 1 must equal the marginal rate of substitution between the two goods for consumer 2. This is the condition for Pareto efficiency in an exchange economy, as illustrated in the Edgeworth-Bowley box shown in Figure 10.5.

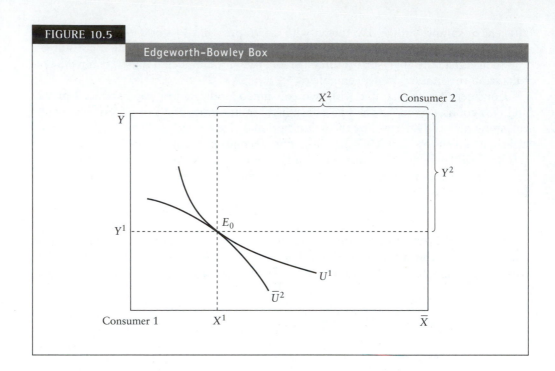

FIGURE 10.5

Edgeworth–Bowley Box

The dimensions of the box are the amounts of the two goods available; particular locations in the box represent how those goods are allocated between the two individuals. The first consumer's indifference curves are drawn with the origin in the lower left-hand corner of the box, while the second consumer's indifference curves are drawn relative to an origin in the upper right-hand corner. A particular indifference curve for consumer 2 is shown, labeled $\overline{U^2}$. The point of tangency E_0 represents an efficient allocation. Consumer 1 gets X^1 units of good X and Y^1 units of good Y, while consumer 2 gets $X^2 = \bar{X} - X^1$ units of good X and $Y^2 = \bar{Y} - Y^1$ units of good Y.

Assuming that the second-order conditions are satisfied, the seven equations (10.52) implicitly define the efficient allocations of goods X and Y to the two consumers as functions of the exogenous variables $\bar{X}$, $\bar{Y}$, and $\overline{U^2}$. In particular, by varying $\overline{U^2}$, we can trace out all of the efficient allocations possible in the Edgeworth-Bowley box by finding the efficient allocation on each indifference curve of consumer 2. The locus of all efficient points is called the **contract curve,** which is shown in Figure 10.6 with three representative efficient points and the associated indifference curves for both consumers. The subscripts on the indifference curves indicate the efficient point to which they correspond.

One of the fundamental results of welfare economics is that perfect competition will lead to Pareto efficiency. For this result to be true, it must be that when consumers maximize utility subject to exogenous prices, the resulting equilibrium is on the contract curve. Problems 10.18–10.20 explore this.

Adding a third consumer to the model adds two more choice variables, X^3 and Y^3, and one more constraint, $U^3(X^3, Y^3) \geq \overline{U^3}$, with an associated Lagrange multiplier. This makes the Lagrangian for the problem

$$\mathscr{L} = U^1(X^1, Y^1) + \lambda_1(U^2(X^2, Y^2) - \overline{U^2}) + \lambda_2(U^3(X^3, Y^3) - \overline{U^3})$$
$$+ \lambda_3(\bar{X} - X^1 - X^2 - X^3) + \lambda_4(\bar{Y} - Y^1 - Y^2 - Y^3),$$

(10.54)

FIGURE 10.6

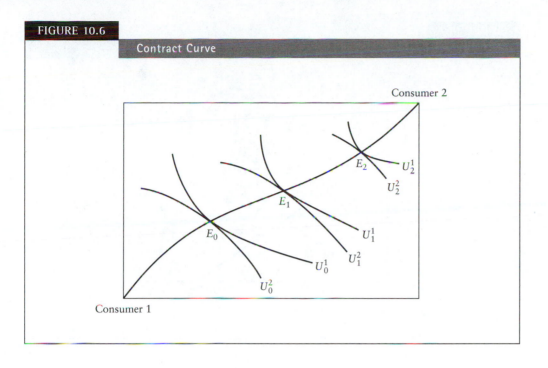

where the Lagrange multipliers have been renumbered, and makes the first-order conditions

$$\frac{\partial \mathcal{L}}{\partial X^1} = \frac{\partial U^1}{\partial X^1} - \lambda_3 = 0 \qquad \frac{\partial \mathcal{L}}{\partial X^2} = \lambda_1 \frac{\partial U^2}{\partial X^2} - \lambda_3 = 0 \qquad \frac{\partial \mathcal{L}}{\partial X^3} = \lambda_2 \frac{\partial U^3}{\partial X^3} - \lambda_3 = 0$$

$$\frac{\partial \mathcal{L}}{\partial Y^1} = \frac{\partial U^1}{\partial Y^1} - \lambda_4 = 0 \qquad \frac{\partial \mathcal{L}}{\partial Y^2} = \lambda_1 \frac{\partial U^2}{\partial Y^2} - \lambda_4 = 0 \qquad \frac{\partial \mathcal{L}}{\partial Y^3} = \lambda_2 \frac{\partial U^3}{\partial Y^3} - \lambda_4 = 0$$

$$\frac{\partial \mathcal{L}}{\partial \lambda_1} = U^2(X^2, Y^2) - \overline{U^2} = 0 \qquad \frac{\partial \mathcal{L}}{\partial \lambda_2} = U^3(X^3, Y^3) - \overline{U^3} = 0$$

$$\frac{\partial \mathcal{L}}{\partial \lambda_3} = \bar{X} - X_1 - X_2 - X_3 = 0 \qquad \frac{\partial \mathcal{L}}{\partial \lambda_4} = \bar{Y} - Y_1 - Y_2 - Y_3 = 0.$$

$$(10.55)$$

Eliminating the Lagrange multipliers from the first six of these conditions yields tangency conditions between pairs of indifference curves:

$$\frac{\partial U^1/\partial X^1}{\partial U^1/\partial Y^1} = \frac{\partial U^2/\partial X^2}{\partial U^2/\partial Y^2} = \frac{\partial U^3/\partial X^3}{\partial U^3/\partial Y^3}. \qquad (10.56)$$

The extension to many consumers and many goods is straightforward. This is another case when the geometric illustration of the solution, which can be done only for the special case that can be drawn in two dimensions, can be extended to the general case.

10.8   INTERTEMPORAL CONSUMPTION

In this section we use the utility-maximization model to investigate intertemporal consumption choices. We begin with a simple, two-period model with equal weight

FIGURE 10.7

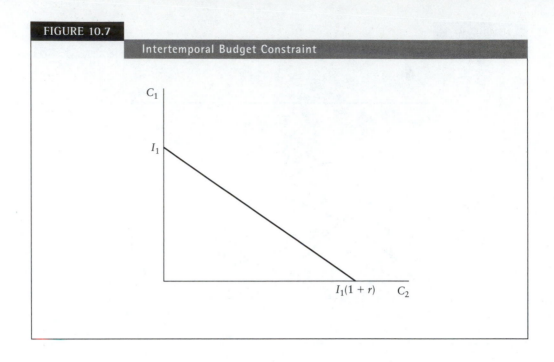

Intertemporal Budget Constraint

given in the utility function to future and present consumption. Then we extend the model to many periods and to include a subjective rate of time preference.

In the simplest two-period model, utility depends on (aggregate) consumption in the two periods, C_1 and C_2. A simple utility function that exhibits no time preference is $U(C_1, C_2) = C_1 C_2$. We will assume that income is earned only in the first period, in amount I_1. Economists often use this sort of model (albeit with time preference) to study saving; it serves as the basis for overlapping-generations models, for example, in the analysis of Social Security. The consumer's problem is to maximize utility by allocating consumption across the two periods.

Consumption in the second period is possible only by saving in the first period, that is, by consuming less than I_1. The maximum amount of consumption possible in the second period is $I_1(1 + r)$, where r is the (exogenous) interest rate; this amount of consumption in the second period is possible only if *no* income is consumed in the first period. At the other extreme, the consumer could consume *all* of the first-period income I_1 and consume nothing in the second period. The consumer can choose any combination of consumption in the two periods as long as their sum does not exceed I_1 plus whatever interest is earned on the amount saved. The budget constraint facing the consumer is thus that the **present discounted value**[10] of present and future consumption not exceed I_1; that is,

$$C_1 + \frac{C_2}{1 + r} \le I_1. \tag{10.57}$$

This intertemporal budget constraint is shown in Figure 10.7.

---

[10]The *present discount value of a dollar* received at a future date is the amount that, if received today and saved, would accumulate (counting principal and interest) to one dollar at the future date. Equivalently, it is the maximum amount that could be borrowed today with the principal and interest of the loan repaid at the future date with one dollar.

The Lagrangian for this constrained maximization problem is

$$\mathscr{L}(C_1, C_2, \lambda, I_1, r) = U(C_1, C_2) + \lambda \left(I_1 - C_1 - \frac{C_2}{1+r} \right) = C_1 C_2 + \lambda \left(I_1 - C_1 - \frac{C_2}{1+r} \right).$$

(10.58)

The three first-order conditions are

$$\frac{\partial \mathscr{L}}{\partial C_1} = C_2 - \lambda = 0, \quad \frac{\partial \mathscr{L}}{\partial C_2} = C_1 - \frac{\lambda}{1+r} = 0, \quad \text{and} \quad \frac{\partial \mathscr{L}}{\partial \lambda} = I_1 - C_1 - \frac{C_2}{1+r} = 0$$

(10.59)

while the second-order condition is that the determinant of the bordered Hessian of the problem be positive:[11]

$$|\bar{H}| = \begin{vmatrix} 0 & 1 & -1 \\ 1 & 0 & -\dfrac{1}{1+r} \\ -1 & -\dfrac{1}{1+r} & 0 \end{vmatrix} = -\left(-\frac{1}{1+r} \right) - \left(-\frac{1}{1+r} \right) = \frac{2}{1+r} > 0.$$

(10.60)

Eliminating the Lagrange multiplier from the first two of the first-order conditions (10.59) gives us the standard equality between the marginal rate of substitution for this utility function and the ratio of prices of first- and second-period consumption. (In this model, this ratio depends only on the interest rate: a higher interest rate makes future consumption relatively cheaper since less current consumption has to be foregone in order to finance a given amount of future consumption.) Thus we have

$$\frac{C_2}{C_1} = \frac{1}{1/(1+r)} = 1 + r.$$

(10.61)

A slight rearrangement of equation (10.61) yields $C_2 = C_1(1+r)$. Since first- and second-period consumption get equal weight in the utility function we are using, the consumer will maximize utility by equating the present discounted values of the two periods' consumption.[12] In fact, solving the first-order conditions for the two periods' consumption yields

$$C_1^* = \frac{I_1}{2} \quad \text{and} \quad C_2^* = \frac{I_1(1+r)}{2}$$

(10.62)

so with this utility function, the interest rate has no effect on first-period consumption (and therefore on saving, which equals first-period income minus first-period consumption).

It should be clear from equation (10.61) that, since the ratio of optimal consumption in the two periods does not depend on income, the timing of when income is received does not affect the time path of consumption. Only the present discounted value

---

[11] Since we have only two choice variables (except for the Lagrange multiplier), we need to consider only the second-order border-preserving principal minor (which is the determinant of the bordered Hessian).

[12] This somewhat odd result—that consumption in the "retirement" period is greater than consumption in the "working" period—is a result of using a utility function with no subjective rate of time preference.

of the income stream matters.[13] This point is explored further in the context of the two-period model in Problem 10.21.

Next we will extend our model to more than two time periods and show that the invariance of consumption to the timing of income still holds. Suppose that the consumer lives for T time periods, indexed by the subscript t; thus C_t is consumption in time t. We will, for the time being, continue to use the multiplicative utility function with no time preference:

$$U(C_1, C_2, \ldots, C_T) = C_1 C_2 \cdots C_T = \prod_{t=1}^{T} C_t. \tag{10.63}$$

Income is earned in each period; assuming perfect capital markets and a constant interest rate, the present discounted value of the income stream is

$$I_1 + \frac{I_2}{1+r} + \cdots + \frac{I_T}{(1+r)^{T-1}} = \sum_{t=1}^{T} \frac{I_t}{(1+r)^{t-1}} \tag{10.64}$$

while the present discounted value of the stream of present and future consumption is

$$C_1 + \frac{C_2}{1+r} + \cdots + \frac{C_T}{(1+r)^{T-1}} = \sum_{t=1}^{T} \frac{C_t}{(1+r)^{t-1}}. \tag{10.65}$$

With perfect capital markets, the consumer can borrow and save across time periods in many ways as long as the present discounted value of consumption is less than the present discounted value of income. The Lagrangian for the intertemporal utility-maximization problem is

$$\mathcal{L} = \prod_{t=1}^{T} C_t + \lambda \left(\sum_{t=1}^{T} \frac{I_t}{(1+r)^{t-1}} - \sum_{t=1}^{T} \frac{C_t}{(1+r)^{t-1}} \right). \tag{10.66}$$

The first-order conditions are

$$\frac{\partial \mathcal{L}}{\partial C_t} = \prod_{s \neq t} C_s - \frac{\lambda}{(1+r)^{t-1}} = 0, \quad t = 1, \ldots, T$$

$$\frac{\partial \mathcal{L}}{\partial \lambda} = \sum_{t=1}^{T} \frac{I_t}{(1+r)^{t-1}} - \sum_{t=1}^{T} \frac{C_t}{(1+r)^{t-1}} = 0. \tag{10.67}$$

Taking the first-order conditions for any two consecutive time periods t and $t+1$ and eliminating the Lagrange multiplier, we obtain

$$\frac{C_t}{(1+r)^{t-1}} = \frac{C_{t+1}}{(1+r)^t} \quad \text{or} \quad \frac{C_{t+1}}{C_t} = 1+r, \tag{10.68}$$

which is the same result obtained in the two-period model. It also would be the result if all income were earned in the first period. Any increase in income increases all periods' consumption by the same percentage (so that consumption still grows at rate r). Macroeconomists refer to this result as **consumption smoothing**: the path of consumption over time will be smoother than the time path of income.

---

[13]This result depends on the assumptions of perfect capital markets (borrowing and lending are done at the same interest rate, and—other than the consumer's limited income—there are no limits on either borrowing or saving) and no uncertainty.

The effect on the time path of consumption of an increase in the interest rate is affected by the time at which income is earned, however. Let us return to the two-period model to explore this issue. (The application to a many-period model is left as an end-of-chapter problem.)

In the two-period model when income is earned in the second period as well as the first, the first-order conditions (10.59) become

$$\frac{\partial \mathcal{L}}{\partial C_1} = C_2 - \lambda = 0, \quad \frac{\partial \mathcal{L}}{\partial C_2} = C_1 - \frac{\lambda}{1+r} = 0, \quad \text{and}$$
$$\frac{\partial \mathcal{L}}{\partial \lambda} = I_1 + \frac{I_2}{1+r} - C_1 - \frac{C_2}{1+r} = 0. \tag{10.69}$$

The second-order condition is satisfied for this problem (as you can confirm), so the solution to the first-order conditions (10.69) yields the optimal values of the two periods' consumption (and the Lagrange multiplier) as functions of the exogenous variables r, I_1, and I_2. We can solve equations (10.69) by successive substitution, Cramer's rule, or matrix inversion to get

$$C_1^* = \frac{I_1(1+r) + I_2}{2(1+r)} = \frac{I_1}{2} + \frac{I_2}{2(1+r)}, \quad C_2^* = \frac{I_1(1+r) + I_2}{2}, \quad \text{and}$$
$$\lambda^* = \frac{I_1(1+r) + I_2}{2} \tag{10.70}$$

from which we can directly obtain

$$\frac{\partial C_1^*}{\partial r} = -\frac{I_2}{2(1+r)^2}. \tag{10.71}$$

As long as second-period income is positive, $\partial C_1^*/\partial r < 0$.

The economic explanation of this result is that when the interest rate increases, the relative price of future consumption declines (the opportunity cost of current consumption rises). That is, the consumer does not have to give up as much first-period consumption to finance the same increase in future consumption because interest earnings will be higher (if the consumer chooses higher current consumption, the drop in future consumption is larger than when the interest rate was lower). This substitution effect would lead to less first-period consumption (and hence more saving). But the income effect of the increased consumption possibilities due to the higher return on saving counteracts the substitution effect. For the particular utility function we have used, the income effect exactly offsets the substitution effect when there is no second-period income. When there is second-period income, it becomes less valuable when the interest rate increases because, with a higher interest rate, borrowing costs are higher. Thus the amount of first-period consumption that can be financed by borrowing against future income is smaller. So when there is second-period income, there is an additional income effect, which acts to decrease first-period consumption (and increase saving).

Figure 10.8 illustrates the effect of an increase in the interest rate for the cases of no second-period income and positive second-period income. In Figure 10.8a there is no second-period income; so when the interest rate increases, the intertemporal budget constraint swivels around the vertical intercept. Equations (10.70) indicate that for the utility function we are using, first-period consumption is unaffected but second-period consumption increases. The two equilibria (before and after the interest rate increases) are labeled points E and E'. In Figure 10.8b there is second-period income, so second-period consumption is possible even if no first-period income is saved. When

FIGURE 10.8

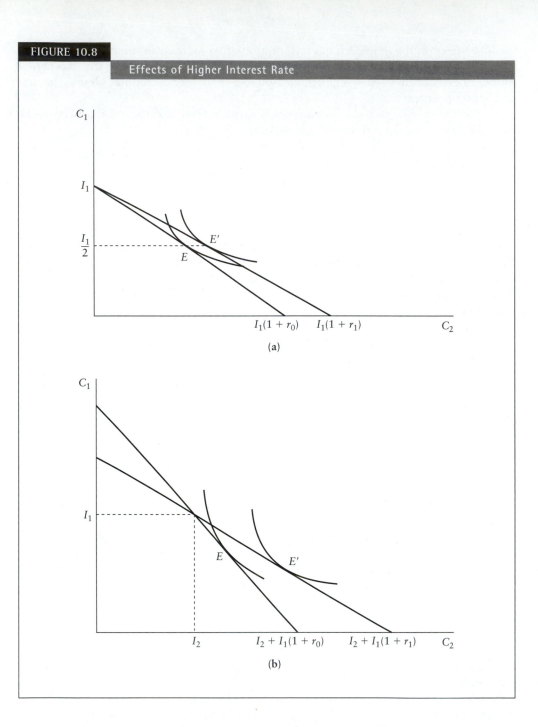

(a)

(b)

the interest rate increases, the intertemporal budget constraint becomes flatter, as before. But the consumer can always choose the combination of first- and second-period consumption that corresponds to no borrowing or saving, so the intertemporal budget constraint still goes through that point. The maximum amount of first-period consumption possible is now lower (due to higher borrowing costs) while the maximum amount of second-period consumption is higher. Equation (10.71) shows that for the multiplicative utility function with no subjective rate of time preference, first-period consumption will decrease when the interest rate rises. This is illustrated in Figure 10.8b as the equilibrium moves from point E to point E'.

We conclude by incorporating in the utility function a preference for present consumption over future consumption. For the two-period model the utility function is now

$$U(C_1, C_2) = C_1 C_2^{\rho} \quad \text{where } 0 < \rho < 1 \tag{10.72}$$

and the first-order conditions for the two-period intertemporal consumption problem with all income earned in period 1 are

$$\frac{\partial \mathscr{L}}{\partial C_1} = C_2^{\rho} - \lambda = 0, \quad \frac{\partial \mathscr{L}}{\partial C_2} = \rho C_1 C_2^{\rho-1} - \frac{\lambda}{1+r} = 0, \quad \text{and} \quad \frac{\partial \mathscr{L}}{\partial \lambda} = I_1 - C_1 - \frac{C_2}{1+r} = 0. \tag{10.73}$$

Eliminating the Lagrange multiplier from the first two of these conditions now yields

$$\frac{C_2}{\rho C_1} = \frac{1}{1/(1+r)} = 1 + r \quad \text{so} \quad \frac{C_2}{C_1} = \rho(1+r). \tag{10.74}$$

The stronger is the rate of time preference, the closer ρ is to zero and the higher is present compared to future consumption.

10.9   TRANSACTIONS DEMAND FOR MONEY

In the 1950s William Baumol[14] and James Tobin[15] developed a simple utility-maximization model to explain the transactions demand for money. It was based on a theory of how firms decided how much inventory to hold; the analogy was that people hold an inventory of money to finance their transactions each month. In this application we present a simple Baumol-Tobin model of an individual's transactions demand for money and derive the income and interest elasticities of money demand.

Our representative individual has a certain amount of transactions that she plans on making each month. If her planned real consumption is y and the aggregate price level is P then the dollar amount of transactions she plans is Py. We assume that these transactions occur evenly throughout the month. Our individual needs to allocate her wealth between two assets: a savings account that earns interest at rate i per month but which cannot be used to pay for transactions, and cash, which is needed to make purchases but which earns no interest. To make the model simple, we assume that interest is paid at the end of the month, based on the average monthly balance in her savings account. If money holdings average M, therefore, the interest foregone equals iM. Our representative individual therefore would like to hold as little money as possible, so that she earns interest on her assets, but she needs money to make her purchases. Every time she withdraws cash from her savings account she must pay a fixed real cost of f.[16]

If, at the beginning of the month, she withdraws enough cash to pay for all of the month's transactions, she has to pay the fixed cost only once, but she loses interest.

[14]William Baumol, "The Transactions Demand for Cash: An Inventory Theoretic Approach," *Quarterly Journal of Economics*, November 1952.

[15]James Tobin, "The Interest-Elasticity of Transactions Demand for Cash," *Review of Economics and Statistics* 38, August 1956.

[16]This might be a bank fee or it might represent the time cost of a trip to the bank.

Since she plans to make Py dollars' worth of transactions during the month, if she withdraws that amount at the beginning of the month and spends it evenly throughout the month, her average holdings of money will be $Py/2$. If, on the other hand, she withdraws at the beginning of the month only enough to last halfway through the month, and then withdraws the same amount again halfway through the month, she will still be able to make the same amount of transactions, but her average money holdings will be only $Py/4$. In general, if she makes n evenly spaced withdrawals, each of amount W, her average money holdings will be $M = W/2$, but she will be able to finance $nW = 2nM$ dollars' worth of transactions. Meanwhile, her total (nominal) costs associated with her money demand will be $nPf + iM$.[17]

The individual's problem is therefore to choose the number of withdrawals and the size of each withdrawal, or equivalently her average money balance, to minimize her total costs, subject to the constraint that she wants to be able to make Py dollars' worth of transactions. Since our purpose is to derive her demand for money, we will use n and M as choice variables rather than n and W. The Lagrangian for this problem is

$$\mathscr{L}(M, n, P, f, i, \lambda, y) = nPf + iM + \lambda(Py - 2nM) \tag{10.75}$$

and the first-order conditions are

$$\frac{\partial \mathscr{L}}{\partial n} = Pf - \lambda(2M) = 0$$

$$\frac{\partial \mathscr{L}}{\partial M} = i - \lambda(2n) = 0 \tag{10.76}$$

$$\frac{\partial \mathscr{L}}{\partial \lambda} = Py - 2nM = 0.$$

Since this is a minimization problem with two choice variables (not counting the Lagrange multiplier) and one constraint, the second-order condition is that the determinant of the bordered Hessian must be negative:

$$|\bar{H}| = \begin{vmatrix} 0 & -2\lambda & -2M \\ -2\lambda & 0 & -2n \\ -2M & -2n & 0 \end{vmatrix} = 2\lambda(-4nM) - 2M(4\lambda n) = -16\lambda nM < 0. \tag{10.77}$$

The first-order conditions (10.76) are nonlinear in the choice variables but we can solve for the optimal money demand by successive substitution. First, we use the first two equations to eliminate the Lagrange multiplier. The result is, as always in constrained optimization problems, a tangency condition:

$$\frac{i}{Pf} = \frac{\lambda 2n}{\lambda 2M} = \frac{n}{M}. \tag{10.78}$$

The left-hand side of equation (10.78) is the absolute value of the slope of a level curve of the objective function, while the right-hand side is the absolute value of the slope of a level curve of the constraint function. We can now solve for the money demand equation by using the constraint to eliminate n. From the constraint (the third equation in (10.76)),

$$n = \frac{Py}{2M}. \tag{10.79}$$

---

[17]Recall that f is a real cost.

Substituting into equation (10.78) and solving for M, we obtain

$$\frac{i}{Pf} = \frac{n}{M} = \frac{Py/2M}{M} = \frac{Py}{2M^2} \quad \text{so} \quad M^* = P\sqrt{\frac{fy}{2i}}. \qquad \textbf{(10.80)}$$

This has become known as the **square-root law** of the transactions demand for money.

To derive the income and interest elasticities of money demand, we begin by finding the comparative static derivatives of M^* with respect to y and i.[18] These derivatives are found directly from equation (10.80):

$$\frac{\partial M^*}{\partial y} = \frac{P}{2}\sqrt{\frac{f}{2iy}} \quad \text{and} \quad \frac{\partial M^*}{\partial i} = -\frac{P}{2i}\sqrt{\frac{fy}{2i}}. \qquad \textbf{(10.81)}$$

The income elasticity of money demand is therefore

$$\eta_y = \frac{\partial M^*}{\partial y}\frac{y}{M} = \left(\frac{P}{2}\sqrt{\frac{f}{2iy}}\right)\left(\frac{y}{P\sqrt{fy/2i}}\right) = \left(\frac{P}{2}\sqrt{\frac{f}{2iy}}\right)\left(\frac{y}{P}\sqrt{\frac{2i}{fy}}\right) = \frac{1}{2} \quad \textbf{(10.82)}$$

and the interest elasticity of money demand is

$$\eta_i = \frac{\partial M^*}{\partial i}\frac{i}{M} = \left(-\frac{P}{2i}\sqrt{\frac{fy}{2i}}\right)\left(\frac{i}{P\sqrt{fy/2i}}\right) = \left(-\frac{P}{2i}\sqrt{\frac{fy}{2i}}\right)\left(\frac{i}{P}\sqrt{\frac{2i}{fy}}\right) = -\frac{1}{2}. \quad \textbf{(10.83)}$$

A further important result of equation (10.80) is that money demand is proportional to the price level P. Said another way, there is no money illusion and the demand for *real balances* (the demand for purchasing power) is equal to

$$\frac{M^*}{P} = \sqrt{\frac{fy}{2i}}. \qquad \textbf{(10.84)}$$

The demand for real balances is an increasing function of income and a decreasing function of the interest rate. This provides the microfoundations for the usual assumptions about money demand (also called the *demand for liquidity*) in macroeconomic models.

10.10 MACROECONOMIC TRADEOFFS

In this application we explore the tradeoff between unemployment and inflation facing macroeconomic policymakers. The federal government would like to keep both inflation and unemployment low, but in the short run if monetary or fiscal policy is used to lower either inflation or unemployment, the other will rise, a relationship captured by the short-run Phillips curve. We will model the government's problem as one of minimizing a loss function, subject to the Phillips curve as a constraint. Then we will investigate the comparative static effects of changes in two parameters: one governing the relative weights of inflation and unemployment in the loss function and the other measuring the steepness of the Phillips curve.

---

[18]Although y more accurately represents consumption, not income, when put into a macroeconomic context money demand is usually thought of as a function of income. Each individual needs only enough money to pay for her consumption purchases, but for the economy as a whole saving equals investment, which comprises purchases by firms; so if they, like individuals, need money to make purchases, money demand should depend on income, not just personal consumption.

Let

$$L(u, \pi, \alpha) \tag{10.85}$$

be a loss function measuring how unhappy policymakers are with a particular combination of the unemployment rate u and the inflation rate π, with α being an exogenous parameter affecting the relative importance of inflation (versus unemployment) in the loss function. A quadratic loss function, $L(u, \pi, \alpha) = (u + \alpha \pi)^2$, is frequently used in economic analysis; we will leave the function general and explore the quadratic loss function in Problem 10.29. We will assume, however, that all three first derivatives of the loss function are positive and, further, that an increase in α increases the marginal loss associated with inflation: $\partial^2 L / \partial \pi \partial \alpha \equiv L_{\pi\alpha} > 0$.

The government cannot pick whatever combination of unemployment and inflation it wants: in the short run it is constrained by the short-run Phillips curve showing the relationship between inflation and unemployment

$$\pi = g(u, \beta) \tag{10.86}$$

where β is an exogenous parameter affecting the steepness of the Phillips curve. Since in the short run an increase in unemployment caused by demand-side fiscal or monetary policy will cause a movement along the Phillips curve to a lower inflation rate, $\partial g / \partial u \equiv g_u < 0$. Two typical Phillips curves are shown in Figure 10.9. We will assume that an increase in β means that the economy is more susceptible to inflation, making the Phillips curve shift upward, so $\partial g / \partial \beta \equiv g_\beta > 0$, and become steeper (the slope becomes more negative), so $\partial^2 g / \partial u \partial \beta \equiv g_{u\beta} < 0$. Otherwise we will leave the form of the function g general. So, in Figure 10.9, the curve labeled II corresponds to a higher value of β than does the curve labeled I.

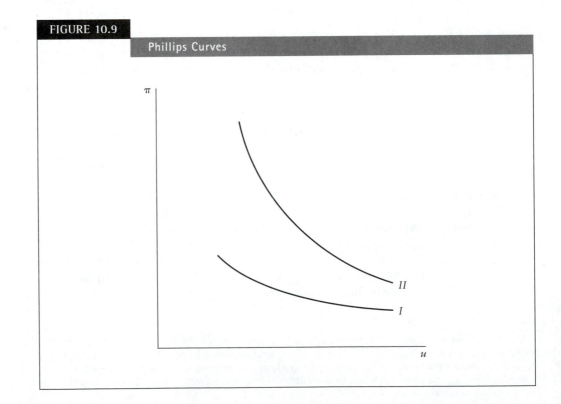

FIGURE 10.9

Phillips Curves

The government's problem is to choose unemployment and inflation to minimize the loss function (10.85) subject to the constraint (10.86). Unlike most of the constraints considered in Chapter 9 and the other applications in this chapter, (10.86) is a true equality constraint, so it is unclear whether the Lagrangian should be written as

$$\mathscr{L}(u, \pi, \alpha, \lambda, \beta) = L(u, \pi, \alpha) + \lambda(\pi - g(u, \beta)) \tag{10.87}$$

or as

$$\mathscr{L}(u, \pi, \alpha, \lambda, \beta) = L(u, \pi, \alpha) + \lambda(g(u, \beta) - \pi). \tag{10.88}$$

The only difference will be in the interpretation of the Lagrange multiplier. We will use the Lagrangian (10.88) so that λ will be interpreted as the marginal loss associated with higher inflation.

The first-order conditions for the government's constrained minimization problem are

$$\frac{\partial \mathscr{L}}{\partial u} = L_u + \lambda g_u = 0$$

$$\frac{\partial \mathscr{L}}{\partial \pi} = L_\pi - \lambda = 0 \tag{10.89}$$

$$\frac{\partial \mathscr{L}}{\partial \lambda} = g(u, \beta) - \pi = 0,$$

the second equation of which confirms our interpretation of λ. By eliminating the Lagrange multiplier from the first two equations, we derive the tangency condition

$$\frac{L_u}{L_\pi} = -g_u \tag{10.90}$$

where, as usual, the left-hand side represents the absolute value of the slope of a level curve of the objective (loss) function and the right-hand side represents the absolute value of the slope of the constraint (the Phillips curve). Combining equation (10.90) with the third of the first-order conditions (10.89), we get the result that the optimal combination of unemployment and inflation must be the point on the Phillips curve that is tangent to a level curve of the loss function. Worded another way, the government must choose the point on the Phillips curve such that the tradeoff they are willing to make between marginal changes in unemployment and inflation just equals the tradeoff the Phillips curve allows. The situation is illustrated in Figure 10.10, in which the curve labeled L_0 is a level curve of the loss function and the point labeled E_0 represents the optimal combination of unemployment and inflation consistent with the Phillips curve shown.

The level curve L_0 in Figure 10.10 is concave to the origin. This shape is required by the second-order condition[19] of the government's constrained minimization problem, which implies that the loss function is quasiconvex (see Theorem 9.3). Stated in terms of the bordered Hessian of the Lagrangian (10.88), the second-order condition is

---

[19]Since there are only two choice variables (other than the Lagrange multiplier) and one constraint, there is only one second-order condition, given in equation (10.91).

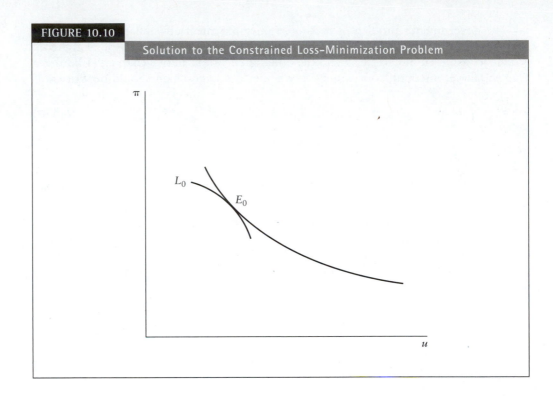

FIGURE 10.10

Solution to the Constrained Loss-Minimization Problem

that the determinant of the bordered Hessian must be negative:

$$|\bar{H}| = \begin{vmatrix} L_{uu} + \lambda g_{uu} & L_{u\pi} & g_u \\ L_{\pi u} & L_{\pi\pi} & -1 \\ g_u & -1 & 0 \end{vmatrix}$$

$$= g_u(-L_{u\pi} - L_{\pi\pi}g_u) + (-(L_{uu} + \lambda g_{uu}) - (L_{\pi u}g_u)) < 0.$$

(10.91)

Next, we investigate the comparative static effects of an increase in β, which, as explained previously, shifts the Phillips curve up and makes it steeper. The total differentials of the first-order conditions (10.89), written in matrix notation, are

$$\bar{H} \begin{bmatrix} du \\ d\pi \\ d\lambda \end{bmatrix} = \begin{bmatrix} -L_{u\alpha}d\alpha - \lambda g_{u\beta}d\beta \\ -L_{\pi\alpha}d\alpha \\ -g_\beta d\beta \end{bmatrix}.$$

(10.92)

Since $d\beta$ appears in more than one row of the right-hand side of (10.92), we know immediately that the second-order condition will not be sufficient to sign comparative static derivatives with respect to β. However, we have imposed some additional structure on the model already, by the assumptions we made earlier about the first and second derivatives of the loss function $L(u, \pi, \alpha)$ and the constraint function $g(u, \beta)$. Will these assumptions be sufficient to sign the comparative static effects of an increase in β? Setting $d\alpha = 0$ and using Cramer's rule to solve the system of equations (10.92) for du, we get

$$du = \frac{\begin{vmatrix} -\lambda g_{u\beta}d\beta & L_{u\pi} & g_u \\ 0 & L_{\pi\pi} & -1 \\ -g_\beta d\beta & -1 & 0 \end{vmatrix}}{|\bar{H}|} = \frac{-g_\beta d\beta(-L_{u\pi} - L_{\pi\pi}g_u) + \lambda g_{u\beta}d\beta}{|\bar{H}|}.$$

(10.93)

We have already assumed that $g_\beta > 0$ and that $g_{u\beta} < 0$. But without further assumptions about second derivatives of the loss function, the numerator of the right-hand side of equation (10.93) cannot be signed. This indeterminacy is illustrated in Figure 10.11. The original equilibrium is the same as that shown in Figure 10.10 and is labeled E_0. When β increases, the Phillips curve shifts to the curve labeled II and the new tangency is on a higher (worse) level curve of the loss function, L_1. But, without knowing how the shape of the level curves of the loss function change as we move up and to the right

FIGURE 10.11

Possible Effects on Unemployment of an Increase in β

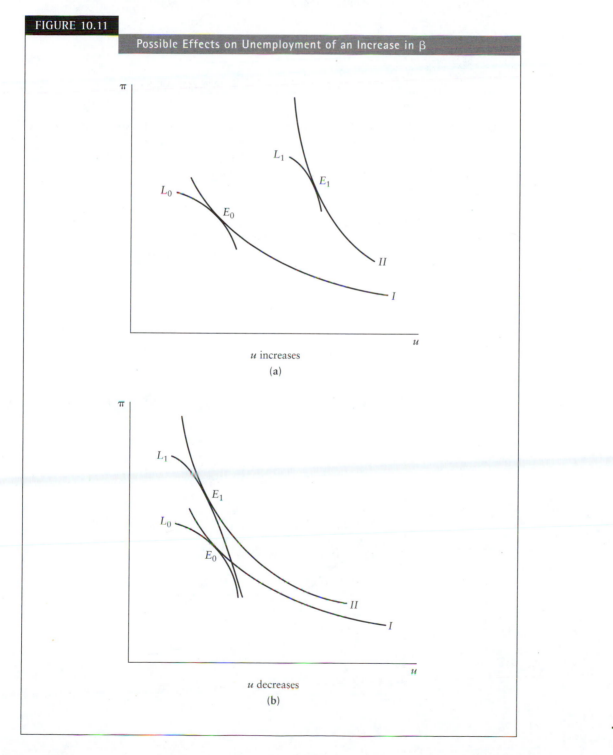

in the diagram (in particular, what happens to $L_{u\pi}$ and $L_{\pi\pi}$), it is impossible to know whether the new equilibrium is at a higher unemployment rate, as shown in Figure 10.11a, or a lower unemployment rate, as shown in Figure 10.11b.

To find the comparative static effect on the optimal inflation rate of an increase in β, we use Cramer's rule to solve the system of equations (10.92) for $d\pi$ (after setting $d\alpha = 0$) and obtain

$$d\pi = \frac{\begin{vmatrix} L_{uu} + \lambda g_{uu} & -\lambda g_{u\beta}d\beta & g_u \\ L_{\pi u} & 0 & -1 \\ g_u & -g_\beta d\beta & 0 \end{vmatrix}}{|\bar{H}|}$$ (10.94)

$$= \frac{g_u(-L_{\pi u}g_\beta d\beta) + (-g_\beta d\beta(L_{uu} + \lambda g_{uu}) + \lambda g_u g_{u\beta}d\beta)}{|\bar{H}|}.$$

As was the case with the unemployment rate, we are unable to sign the comparative static effect on the inflation rate without further assumptions, in this case about both the loss and constraint functions. The new optimal inflation rate might be higher than the original inflation rate, as shown in both panels of Figure 10.11, or it might be lower, as shown in Figure 10.12.

All of the cases illustrated in Figures 10.11 and 10.12 show a move to a higher (and therefore worse) level curve of the loss function. This is one comparative static effect that we can show unambiguously, by using the envelope theorem. By the envelope theorem, the effect of an increase in β on the optimal value of the loss function equals the partial derivative of the Lagrangian (10.88) with respect to β. That derivative is unambiguously positive:

$$\frac{\partial \mathcal{L}}{\partial \beta} = \lambda g_\beta > 0$$ (10.95)

since both λ and g_β are positive.

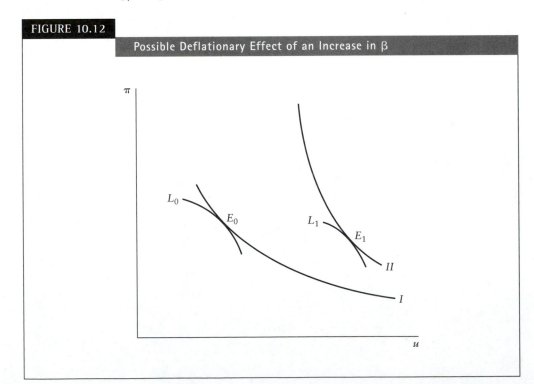

FIGURE 10.12

Possible Deflationary Effect of an Increase in β

To summarize, if the Phillips curve shifts up and becomes steeper, the economy is worse off, as measured by the increase in the value of the government's loss function. This result is intuitive. But it might also seem intuitive that another result of this shift would be that the government would be forced to accept a higher inflation rate; this result is not true in general. Similarly, while it might seem that the government would respond to the shifted Phillips curve by trying harder to fight inflation, and therefore allowing more unemployment, this result is not true in general either. Problem 10.29 investigates whether these intuitive results occur with a quadratic loss function.

The comparative static effects of an increase in the parameter α are also ambiguous. Setting $d\beta = 0$ and solving the system of equations (10.92) for du, we obtain

$$du = \frac{\begin{vmatrix} -L_{u\alpha}\,d\alpha & L_{u\pi} & g_u \\ -L_{\pi\alpha}\,d\alpha & L_{\pi\pi} & -1 \\ 0 & -1 & 0 \end{vmatrix}}{|\bar{H}|} = \frac{L_{u\alpha}\,d\alpha + g_u L_{\pi\alpha}\,d\alpha}{|\bar{H}|}. \qquad (10.96)$$

The second-order condition is that the denominator on the right-hand side of equation (10.96) is negative. We have already assumed that $g_u < 0$ and that $L_{\pi\alpha} > 0$. So if we are willing to assume that $L_{u\alpha} < 0$, the comparative static effect on the unemployment rate would be unambiguously positive, which is intuitively appealing: a change that makes inflation worse (in terms of its effect on the government's loss function) might lead the government to fight inflation harder, leading to a higher unemployment rate. Unfortunately, there is no reason to suspect that whatever change increases the government's marginal loss associated with inflation (holding constant the unemployment rate) would make the government's marginal loss associated with unemployment (holding constant the inflation rate) lower. So the end result of this change in the loss function might be to either increase or decrease the government's ultimate choice of the unemployment rate.

This application has illustrated an important result of mathematical economics models: often, the second-order conditions will not be sufficient to sign comparative static derivatives. More assumptions need to be made about the form of the objective function, constraint, or both. Economists often use two strategies in this situation. The first is to make additional, seemingly reasonable assumptions about the objective function and constraint and then see whether they are sufficient to sign the comparative static derivatives. The second is to identify what assumptions are sufficient for getting the comparative static derivatives to have a particular sign and then investigate whether those assumptions seem realistic. In either case, figuring out why the comparative static effects are ambiguous is usually a worthwhile exercise in economic reasoning.

Problems

10.1 For each of the following production functions, find and sign (if possible) the derivatives of the conditional and unconditional demand for labor functions with respect to the wage rate and the rental rate of capital. (R stands for another productive resource, say, land.) Assume all parameters are positive and less than 1.

(a) $\quad F(L, K) = L^\alpha K^{1-\alpha}$

(b) $\quad F(L, K) = A(\alpha L^\rho + (1-\alpha) K^\rho)^{1/\rho}$

(c) $F(L, K, R) = L^\alpha K^\beta R^\gamma$

(d) $F(L, K, R) = \alpha \ln L + \beta \ln K + \gamma \ln R + \delta(\ln L)(\ln K) + \phi(\ln L)(\ln R) + \theta(\ln K)(\ln R)$

10.2   For the production functions listed in Problem 10.1, find and sign (if possible) each of the following:

(a)   The derivative of marginal cost with respect to w

(b)   The derivative of marginal cost with respect to output

(c)   The derivative of the conditional demand for labor with respect to output

(d)   The derivative of the unconditional demand for labor with respect to output

(e)   The derivative of the unconditional demand for labor with respect to the price of the firm's output

(f)   The slope of the firm's supply curve (that is, $\partial Q^* / \partial P$)

(g)   The derivative of the profit-maximizing output level with respect to w

(h)   The effect on the firm's profit of an increase in w

10.3   For the following utility functions, find and sign (if possible) the derivatives of the Hicksian (compensated) and Marshallian (ordinary) demands for good 1 with respect to the prices of each good. Assume all parameters are positive and less than 1.

(a) $U(x_1, x_2) = x_1^\alpha x_2^\beta$

(b) $U(x_1, x_2) = \alpha \ln x_1 + \beta \ln x_2$

(c) $U(x_1, x_2) = (\alpha x_1^\rho + (1 - \alpha) x_2^\rho)^{1/\rho}$

(d) $U(x_1, x_2) = \alpha \ln x_1 + \beta \ln x_2 + \gamma(\ln x_1)(\ln x_2)$

10.4   For the utility functions listed in Problem 10.3, find and sign (if possible) the derivative of the Marshallian demand for good 1 with respect to income. (Why don't we find the corresponding derivative of the Hicksian demand for good 1?)

10.5   For a perfectly competitive firm with production function $F(L, K)$, does the derivative of the conditional demand for labor with respect to the rental rate of capital equal the derivative of the conditional demand for capital with respect to the wage? Explain in terms of factor substitution effects.

10.6   Would the answer to Problem 10.5 change if we were considering unconditional instead of conditional input demands? (The economic explanation of this result will have to include consideration of the output effect as well as the factor substitution effect.) Why is this result important in terms of defining whether two inputs are substitutes or complements?

10.7   Consider a firm with production function $F(L, K)$ that is a perfect competitor in input markets but a monopolist in the output market. Find, and compare to a perfectly competitive firm,

(a)   The first- and second-order conditions for the firm's profit-maximization problem

(b)   The derivative of the unconditional demand for capital function with respect to the wage rate

(c) The effect on the profit-maximizing output level of an increase in the wage rate

(d) The effect on profit of an increase in the wage rate

10.8 A college is trying to decide what combination of three fuel sources (oil (O), wood (W), and coal (C)) to use in its heating plant. E_0 is the amount of energy the college wants to produce, $F(O, W, C)$ is the energy production function, and the college is a perfect competitor in the fuel markets (so its decisions do not affect the three fuel prices).

(a) Find the first- and second-order conditions for the college's cost-minimization problem and give an economic interpretation of the first-order conditions.

(b) Find and sign (if possible) the comparative static derivatives showing how the college's (conditional) demands for each fuel are affected by a government subsidy for wood. (The subsidy effectively lowers the price of wood.) Explain the economics of your answers.

(c) Find and sign (if possible) the comparative static derivative showing how the total cost of producing E_0 is affected by a government subsidy for wood.

(d) If all three fuel prices increase by the same percentage, find an expression showing the change in the college's overall cost of producing E_0. Does the overall cost change by the same proportion as the three fuel prices?

10.9 Find the first- and second-order conditions for the problem of maximizing the profit of a perfectly competitive firm with many inputs. Do all unconditional input demand functions slope downward?

10.10 Does the result of Problem 10.5 generalize? That is, for a perfectly competitive firm with many inputs, does the derivative of the conditional demand for input i with respect to the price of input j equal the derivative of the conditional demand for input j with respect to the price of input i?

10.11 Does the result of Problem 10.6 generalize? That is, for a perfectly competitive firm with many inputs, does the derivative of the unconditional demand for input i with respect to the price of input j equal the derivative of the unconditional demand for input j with respect to the price of input i?

10.12 Assuming that consumers are price takers (that is, they consider the prices of the goods they buy to be exogenous) and that there are many goods, does the derivative of the Hicksian (compensated) demand function for good i with respect to the price of good j equal the derivative of the Hicksian demand function for good j with respect to the price of good i?

10.13 A monopolist produces output (Q) but, as a byproduct, also produces pollution (e). The monopolist has to pay a pollution tax equal to t per unit of output, so the total costs of the monopolist equal its production costs, $C(Q)$, plus the total pollution taxes, te.

(a) Write out the Lagrangian for the firm's problem of choosing Q and e to maximize profits, subject to the constraint that $e - F(Q) \geq 0$, where F is the pollution production function showing how much pollution results from any given output level.

(b) Find the first- and second-order conditions and give an economic interpretation of each of the first-order conditions.

(c) Find and sign (if possible) the comparative static derivative showing how the firm's profit-maximizing output level changes when the pollution tax increases. Explain the economics.

(d) Find and sign (if possible) the comparative static derivative showing how the firm's profits are affected by an increase in the pollution tax.

(e) Find and sign (if possible) the comparative static derivative showing how the firm's price changes when the pollution tax increases. Explain the economics.

10.14 A visitor to a recreation site has the utility function $U(C, v)$ where v is visits to the site and C is consumption of all other goods. For convenience, set the price of C equal to 1; the recreation site has an entrance fee P. So the visitor's budget constraint is $C + Pv \leq I$, where I is (exogenous) income.

(a) Write out the Lagrangian and find the first-order conditions for the visitor's utility-maximization problem. Give an economic interpretation for the first-order conditions.

(b) Show that $\partial v^*/\partial P$ is not signable in general.

(c) Show that if visits are a normal good ($\partial v^*/\partial I > 0$), then $\partial v^*/\partial P < 0$.

10.15 Redo the analysis from Section 10.5 by choosing C, L, and l to maximize $U(C, l)$ subject to two constraints: $C \leq wL + I$ and $l = T - L$, where l is leisure. In particular, derive the derivative of the labor supply function with respect to w and show that it is equivalent to equation (10.45).

10.16 For the following utility functions, find $\partial L^*/\partial w$ and $\partial L^*/\partial I$. How large is the substitution effect of an increase in the wage rate?

(a) $U(C, T - L) = C^\alpha (T - L)^\beta$

(b) $U(C, T - L) = A(\alpha C^\rho + (1 - \alpha)(T - L)^\rho)^{1/\rho}$

(c) $U(C, T - L) = \gamma \ln(C - \alpha) + \beta \ln(T - L)$

10.17 For the utility functions in Problem 10.16, find, sign (if possible), and give an economic interpretation for $\partial L^*/\partial \alpha$.

10.18 Suppose there are two consumers with identical utility functions but different incomes. Show that if each consumer maximizes utility, the conditions for Pareto efficiency will be satisfied.

10.19 Would your answer to Problem 10.18 change if the two consumers had different utility functions?

10.20 Would your answer to Problems 10.18 and 10.19 change if there were many consumers?

10.21 In a two-period intertemporal consumption model, income is earned only in the second period. Assuming perfect capital markets, find the first- and second-order conditions for the problem of maximizing $U(C_1, C_2) = C_1 C_2$ subject to the intertemporal budget constraint. Show that these conditions are identical to the results derived in Section 10.8 if the present discounted value of income is the same in both cases.

10.22 Use the first-order conditions (10.59) to prove the consumption-smoothing result that if second-period income increases, consumption in both periods will increase.

10.23 Use the first-order conditions (10.59) to derive the effects on saving and second-period consumption of an increase in the interest rate.

10.24 Do the results of equation (10.71) hold for all utility functions of the form $U(C_1, C_2)$? That is, are the first- and second-order conditions sufficient to show that $\partial C_1^*/\partial r < 0$ in a two-period model when income is earned in both periods?

10.25 Solve the first-order conditions (10.73) and confirm that if ρ increases, first-period consumption falls while second-period consumption rises. Explain the economic reasoning for these results.

10.26 Use the first-order conditions (10.67) to show that $C_t = C_1(1+r)^{t-1}$ in the model with many time periods. Then solve for C_1^* and show that $\partial C_1^*/\partial r < 0$. Explain the economic reasoning.

10.27 In the Baumol-Tobin model of Section 10.9, derive, sign if possible, and give economic explanations for each of the following:

(a) $\partial M^*/\partial f_0$

(b) $\partial^2 M^*/\partial i\, \partial P$

(c) $\partial^2 M^*/\partial y\, \partial P$

(d) $\partial^2 M^*/\partial y\, \partial i$

(e) The effect on the representative individual's equilibrium (nominal) costs associated with her money demand of

  (i) A higher fixed cost of converting other assets into money

  (ii) A higher interest rate

  (iii) A higher price level

10.28 In the Baumol-Tobin model of Section 10.9, suppose that the real cost of converting other assets into money depends on the size of the transaction as well as a fixed cost: $f = f_0 + \alpha W$, where $\alpha > 0$.

(a) Find and sign (if possible) $\partial M^*/\partial \alpha$ and $\partial M^*/\partial f_0$ and give economic explanations.

(b) Find $\partial M^*/\partial i$ and $\partial M^*/\partial y$. Do they have the same signs as in the simple Baumol-Tobin model?

10.29 For the model of Section 10.10, if the loss function is $L(u, \pi, \alpha) = (u + \alpha\pi)^2$ find, sign if possible, and give an economic explanation for

(a) How an increase in α affects the slope of the level curve of the loss function

(b) $\partial u^*/\partial \alpha$

(c) $\partial u^*/\partial \beta$

(d) $\partial \pi^*/\partial \alpha$

(e) $\partial \pi^*/\partial \beta$

(f) $\partial L^*/\partial \alpha$

(g) $\partial L^*/\partial \beta$

10.30 For the model of Section 10.10, suppose the Phillips curve relationship is given by $\pi = g(u, \beta, \gamma)$, where γ is a positive parameter that causes a parallel shift in the Phillips curve: $\partial g/\partial \gamma > 0$ but $\partial^2 g/\partial u\, \partial \gamma = 0$. Find, sign if possible, and give an economic explanation for the effects of an increase in γ on the equilibrium rates of inflation and unemployment.

10.31 Do Problem 10.30 using the loss function $L(u, \pi, \alpha) = (u + \alpha\pi)^2$.

10.32 Do Problems 10.29 and 10.30 using the loss function $L(u, \pi, \alpha) = u^2 + \alpha\pi^2$.

Optimization with Inequality Constraints: Theory

11.1   INTRODUCTION

Almost all constraints encountered in economic analysis are inequality constraints, although they are usually treated as if they were equalities. For example, an individual may choose not to spend all her income, so her budget constraint is an inequality. But in the usual (one-period) utility-maximization problem, saving income does not increase utility, so the consumer will spend all her income. Thus treating the budget constraint as an equality will lead to the same result as treating it as an inequality. In Chapters 9 and 10 we set up constraints as (weak) inequalities. But in using the Lagrangian method, we assume that the constraints are binding; that is, they are satisfied as equalities. If leaving the constraint slack (not binding) leads to a higher value of the objective function in a maximization problem (or a lower value of the objective function in a minimization problem), the Lagrangian method cannot find that solution.

For the most part, we have also assumed interior solutions so far. But border solutions are clearly an important part of real economic decision making: consumers routinely decide not to purchase some goods; firms do not use positive quantities of every possible input, and so on. Dealing with border solutions is also a case of incorporating inequality constraints. For example, in a utility maximization problem $x_i \geq 0$ for each good i.

For most optimization problems in economic theory, treating inequality constraints as if they were equalities and ignoring border solutions does not lead to erroneous conclusions. The assumptions we normally make about the functions we deal with (such as utility functions and production functions) are usually enough to rule out border solutions and nonbinding constraints. By incorporating the possibilities of border solutions and nonbinding constraints in our formal analysis, however, we can see exactly what assumptions are necessary to rule out these possibilities and under what conditions they may be important.

In this chapter we introduce the **Kuhn-Tucker conditions** for optimization with inequality constraints. First we consider a one-variable maximization problem where the choice variable is constrained to be nonnegative. Then we impose a functional constraint while ignoring the nonnegativity constraint. Next we combine the findings of the previous sections to formulate the Kuhn-Tucker conditions and generalize them to a many-variable maximization problem with many constraints. We describe in passing the necessary modifications for minimization problems. (This is not because minimization problems are less important but rather because the modifications are minor and do not need lengthy explanations.) Section 11.5 contains some worked-out examples of applying the Kuhn-Tucker conditions. We then introduce linear programming as a special case of the preceding analysis and discuss duality in linear programming problems. The chapter concludes with a summary section and some problems based on the chapter material.

11.2   ONE–VARIABLE OPTIMIZATION WITH A NONNEGATIVITY CONSTRAINT

In Chapter 1 we described the first- and second-order conditions for a one-variable optimization problem. The value of x that optimizes the function $f(x)$ must necessarily satisfy the first-order condition $f'(x) = 0$. For a maximum the second-order condition is $f''(x) < 0$, while for a minimum the second-order condition is $f''(x) > 0$. These conditions do not ensure, however, that the value of x that satisfies them is nonnegative. For most economic problems, negative values of x are not meaningful, which is why we must examine border conditions ($x = 0$, in this case) as well as the solutions to the first-order conditions. Figure 11.1 shows the two possible solutions to a one-variable maximization problem: either the first-order condition is satisfied for a positive value of x (this is called an interior solution), or the maximal value of the function is obtained where $x = 0$ (this is called a border, or corner, solution).[1] Since either $f'(x) = 0$ while $x > 0$ or $f'(x) \leq 0$ while $x = 0$, the necessary conditions can be summarized as

$$x \geq 0, \quad f'(x) \leq 0, \quad \text{and} \quad xf'(x) = 0. \tag{11.1}$$

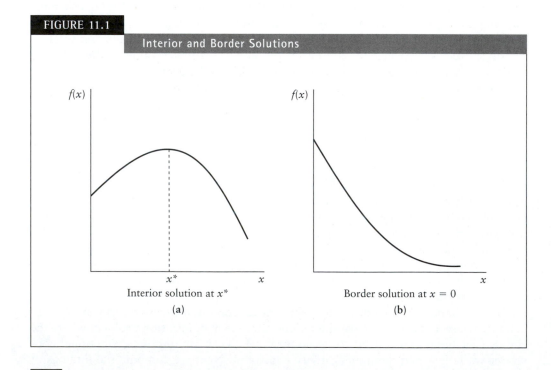

FIGURE 11.1

Interior and Border Solutions

Interior solution at x^*

(a)

Border solution at $x = 0$

(b)

---

[1]It is also possible for the first-order condition to be satisfied where $x = 0$; we ignore this case because the necessary conditions that we discuss in the text cover this possibility.

The conditions for a minimization problem are the same except that $f'(x) \geq 0$. The third of the conditions (11.1) is an example of a **complementary slackness condition** requiring that one of two weak inequalities hold as an equality—both of the inequalities cannot be slack. These complementary slackness conditions comprise an important part of the necessary conditions for inequality-constrained optimization problems.

Conditions (11.1) are usually applied in the following way. First, we ignore the nonnegativity constraint, solving the problem by finding a value x^* for which $f'(x^*) = 0$. If x^* is nonnegative, it is a possible solution to the maximization problem. Next, we check the border solution $x = 0$. If the function is decreasing at $x = 0$, that is, if $f'(0) < 0$, then $x = 0$ is a possible solution to the maximization problem. As always, if there is more than one possible solution, we must evaluate the function $f(x)$ at each to see which gives the global maximum. This procedure generalizes to more complex problems: the set of possible solutions is broken down into various cases, depending on what combinations of constraints are binding. For each case, we either find or rule out a solution. If we find more than one possible solution, we must evaluate the objective function for each to see which is the global maximum.

So far we have ignored second-order conditions. For positive values of x, the usual second-order conditions are sufficient to guarantee a local maximum. The necessary conditions (11.1) are themselves sufficient to guarantee that if $f'(0) < 0$, then $x = 0$ is a local (constrained—the objective function could take a larger value if x were allowed to be negative) maximum. If, as we often assume in economics, the function $f(x)$ is globally concave, we explore two possibilities: if there is a positive solution x^*, it is a global maximum; if there is no positive solution, either $x = 0$ is the global maximum or there is no maximum (for example, if $f(x) = \sqrt{x}$). Because we so often assume that objective functions in economics are globally concave, we can usually ignore nonnegativity constraints; we need to consider corner solutions only if we find a negative value for the choice variable as the solution for the unconstrained maximization problem.

11.3   ONE–VARIABLE OPTIMIZATION WITH ONE INEQUALITY CONSTRAINT

In this section we ignore nonnegativity constraints and focus on constraints of the form $g(x) \leq g_0$. These are the types of constraints encountered in most economics problems (see Chapters 9 and 10). In the previous chapters we simply assumed that the constraints would be binding. Here we allow for them to be slack.

Figure 11.2 shows the possible solutions to the problem of maximizing $f(x)$ subject to the constraint $g(x) \leq g_0$. The typical case is illustrated by Figure 11.2a, where the constraint ends up being binding. In Figure 11.2b the constraint is nonbinding: the maximal value of $f(x)$ occurs at a value of x for which $g(x) < g_0$.

Consider now the Lagrangian for the constrained maximization problem:

$$\mathscr{L}(x, \lambda, g_0) = f(x) + \lambda(g_0 - g(x)). \tag{11.2}$$

In Chapter 9, we justified maximizing the Lagrangian as being equivalent to maximizing the objective function $f(x)$ since the Lagrangian equals $f(x^*)$ plus zero if the constraint is binding at the solution x^*. Now, however, we want to allow for the possibility that the constraint is slack, in which case $g_0 - g(x) > 0$. In order for the Lagrangian to equal the objective function, the Lagrange multiplier λ must be zero. Thus the values of x and λ that maximize the objective function subject to the constraint

FIGURE 11.2

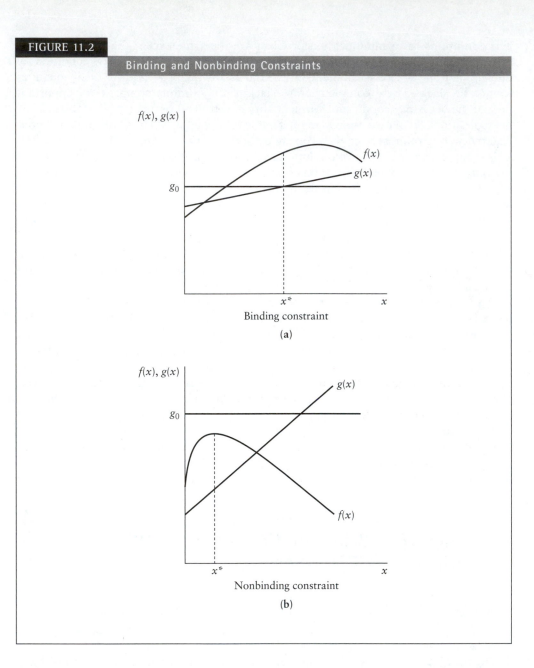

Binding constraint

(a)

Nonbinding constraint

(b)

must satisfy one of two conditions: either λ is positive[2] and the constraint is binding, or λ is zero and the constraint is slack. Since the derivative of the Lagrangian with respect to λ is $(g_0 - g(x))$, we write these two conditions as two nonnegativity conditions plus a complementary slackness condition:

$$\lambda \geq 0, \quad \frac{\partial \mathscr{L}}{\partial \lambda} \geq 0, \quad \text{and} \quad \lambda \frac{\partial \mathscr{L}}{\partial \lambda} = 0. \tag{11.3}$$

---

[2]Recall from Chapter 9 that, because of the way we have set up the Lagrangian, the Lagrange multiplier measures how the optimal value of the objective function is affected when the constraint is relaxed marginally. If a constraint is binding, there are some values for the choice variable(s) that would increase the value of the objective function, but they are ruled out because they would violate the constraint. Relaxing the constraint increases the set of possible values for the choice variable(s), some of which will increase the value of the objective function. Thus, if the Lagrange multiplier is nonzero, it must be positive. We discuss the interpretation of the Lagrange multiplier in more detail in Chapter 13.

In addition, of course, the derivative of the Lagrangian with respect to x must equal zero.[3] For a minimization problem the necessary conditions (11.3) are modified by setting $\partial \mathcal{L}/\partial \lambda \leq 0$ so that the constraint is nonpositive.

We are thus left with two possibilities: either the constraint is binding, in which case the Lagrangian analysis of Chapter 9, including the second-order conditions described there, is appropriate, or the constraint is nonbinding, in which case the analysis of Chapter 7 is appropriate. As is the case with nonnegativity constraints, conditions (11.3) are usually used to identify the possible cases and to rule out possible solutions if they do not satisfy the conditions (11.3).

11.4 THE KUHN–TUCKER CONDITIONS

There is an evident similarity between conditions (11.1) and (11.3). Indeed, by combining the two we get the Kuhn-Tucker conditions, named after the two coauthors of the pioneering work in the field of inequality-constrained maximization problems. The only alteration we must make is to replace $f'(x)$ with $\partial \mathcal{L}/\partial x$ in conditions (11.1). The Kuhn-Tucker conditions apply to the problem of maximizing $f(x)$ subject to the constraints $x \geq 0$ and $g(x) \leq g_0$. The Lagrangian (11.2) is formed and the necessary conditions are as follows:

Kuhn Tucker Conditions (one variable, one constraint)

The necessary[4] conditions for the problem of maximizing the Lagrangian (11.2) subject to the constraints that $x \geq 0$ and $g(x) \leq g_0$ are

$$\text{(a)} \quad \frac{\partial \mathcal{L}}{\partial x} \leq 0 \qquad \text{(d)} \quad \frac{\partial \mathcal{L}}{\partial \lambda} \geq 0$$

$$\text{(b)} \quad x \geq 0 \qquad \text{(e)} \quad \lambda \geq 0 \qquad \qquad \textbf{(11.4)}$$

$$\text{(c)} \quad x\frac{\partial \mathcal{L}}{\partial x} = 0 \qquad \text{(f)} \quad \lambda\frac{\partial \mathcal{L}}{\partial \lambda} = 0.$$

For minimization problems, the inequalities in conditions (11.4)(a) and (11.4)(d) are reversed. Conditions (11.4)(c) and (11.4)(f) are the familiar complementary slackness conditions. The possible cases for which solutions must be checked are as follows:

1. $x = 0$ and $\lambda = 0$ (a border solution and a nonbinding constraint),
2. $x = 0$ and $\lambda > 0$ (a border solution and a binding constraint),
3. $x > 0$ and $\lambda = 0$ (an interior solution and a nonbinding constraint), and
4. $x > 0$ and $\lambda > 0$ (the usual case of an interior solution and a binding constraint).

Notice that for case (4) the Kuhn-Tucker conditions collapse to the familiar Lagrangian first-order conditions. Cases (1) and (2) can be ruled out if $\partial \mathcal{L}/\partial x > 0$ when evaluated at $x = 0$ (for a maximization problem; the inequality is reversed for minimization problems), and case (3) can be ruled out if $\partial \mathcal{L}/\partial \lambda < 0$ when evaluated at $\lambda = 0$. If the solutions to the Lagrangian first-order conditions yield negative values for x, cases (3) and (4) can be ruled out and a border solution should be investigated. Similarly, if the solution to the Lagrangian first-order conditions yields a negative value for the Lagrange

---

[3]If the constraint is binding, as in Chapter 9, we also need the constraint qualification that the optimal value of x is not a critical value of the constraint.

[4]Assuming that the constraint qualification (defined in Chapter 9) is satisfied if the constraint is binding.

Optimization with Inequality Constraints: Theory

multiplier, cases (2) and (4) are ruled out and we have an indication that the constraint is nonbinding.

If the objective function $f(x)$ is (globally) quasiconcave and the constraint function $g(x)$ is (globally) quasiconvex, any solution that exists with positive x and λ is the global maximum.

The Kuhn-Tucker conditions and solution technique extend immediately to the general case of many choice variables and many inequality constraints. (If any constraints are true equalities, then the cases in which those constraints would be nonbinding do not need to be considered.)

Kuhn Tucker Conditions (many variables, many constraints)

The necessary[5] conditions for the problem of maximizing the Lagrangian

$$\mathscr{L}(\mathbf{x}, \lambda_1, \ldots, \lambda_m) = f(\mathbf{x}) + \lambda_1(g_1 - g^1(\mathbf{x})) + \cdots + \lambda_m(g_m - g^m(\mathbf{x})) \quad \textbf{(11.5)}$$

subject to the constraints that $x_i \geq 0$ for $i = 1, \ldots, n$ and $g^j(\mathbf{x}) \leq g_j$ for $j = 1, \ldots, m$ are

$$
\begin{array}{llll}
\text{(a)} & \dfrac{\partial \mathscr{L}}{\partial x_i} \leq 0, & i = 1, \ldots, n & \text{(d)} \quad \dfrac{\partial \mathscr{L}}{\partial \lambda_j} \geq 0, \quad j = 1, \ldots, m \\[2mm]
\text{(b)} & x_i \geq 0, & i = 1, \ldots, n & \text{(e)} \quad \lambda_j \geq 0, \quad j = 1, \ldots, m \qquad \textbf{(11.6)} \\[2mm]
\text{(c)} & x_i \dfrac{\partial \mathscr{L}}{\partial x_i} = 0, & i = 1, \ldots, n & \text{(f)} \quad \lambda_j \dfrac{\partial \mathscr{L}}{\partial \lambda_j} \geq 0, \quad j = 1, \ldots, m
\end{array}
$$

For minimization problems, the inequalities in conditions (11.6)(a) and (11.6)(d) are reversed. If the objective function is globally quasiconcave (quasiconvex) and all constraint functions are quasiconvex (quasiconcave), then if a solution exists in which all choice variables and all Lagrange multipliers are positive, it is the global maximum (minimum). The usual procedure in such cases is to solve the Lagrangian for the problem, assuming all choice variables are positive and all constraints hold with equality; then check to see if the optimal values of any choice variables or Lagrange multipliers are nonpositive. If so, border solutions for the relevant choice variables should be considered and the constraints corresponding to the nonpositive Lagrange multipliers should be ignored.

11.5   APPLICATIONS OF KUHN–TUCKER CONDITIONS: EXAMPLES

In this section we will work through four examples illustrating the application of the Kuhn-Tucker conditions. The first is a constrained maximization problem that has an interior solution with a binding constraint. Second, we will relax the constraint so that it ends up being nonbinding at the values of the choice variables that maximize the objective function. Third, we will solve an unconstrained maximization problem that

---

[5]When there are many constraints, we need a generalization of the constraint qualification: the matrix of first derivatives of the binding constraints with respect to the choice variables, evaluated at the optimal values of the choice variables, must be of full rank. That is, all rows and columns are independent.

has a border solution. Last, we will work through a constrained maximization problem that ends up with a slack constraint and a border solution.

11.5.1 Binding Constraint, Interior Solution

Suppose we want to choose x and y to maximize the objective function

$$f(x, y) = xy(9 - x - y) \tag{11.7}$$

subject to the constraints $x + y \leq 5$, $x \geq 0$, and $y \geq 0$. The Lagrangian for this problem is

$$\mathcal{L}(x, y, \lambda) = xy(9 - x - y) + \lambda(5 - x - y). \tag{11.8}$$

Assuming for the moment that both choice variables will be positive and that the constraint is binding, this is a straightforward Lagrangian problem; the first-order conditions are

$$\frac{\partial \mathcal{L}}{\partial x} = y(9 - x - y) - xy - \lambda = 0$$

$$\frac{\partial \mathcal{L}}{\partial y} = x(9 - x - y) - xy - \lambda = 0 \tag{11.9}$$

$$\frac{\partial \mathcal{L}}{\partial \lambda} = 5 - x - y = 0.$$

The first two equations imply that $y(9 - x - y) = x(9 - x - y)$, which has two possible solutions: either $x = y$ or $x + y = 9$. The latter, however, violates the constraint, so we are left with $x = y$, which implies (using the third equation in (11.9)) $x^* = y^* = 2.5$. Substituting these values into either of the first two equations in (11.9), we obtain $\lambda^* = 3.75$. Since the solutions for x, y, and λ are all positive while the derivatives of the Lagrangian with respect to all three variables are zero, the Kuhn-Tucker conditions are satisfied. Since the second-order conditions can be shown to hold, this is the (constrained) maximum of the objective function. The positive value of the Lagrange multiplier indicates that the constraint is binding at the optimal values of the choice variables. It turns out that the objective function in this example is not globally quasiconcave,[6] so we should check border solutions where either x or y is zero (or both are zero); it is easy to see that if either x or y is zero, the value of the objective function will also be zero, which is less than the value obtained when $x^* = y^* = 2.5$.

11.5.2 Nonbinding Constraint, Interior Solution

Now consider the problem of maximizing the same objective function (11.7), but now the constraints are that $x + y \leq 8$, $x \geq 0$, and $y \geq 0$. So the only difference between the two problems is that we have relaxed the first constraint. The Lagrangian for this problem is

$$\mathcal{L}(x, y, \lambda) = xy(9 - x - y) + \lambda(8 - x - y). \tag{11.10}$$

---

[6]The determinant of the bordered Hessian is positive when evaluated at the *optimal* values of x, y, and λ, but it is not positive for *all* values of x, y, and λ.

Assuming for the moment that both choice variables will be positive and that the constraint will be binding, the first-order conditions are

$$\frac{\partial \mathcal{L}}{\partial x} = y(9 - x - y) - xy - \lambda = 0$$

$$\frac{\partial \mathcal{L}}{\partial y} = x(9 - x - y) - xy - \lambda = 0 \tag{11.11}$$

$$\frac{\partial \mathcal{L}}{\partial \lambda} = 8 - x - y = 0,$$

which can be solved as in the previous section to get $x^* = 4$, $y^* = 4$, and $\lambda^* = -12$. But the negative Lagrange multiplier violates the Kuhn-Tucker conditions and indicates that the constraint will be slack at the true optimal values of x and y. Thus we examine the case of $x > 0$, $y > 0$, and $\lambda = 0$. Setting $\lambda = 0$ and solving the first two of the first-order conditions (11.11) for x and y (the third first-order condition is ignored because the Kuhn-Tucker conditions do not require that the constraint hold with equality if $\lambda = 0$), we obtain $x^* = y^* = 3$. This satisfies the constraint $(\partial \mathcal{L}/\partial \lambda \geq 0)$ since $x^* + y^* = 3 + 3 = 6 \leq 8$. The second-order conditions can be shown to hold and the objective function takes a greater value than if either x or y were to equal zero. In this example, the unconstrained maximum of the function occurs at values of the choice variables that satisfy the inequality constraint, so the constrained and unconstrained solutions coincide.

11.5.3   Border Solution, No Constraint

Now suppose we want to maximize the function

$$f(x, y) = 10 - x^2 - y^2 + 2y - 2x \tag{11.12}$$

subject only to nonnegativity constraints on x and y. The first-order conditions for the maximization problem are

$$-2x - 2 = 0$$
$$-2y + 2 = 0 \tag{11.13}$$

which yield the solutions $x^* = -1$ and $y^* = 1$. This obviously violates the nonnegativity constraint on x and suggests that we consider a border solution where $x = 0$. The Kuhn-Tucker conditions require that $x \geq 0$, $f_x \leq 0$, and $xf_x = 0$. When evaluated at $x = 0$, $f_x < 0$ for our example, so the Kuhn-Tucker conditions are satisfied. The reader can confirm that the second-order conditions are also satisfied. In our example, the solution for y is invariant to the choice of x, although this is not typical: usually, when a border condition for one variable is examined, it will alter the optimal values of other choice variables. So the optimal solution to our problem is $x^* = 0$ and $y^* = 1$.

11.5.4   Border Solution, Nonbinding Constraint

As an example of a problem with a border solution and a nonbinding constraint, we will simply add a nonbinding constraint to the problem of the previous section. Consider the problem of maximizing the objective function (11.12) subject to the constraints $x \geq 0$, $y \geq 0$, and $x + y \leq 5$. The Lagrangian for this problem is

$$\mathcal{L}(x, y, \lambda) = 10 - x^2 - y^2 + 2y - 2x + \lambda(5 - x - y). \tag{11.14}$$

Assuming for the moment that both choice variables will be positive and that the constraint will be binding, the first-order conditions are

$$-2x - 2 - \lambda = 0$$
$$-2y + 2 - \lambda = 0 \qquad \textbf{(11.15)}$$
$$5 - x - y = 0.$$

Using the first two first-order conditions to eliminate λ, we can solve for y as a function of x: $y = 2 + x$. Substituting this into the constraint (the third equation in (11.15)), we obtain the solutions $x^* = 1.5$ and $y^* = 3.5$. Using either of the first two equations in (11.15), we find that this implies $\lambda^* = -5$, which violates the Kuhn-Tucker conditions and suggests that the objective function can obtain a higher value if we leave the constraint slack. By setting $\lambda = 0$ and ignoring the constraint, we return to the problem of the previous section, in which we found that the optimal solution was a border solution where $x^* = 0$ and $y^* = 1$.

We conclude this example by confirming that, when evaluated at the optimal values of x, y, and λ, all of the Kuhn-Tucker conditions (11.6) are satisfied. When evaluated at $x = 0$, $y = 1$, and $\lambda = 0$, $\partial \mathscr{L}/\partial x = -2 < 0$, $\partial \mathscr{L}/\partial y = -2(1) + 2 = 0$, and $\partial \mathscr{L}/\partial \lambda = 5 - 1 = 4 > 0$. These do indeed satisfy the Kuhn-Tucker conditions.

11.6   AN INTRODUCTION TO LINEAR PROGRAMMING

If the objective function and all constraints are linear, then the problem of maximizing the Lagrangian (11.5) subject to inequality constraints is a **linear programming** problem. The solution to a linear programming problem is found by applying the Kuhn-Tucker conditions to the special case of linear functions and constraints. The solution method involves evaluating the objective function for all **feasible combinations** of choice variables, that is, combinations that satisfy all the inequality constraints (including nonnegativity constraints). We will illustrate the solution graphically for a problem with two choice variables and two constraints (in addition to the nonnegativity constraints). In the next chapter we will look at an example with two choice variables and several constraints. More complex problems are solved using computational methods; the most common is called the **simplex method**,[7] which is available in many computer software packages.

Consider the simple example of maximizing the function $f(x, y) = x + y$ subject to the constraints that $x \geq 0$, $y \geq 0$, $x + 2y \leq 5$, and $6x + 3y \leq 18$. Figure 11.3 shows the **feasible set** of (x, y) combinations as the shaded region. We find the solution to the maximization problem by finding the highest level curve of the objective function that is in the feasible set. Some level curves of the function $f(x, y) = x + y$ are shown in Figure 11.3, where the solution is clearly seen to be at the point labeled A. Because both constraints are binding at this point, we can use the equations of the two constraints to solve for the optimal values of the choice variables, $x^* = 7/3$ and $y^* = 4/3$.

The solution to the linear programming problem is a tangency condition between the border of the feasible set and a level curve of the objective function. In this respect the solution of the linear programming problem is like the solution to a Lagrangian problem of maximizing an objective function subject to a constraint. But because the border of the feasible set is not differentiable at point A, the usual characterization of that tangency as the equality between slopes of level curves is not possible. The solution

---

[7]The simplex method is a procedure for identifying **nodes**, or **kink points**, of the feasible set and identifying for a given node whether there is a direction to move that will increase the value of the objective function without violating any constraints.

FIGURE 11.3

Feasible Set and Solution to the Maximization Problem

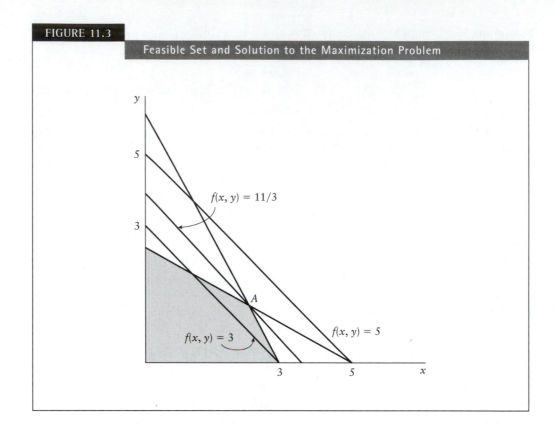

to the problem is therefore found by identifying all of the kink points (nodes) of the border of the feasible set (including the points where one of the choice variables is zero) and evaluating the objective function for each of them. Whichever kink point gives the highest value of the objective function yields the solution to the maximization problem.[8]

The optimal values of the Lagrange multipliers may be of interest as well, since they measure shadow prices (or imputed values) of the constraints (see Chapter 9). Once we have found the optimal values of the choice variables, we can find the optimal values of the Lagrange multipliers using the complementary slackness conditions. First, we identify which (if any) constraints are slack. The Lagrange multipliers associated with these constraints will equal zero: for a constraint that does not bind, relaxing the constraint will not enable the objective function to reach a higher value, so no agent will be willing to pay a positive amount to get the constraint relaxed. For the remaining, binding constraints, we can derive the associated Lagrange multipliers from the derivatives of the Lagrangian with respect to the choice variables, since the complementary slackness condition (11.6)(c) indicates that as long as the choice variables are positive, these derivatives must equal zero. But if the optimal value of a choice variable is zero (if we have a border solution), the derivative of the Lagrangian with respect to that variable may not equal zero, so we do not use that derivative in solving for the Lagrange multipliers.

---

[8]It is possible that the level curves of the objective function are parallel to one segment of the feasible set. In this case, all points on that segment are solutions to the constrained maximization problem. Since two kink points will yield the same (highest) value for the objective function, it is still sufficient to consider only kink points in order to identify these cases.

For our example, the Lagrangian is

$$\mathscr{L}(x, y, \lambda_1, \lambda_2) = x + y + \lambda_1(5 - x - 2y) + \lambda_2(18 - 6x - 3y). \qquad \textbf{(11.16)}$$

Neither choice variable is zero, so we can solve for the Lagrange multipliers using the two equations

$$\frac{\partial \mathscr{L}}{\partial x} = 1 - \lambda_1 - 6\lambda_2 = 0$$

$$\frac{\partial \mathscr{L}}{\partial y} = 1 - 2\lambda_1 - 3\lambda_2 = 0 \qquad \textbf{(11.17)}$$

which can be solved for the optimal values of the Lagrange multipliers: $\lambda_1^* = 1/3$ and $\lambda_2^* = 1/9$.

11.7 DUALITY IN LINEAR PROGRAMMING

We can analyze every linear programming problem in two equivalent ways. For every maximization problem, there is an associated minimization problem that yields the same solution, and every minimization problem has an equivalent maximization problem. It is traditional to call the original problem (usually a maximization problem) the **primal problem** and the associated (usually minimization) problem the **dual**. The choice variables in the primal problem become the Lagrange multipliers in the dual, and the Lagrange multipliers in the primal become the choice variables in the dual. The dual therefore directly solves for the shadow prices of constraints in the primal, which can be convenient if those shadow prices are of primary importance. More important, the primal and dual yield the same solutions for the choice variables, Lagrange multipliers, and objective function. Since linear programming problems are solved by searching the kink points of the feasible set—and since the more constraints there are, the more kink points there are—it is usually helpful to convert any linear programming problem into whichever version (the primal or dual) has fewer constraints.

We begin by considering the linear programming problem used in the previous section. Since it has two choice variables and two constraints, both the primal and dual problem can be solved graphically. We then discuss a more general linear programming problem.

The problem analyzed in the previous section is

$$
\begin{aligned}
\text{Maximize} \quad & f(x, y) = x + y \\
\text{subject to} \quad & x + 2y \le 5 \\
& 6x + 3y \le 18 \\
& x \ge 0 \\
& y \ge 0
\end{aligned}
\qquad \textbf{(11.18)}
$$

which can be restated as

Maximize[9] $\mathscr{L}(x, y, \lambda_1, \lambda_2) = x + y + \lambda_1(5 - x - 2y) + \lambda_2(18 - 6x - 3y)$ $\qquad$ **(11.19)**
subject to $\qquad\qquad x \ge 0, y \ge 0.$

---

[9]Recall from Chapter 9 that the Lagrangian is maximized with respect to the choice variables x and y but is minimized with respect to the Lagrange multipliers.

The Kuhn-Tucker conditions for this problem are

$$x \geq 0, \quad \frac{\partial \mathcal{L}}{\partial x} = 1 - \lambda_1 - 6\lambda_2 \leq 0, \quad x\frac{\partial \mathcal{L}}{\partial x} = 0$$

$$y \geq 0, \quad \frac{\partial \mathcal{L}}{\partial y} = 1 - 2\lambda_1 - 3\lambda_2 \leq 0, \quad y\frac{\partial \mathcal{L}}{\partial y} = 0$$

$$\lambda_1 \geq 0, \quad \frac{\partial \mathcal{L}}{\partial \lambda_1} = 5 - x - 2y \geq 0, \quad \lambda_1\frac{\partial \mathcal{L}}{\partial \lambda_1} = 0 \tag{11.20}$$

$$\lambda_2 \geq 0, \quad \frac{\partial \mathcal{L}}{\partial \lambda_2} = 18 - 6x - 3y \geq 0, \quad \lambda_2\frac{\partial \mathcal{L}}{\partial \lambda_2} = 0.$$

We saw in the previous section that the solution to this problem is $x^* = 7/3$, $y^* = 4/3$, $\lambda_1^* = 1/3$, and $\lambda_2^* = 1/9$.

To solve the dual to this programming problem, we first write the Lagrangian (11.19) with the Lagrange multipliers λ_1 and λ_2 acting as choice variables and x and y serving as Lagrange multipliers:

$$\mathcal{L}(x, y, \lambda_1, \lambda_2) = x + y + \lambda_1(5 - x - 2y) + \lambda_2(18 - 6x - 3y)$$
$$= x + y + 5\lambda_1 - x\lambda_1 - 2y\lambda_1 + 18\lambda_2 - 6x\lambda_2 - 3y\lambda_2$$
$$= 5\lambda_1 + 18\lambda_2 + x - x\lambda_1 - 6x\lambda_2 + y - 2y\lambda_1 - 3y\lambda_2 \tag{11.21}$$
$$\mathcal{L}(\lambda_1, \lambda_2, x, y) = 5\lambda_1 + 18\lambda_2 + x(1 - \lambda_1 - 6\lambda_2) + y(1 - 2\lambda_1 - 3\lambda_2)$$

which is minimized subject to nonnegativity constraints on λ_1 and λ_2. The Kuhn-Tucker conditions for this minimization problem are

$$\lambda_1 \geq 0, \quad \frac{\partial \mathcal{L}}{\partial \lambda_1} = 5 - x - 2y \geq 0, \quad \lambda_1\frac{\partial \mathcal{L}}{\partial \lambda_1} = 0$$

$$\lambda_2 \geq 0, \quad \frac{\partial \mathcal{L}}{\partial \lambda_2} = 18 - 6x - 3y \geq 0, \quad \lambda_2\frac{\partial \mathcal{L}}{\partial \lambda_2} = 0$$

$$x \geq 0, \quad \frac{\partial \mathcal{L}}{\partial x} = 1 - \lambda_1 - 6\lambda_2 \leq 0, \quad x\frac{\partial \mathcal{L}}{\partial x} = 0 \tag{11.22}$$

$$y \geq 0, \quad \frac{\partial \mathcal{L}}{\partial y} = 1 - 2\lambda_1 - 3\lambda_2 \leq 0, \quad y\frac{\partial \mathcal{L}}{\partial y} = 0$$

which are identical to (11.20). Thus the same values of x, y, λ_1, and λ_2 will solve both the primal and dual problems. Another feature of duality in linear programming is that the objective function in the dual problem will have the same value at the optimum as the objective function in the primal problem.

We illustrate duality further by solving the minimization problem graphically and comparing it with the solution to (11.19) shown in Figure 11.3. The feasible set defined by the two constraints in (11.21) is the shaded region in Figure 11.4, where λ_1 and λ_2 are on the axes. A few level curves of the objective function $g(\lambda_1, \lambda_2) = 5\lambda_1 + 18\lambda_2$ are also shown. Since this is a minimization problem, the solution is the point of the feasible set that is on the lowest level curve of the objective function. This point is labeled point B and corresponds to $\lambda_1^* = 1/3$ and $\lambda_2^* = 1/9$. Thus we get the same solution as that derived from the Kuhn-Tucker conditions (11.20) for the primal problem.

For this simple example, it was quite easy to solve the Kuhn-Tucker conditions for the optimal values of x, y, λ_1, and λ_2. But for more complex problems, it may be considerably easier to solve the primal than the dual problem, or *vice versa*, depending on which has fewer constraints.

We now consider a general linear programming problem with n choice variables and m constraints. We choose the maximization version to be the primal problem. The

FIGURE 11.4

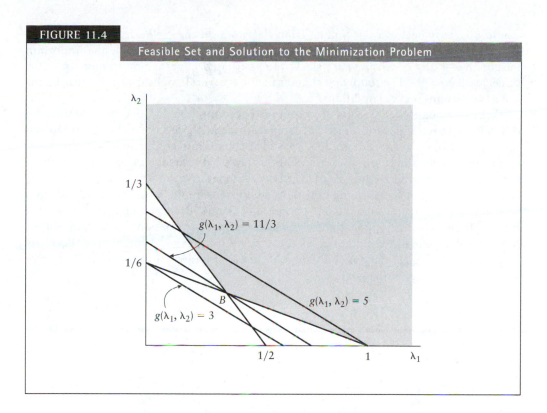

objective function is $f(\mathbf{x}) = \mathbf{a'x}$ where $\mathbf{a'}$ is a row vector of n constants and $\mathbf{x}$ is the (column) vector of the n choice variables. This objective function is maximized subject to the nonnegativity constraints $x_i \geq 0, i = 1, \ldots, n$, and m other linear constraints $B\mathbf{x} \leq \mathbf{b}_0$ where $\mathbf{b}_0$ is a column vector of m constants and B is an $(m \times n)$ matrix of constants; each row of B contains the coefficients on the choice variables for one of the constraints. The Lagrangian for this problem, written in matrix notation, is

$$\mathscr{L}(\mathbf{x}, \boldsymbol{\lambda}) = \mathbf{a'x} + \boldsymbol{\lambda}'(\mathbf{b}_0 - B\mathbf{x}) \tag{11.23}$$

where $\boldsymbol{\lambda}'$ is a row vector of m Lagrange multipliers. The simplex method will yield the solution for the optimum values of the choice variables $\mathbf{x}$, and the Kuhn-Tucker conditions (11.20) could in principle be used to find the optimal values for the Lagrange multipliers.

The dual to this maximization problem is to minimize a function $g(\boldsymbol{\lambda}) = \mathbf{b}_0'\boldsymbol{\lambda}$ subject to $\lambda_j \geq 0$, $j = 1, \ldots, m$ and the n other constraints $B'\boldsymbol{\lambda} \geq \mathbf{a}$. This has the Lagrangian

$$\mathscr{L}(\boldsymbol{\lambda}, \mathbf{x}) = \mathbf{b}_0'\boldsymbol{\lambda} + \mathbf{x}'(\mathbf{a} - B'\boldsymbol{\lambda}) \tag{11.24}$$

and the simplex method can be used to derive the same values of $\boldsymbol{\lambda}$ as those derived from maximizing (11.23).

SUMMARY

The advantage of the Kuhn-Tucker theory (including linear programming as a special case) over the Lagrangian method is that it allows inequality constraints, including nonnegativity constraints, to be formally taken into consideration. The Kuhn-Tucker conditions provide a way of finding if border solutions are relevant and if any constraints can be ignored because they are nonbinding. They also provide the important insight that if any constraints are slack, the associated Lagrange multipliers (shadow prices) will be zero. The following chapter includes several applications illustrating the use of the Kuhn-Tucker conditions.

When inequality constraints are taken into account, comparative static results are difficult to derive in general, since the comparative static effects of a change in any parameter will usually differ depending on which constraints are binding and which are slack. Ordinarily, the Kuhn-Tucker conditions are used to define the conditions under which the Lagrangian method is appropriate. Comparative statics results are then derived, as in Chapter 9, assuming that those conditions apply; when border conditions or slack constraints are encountered, comparative static results are analyzed on a case-by-case basis. The first application of the next chapter provides an illustration: the effects of a change in one good's price on the demand for that good and another good are analyzed, allowing for the possibility of a border solution where the consumer chooses not to purchase any of the good.

Problems

11.1 Maximize the following functions subject to nonnegativity constraints on the choice variables:

(a) $f(x, y) = xy - 0.5x^2 - y^2 - 2y$

(b) $f(x, y) = 3(xy)^{1/3} - 2x - y$

(c) $f(x, y) = 2x - x^2 - 4y - y^2$

(d) $f(x, y) = 2x - x^2 + 5y - y^2 - (x + y)^2$

(e) $f(x, y) = \ln(x + 2) + \ln(y + 1) - x - 0.5y$

(f) $f(x, y, z) = (2 - x)x + (1 - y)y + (3 - z)z - (x + y + z)^2$

(g) $f(x, y, z) = -x - 2y + 3z + \ln(xy) - z^2$

11.2 Minimize the following functions subject to nonnegativity constraints on the choice variables:

(a) $f(x, y) = 3x^2 + 2y^2 + 4xy - 4x - 2y$

(b) $f(x, y) = x^2 + 2y^2 - xy - x - y$

(c) $f(x, y) = x^2 + y^2 + 4x - 8y + 20$

(d) $f(x, y, z) = 3x^2 + 2y^2 + z^2 - x - y + z - xz + yz$

(e) $f(x, y, z) = x^4/8 + y^3/3 + 3z^2 - xy - z$

11.3 Redo Problem 11.1, adding the constraint $6x + 2y \leq 4$.

11.4 Redo Problem 11.2, adding the constraint $x + 2y \geq 1$.

11.5 Write the Kuhn-Tucker conditions and find the optimal values of the choice variables and Lagrange multipliers for the following problems:

(a) Maximize $3x + 6y$ subject to the constraints $2x + 3y \leq 3$, $x + 4y \leq 2$, $x \geq 0$, $y \geq 0$.

(b) Maximize $50x + 10y$ subject to the constraints $x - y \leq 3$, $5x + 2y \leq 20$, $x \geq 0$, $y \geq 0$.

(c) Minimize $10x + 6y$ subject to the constraints $20x + 10y \geq 100$, $10x + 10y \geq 80$, $4x + 8y \geq 40$, $x \geq 0$, $y \geq 0$.

11.6 Set up and write out the Kuhn-Tucker conditions for the dual to each part of Problem 11.5.

12

Optimization with Inequality Constraints: Applications

12.1 INTRODUCTION

In Chapter 11 we introduced and explained the Kuhn-Tucker conditions. The main advantage of Kuhn-Tucker analysis compared to the Lagrangian analysis used in Chapter 9 is the explicit recognition of the possibility of slack constraints and border solutions. This chapter contains several applications of Kuhn-Tucker analysis. The first is an illustration using a very simple two-good utility-maximization problem. Here we focus on the role of various assumptions in ruling out slack constraints and identifying conditions leading to border solutions. The difficulty of getting comparative static results in Kuhn-Tucker analysis is also addressed. Section 12.3 offers a simple version of the classic diet problem, in which the cost of achieving given nutritional goals is minimized, as an example of a linear programming problem with two variables and several constraints. Another classic example of an economics optimization problem with inequality constraints follows in which a monopolist maximizes revenue instead of profit. Section 12.5 returns to the labor supply problem investigated in Chapter 6 and uses the Kuhn-Tucker conditions to provide more insight. The following section also revisits a problem analyzed in an earlier chapter: intertemporal consumption, which was discussed in Chapter 10. In this chapter we look at liquidity constraints in the intertemporal consumption problem.

12.2 UTILITY MAXIMIZATION WITH TWO GOODS

In the usual utility-maximization problem we ignore the fact that there two types of inequality constraints: nonnegativity constraints for the consumption of each good, and the budget constraint, which we usually treat as an equality. In this application we allow the possibility that each of these inequality constraints might be either binding or slack. In the process we will illustrate many of the concepts discussed in Chapter 11. For convenience we consider only the two-good case; most of what we say generalizes to the case of many goods.

The two-good utility-maximization problem can be formalized as follows. The problem is to choose consumption levels of the two goods, x and y, to maximize a (quasiconcave) utility function $U(x, y)$ subject to three inequality constraints: $x \geq 0$,

$y \geq 0$, and $p_x x + p_y y \leq I$, where I is income and p_x and p_y are the prices of the two goods. The Lagrangian for this problem is

$$\mathcal{L}(x, y, \lambda) = U(x, y) + \lambda(I - p_x x - p_y y). \qquad \text{(12.1)}$$

The Kuhn-Tucker conditions for this problem (see conditions (11.6)) are

(a) $\quad \dfrac{\partial \mathcal{L}}{\partial x} = \dfrac{\partial U}{\partial x} - \lambda p_x \leq 0 \quad$ and $\quad \dfrac{\partial \mathcal{L}}{\partial y} = \dfrac{\partial U}{\partial y} - \lambda p_y \leq 0$

(b) $\quad x \geq 0 \quad$ and $\quad y \geq 0$

(c) $\quad x\dfrac{\partial \mathcal{L}}{\partial x} = 0 \quad$ and $\quad y\dfrac{\partial \mathcal{L}}{\partial y} = 0$

(d) $\quad \dfrac{\partial \mathcal{L}}{\partial \lambda} = I - p_x x - p_y y \geq 0$ $\qquad\qquad$ **(12.2)**

(e) $\quad \lambda \geq 0$

(f) $\quad \lambda\dfrac{\partial \mathcal{L}}{\partial \lambda} = 0.$

The special case of these conditions that we usually analyze is one in which the nonnegativity constraints are slack (the consumption of both goods is positive) and the income constraint is binding (it holds as an equality). We analyzed a more general version (with many choice variables) of this case in Chapter 9. Here we will focus on the other cases.

First, consider the case when the budget constraint is slack, that is, condition (12.2) (d) is a strict inequality. From the complementary slackness condition (12.2) (f), this strict inequality implies that λ is zero; from condition (12.2) (a), the marginal utilities of both goods x and y must therefore be nonpositive. In other words, the only way the budget constraint can be slack is if neither good provides positive marginal utility (at the consumption levels that maximize utility). As long as increased consumption of at least one good will increase utility, a utility-maximizing consumer will exhaust all of her income. This is why the possibility of a slack budget constraint is ordinarily ignored in utility-maximization problems. (Of course, if saving is allowed, *current* income is not always exhausted.) In fact, we usually assume at the outset that increased consumption of every good always provides positive utility (this assumption is sometimes called the "more is better" assumption). As we have just shown, with that assumption, the income constraint cannot be slack at the optimum.

Now let us consider the nonnegativity constraints. We will consider the case of $x = 0$; the case of $y = 0$ is totally analogous.[1] If $x = 0$, the complementary slackness condition in (12.2)(c) can be satisfied even if $\partial \mathcal{L}/\partial x < 0$. If $\partial \mathcal{L}/\partial x < 0$, then condition (12.2)(a) implies that $\partial U/\partial x - \lambda p_x < 0$. Since $x = 0$, the consumer spends all her budget on good y, so $y = I/p_y$. Since $y > 0$, conditions (12.2)(a) and (c) imply that $\partial U/\partial y - \lambda p_y = 0$. Putting together the two parts of condition (12.2)(a) and eliminating λ, we obtain the condition that

$$\frac{\partial U/\partial x}{\partial U/\partial y} < \frac{p_x}{p_y}, \qquad \text{(12.3)}$$

which is to say that the marginal rate of substitution between the two goods, evaluated at $x = 0$ and $y = I/p_y$, is less than their price ratio. This situation is illustrated in

---

[1] Both goods cannot be zero at the optimum because if both x and y are zero, the budget constraint is slack (condition (12.2)(d) is a strict inequality). But we have just argued that this is impossible except when no good has positive marginal utility. So if one of the goods is not purchased, the other must be consumed in a positive amount; all the consumer's income will be spent on one good.

FIGURE 12.1

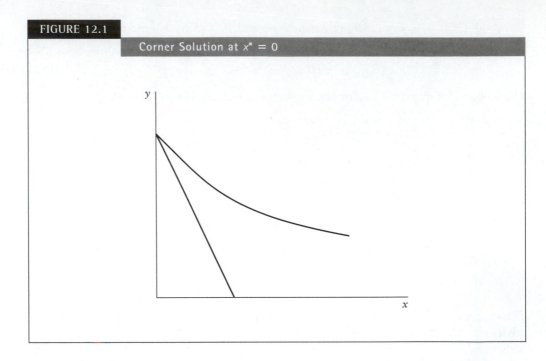

Corner Solution at $x^* = 0$

FIGURE 12.2

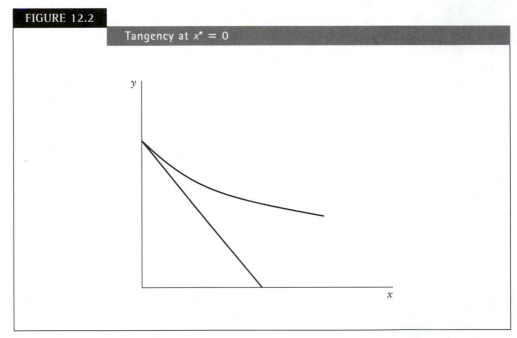

Tangency at $x^* = 0$

Figure 12.1. Since the marginal rate of substitution is the absolute value of the slope of the consumer's indifference curve and the price ratio is the absolute value of the slope of the budget line,[2] condition (12.3) says that, at the point where the budget line intersects the y-axis, the budget line must be steeper than the indifference curve going through that point. This is the condition necessary for the optimal consumption of a good to be zero.

It is possible that both $x = 0$ and $\partial \mathcal{L}/\partial x = 0$, as illustrated in Figure 12.2. In this case, the marginal rate of substitution equals the price ratio, and the consumer's highest

---

FIGURE 12.3

Moving from a Corner Solution to an Interior Solution

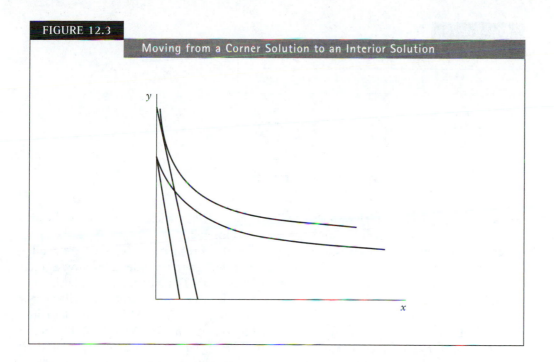

indifference curve is tangent to the budget constraint just at $x = 0$. A slight decrease in the (absolute value of the) slope of the budget constraint would induce a positive consumption of x, whereas a slightly steeper budget constraint would induce the consumer to stay at the corner solution with $\partial \mathcal{L}/\partial x < 0$.

Following this line of reasoning, we can see the comparative static effects of changing any of the parameters when a nonnegativity constraint is binding by examining Figures 12.1 and 12.2. (If the nonnegativity constraint is slack, we can use the analysis of Chapter 9 to derive comparative static results.) If the consumer's marginal rate of substitution is strictly greater than the price ratio, the optimal values of x and y are

$$x^* = 0 \quad \text{and} \quad y^* = \frac{I}{p_y}. \tag{12.4}$$

To find the effects of changing one or more of the parameters p_x, p_y, and I, it is necessary to consider two cases. First, if the marginal rate of substitution, evaluated at $x = 0$ and $y = I/p_y$ (that is, at the point where the budget constraint hits the y-axis), is still less than the price ratio after the parameter changes, we can find effects on the optimal values of x and y by differentiating (12.4). Changes in the parameters will have no effect on x, which will continue to be zero; the optimal consumption of y will change if income or the price of y changes because the maximum amount of y that can be purchased will change in this case. The second possibility is that the change in the parameters may cause the marginal rate of substitution, evaluated at $x = 0$ and $y = I/p_y$, to become greater than the price ratio. In this case the optimal consumption of x will become positive and hence the consumption of y will change as well. An example of this possibility is shown in Figure 12.3.

12.3 TWO–GOOD DIET PROBLEM

A classic example of the use of inequality constraints in economics is the diet problem in which the object is to find the least-costly combination of various food items that together satisfy or exceed a collection of nutritional criteria. Analysis of the diet

TABLE 12.1	RDA per Serving	
	Milk (%)	Cereal (%)
Vitamin A	6	30
Vitamin D	25	25
Calcium	30	15
Iron	0	45

problem in economics, which originated with Stigler's 1945 article,[3] continues to be an important part of some problems in economic development.

In this example we will investigate a simplified diet problem. Two goods are available for consumption: milk and cereal. A serving of each good provides a fraction of the recommended daily allowance (RDA) of several nutrients. To make the example simple but still illustrate a linear programming problem with several constraints, we will focus on four nutrients: vitamin A, vitamin D, calcium, and iron. Table 12.1 gives the percentages of the RDA of each nutrient provided by one serving of each good. The cost of a serving of milk is 12 cents and the cost of a serving of cereal is 24 cents. The diet problem is to find the minimum-cost combination of milk and cereal consumption that will provide at least 100% of the RDA of each nutrient.

The formal statement of the problem in the form of a linear programming problem is as follows:

$$
\begin{aligned}
\text{Minimize} \quad & 12M + 24C \\
\text{subject to} \quad & 6M + 30C \geq 100 \\
& 25M + 25C \geq 100 \\
& 30M + 15C \geq 100 \\
& 0M + 45C \geq 100 \\
& M \geq 0, C \geq 0
\end{aligned}
\tag{12.5}
$$

In Figure 12.4, where the various constraints are graphed, the feasible set is shaded. It is clear that the vitamin D constraint is not binding, since the minimum levels of milk and cereal necessary to achieve other nutritional requirements ensure that the vitamin D requirement is also met.

In Figure 12.5 a few level curves of the objective function are added to the graph of the feasible set. The minimum-cost way of satisfying all nutritional requirements is clearly at the kink point where the vitamin A and calcium constraints meet. We can discover the amounts of calcium and milk consumed at this point by treating both of these constraints as equalities and solving them as a system of two equations in two unknowns:

$$
\begin{aligned}
6M + 30C &= 100 \\
30M + 15C &= 100.
\end{aligned}
\tag{12.6}
$$

[3]Stigler, G.J. 1945. "The cost of subsistence," *Journal of Farm Economics* 27:303–12. More recent examples of the diet problem are Silberberg, E. 1985, "Nutrition and the Demand for Tastes," *Journal of Political Economy* 93:881–900; and Tiefenthaler, J. 1995, "Deviations from the least-cost diets for infants," *Journal of Population Economics* 8:281–300.

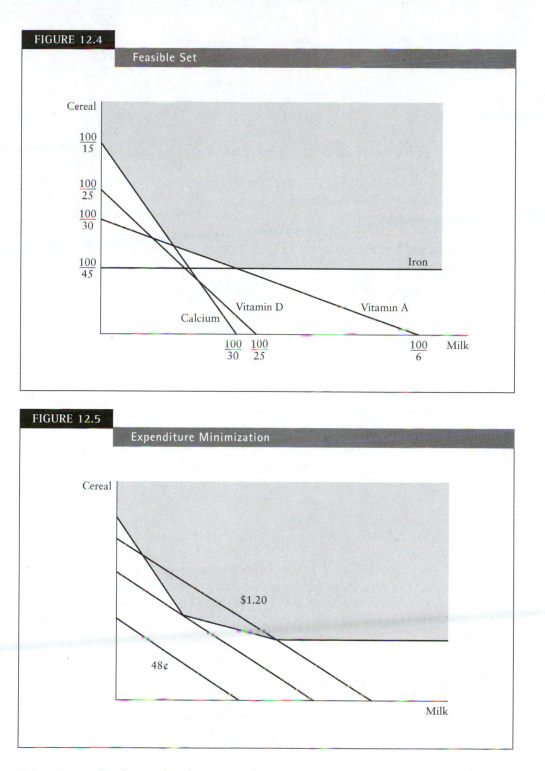

FIGURE 12.4

Feasible Set

Cereal

$\frac{100}{15}$

$\frac{100}{25}$

$\frac{100}{30}$

$\frac{100}{45}$

Iron

Vitamin D   Vitamin A

Calcium

$\frac{100}{30}$ $\frac{100}{25}$ $\frac{100}{6}$ Milk

FIGURE 12.5

Expenditure Minimization

Cereal

$1.20

48¢

Milk

Using Cramer's rule to solve this system for M, we have

$$M^* = \frac{\begin{vmatrix} 100 & 30 \\ 100 & 15 \end{vmatrix}}{\begin{vmatrix} 6 & 30 \\ 30 & 15 \end{vmatrix}} = \frac{-1500}{-810} = \frac{50}{27}.$$ **(12.7)**

Solving for C, we get

$$C^* = \frac{\begin{vmatrix} 6 & 100 \\ 30 & 100 \end{vmatrix}}{\begin{vmatrix} 6 & 30 \\ 30 & 15 \end{vmatrix}} = \frac{-2400}{-810} = \frac{80}{27}. \qquad (12.8)$$

So the solution to the diet problem is to consume nearly two servings of milk and nearly three servings of cereal.

We can use this solution with the Kuhn-Tucker conditions to solve for the values of the Lagrange multipliers associated with the four nutritional constraints. At the optimal solution, the vitamin D and iron constraints are slack. Therefore their Lagrange multipliers are zero. The other two multipliers can be found from the relevant complementary slackness conditions: since M and C are positive, the derivatives of the Lagrangian with respect to M and C must each be zero. For this example the Lagrangian is

$$\mathcal{L} = 12M + 24C + \lambda_1(100 - 6M - 30C) + \lambda_2(100 - 25M - 25C) \\ + \lambda_3(100 - 30M - 15C) + \lambda_4(100 - 0M - 45C). \qquad (12.9)$$

The derivatives with respect to M and C are

$$\frac{\partial \mathcal{L}}{\partial M} = 12 - 6\lambda_1 - 25\lambda_2 - 30\lambda_3 - 0\lambda_4 = 12 - 6\lambda_1 - 30\lambda_3 = 0 \\ \frac{\partial \mathcal{L}}{\partial C} = 24 - 30\lambda_1 - 25\lambda_2 - 15\lambda_3 - 45\lambda_4 = 24 - 30\lambda_1 - 15\lambda_3 = 0 \qquad (12.10)$$

where we have used the fact that both λ_2 and λ_4 are zero since they are the Lagrange multipliers for slack constraints. Solving equations (12.10) for λ_1 and λ_3 yields

$$\lambda_1^* = \frac{90}{135} = \frac{2}{3} \quad \text{and} \quad \lambda_3^* = \frac{36}{135} = \frac{4}{15}. \qquad (12.11)$$

These are the shadow prices of the vitamin A and calcium constraints, respectively. For example, if the RDA of vitamin A were to be increased by 1%, an extra two-thirds cent expenditure would be required. It is more expensive (at the margin) to satisfy the vitamin A constraint than it is to satisfy the calcium constraint, while the vitamin D and iron constraints do not themselves add any expense, since they are satisfied in the course of buying (and eating) enough milk and cereal to satisfy the other nutritional constraints.

12.4 SALES MAXIMIZATION

In a pure monopoly, profits can persist indefinitely since there are no new entrants into the industry to drive price (and therefore profit) down. Because of this, many economists believe that the managers of monopolies may pursue goals other than profit maximization. The pursuit of other goals is especially likely in modern corporations, whose managers are not typically owners; such managers may have different objectives than the stockholders. One possible goal for the managers to pursue is to maximize sales (revenues). Usually, however, economists assume that the managers are constrained to achieve at least some target level of profits; if profits fall too far, stockholders will get upset and change management. In this example we will investigate the case in which a monopolist is maximizing revenue subject to a profit constraint.

The monopolist uses labor (L) and capital (K) to produce output (Q) according to the production function $Q = F(L, K)$. The monopolist faces the inverse demand function $P = P(Q)$, so revenues equal $R(Q) = P(Q)Q$. Substituting in the production function, we get revenues as a function of labor and capital:

$$R(Q) = R(F(L, K)). \tag{12.12}$$

We will assume that the firm, though a monopolist in the output market, faces competitive input markets. Thus it treats the wage rate (w) and the rental rate of capital (r) as parameters and its costs are equal to $wL + rK$. The stockholders of this firm require that profits be at least equal to π_0. Thus the profit constraint takes the form

$$R(F(L, K)) - wL - rK \geq \pi_0. \tag{12.13}$$

In order to set up the problem in the form discussed in Chapter 11 (the Lagrangian is given in (11.5) and the Kuhn-Tucker conditions are given in (11.6)) we will rewrite the profit constraint as

$$\pi_0 - (R(F(L, K)) - wL - rK) \leq 0. \tag{12.14}$$

The manager's problem is to choose labor and capital to maximize revenues (12.12) subject to the profit constraint (12.14) (and nonnegativity constraints on L and K). The Lagrangian for this problem[4] is

$$\mathcal{L}(L, K, \lambda) = R(F(L, K)) + \lambda(R(F(L, K)) - wL - rK - \pi_0) \tag{12.15}$$

and the Kuhn-Tucker conditions are

(a) $\dfrac{\partial \mathcal{L}}{\partial L} = R'F_L + \lambda(R'F_L - w) \leq 0$ and $\dfrac{\partial \mathcal{L}}{\partial K} = R'F_K + \lambda(R'F_K - r) \leq 0$

(b) $L \geq 0$ and $K \geq 0$

(c) $L\dfrac{\partial \mathcal{L}}{\partial L} = 0$ and $K\dfrac{\partial \mathcal{L}}{\partial K} = 0$

(d) $\dfrac{\partial \mathcal{L}}{\partial \lambda} = R(F(L, K) - wL - rK - \pi_0) \geq 0$

(e) $\lambda \geq 0$

(f) $\lambda\dfrac{\partial \mathcal{L}}{\partial \lambda} = 0$

$$\tag{12.16}$$

where $F_L = \partial F/\partial L$ and $F_K = \partial F/\partial K$. We begin analyzing these conditions by assuming that the profit constraint is binding and that both L and K are positive. The complementary slackness conditions (12.16)(c) and (f) therefore imply that the conditions (12.16)(a) and (d) hold as equalities. That is, the usual Lagrangian method is appropriate. In this case, the conditions in (12.16)(a) can be rewritten as

$$\left(\frac{1+\lambda}{\lambda}\right) R'F_L = w \quad \text{and} \quad \left(\frac{1+\lambda}{\lambda}\right) R'F_K = r. \tag{12.17}$$

The first thing to note is that the two conditions in (12.17) together imply that $F_L/F_K = w/r$, which is to say that the marginal rate of technical substitution equals the input price ratio. This is not surprising, since this is the condition for cost minimization

---

[4]Recall that, when formulating the Lagrangian, we write the constraint so that it is nonnegative.

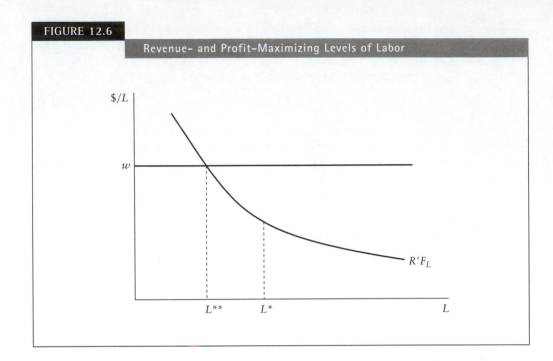

FIGURE 12.6

Revenue- and Profit-Maximizing Levels of Labor

(see Chapter 10), which the firm will continue to do even if it is not maximizing profit; minimizing costs makes it easier to satisfy the profit constraint.

The second implication of (12.17) is that the firm will expand beyond the profit-maximizing output level. Although we can see this from either of the conditions in (12.17), we will focus on the labor condition. Since $\lambda > 0$, $(1 + \lambda)/\lambda > 1$. Thus $R'F_L$ is less than the wage rate. Since R' is marginal revenue and F_L is the marginal product of labor, $R'F_L$ represents the addition to the firm's revenue of the marginal worker, which is called the **marginal revenue product of labor.** Because the marginal revenue product of labor is less than the wage rate, the marginal worker is not adding as much to revenue as she is costing. But she is still adding to the firm's revenue, since the marginal revenue product is greater than zero. Figure 12.6 illustrates the situation. We assume that the firm is using its optimal level of capital; Figure 12.6 shows the level of labor L^* that maximizes revenue subject to the profit constraint and also shows the profit-maximizing level of labor, L^{**}. The role of the profit constraint is made clearer when marginal revenue (R') and marginal cost (w/F_L) are graphed against quantity, shown in Figure 12.7. The profit-maximizing quantity, which corresponds to the use of L^{**} units of labor, is where marginal revenue equals marginal cost. A revenue-maximizing firm with a profit constraint will expand output until profits fall to the minimum allowable level; this corresponds to L^* units of labor.

We now consider the case in which the profit constraint is slack. By the complementary slackness condition (12.16)(f), this implies that $\lambda = 0$, which in turn implies that the conditions in (12.16)(a) simplify to

$$R'F_L = R'F_K = 0. \tag{12.18}$$

Again, it is clear that output is expanded beyond the profit-maximizing level, because the firm hires both labor and capital up to the points at which their marginal revenue products are zero. Assuming, as is often done, that the marginal product of labor is always positive, this implies that marginal revenue must be zero.

FIGURE 12.7

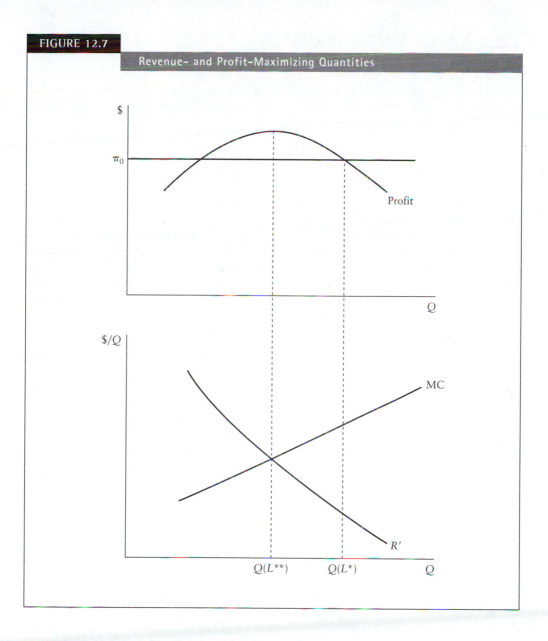

In this case there is no longer any guarantee that the firm will be minimizing cost. Since the profit constraint is no longer binding, the firm does not necessarily need to minimize costs of production: failure to do so will only affect profit, not revenues.[5] When the profit constraint is binding, the firm has an incentive to minimize costs of production because reducing the costs of production enables the firm to increase production and revenue further without falling below the targeted profit level.

We conclude this section by noting that in a two-good model, we usually rule out border conditions in which either L or K is zero by assuming that positive amounts of both inputs are necessary for production. If there were many goods, however, we might well have a solution in which some of them are not used.

---

[5]If the firm is too inefficient, however, profits will fall below the target level and the profit constraint will become binding again.

12.5   LABOR SUPPLY REVISITED

This exercise revisits the labor supply example of Chapter 6 in which the possibility of border constraints was addressed. Here we use the Kuhn-Tucker conditions to explore the possibility more formally. We investigate the labor-leisure choice when utility is given by the Stone-Geary utility function

$$U(C, T - L) = a \ln(C - C_0) + (1 - a) \ln(T - L) \tag{12.19}$$

where a is an exogenous parameter between 0 and 1, C is consumption, C_0 is the subsistence level of consumption, T is the total time available, and L is the amount of time spent working. Thus $T - L$ is leisure time and T is defined in such a way that the subsistence level of leisure is zero. We will also for convenience measure consumption so that $C_0 = 0$ and the price of consumption is normalized to 1. The budget constraint for this problem is that consumption expenditures must not exceed the sum of wage income wL and nonwage income I: $C \leq wL + I$. Individuals choose consumption C and labor supply L to maximize utility subject to this budget constraint, the constraint that the amount of time spent working cannot exceed the total time available: $L \leq T$, and nonnegativity constraints on C and L.

The Lagrangian for this problem is

$$\mathcal{L}(C, L, \lambda_1, \lambda_2) = a \ln C + (1 - a) \ln(T - L) + \lambda_1(wL + I - C) + \lambda_2(T - L) \tag{12.20}$$

and the Kuhn-Tucker conditions are

(a) $\dfrac{\partial \mathcal{L}}{\partial C} = \dfrac{a}{C} - \lambda_1 \leq 0$ and $\dfrac{\partial \mathcal{L}}{\partial L} = -\left(\dfrac{1 - a}{T - L}\right) + \lambda_1 w - \lambda_2 \leq 0$

(b) $C \geq 0$ and $L \geq 0$

(c) $C\dfrac{\partial \mathcal{L}}{\partial C} = 0$ and $L\dfrac{\partial \mathcal{L}}{\partial L} = 0$

(d) $\dfrac{\partial \mathcal{L}}{\partial \lambda_1} = wL + I - C \geq 0$ and $\dfrac{\partial \mathcal{L}}{\partial \lambda_2} = T - L \geq 0$ $\qquad$ (12.21)

(e) $\lambda_1 \geq 0$ and $\lambda_2 \geq 0$

(f) $\lambda_1\dfrac{\partial \mathcal{L}}{\partial \lambda_1} = 0$ and $\lambda_2\dfrac{\partial \mathcal{L}}{\partial \lambda_2} = 0.$

In Chapter 6 we investigated the case where the nonnegativity constraints in condition (12.21)(b) and the last part of condition (12.21)(d) are slack and the budget constraint in the first part of (12.21)(d) is binding. Here we are interested in considering border solutions and a slack budget constraint.

First, we rule out a slack budget constraint. If the budget constraint is slack, λ_1 must be zero by the complementary slackness condition (12.21)(f). But then condition (12.21)(a) implies that $a/C \leq 0$, which violates condition (12.21)(b). As long as consumption has a positive marginal utility, all income must be spent.

Next we consider the cases in which consumption just equals the subsistence level (which in our case is zero) and where all available time is spent working. In the former case, since $C = 0$, the inequality in the first part of condition (12.21)(a) can be strict. But if $C = 0$, then the first term of that condition is infinitely large. Thus the left-hand side of the inequality cannot be negative, which is a contradiction. Since the marginal utility of consumption is infinitely large at the subsistence level, consumption will always exceed the subsistence level. Similarly, at least some time must be spent in leisure because the marginal utility of leisure is infinite if no leisure is being taken. If some positive time is taken in leisure, the time constraint is slack and the Lagrange multiplier associated with it, λ_2, must therefore be zero by the second part of condition (12.21)(f).

We now consider the interesting case of no labor supply. If $L = 0$, the second inequality in condition (12.21)(a) can be strict. This implies that $\lambda_1 w \leq (1 - a)/(T - L)$, evaluated at $L = 0$ and at the optimal level of consumption, which would be equal to I since there is no labor income. Since C is positive, the complementary slackness condition (12.21)(c) implies that the first part of condition (12.21)(a) is an equality, so $\lambda_1 = a/C$. Combining this with the other results in this paragraph, we will have a border solution (labor supply will be zero) if

$$ w < \left(\frac{1 - a}{a} \right) \left(\frac{I}{T} \right). \tag{12.22} $$

In Chapter 6 [equation (6.76)] we found that (if $C_0 = 0$) solving the Lagrangian yields a labor supply function $L^* = aT - (1 - a)I/w$. Comparing this equation with equation (12.22), we observe that if the wage is low enough that the labor supply border constraint is binding, then the solution we obtain for L by solving the problem as a Lagrangian will be negative. This illustrates the general point that, when objective functions are quasiconcave and constraints are quasiconvex, we need to consider border solutions only when the optimal values of the choice variables from a Lagrangian problem are negative.

12.6 INTERTEMPORAL CONSUMPTION WITH LIQUIDITY CONSTRAINTS

In Section 10.5 we analyzed the intertemporal consumption choice assuming perfect capital markets. But it is widely recognized that capital markets are not perfect. For example, many consumers face **liquidity constraints**: they would like to borrow but are not able to. In this section we modify the analysis of Chapter 10 to incorporate the possibility of liquidity constraints, that is, constraints on borrowing. We will analyze the simplest case, a two-period model in which no borrowing is allowed, and assume a very simple utility function with no subjective rate of time preference. We will relax some of these simplistic assumptions in end-of-chapter problems.

In a two-period model, utility is a function of consumption in the two periods, $U(C_1, C_2)$. We will use a particularly simple utility function, $U(C_1, C_2) = C_1 C_2$, and assume that income is earned in each period, I_1 and I_2. If the consumer is not allowed to borrow, $C_1 \leq I_1$. If any first-period income is saved, it (and the interest earned) can be used to increase second-period consumption above second-period income; that is, $C_2 > I_2$. Figure 12.8 shows the intertemporal budget constraint, assuming no borrowing. Because our emphasis is on saving and borrowing, it will be convenient to write consumption in the first period as first-period income minus saving S: $C_1 = I_1 - S$. If consumers are not allowed to borrow, $S \geq 0$. The other constraint[6] facing the consumer is that second-period consumption cannot exceed the sum of second-period income, the amount of first-period income saved, and the interest earned on the saving: $C_2 \leq I_2 + S(1 + r)$, where r is the interest rate.

---

[6]Formally, we also have the constraint that $S \leq I_1$. But this constraint will never be binding, since the marginal utility of first-period consumption is positive when evaluated at $C_1 = 0$. So all cases where this constraint binds can be ruled out.

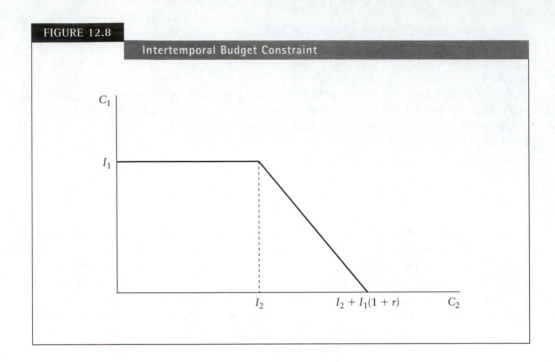

FIGURE 12.8

Intertemporal Budget Constraint

The intertemporal consumption problem allowing for liquidity constraints is to choose S and C_2 to maximize utility subject to the intertemporal budget constraint

$$C_2 \leq I_2 + S(1+r) \tag{12.23}$$

and the nonnegativity constraints $S \geq 0$ and $C_2 \geq 0$.[7] The Lagrangian for this problem is

$$\mathscr{L}(S, C_2, \lambda) = (I_1 - S)C_2 + \lambda(I_2 + S(1+r) - C_2). \tag{12.24}$$

The Kuhn-Tucker conditions are

(a) $\quad \dfrac{\partial \mathscr{L}}{\partial S} = -C_2 + \lambda(1+r) \leq 0 \quad$ and $\quad \dfrac{\partial \mathscr{L}}{\partial C_2} = I_1 - S - \lambda \leq 0$

(b) $\quad S \geq 0 \quad$ and $\quad C_2 \geq 0$

(c) $\quad S\dfrac{\partial \mathscr{L}}{\partial S} = 0 \quad$ and $\quad C_2\dfrac{\partial \mathscr{L}}{\partial C_2} = 0$

(d) $\quad \dfrac{\partial \mathscr{L}}{\partial \lambda} = I_2 + S(1+r) - C_2 \geq 0$

(e) $\quad \lambda \geq 0$

(f) $\quad \lambda\dfrac{\partial \mathscr{L}}{\partial \lambda} = 0.$

(12.25)

Assuming for the time being that both S and C_2 are positive and that the constraint (12.23) is binding (so $\lambda > 0$), the complementary slackness conditions (12.25)(c) and (f)

---

[7]Strictly speaking, we should add the constraint that $S \leq I_1$; but, as explained in Footnote 6, this constraint cannot be binding.

reduce to the first-order conditions for the problem of maximizing the Lagrangian (12.24):

$$\frac{\partial \mathcal{L}}{\partial S} = -C_2 + \lambda(1 + r) = 0$$

$$\frac{\partial \mathcal{L}}{\partial C_2} = I_1 - S - \lambda = 0 \tag{12.26}$$

$$\frac{\partial \mathcal{L}}{\partial \lambda} = I_2 + S(1 + r) - C_2 = 0.$$

Clearly, $\lambda^* = I_1 - S$, which equals first-period consumption, so it is positive. Thus the Kuhn-Tucker conditions (12.25)(d)–(f) will be satisfied with the second-period consumption constraint binding: C_2 will exhaust all available wealth. Said another way, nothing will be saved in period 2; without additional periods, saving in period 2 is useless. Since the marginal utility of consumption is positive, utility can always be increased until no more consumption can be financed.

Substituting $\lambda^* = I_1 - S$ into the first of the first-order conditions (12.26) and combining with the third first-order condition, we obtain two equations in the two unknowns

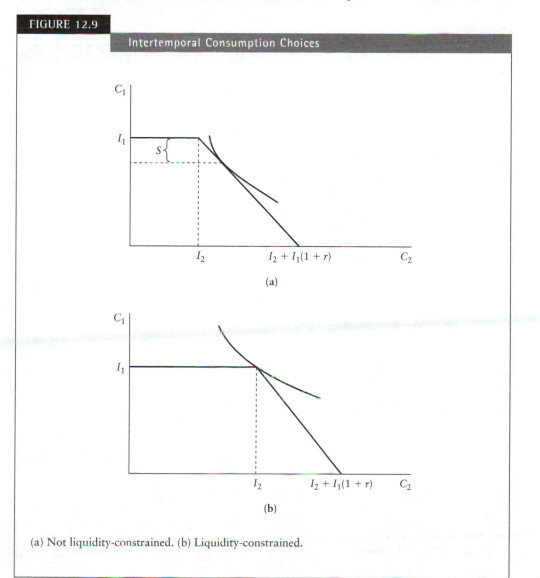

FIGURE 12.9

Intertemporal Consumption Choices

(a)

(b)

(a) Not liquidity-constrained. (b) Liquidity-constrained.

S and C_2, which can be solved either by substitution or Cramer's rule to get

$$C_2^* = \frac{I_1(1+r) + I_2}{2} \quad \text{and} \quad S^* = \frac{I_1(1+r) - I_2}{2(1+r)} \tag{12.27}$$

which implies

$$C_1^* = \frac{I_1(1+r) + I_2}{2(1+r)}. \tag{12.28}$$

Thus, as long as we have an internal solution, we have the same result as that in Section 10.5, where $C_1 = C_2/(1+r)$.

Second-period consumption C_2^* will always be positive because the marginal utility of C_2 is positive when evaluated at $C_2 = 0$. But if I_2 is large enough, specifically if $I_2 > I_1(1+r)$, the consumer will not want to save, preferring instead to borrow against future income ($S^* < 0$). If no borrowing is allowed, the consumer is liquidity-constrained and $S^* = 0$. In this case, $C_1^* = I_1$ and $C_2^* = I_2$. The two cases are illustrated in Figure 12.9. In Figure 12.9a, the consumer has chosen to save a portion of first-period

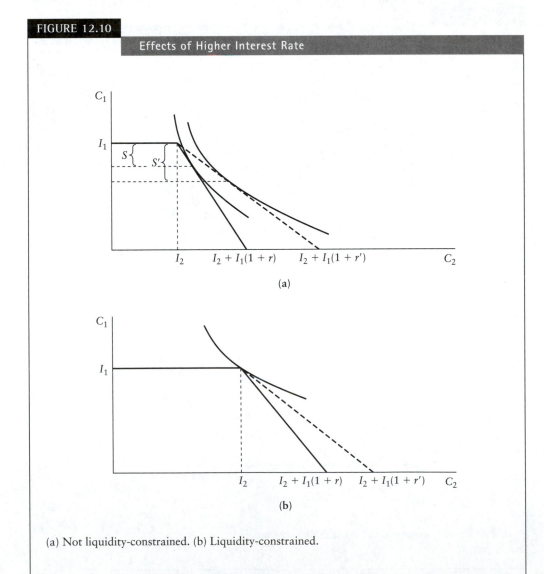

FIGURE 12.10

Effects of Higher Interest Rate

(a)

(b)

(a) Not liquidity-constrained. (b) Liquidity-constrained.

income, since I_2 is relatively low. In Figure 12.9b, I_2 is high enough that the consumer chooses the kink point of the budget constraint and is liquidity-constrained.

The most important policy implication of liquidity constraints has to do with the responsiveness of saving to the interest rate. As explored in Problem 12.16, we can use equations (12.27) to show that, in our example, an increase in the interest rate will increase saving if the equilibrium has positive saving.[8] This is illustrated in Figure 12.10a. But if the consumer is liquidity-constrained, an increase in the interest rate will lead to saving only if it makes $I_2 < I_1(1 + r)$, moving the consumer from a corner solution ($S^* = 0$) to an interior solution ($S^* > 0$). Figure 12.10b illustrates a situation in which the interest rate increases but the utility-maximizing choice of C_1 and C_2 (or, equivalently, S and C_2) does not change. Thus the impact on saving of government policy actions that change interest rates will differ depending on what fraction of consumers are liquidity-constrained.

Problems

12.1 For each of the utility functions given below, find the circumstances (if any) under which the optimal consumption of good x would be zero:

(a) $U(x, y) = a \ln x + b \ln y$ where $a, b > 0$

(b) $U(x, y) = \ln(x + a) + \ln(y + b)$ where $a, b > 0$

(c) $U(x, y) = A(ax^\rho + (1 - a)y^\rho)^{1/\rho}$ where $A > 0$, $0 < a < 1$, $0 < \rho < 1$

12.2 Suppose that rice and fish are the only foods available. It is necessary to obtain at least 1000 calories, 1.25 grams of sodium, 25 grams of protein, and 50 grams of carbohydrates. A serving of fish costs 30 cents and provides 70 calories, 0.25 grams of sodium, 15 grams of protein, and no carbohydrates. A serving of rice costs 7 cents and provides 120 calories, no sodium, 3 grams of protein, and 26 grams of carbohydrates.

(a) What is the minimum-cost way of satisfying all nutritional requirements?

(b) What is the shadow price of the calorie constraint? Explain in words what this shadow price measures.

12.3 A small company manufactures two different kinds of computers: a basic model (B) and an upscale model with several added components (U). The basic model yields the firm $100 per unit in profit, whereas the upscale model yields $250 per unit in profit. The firm employs five workers who assemble the computers and two workers who run diagnostic tests on the computers before they are sold. Each employee works 40 hours per week. It takes one hour to assemble the basic model and three hours to assemble the upscale model; thus the number of computers of each type assembled each week must satisfy the constraint $B + 3U \leq 200$. It takes one hour to run diagnostic tests on the upscale model but only 0.5 hour to run diagnostic tests on the basic model. Thus the number of computers of each type that can be tested in a week must satisfy the constraint $0.5B + U \leq 80$.

(a) How many computers of each type should the firm make to maximize profits subject to constraints on assembly and testing time?

(b) Would adding more assembly workers or more testers add more to profits?

12.4 For the firm in Problem 12.3, suppose that the profit per unit of upscale computers is some parameter π.

(a) For what values of π would the firm choose to produce only upscale computers? [*Hint*: Either compare the firm's profits for all three possible solutions to this problem—two corner solutions plus the interior solution—or identify which values of π would make one of the constraints slack.]

(b) For what values of π would the firm choose to produce only basic computers?

(c) Explain the economic reasoning behind your answers to parts (a) and (b).

12.5 For the firm in Problem 12.3, suppose that the price of the firm's basic model is given by the inverse demand function $P_B = 100 - B$ and the price of the upscale model is given by $P_U = 500 - U$. The upscale model is produced with constant marginal cost of \$25 per unit while the basic model is produced with constant marginal cost of \$10 per unit. There are no fixed costs, so the firm's profits are $\Pi = (500 - U)U + (100 - B)B - 25U - 10B$. How many of each kind of computer should the firm produce to maximize profits subject to the assembly and testing time constraints?

12.6 Redo Problem 12.5 if the inverse demand function for upscale computers is $P_U = 600 - U$.

12.7 A firm is deciding on its advertising strategy. Its options are to advertise on radio, which costs \$100 per ad, or in the newspaper, which costs \$50 per ad, or both. The firm wants to reach at least 55,000 families with incomes over \$50,000 and at least 100,000 people between the ages of 18 and 30. Every radio ad reaches 3,000 families with income over \$50,000 and 12,000 people between the ages of 18 and 30. Every newspaper ad reaches 5,000 families with incomes over \$50,000 and 5,000 people between the ages of 18 and 30. Find the minimum-cost advertising strategy for the firm.

12.8 Consider a two-period problem of production smoothing. A firm knows it will sell 50 units of its product in the first period and 150 units in the second period. It can produce in both periods, but it can also produce extra in the first period and hold the excess production as inventory so that fewer units need to be produced in the second period. Production costs in each period are $C_i = 2Q_i^2$, where Q_i is the quantity produced in period i. The inventory costs equal $200(Q_1 - 50)$. What combination of Q_1 and Q_2 will minimize costs subject to the constraints that the firm must produce at least 50 units in period 1 and have at least 150 units to sell in period 2? [*Hint*: Write the second constraint as $Q_1 + Q_2 \geq 200$.]

12.9 Redo Problem 12.8, if inventory costs are $600(Q_1 - 50)$.

12.10 If, in Problem 12.8, inventory costs equal $\theta(Q_1 - 50)$, for what values of θ would the firm choose to hold no inventories?

12.11 Now suppose that the firm of Problem 12.10 can sell its product for $15 per unit in period 1 and $20 per unit in period 2.

  (a) If $\theta = 200$, what are the profit-maximizing choices of Q_1 and Q_2?

  (b) For what values of θ would the firm choose to hold no inventories?

12.12 Suppose a monopolist wants to maximize revenues subject to a profit constraint. There are three inputs available: capital and two kinds of labor, one more expensive than the other. The firm needs capital and at least one (but not necessarily both) kinds of labor to produce its product. If the profit constraint is binding,

  (a) under what conditions would the firm choose not to employ any of the more expensive labor?

  (b) would the firm ever choose to employ *only* the more expensive labor? Explain.

12.13 For the model of Section 12.5, find the conditions (if any) under which labor supply would be zero given the following utility functions:

  (a) $U(C, T - L) = C^\alpha (T - L)^\beta$

  (b) $U(C, T - L) = (\alpha C^\rho + (1 - \alpha)(T - L)^\rho)^{1/\rho}$

12.14 Redo the analysis of Section 12.6 with C_1 as a choice variable instead of S.

12.15 For the model of Section 12.6, suppose that utility is equal to $C_1 C_2^\rho$, where $0 < \rho < 1$. Is it more or less likely to have liquidity constraints as ρ increases?

12.16 Using equations (12.26), show that $\partial S^*/\partial r > 0$ when $S^* > 0$.

12.17 For the model of Section 12.6, suppose it is possible to borrow up to a limit B, where $B < I_2/(1 + r)$. Find the conditions under which the consumer will be liquidity-constrained.

13

Value Functions and the Envelope Theorem: Theory

13.1   INTRODUCTION

Most economic analysis is comparative statics. The object is to determine the effects on the equilibrium values of endogenous variables of changes in the values of parameters. The purpose of this chapter is to develop value functions and the envelope theorem as tools for comparative static analysis.

There are three kinds of endogenous variables in economics. Of these, choice variables are fundamental because most economic agents assign values to choice variables in order to maximize or minimize the value of some objective function. A second kind of endogenous variable is the optimal value of an agent's objective function. The third kind of endogenous variable is an equilibrium price or quantity in a market. In this chapter we focus on the second kind of endogenous variable. We will explain how value functions and the envelope theorem are used to obtain comparative static results for the optimal values of objective functions.

A *value function* expresses the *optimal*[1] value of an objective function as a function of the parameters that define an economic agent's environment. There are value functions for both unconstrained and constrained optimization problems.

Consider the following unconstrained optimization problem. A monopolist wants to maximize her rate of profit, which is a function of her rate of output and the parameters of her demand and cost functions. Her optimal choice for the rate of output is therefore a function of those parameters. The monopolist's value function expresses the maximal rate of profit as a function of the demand and cost parameters by substituting her optimal choice function into her profit function. We will develop a specific example of a value function for a monopolist in Section 13.2.

The *envelope theorem* is a comparative static relationship between an objective function and its associated value function. For an unconstrained maximization problem, the envelope theorem states that the partial derivative of the value function with

---

[1]In earlier chapters we sometimes called the optimal value of an objective function its *equilibrium* value. An economic agent is in equilibrium when he has assigned optimal values to his choice variables, thereby achieving the optimal value of his objective function.

respect to any parameter is equal to the partial derivative of the associated objective function with respect to that parameter. By taking the partial, rather than the total, derivative of the objective function, we hold the choice variables constant at their optimal values. The envelope theorem for constrained maximization states that the partial derivative of the value function with respect to any parameter is equal to the partial derivative of the objective function with respect to that parameter plus the product of the Lagrange multiplier and the partial derivative of the constraint function with respect to that parameter. Again, the choice variables are held constant at their optimal values.

These are remarkable results. The envelope theorems simplify comparative static analyses of value functions by enabling us to ignore the fact that changing the value of a parameter will (in general) change the optimal values of the choice variables.

In Section 13.2 we develop value functions for unconstrained optimization problems. Then in Section 13.3 we present the envelope theorem for unconstrained optimization. In Sections 13.4 and 13.5 we develop value functions and the associated envelope theorem for constrained optimization problems. We conclude with Section 13.6, in which we use the envelope theorem for constrained problems to obtain an economic interpretation of Lagrange multipliers. This interpretation is more sophisticated than the interpretation we gave in Chapter 9 before we had the envelope theorem.

13.2   VALUE FUNCTIONS FOR UNCONSTRAINED PROBLEMS

A **value function** is an objective function in which the economic agent's choice variables have been assigned their optimal values as functions of the parameters that define that agent's environment. Let $z = f(x, y; \boldsymbol{\alpha})$ be an objective function, in which x and y are the agent's choice variables and $\boldsymbol{\alpha}$ is a vector of parameters that (together with the form of the function f) define the agent's environment. Let $x^* = x^*(\boldsymbol{\alpha})$ and $y^* = y^*(\boldsymbol{\alpha})$ be the agent's optimal choices. Then the agent's value function is $V(\boldsymbol{\alpha}) = f(x^*(\boldsymbol{\alpha}), y^*(\boldsymbol{\alpha}); \boldsymbol{\alpha})$.

Let us reconsider the monopolist first analyzed in Chapter 2. Suppose that the monopolist faces the linear (inverse) demand function

$$P = a - bQ \tag{13.1}$$

in which P is the price, Q is the rate of output, and the parameters a and b are positive. Let marginal cost be the positive constant c, and assume that $c < a$.

The monopolist wants to maximize profit. Her objective function is

$$\pi(Q; a, b, c) = (P - c)Q = (a - c)Q - bQ^2. \tag{13.2}$$

The optimal value[2] for Q is

$$Q^*(a, b, c) = \frac{a - c}{2b}. \tag{13.3}$$

---

[2]Setting the derivative of (13.2) with respect to $Q = 0$ and solving for Q produces $Q^* = (a - c)/2b$ in (13.3). This solution is unique, and the second-order condition for a maximum is satisfied. There are, however, two further second-order conditions. The monopolist will not produce a negative rate of output, nor will she produce at a loss. Both these conditions are met by requiring that $Q^* = 0$ if $a < c$.

Define the monopolist's value function, V, by

$$\pi^* = V(a, b, c) = \pi(Q^*(a, b, c); a, b, c)$$

$$= (a - c)\frac{a - c}{2b} - b\left(\frac{a - c}{2b}\right)^2$$

$$= \frac{(a - c)^2}{4b}.$$

(13.4)

The value function $V(a, b, c)$ provides the monopolist's *maximal* rate of profit as a function of the demand parameters a and b and the cost parameter c. The function $V(a, b, c)$ incorporates optimal behavior because the value for the choice variable Q in the profit function π is the optimal value Q^* as defined in (13.3).

We now examine the relationship between the monopolist's objective function and her associated value function. It is clear from the value function in (13.4) that the monopolist's maximal rate of profit depends partly on the parameter a, which is the vertical intercept of her (inverse) demand function. Let a' and a'' be fixed values for a, and assume that $c < a' < a''$. For each of these values for a we plot the objective function (13.2) as one of the parabolas shown in Figure 13.1a. The independent variable for the objective function is the rate of output Q. As we can see, the values of the demand parameters a and b and the cost parameter c determine the shape and the position of the two versions of the objective function.

In Figure 13.1a we show the monopolist's optimal rates of output for the cases in which $a = a'$ or $a = a''$. The associated maximal rates of profit are indicated on the vertical axis. We can obtain these rates of profit either by substituting the optimal rates of output into the objective function (13.2), or more directly by substituting the alternative values for a into the value function (13.4).

In Figure 13.1b we plot the monopolist's value function. By considering the parameter a as a variable while holding the values of parameters b and c fixed, we can plot the value function in two dimensions. The independent variable a is plotted horizontally; the

FIGURE 13.1

Two Objective Functions and the Value Function for a Monopolist

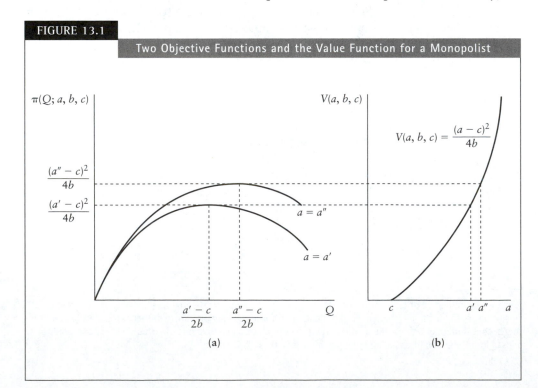

(a)　　　　(b)

corresponding level of the value function is plotted vertically. For each value of a, the height of the value function in Figure 13.1b is the maximal height of the objective function defined by that value of a in Figure 13.1a.

The value function is the segment of the parabola $V(a, b, c) = (a - c)^2/4b$ for $c < a$. Larger values for a shift the graph of the monopolist's demand function upward. Her average (and marginal) cost function does not shift because it depends only on c, which we hold constant. A larger value for a enables the monopolist to obtain a higher rate of profit.[3]

Alternatively, we may define the profit function (13.2) as the **direct objective function** because the argument of that function is the choice variable Q. By choosing the value for Q, the agent chooses the level of profit "directly." From this point of view, the value function (13.4) is defined as the **indirect objective function**; that is, it transforms values for the parameters a, b, and c "indirectly" into a value for the rate of profit by first using (13.3) to determine the optimal choice for Q, then using this value for Q to determine the optimal rate of profit. The indirect objective function (the value function) incorporates optimal behavior into the objective function. Therefore, the indirect objective function indicates the *maximal* rate of profit that the monopolist can obtain for any set of values for the parameters. The direct objective function is defined for *all* feasible values for the choice variable. Therefore, a value of the direct objective function need not be the maximal rate of profit.

13.3 THE ENVELOPE THEOREM FOR UNCONSTRAINED OPTIMIZATION

13.3.1 General Discussion

The **envelope theorem** is a comparative static relationship between the derivatives of an objective function and the derivatives of the associated value function. The envelope theorem for unconstrained optimization states that when computing the marginal effect of a parameter on the maximal value of the objective function, we can treat the choice variables as constants. We will now present the envelope theorem formally.

Let $z = f(x, y; \boldsymbol{\alpha})$ be an objective function in which x and y are the choice variables and $\boldsymbol{\alpha}$ is a vector of parameters. The agent maximizes the value of z by choosing values for x and y according to the optimal choice functions $x^* = x^*(\boldsymbol{\alpha})$ and $y^* = y^*(\boldsymbol{\alpha})$. Let z^* be the corresponding maximal value for z. Finally, let

$$z^* = V(\boldsymbol{\alpha}) = f(x^*(\boldsymbol{\alpha}), y^*(\boldsymbol{\alpha}); \boldsymbol{\alpha}) \tag{13.5}$$

be the agent's value function.

THEOREM ## 13.1

The Envelope Theorem for Unconstrained Optimization

Let α_i be one of the parameters in $\boldsymbol{\alpha}$. The relationship between the partial derivatives of the value function $V(\boldsymbol{\alpha})$ and the partial derivatives of the associated objective function $f(x, y; \boldsymbol{\alpha})$, all taken with respect to α_i, is

$$\frac{\partial z^*}{\partial \alpha_i} = \frac{\partial}{\partial \alpha_i} V(\boldsymbol{\alpha}) = \frac{\partial}{\partial \alpha_i} f(x^*, y^*; \boldsymbol{\alpha}). \tag{13.6}$$

[3]If a were less than c, the monopolist's entire demand function would lie below her average cost function. For these values of a her value function would coincide with the horizontal axis because her optimal output and rate of profit would be zero.

Proof

The expression for $\partial V(\boldsymbol{\alpha})/\partial \alpha_i$ is

$$
\begin{aligned}
\frac{\partial}{\partial \alpha_i} V(\boldsymbol{\alpha}) &= \frac{\partial}{\partial \alpha_i} f(x^*(\boldsymbol{\alpha}), y^*(\boldsymbol{\alpha}); \boldsymbol{\alpha}) \\
&= \frac{\partial}{\partial x} f(x^*, y^*; \boldsymbol{\alpha}) \frac{\partial x^*}{\partial \alpha_i} + \frac{\partial}{\partial y} f(x^*, y^*; \boldsymbol{\alpha}) \frac{\partial y^*}{\partial \alpha_i} + \frac{\partial}{\partial \alpha_i} f(x^*, y^*; \boldsymbol{\alpha}) \\
&= 0 \frac{\partial x^*}{\partial \alpha_i} + 0 \frac{\partial y^*}{\partial \alpha_i} + \frac{\partial}{\partial \alpha_i} f(x^*, y^*; \boldsymbol{\alpha}) \\
&= \frac{\partial}{\partial \alpha_i} f(x^*, y^*; \boldsymbol{\alpha}).
\end{aligned}
\tag{13.7}
$$

By the definition of a value function, all the partial derivatives in (13.7) are evaluated at the optimal values for x and y. The first-order conditions for an optimum require that the values of the partial derivatives with respect to the choice variables be equal to zero.[4] This concludes the proof. ∎

The envelope theorem for unconstrained optimization states that the marginal effect on the value function of any parameter α_i is equal to the marginal effect of that parameter on the objective function, holding the choice variables x and y constant at their optimal values.

We offer three comments to prepare for the economic interpretation of the envelope theorem. First, in general, the values of the partial derivatives of a function depend on the values of all the variables and parameters in that function, not just the variable with respect to which the partial derivative is taken. Second, under the envelope theorem all the partial derivatives are evaluated at the optimal values of the choice variables. To emphasize these facts we write, for example, $\partial f(x^*, y^*; \boldsymbol{\alpha})/\partial x$, rather than simply $\partial f/\partial x$, in (13.7). By the first-order conditions for an unconstrained maximum, these partial derivatives with respect to the choice variables are equal to zero.

The third comment is based on the distinction between a shift of the graph of a function and a movement along that function. The value of $\partial f(x^*, y^*; \boldsymbol{\alpha})/\partial \alpha_i$ is the marginal effect on the value of z^* caused by the marginal effect of the parameter α_i in *shifting* the graph of the objective function f. By contrast, the product of $\partial f(x^*, y^*; \boldsymbol{\alpha})/\partial x$ and $\partial x^*/\partial \alpha_i$ is the marginal effect on z^* caused by the *movement along* the objective function as the value of x^* responds to the change in the value of α_i. The envelope theorem states that the marginal effect on z^* of a change in the value of any parameter is equal to the marginal effect of that parameter in shifting the graph of the objective function; we may ignore the movements along the objective function caused by the dependence of the choice variables on that parameter. The values of x^* and y^* do respond to a change in the value of α_i, but at the optimal point $x = x^*$ and $y = y^*$, the slopes of the objective function with respect to x and y are equal to zero.

13.3.2 An Example

We now apply the envelope theorem to the example of the monopolist in Section 13.2. We want to determine the marginal effect on the monopolist's maximal profit of an upward shift of her (inverse) demand function.

---

[4]Although we are discussing unconstrained optimization problems in this section, most "unconstrained" problems in economics have nonnegativity constraints for the choice variables. We assume here that the optimal solution is an interior solution. Then the partial derivatives of the objective function with respect to the choice variables are equal to zero when the choice variables are assigned their optimal values.

The relevant parameter is a, the vertical intercept of the demand function. From (13.2) the monopolist's objective function is her profit function, $\pi(Q; a, b, c) = (a - c)Q - bQ^2$. We have the partial[5] derivative $\partial\pi/\partial a = Q$. The value of this derivative at the optimal rate of output is $Q^* = (a - c)/2b$. From (13.4) the value function is $V(a, b, c) = (a - c)^2/4b$. The partial derivative is $\partial V/\partial a = (a - c)/2b = Q^*$. Then

$$\frac{\partial\pi^*}{\partial a}(Q^*(a, b, c), a, b, c) = \frac{\partial V}{\partial a}(a, b, c) = Q^* = \frac{a - c}{2b}. \tag{13.8}$$

This application of the envelope theorem says that if the graph of the monopolist's (inverse) demand function shifts upward by \$1, then her maximal profit will increase by \$1Q. We can interpret this result geometrically by examining Figure 13.2.

The monopolist's initial situation is described by the solid lines that plot her demand, marginal revenue, and marginal cost functions. The monopolist operates at point A on her demand function. Her optimal rate of output is $Q^* = (a - c)/2b$ and

FIGURE 13.2

The Envelope Theorem: Effect on a Monopolist's Profit of a Shift in the Demand Function

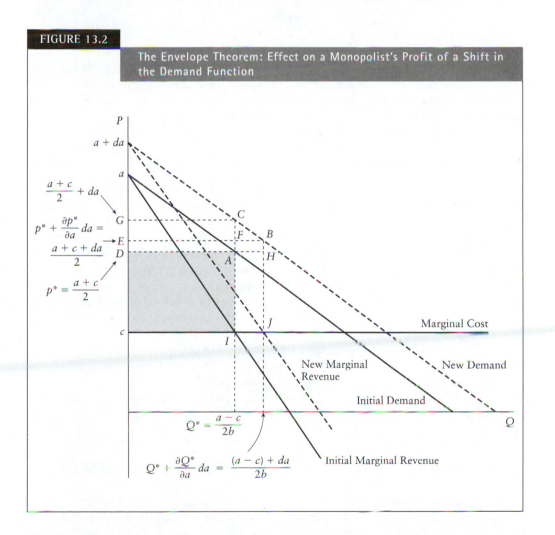

---

[5]The total derivative $d\pi/da$, evaluated at the optimal output, is $(\partial\pi/\partial Q)(\partial Q^*/\partial a) + \partial\pi/\partial a = (0)(1/2b) + Q^* = (a - c)/2b$. The total and partial derivatives of the objective function, evaluated at the optimal values of the choice variables, are equal according to the envelope theorem.

her optimal price is $P^* = a - bQ^* = (a + c)/2$. Then her maximal rate of profit is $\pi^* = (P^* - c)Q^* = (a - c)^2/4b$, which is the shaded area $IcDA$ in Figure 13.2.

If we increase a by the differential da and hold b constant, the monopolist's demand function shifts to the new demand function indicated by a dashed line in Figure 13.2. The two demand functions are parallel; the vertical distance between them is da.

The shift of the graph of the demand function changes the monopolist's optimal rate of output, price, and rate of profit. The marginal effect on her output is $dQ = (\partial Q^*/\partial a)\, da = (1/2b)\, da$. The effect on her price is $dP = (\partial P^*/\partial a)\, da = [1 - b(\partial Q^*/\partial a)]\, da = (1/2)\, da$. The monopolist moves from point A on her initial demand function to point B on her new demand function. The change in her profit is the sum of areas $IAHJ$, $ADEF$, and $AFBH$. These areas are

$$\text{Area } IAHJ = \left(\frac{a + c}{2} - c \right) \frac{\partial Q^*}{\partial a}\, da = \left(\frac{a - c}{2} \right)\left(\frac{1}{2b} \right) da,$$

$$\text{Area } ADEF = Q^* \frac{\partial P^*}{\partial a}\, da = \left(\frac{a - c}{2b} \right)\left(\frac{1}{2} \right) da, \qquad \textbf{(13.9)}$$

$$\text{Area } AFBH = \left(\frac{\partial P^*}{\partial a}\, da \right)\left(\frac{\partial Q^*}{\partial a}\, da \right) = \frac{1}{4b}\, (da)^2.$$

The sum of areas $IAHJ$ and $ADEF$ is $[(a - c)/2b]\, da$. This is equal to the area $ADGC$, which is the increase in profit that the monopolist could obtain simply by holding her output constant at $Q^* = (a - c)/2b$ and increasing her price by the full amount of the upward shift of the graph in her demand function, da.

Using (13.2) we can write the monopolist's objective function as

$$\pi(Q; a, b, c) = (P - c)Q = (a - bQ - c)Q. \qquad \textbf{(13.10)}$$

Holding b and c constant, increase a by da. If output is held constant at the optimal rate for the original demand function, the monopolist will move from point A to point C. Profit will increase by the amount $d\pi = Q^*\, da = [(a - c)/2b]\, da$.

Moving from point A to point B requires adjusting the value of the choice variable Q to the change in the value of the parameter a. Moving from point A to point C makes no adjustment to the choice variable. We conclude that, except for an approximation error, the marginal effect on the monopolist's maximal profit of a shift of the graph of her demand function is the same whether or not we allow for an optimal adjustment to her choice variable Q. We have demonstrated the envelope theorem for a parallel upward shift of the graph in a monopolist's demand function.

The approximation error is the area $AFBH$. Any derivative is an instantaneous rate of change. To approximate the effect on the value of a function caused by a change in the value of one of its variables, we multiply a derivative by the change in that variable. The result contains an error that approaches zero as the change in the variable approaches zero.[6] In the present example the error is a multiple of $(da)^2$, which rapidly approaches zero as da approaches zero.

13.4   VALUE FUNCTIONS FOR CONSTRAINED OPTIMIZATION

Value functions for constrained optimization are analogous to those for unconstrained optimization. Let $z = f(x, y; \boldsymbol{\alpha})$ be an objective function in which x and y are choice variables and $\boldsymbol{\alpha}$ is a vector of parameters. Let $h(x, y; \boldsymbol{\alpha}) \geq 0$ define a constraint.

---

[6]Unless the function is linear or constant, in which case the error is zero.

Finally, let $x^*(\alpha)$ and $y^*(\alpha)$ be the agent's constrained optimal choices. Then $V(\alpha) = f(x^*(\alpha), y^*(\alpha); \alpha)$ is the value function for this constrained optimization problem.

We will now develop an example of a value function for a consumer who maximizes utility subject to a budget constraint.

Let the consumer have the Cobb-Douglas utility function

$$U(x, y) = Hx^\gamma y^\varepsilon, \tag{13.11}$$

in which x and y are the rates at which he consumes goods X and Y, H is a positive constant, and the exponents γ and ε are positive constants.

Let P_x and P_y be the constant prices of the two goods, and let I be the consumer's fixed income. The budget constraint is

$$h(x, y; \alpha) \equiv I - g(x, y; P_x, P_y) \geq 0, \quad \text{in which } g(x, y; P_x, P_y) = P_x x + P_y y. \tag{13.12}$$

We know that the budget constraint is binding because the marginal utilities for both goods are positive (for positive rates of consumption). Using the techniques in Chapter 9, we can obtain the consumer's optimal choice functions as

$$x^* = x^*(P_x, P_y, I) = \frac{I}{P_x}\left(\frac{\gamma}{\gamma + \varepsilon}\right),$$
$$y^* = y^*(P_y, P_x, I) = \frac{I}{P_y}\left(\frac{\varepsilon}{\gamma + \varepsilon}\right). \tag{13.13}$$

Now substitute these optimal choice functions into the utility function (13.11). The result is the consumer's value function:

$$V(H, \gamma, \varepsilon, I, P_x, P_y) = H\left[\frac{I}{P_x}\left(\frac{\gamma}{\gamma + \varepsilon}\right)\right]^\gamma \left[\frac{I}{P_y}\left(\frac{\varepsilon}{\gamma + \varepsilon}\right)\right]^\varepsilon$$
$$= H\left(\frac{I}{\gamma + \varepsilon}\right)^{\gamma + \varepsilon}\left(\frac{\gamma}{P_x}\right)^\gamma\left(\frac{\varepsilon}{P_y}\right)^\varepsilon. \tag{13.14}$$

The utility function (13.11) is the *direct* utility function because its arguments are the consumer's choice variables, x and y. The value function (13.14) is the *indirect* utility function because its arguments are the parameters whose values define the consumer's environment. By incorporating optimal behavior into the direct utility function, the value function produces the maximal level of utility permitted by the values of the parameters. Since the arguments of the value function are the parameters of the agent's constrained maximization problem, the value function is a convenient tool for comparative static analyses of the consumer.

13.5 THE ENVELOPE THEOREM FOR CONSTRAINED OPTIMIZATION

The envelope theorem for constrained optimization is analogous to the theorem for unconstrained optimization, except for an adjustment required by the constraint. The adjustment involves the Lagrange multiplier.

Consider the following problem:

$$\begin{array}{ll} \text{Maximize} & z = f(x, y; \alpha) \\ x, y \\ \text{subject to} & g(x, y; \alpha) \leq g_0, \end{array} \tag{13.15}$$

in which $\boldsymbol{\alpha}$ is a vector of parameters and g_0 is a single parameter.[7] At the optimal solution we will have either $g(x^*, y^*; \boldsymbol{\alpha}) = g_0$ or $g(x^*, y^*; \boldsymbol{\alpha}) < g_0$, depending on whether the constraint is binding or slack.

The Lagrange function for this problem is

$$\mathscr{L}(x, y, \lambda; \boldsymbol{\alpha}, g_0) = f(x, y; \boldsymbol{\alpha}) + \lambda[g_0 - g(x, y; \boldsymbol{\alpha})]. \qquad \textbf{(13.16)}$$

To present the envelope theorem it is convenient to incorporate the constraint into a single function. We define the constraint function as

$$h(x, y; g_0, \boldsymbol{\alpha}) = g_0 - g(x, y; \boldsymbol{\alpha}). \qquad \textbf{(13.17)}$$

Then the Lagrange function becomes

$$\mathscr{L}(x, y, \lambda; \boldsymbol{\alpha}, g_0) = f(x, y; \boldsymbol{\alpha}) + \lambda h(x, y; g_0, \boldsymbol{\alpha}). \qquad \textbf{(13.18)}$$

Let $x^*(\boldsymbol{\alpha}, g_0)$, $y^*(\boldsymbol{\alpha}, g_0)$, and $\lambda^*(\boldsymbol{\alpha}, g_0)$ be the optimal values for x, y, and λ. Let z^* be the corresponding maximal value for z. Finally, let

$$z^* = V(\boldsymbol{\alpha}, g_0) = f(x^*(\boldsymbol{\alpha}, g_0), y^*(\boldsymbol{\alpha}, g_0); \boldsymbol{\alpha}) \qquad \textbf{(13.19)}$$

be the agent's value function.

THEOREM 13.2

The Envelope Theorem for Constrained Optimization

Let k represent the parameter g_0 or any of the parameters in $\boldsymbol{\alpha}$. The relationship among the partial derivatives of the value function $V(\boldsymbol{\alpha}, g_0)$, the associated objective function $f(x, y; \boldsymbol{\alpha})$, and the constraint function $h(x, y; g_0, \boldsymbol{\alpha})$ is

$$\frac{\partial z^*}{\partial k} = \frac{\partial}{\partial k} V(\boldsymbol{\alpha}; g_0) = \frac{\partial}{\partial k} f(x^*, y^*; \boldsymbol{\alpha}) + \lambda^* \frac{\partial}{\partial k} h(x^*, y^*; g_0, \boldsymbol{\alpha}). \qquad \textbf{(13.20)}$$

Proof

We consider separately the cases in which the constraint is binding or slack.

Case 1: Binding Constraint

From (13.19) we have

$$\frac{\partial V}{\partial k}(\boldsymbol{\alpha}, g_0) = \frac{\partial f}{\partial x}\frac{\partial x^*}{\partial k} + \frac{\partial f}{\partial y}\frac{\partial y^*}{\partial k} + \frac{\partial f}{\partial k}. \qquad \textbf{(13.21)}$$

If the constraint is binding, the first-order conditions for a maximum of the Lagrangian are

$$\frac{\partial f}{\partial x} + \lambda^* \frac{\partial h}{\partial x} = 0,$$
$$\frac{\partial f}{\partial y} + \lambda^* \frac{\partial h}{\partial y} = 0, \qquad \textbf{(13.22)}$$
$$h = 0.$$

---

[7]The parameter g_0 designates the level of the constraint. For example, in a problem of constrained maximization of utility, the level parameter g_0 would be the consumer's income and $g(x, y; \boldsymbol{\alpha})$ would be total expenditure, or $g(x, y; \boldsymbol{\alpha}) = P_x x + P_y y$. We will discuss this interpretation of g_0 in detail in Section 13.6.

Totally differentiating the condition $h = 0$ with respect to k yields

$$\frac{\partial h}{\partial x}\frac{\partial x^*}{\partial k} + \frac{\partial h}{\partial y}\frac{\partial y^*}{\partial k} + \frac{\partial h}{\partial k} = 0. \tag{13.23}$$

Using the first two conditions in (13.22) and the condition (13.23), we may rewrite (13.21) as

$$\frac{\partial V}{\partial k}(\boldsymbol{\alpha}, g_0) = -\lambda^*\left(\frac{\partial h}{\partial x}\frac{\partial x^*}{\partial k} + \frac{\partial h}{\partial y}\frac{\partial y^*}{\partial k}\right) + \frac{\partial f}{\partial k} = \frac{\partial f}{\partial k} + \lambda^*\frac{\partial h}{\partial k}, \tag{13.24}$$

which establishes (13.20). ∎

Case 2: Slack Constraint

We know from Section 11.4 that if the constraint is slack, $\partial f/\partial x$ and $\partial f/\partial y$ are also equal to zero when evaluated at x^* and y^*, then $\lambda^* = \partial f/\partial x = \partial f/\partial y = 0$ (13.19, 13.21) and $\lambda^* = 0$ establishes (13.20), concluding the proof. ∎

Note that by the first-order conditions (13.22) for constrained optimization, the partial derivatives of the objective function, $\partial f/\partial x$ and $\partial f/\partial y$, are not, in general, equal to zero. This fact is central to the economic interpretation of the Lagrange multiplier.

13.6 ECONOMIC INTERPRETATION OF THE LAGRANGE MULTIPLIER

13.6.1 Using the Envelope Theorem

In a constrained optimization problem the constraint defines the choices available to the agent. To relax the constraint is to expand the agent's set of permissible choices. All the choices permitted under the initial constraint are still available, but the agent may make certain new choices that were not previously allowed. These new choices may enable the agent to achieve a better value for her objective function than she could under the original constraint.

Solving a constrained optimization problem by using a Lagrangian function produces an optimal value for the Lagrange multiplier, as well as optimal values for the choice variables in the original problem. We will now use the envelope theorem and the Kuhn-Tucker complementary slackness condition to show that the optimal value of the Lagrange multiplier is the marginal effect on the value function of relaxing the constraint.

Consider the constrained maximization[8] problem

$$\begin{aligned} \text{Maximize}_{x,y} \quad & z = f(x, y; \boldsymbol{\alpha}) \\ \text{subject to} \quad & h(x, y; g_0, \boldsymbol{\alpha}) \geq 0, \end{aligned} \tag{13.25}$$

in which g_0 is a level parameter and $\boldsymbol{\alpha}$ is a vector of parameters.[9] Let $V(\boldsymbol{\alpha}, g_0)$ be the value function for this problem, and let k be any parameter. By the envelope theorem in (13.20) for constrained optimization,

$$\frac{\partial V}{\partial k} = \frac{\partial f}{\partial k} + \lambda^*\frac{\partial h}{\partial k}, \tag{13.26}$$

in which all partial derivatives are evaluated at x^*, y^*, and λ^*.

---

[8] The case of constrained minimization is analogous.

[9] Usually (but not always), there are also nonnegativity constraints on the choice variables.

To understand the economic significance of the Lagrange multiplier, we first recognize that the constraint function $h(x, y; g_0, \boldsymbol{\alpha})$ measures the quantity of a scarce resource that is *unallocated* (or unconsumed) after the agent chooses values for x and y. The constraint requires the agent to choose values for x and y so that the unallocated quantity of the scarce resource is nonnegative.

If $h(x^*, y^*; g_0, \boldsymbol{\alpha}) = 0$, the constraint is *binding*. The agent cannot achieve a higher value for the objective function because he has exhausted the scarce resource. If $\partial h(x^*, y^*; g_0, \boldsymbol{\alpha})/\partial k > 0$, an increase in the value of k relaxes the constraint by increasing the unallocated quantity of the scarce resource from zero to a positive quantity. The agent can then achieve a higher value of the objective function by new values for x and y that were not previously feasible. Similarly, if $\partial h(x^*, y^*; g_0, \boldsymbol{\alpha})/\partial k < 0$, an increase in k tightens the constraint, forcing the agent to accept a lower value for the objective function.

13.6.2 Three Cases

The complementary slackness conditions we examined in Chapter 11 require $\lambda^* \geq 0$ and $\lambda^* h(x^*, y^*; g_0, \boldsymbol{\alpha}) = 0$. The constraint is $h(x^*, y^*; g_0, \boldsymbol{\alpha}) \geq 0$. Then there are three possibilities: $\lambda^* > 0$ and $h = 0$; $\lambda^* = 0$ and $h > 0$; and $\lambda^* = h = 0$.

Case 1: $\lambda^* > 0$, $h = 0$

The constraint is binding. From (13.26) the marginal effect of k on the value function is the sum of the effect of k in shifting the graph of the objective function plus λ^* times the effect of k in shifting the graph of the constraint function. We multiply $\partial h/\partial k$ by λ^* to convert the units in which the constraint function is measured into units in which the objective function is measured. Thus, it is reasonable to interpret λ^* as the marginal effect on the value function of an increase in the unallocated quantity of the scarce resource. Equivalently, λ^* is the marginal effect on the value function of a relaxation of the constraint.

Case 2: $\lambda^* = 0$, $h > 0$

If $h > 0$, the constraint is nonbinding, or slack. The agent's choice of values for x and y has achieved the unconstrained optimal value for her objective function. Relaxing the constraint function to expand the set of permissible choices provides no benefit. Then $\lambda^* = 0$, and the product term in (13.26) vanishes.

Case 3: $\lambda^* = h = 0$

This is an unusual case. Since $h = 0$, the agent's optimal choices x^* and y^* lie on the boundary of the constraint. The fact that $\lambda^* = 0$ means that the agent would not be able to increase the value of her objective function even if the constraint were relaxed. As in Case 2, the agent has achieved the unconstrained optimal value for her objective function because her constrained and unconstrained optimal choices coincide. Then the second term in (13.26) vanishes, so that the marginal effect of k on the value function is limited to its effect in shifting the graph of the objective function.

13.6.3 An Example Using Utility Maximization

Consider the problem in Section 13.4 in which a consumer maximizes utility subject to a constraint on income. The consumer's value function (repeated from (13.14)) is

$$V(H, \gamma, \varepsilon, I, P_x, P_y) = H\left(\frac{I}{\gamma + \varepsilon}\right)^{\gamma + \varepsilon}\left(\frac{\gamma}{P_x}\right)^{\gamma}\left(\frac{\varepsilon}{P_y}\right)^{\varepsilon}. \tag{13.27}$$

Clearly, the nature of the consumer's utility function requires that he allocate his entire income to purchases of goods X and Y.[10] Then the constraint is binding. Increasing the consumer's income or reducing the prices of the two goods will relax the constraint, permitting him to achieve a higher level of utility. Therefore, the value of λ^* should be positive.

The Lagrangian for the consumer's constrained maximization problem is

$$\mathcal{L}(x, y, \lambda; I, \boldsymbol{\alpha}) = U(x, y; \boldsymbol{\alpha}) + \lambda h(x, y; I, \boldsymbol{\alpha}), \tag{13.28}$$

in which $U(x, y; \boldsymbol{\alpha}) = Hx^\gamma y^\varepsilon$ and $h(x, y; I, \boldsymbol{\alpha}) = M - P_x x - P_y y$. The first-order conditions for a solution are

$$\frac{\partial U}{\partial x} - \lambda^* P_x = 0,$$

$$\frac{\partial U}{\partial y} - \lambda^* P_y = 0, \tag{13.29}$$

$$I - P_x x - P_y y = 0,$$

in which λ^* is the optimal value of λ. Solving for λ, we have

$$\lambda^* = \frac{1}{P_x}\frac{\partial U}{\partial x} = \frac{1}{P_y}\frac{\partial U}{\partial y}. \tag{13.30}$$

The quantity $(1/P_x)$ is the additional quantity of good X that the consumer could purchase if his income were increased by \$1. Then $\lambda^* = (1/P_x)\,\partial U/\partial x = (1/P_y)\,\partial U/\partial y$ is the additional utility that the consumer could obtain by using the extra dollar of income to purchase either more good X or more good Y. Since both $\partial U/\partial x$ and $\partial U/\partial y$ are evaluated at the optimal choices x^* and y^*, the consumer will be indifferent between allocating the extra dollar to purchasing good X or good Y. Then λ^* is obviously the marginal utility of income in this problem.

We will now interpret λ^* further by evaluating the individual marginal effects on the value function of changes in the parameters H, I, and P_x. Consider first the parameter H. From (13.26) and (13.27) we have

$$\begin{aligned}\frac{\partial V}{\partial H}(H, \gamma, \varepsilon, I, P_x, P_y) &= \frac{\partial U}{\partial H} + \lambda^*\frac{\partial h}{\partial H} \\ &= (x^*)^\gamma (y^*)^\beta + \lambda^* \cdot 0 \\ &= (x^*)^\gamma (y^*)^\beta. \end{aligned} \tag{13.31}$$

The Lagrange multiplier is not involved in measuring the effect of H on the value function because H does not appear in the constraint function. Then the marginal effect of H on the value function is limited to its effect in shifting the graph of the utility function. The shift of the graph is evaluated at x^* and y^*.

Next consider the parameter I. An increase in the consumer's income shifts the graph of the constraint. On a graph of the utility-maximization problem, the consumer's budget line will shift outward parallel to itself, expanding the consumer's permissible choices for x and y. Since the constraint is binding at the solution that was optimal before the increase in I, we should expect that the marginal effect of I on the

---

[10]This conclusion is not as restrictive as it might seem. One of the goods could be a financial asset. To allocate some income to purchase this asset is to save, rather than consume.

value function will be positive. From (13.26) and (13.27) we have

$$
\begin{aligned}
\frac{\partial V}{\partial I}(H, \gamma, \varepsilon, I, P_x, P_y) &= \frac{\partial U}{\partial I} + \lambda^* \frac{\partial h}{\partial I} \\
&= 0 + \lambda^* \cdot 1 \\
&= \lambda^*.
\end{aligned}
\tag{13.32}
$$

It is easy to see that λ^* is the marginal utility of income in this problem. The constraint function is $h(x, y; I, \boldsymbol{\alpha}) = I - P_x x - P_y y$. For any choice of values for x and y, the value of h is the amount of income not allocated to the two goods. For the utility function in this problem, $h = 0$ when the consumer makes optimal choices. Then a unit increase in I increases unallocated income from \$0 to \$1. Therefore, the marginal effect of I on V is the marginal effect of I on unallocated income, times the marginal utility of income. That is, $\partial V/\partial I = [\partial h/\partial I]\lambda^* = 1\lambda^* = \lambda^*$.

Finally, consider the parameter P_x. Increasing the price of good X causes the consumer's budget line to rotate about the y-intercept, becoming steeper as P_x increases. This shift of the graph of the budget line will contract the consumer's set of permissible choices. Thus, the marginal effect of P_x on V will be nonpositive. We find

$$
\begin{aligned}
\frac{\partial V}{\partial P_x}(H, \gamma, \varepsilon, I, P_x, P_y) &= \frac{\partial U}{\partial P_x} + \lambda^* \frac{\partial h}{\partial P_x} \\
&= 0 + \lambda^*[-x^*] \\
&= -\lambda^* x^* < 0.
\end{aligned}
\tag{13.33}
$$

At the margin defined by x^* and y^*, unallocated income h is zero. An increase of \$1 in P_x is equivalent to reducing the consumer's unallocated income by \$$x^*$ (from \$0 to $-\$1x^*$). The marginal utility of income is λ^*. Then the marginal effect of P_x on V is $\partial V/\partial P_x = \lambda^*[\partial h/\partial P_x] = \lambda^*[-x^*] = -\lambda^* x^* < 0$.

SUMMARY

In this chapter we developed value functions and the envelope theorem as tools for comparative static analyses. We considered both unconstrained and constrained optimization problems. In each kind of problem the agent's optimal behavior depends on the parameters that define her environment.

A value function is an objective function in which the agent's optimal behavior functions are substituted for her choice variables. Thus, a value function has as its arguments the parameters of the original problem.

The envelope theorem for an unconstrained problem states that the partial derivative of a value function with respect to any parameter is equal to the partial derivative (with respect to that parameter) of the associated objective function, holding the values of the choice variables constant at their optimal values. Thus, the marginal effect of a parameter on the optimal value of an objective function is equal to the effect of that parameter in shifting the graph of that objective function. Movements along the objective function caused by the choice variables responding to the change in the value of the parameter can be ignored. For a constrained problem, the envelope theorem states that the marginal effect on the value function of any parameter k is the marginal effect of k on the objective function plus the product of the Lagrange multiplier and the marginal effect of k on the constraint function.

We concluded by using the envelope theorem to show that the optimal value of the Lagrange multiplier is the marginal effect on the value function of a shift of the graph of the constraint function. If a parameter appears in the constraint, then a change in the value of that parameter will shift the graph of the constraint. This shift of the graph of the constraint will change the unallocated quantity of the constrained resource (which is income in our example of utility maximization). The value of λ^* is the marginal effect on the value function of a change in the unallocated quantity of the constrained resource.

Problems

13.1 A monopolist faces the (inverse) demand function $P = a - bQ$, in which P is the price, Q is the rate of output, and the parameters a and b are positive. The monopolist's cost function is $C(Q) = e + gQ + cQ^2$, in which the parameters e, g, and c are positive. Assume that $g < a$. Derive the monopolist's value function assuming that his objective is to maximize profit.

13.2 For Problem 13.1, use the envelope theorem to measure the marginal effect on the monopolist's maximal profit of a parallel upward shift of his marginal cost function. Illustrate your result with graphical analysis analogous to that in Section 13.2.2.

13.3 Using the envelope theorem, determine the relative values of the parameters such that the monopolist of Problem 13.1 would be indifferent between (a) shifting the (inverse) demand function upward, parallel to itself, by a distance of $1 and (b) reducing fixed costs by $1.

13.4 Derive the value function for a consumer whose utility function is $U(x, y) = a \ln x + b \ln y$, in which a and b are positive constants and x and y are the quantities of goods X and Y consumed. The consumer's income is the constant I, and the prices of the goods are the constants P_x and P_y.

13.5 For the consumer in Problem 13.4, derive expressions for the marginal utility of income and the marginal effects on the prices P_x and P_y on utility.

13.6 A consumer's utility function is $U(x, y) = xy^2$, in which x and y are the quantities of goods X and Y consumed. The consumer's income I is fixed at $900, and the fixed prices of the two goods are $P_x = \$10$ and $P_y = \$6$.

(a) Derive the consumer's value function in terms of the parameters I, P_x, and P_y. Evaluate that value function at the values of the parameters stated in the problem.

(b) Evaluate the marginal utility of income and the marginal effects of the two prices on utility, using the values for income and prices stated in the problem. Show that these marginal effects on utility are compatible with the optimal value of the Lagrange multiplier and the derivatives of the constraint function for this problem.

13.7 Consider the consumer of Section 6.9 who allocates time between labor and leisure to maximize utility according to the Stone-Geary utility function

$$U(C, T - L) = a \ln(C - C_0) + (1 - a) \ln(T - L),$$

in which a is a constant between 0 and 1, C is consumption, C_0 is the (exogenous) subsistence level of consumption, T is the (exogenous) total time available, and L is the amount of time spent working. The consumer chooses values for C and L to maximize utility subject to the budget constraint $C \leq wL + I$, in which w is the (exogenous) wage rate and I is the (exogenous) nonlabor income. The consumer's choice of L is also constrained by $L \leq T$.

Assume that $a = 0.7$, $T = 16$ hours, $w = \$4$ per hour, and $C_0 = 10$ units.

(a) Use the Lagrangian technique to determine the optimal values for C and L as functions of I. Specify the values for I that imply an interior solution.

(b) Derive the consumer's value function for those values of I that imply an interior solution.

Value Functions and the Envelope Theorem: Duality and Other Applications

14.1   INTRODUCTION

In this chapter we will develop the concept of duality as one application of value functions. We will also present a variety of other applications of value functions, several of which involve the envelope theorem.

Duality in economic theory is a relationship between two constrained optimization problems. If one of the problems requires constrained maximization, the other problem will require constrained minimization. The structure and solution of either problem can provide information about the structure and solution of the other problem. In fact, it is sometimes easier to solve one problem by first solving the related problem, then using the solution to the second problem to infer the solution to the first. If two problems are related through duality, we say that each problem is the **dual** of the other.[1]

We saw one example of duality when we considered linear programming in Chapter 11, but let us briefly consider the more general example from the theory of the consumer introduced in Chapter 9. There we examined the dual problems of constrained utility maximization and constrained expenditure minimization for a consumer. The analogous dual problems for a firm are to maximize output subject to a budget constraint and to minimize expenditure on inputs subject to a constraint on the level of output produced. We will examine duality in more detail in this chapter, with particular attention to the interpretations of the Lagrange multipliers.

We will use value functions and the envelope theorem throughout this chapter. In Chapter 13 we defined an agent's value function as her objective function evaluated at the optimal values of the choice variables. Since the agent's optimal choices are functions of those same parameters that define her environment, the arguments (or independent variables) of the value function are the parameters. The envelope theorem permits us to treat the choice variables as constants when evaluating the partial derivative of the

---

[1] Usually, the problem that is stated first is called the *primal* and the related problem is called its *dual*. This is an arbitrary distinction: under the concept of duality each problem is the dual of the other, so that the dual of the "dual" problem is the "primal" problem.

optimal value of the objective function with respect to any parameter. Specifically, for an unconstrained maximization[2] problem the envelope theorem states that the partial derivative of a value function with respect to any parameter k is equal to the partial derivative of the objective function with respect to k. The envelope theorem for a constrained problem adds to the partial derivative of the objective function the product of the Lagrange multiplier and the partial derivative of the constraint function with respect to k. For both constrained and unconstrained problems, the envelope theorem allows us to ignore the effect on the choice variables of a change in a parameter when we conduct comparative static analyses of the optimal values of objective functions.

In Section 14.2 we first use value functions to develop the concept of duality in general terms. We then present a specific example for a consumer who maximizes a Cobb-Douglas utility function subject to a budget constraint. This example will further illustrate the concept of duality for the consumer we first met in Chapter 9. In Section 14.3 we use value functions and the envelope theorem to establish Roy's identity, which states that optimal quantities demanded can be found as ratios of partial derivatives of a value function. If Roy's identity is used to derive quantities demanded by a consumer, the relevant value function is the indirect utility function. If the application is to quantities of inputs demanded by a firm, the relevant value function is the firm's expenditure function.

In Section 14.4 we present Shephard's lemma, which states that the partial derivative of an expenditure function with respect to any price is equal to the optimal value of the corresponding choice variable in the associated constrained minimization problem. Then in Section 14.5 we use the envelope theorem to derive the Slutsky equation, which resolves the effect of a change in the price of good X on the quantity demanded of any good Y (including good X as a special case) into substitution and income effects. We also show that Shephard's lemma is involved in the derivation of the Slutsky equation.

In Section 14.6 we concentrate on cost functions. We first use a three-input Cobb-Douglas production function to derive short- and long-run cost functions, and we show the relationship between these functions and discuss returns to scale. Next we present reciprocity relations, in which the partial derivative of the optimal value of one choice variable with respect to the price of a second choice variable is equal to the partial derivative of the optimal value of the second choice variable with respect to the price of the first choice variable. We then apply Shephard's lemma to cost functions. We conclude Section 14.6 by using value functions and the envelope theorem to show that the graph of a firm's long-run average cost function is an envelope of the graphs of its short-run average cost functions.

In Section 14.7 we analyze the optimal structure of a monopolist's two-part tariff. A two-part tariff consists of an entry fee T and an incremental price P. A consumer must pay the entry fee for the right to purchase the monopolist's good at a price equal to P per unit.[3] We conclude in Section 14.8 with the Ramsey tax problem, which is the question of how the government should structure unit tax rates on consumption goods so as to maximize consumers' aggregate utility subject to a constraint on the minimum amount of tax revenue collected.

---

[2]For convenience, we will continue to conduct the general discussion in terms of maximization problems. Specific applications will include both maximization and minimization problems. Either kind of problem can be constrained or unconstrained.

[3]Sometimes the entry fee is paid once, sometimes annually. Country clubs often use two-part tariffs. The annual dues serve as the entry fee. After paying the fee, a member may purchase meals during the following year at the prices set on the menu. Nonmembers may not purchase meals (except as guests of members). Another popular example is the sale of cameras, and charges for developing film, by a monopolist. The entry fee is the price of the camera. By purchasing the camera, the consumer obtains the right to purchase developed pictures at a cost of the film and processing fees (plus the cost of the time to take the pictures).

14.2 DUALITY

14.2.1 The Nature of Duality

In this section we will develop the general nature of duality by examining the relationship between constrained utility maximization and constrained expenditure minimization for a consumer. The duality relationships we will present are quite general; we use the dual problems of the consumer only to facilitate the exposition. In the next section we derive specific duality relationships for a consumer with a Cobb-Douglas utility function.

A consumer's expenditure function and his indirect utility function are the value functions for dual problems. An **expenditure function** specifies the minimum expenditure required to obtain a fixed level of utility, given the utility function and the prices of the consumption goods. An **indirect utility function** specifies the maximal utility obtainable, given the prices, income, and the utility function.

Each value function specifies the optimal value of an objective function subject to a constraint. If the value of the value function for either problem is substituted for the *level* of the constraint in the other problem, the value function for the second problem will be equal to the level of the constraint in the first problem. Furthermore, the optimal values for the choice variables will be the same for both problems. Finally, the optimal values of the Lagrange multipliers for the two problems will be reciprocals of each other.

Let $U = U(x, y; \boldsymbol{\alpha})$ be a utility function in which x and y are the rates of consumption for goods X and Y and $\boldsymbol{\alpha}$ is a vector of parameters. The consumer's fixed income is I and the fixed prices of the two goods are P_x and P_y. The Lagrangian for this problem is

$$\mathcal{L}(x, y, \lambda; I, P_x, P_y, \boldsymbol{\alpha}) = U(x, y; \boldsymbol{\alpha}) + \lambda[I - P_x x - P_y y], \qquad \textbf{(14.1)}$$

in which λ is the Lagrangian multiplier for the budget constraint. Assume that the constraint is binding and that the optimal solution is an interior one. We showed in Chapter 9 that the first-order conditions are

$$\frac{\partial \mathcal{L}}{\partial x} = \frac{\partial U}{\partial x} - \lambda^* P_x = 0$$

$$\frac{\partial \mathcal{L}}{\partial y} = \frac{\partial U}{\partial y} - \lambda^* P_y = 0 \qquad \textbf{(14.2)}$$

$$\frac{\partial \mathcal{L}}{\partial \lambda} = I - P_x x = P_y y = 0.$$

The consumer will choose an indifference curve that is tangential to the budget line defined by the fixed income I and prices P_x and P_y. In Figure 14.1a we show the solution to the constrained maximization problem. The highest level of utility obtainable, given the budget constraint, is U^*. The consumption bundle is determined by optimal choice functions $x^* = x^*(I, P_x, P_y, \boldsymbol{\alpha})$ and $y^* = y^*(I, P_x, P_y, \boldsymbol{\alpha})$. These choice functions are *Marshallian* demand functions because they specify the utility-maximizing rates of consumption for a fixed level of income and fixed prices.[4]

---

[4]In Section 14.2.2 we will derive a set of Marshallian demand functions using a specific utility function.

FIGURE 14.1

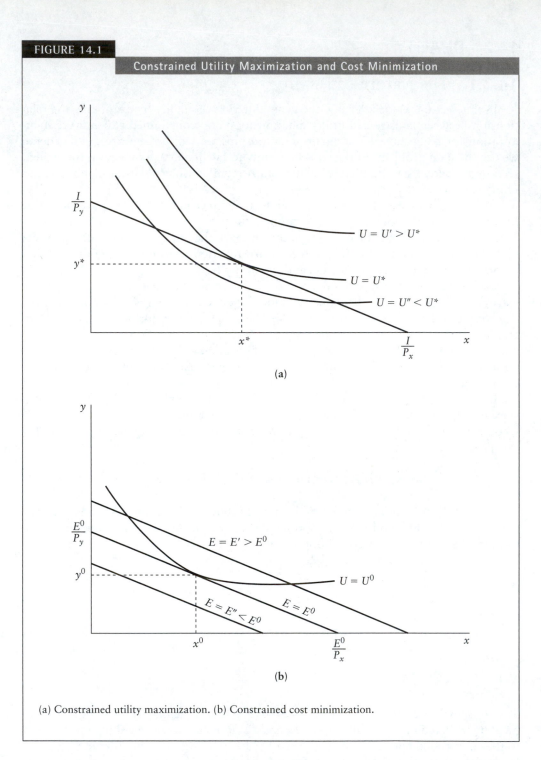

(a) Constrained utility maximization. (b) Constrained cost minimization.

The value function for this constrained maximization problem is the consumer's indirect utility function $V(I, P_x, P_y, \boldsymbol{\alpha})$, which satisfies

$$V(I, P_x, P_y, \boldsymbol{\alpha}) = U(x^*, y^*; \boldsymbol{\alpha}) = U^*. \qquad \textbf{(14.3)}$$

Now consider the dual problem of constrained cost minimization. Let the consumer minimize his expenditure on goods X and Y subject to the constraint that he obtains a level of utility at least equal to U^0. The Lagrangian[5] is now

$$\mathcal{L}(x, y, \mu; U^0, P_x, P_y, \alpha) = P_x x + P_y y + \mu[U^0 - U(x, y; \alpha)], \qquad \textbf{(14.4)}$$

in which μ is the Lagrangian multiplier for the utility constraint. Again assuming that the constraint is binding and the solution is interior, the first-order conditions are

$$\frac{\partial \mathcal{L}}{\partial x} = P_x - \mu \frac{\partial U}{\partial x} = 0$$

$$\frac{\partial \mathcal{L}}{\partial y} = P_y - \mu \frac{\partial U}{\partial y} = 0 \qquad \textbf{(14.5)}$$

$$\frac{\partial \mathcal{L}}{\partial \mu} = U^0 - U(x, y; \alpha) = 0.$$

These conditions require the consumer to choose a budget line that is tangential to the indifference curve for the fixed utility U^0. In Figure 14.1b we see the solution to the constrained minimization problem. The consumption bundle is determined by optimal choice functions $x^0 = x^0(U^0, P_x, P_y, \alpha)$ and $y^0 = y^0(U^0, P_x, P_y, \alpha)$. These are called *Hicksian* functions because utility, rather than income, is constant.

The value function for the constrained minimization-problem is the consumer's expenditure function

$$E(U, P_x, P_y, \alpha) = P_x x^0(U, P_x, P_y, \alpha) + P_y y^0(U, P_x, P_y, \alpha). \qquad \textbf{(14.6)}$$

For the example in Figure 14.1b the expenditure function satisfies

$$E(U^0, P_x, P_y, \alpha) = P_x x^0(U^0, P_x, P_y, \alpha) + P_y y^0(U^0, P_x, P_y, \alpha) = E^0. \qquad \textbf{(14.7)}$$

The problems of constrained cost minimization and constrained utility maximization are analogous. The two prices are parameters in both problems. The fixed level of *income* is the third parameter in the utility-maximization problem; the fixed level of *utility* is the third parameter in the expenditure-minimization problem.

Three duality relationships connect the cost minimization and the utility maximization problems. Let I be the consumer's income and $V(P_x, P_y, I, \alpha) = U^*$ be the associated maximal level of utility, given the prices and other parameters. Now let U^* be the fixed (target) level, U^0, of utility for the cost-minimization problem. Then we have the following **duality relationships:**

1. The solution (x^*, y^*) to the utility-maximization problem and the solution (x^0, y^0) to the cost-minimization problem coincide,[6] as

$$x^*(P_x, P_y, I) = x^0(P_x, P_y, U^0)$$

$$y^*(P_x, P_y, I) = y^0(P_x, P_y, U_0). \qquad \textbf{(14.8)}$$

---

[5]To avoid excessive notation we will use the same symbol $\mathcal{L}$ for different Lagrangian functions if the context is sufficient to avoid confusion. We do, however, need a unique symbol for the Lagrange multiplier for each problem because, in general, the optimal values of λ and μ will not be the same.

[6]While the solution *values* for x^* in the maximization problem is equal to the optimal value x^0 in the minimization problem, and similarly the values for y^* and y^0 are equal, the optimal choice *functions* are different. The function for x^* has I as an argument; the function for x^0 has U^* as an argument. In the following sections we will analyze a specific example that will show the difference between the functions $x^*(P_x, P_y, I)$ and $x^0(P_x, P_y, U^*)$.

2. The value for the value function for either problem will be equal to the level of the constraint in the other problem; that is,

$$E(P_x, P_y, U^*) = P_x x^0 + P_y y^0 = I$$
$$V^*(P_x, P_y, I) = U(x^*, y^*) = U^*. \tag{14.9}$$

In words, (14.9) states that if the prices are fixed at P_x and P_y, and if the consumer's income is equal to the *minimal* expenditure required to obtain the level of utility U^*, the *maximal* utility he can obtain is U^*. The converse also holds.

3. Finally, the optimal values of the two Lagrangian multipliers will be reciprocals, as

$$\mu^0 = \frac{1}{\lambda^*}. \tag{14.10}$$

We can obtain the last result by comparing the first-order conditions for the two problems. From (14.2) we have $\lambda^* = (\partial U/\partial x)/P_x = (\partial U/\partial y)/P_y$, with both marginal utilities evaluated at x^* and y^*. From (14.5) we have $\mu^0 = P_x/(\partial U/\partial x) = P_y/(\partial U/\partial y)$, with the marginal utilities evaluated at x^0 and y^0. Since $x^* = x^0$ and $y^* = y^0$, then $\lambda^* = 1/\mu^0$.

The economic interpretation of $\lambda^* = 1/\mu^0$ is based on the reciprocity of marginal benefit and marginal cost. Each of these Lagrangian multipliers measures the marginal effect on a value function of a change in the level of a constraint. If the problem is to maximize utility subject to a budget constraint, the level of the constraint is the level of income, so that λ^* is the marginal utility of income. In the dual problem, the level of the constraint is the target level utility, U^0, so that μ^0 is the marginal cost of utility. But the marginal cost of utility is the reciprocal of the marginal utility of income.

For example, if an extra \$1 of income, optimally spent, will enable the consumer to obtain an extra 4 units of utility, then (approximately) an extra \$0.25 in income will provide an extra 1 unit of utility. Hence, the marginal cost of utility is \$0.25, which is \$(1/4). Thus, $\mu^0 = \$0.25 = \$(1/4) = 1/\lambda^*$.

14.2.2 Duality for a Consumer with Cobb-Douglas Utility

In this section we will illustrate duality relationships for a consumer by solving constrained utility-maximization and constrained cost-minimization problems using a Cobb–Douglas utility function. Let the utility function be

$$U(x, y) = Ax^\alpha y^\beta, \tag{14.11}$$

in which A, α, and β are positive constants. The consumer's income is fixed at I. Using (14.2), the first-order conditions[7] for the solution to the constrained utility-maximization problem are

$$\frac{\partial U}{\partial x} - \lambda^* P_x = A\alpha x^{\alpha-1} y^\beta - \lambda^* P_x = 0$$
$$\frac{\partial U}{\partial y} - \lambda^* P_y = A\beta x^\alpha y^{\beta-1} - \lambda^* P_y = 0 \tag{14.12}$$
$$I - P_x x - P_y y = 0.$$

---

[7]The specific utility function in this example ensures that the constraint is binding and the solution is interior.

Rewriting the first two equations in (14.12) and then dividing the first equation by the second produces

$$\frac{\alpha y^*}{\beta x^*} = \frac{P_x}{P_y}, \tag{14.13}$$

in which x^* and y^* are the optimal values for x and y in the utility-maximization problem. Solving (14.13) for y^* and using the fact that $P_x x^* = I - P_y y^*$ from the budget constraint, we can obtain the demand function for good Y, and by symmetry the demand function for good X. The two functions are

$$\begin{aligned}
x^* &= x^*(I, P_x, \alpha, \beta) = \frac{I}{P_x}\left(\frac{\alpha}{\alpha+\beta}\right) \\
y^* &= y^*(I, P_y, \alpha, \beta) = \frac{I}{P_y}\left(\frac{\beta}{\alpha+\beta}\right).
\end{aligned} \tag{14.14}$$

The functions in (14.14) are the Marshallian demand functions, which specify the utility-maximizing consumption bundle, given the income, the prices, and the utility function.[8] The consumer's maximal utility is given by his value function

$$\begin{aligned}
U^* &= V(I, P_x, P_y, \alpha, \beta) = A(x^*)^\alpha (y^*)^\beta \\
&= A\left(\frac{I}{P_x}\frac{\alpha}{\alpha+\beta}\right)^\alpha \left(\frac{I}{P_y}\frac{\beta}{\alpha+\beta}\right)^\beta = A\left(\frac{I}{\alpha+\beta}\right)^{\alpha+\beta}\left(\frac{\alpha}{P_x}\right)^\alpha \left(\frac{\beta}{P_y}\right)^\beta.
\end{aligned} \tag{14.15}$$

Using either of the first two conditions in (14.12) and the value function (14.15), we can see that the consumer's marginal utility of income evaluated at the optimal consumption bundle is the optimal value of the Lagrangian multiplier. We have

$$\begin{aligned}
\lambda^* &= \frac{\frac{\partial U}{\partial x}}{P_x} = \frac{A\alpha(x^*)^{\alpha-1}(y^*)^\beta}{P_x} = \frac{A\alpha\left(\frac{I}{P_x}\frac{\alpha}{\alpha+\beta}\right)^{\alpha-1}\left(\frac{I}{P_y}\frac{\beta}{\alpha+\beta}\right)^\beta}{P_x} \\
&= \frac{A\alpha\left(\frac{I}{\alpha+\beta}\right)^{\alpha+\beta-1}\left(\frac{\alpha}{P_x}\right)^{\alpha-1}\left(\frac{\beta}{P_y}\right)^\beta}{P_x} = A\left(\frac{I}{\alpha+\beta}\right)^{\alpha+\beta-1}\left(\frac{\alpha}{P_x}\right)^\alpha \left(\frac{\beta}{P_y}\right)^\beta = \frac{\partial U^*}{\partial I}.
\end{aligned} \tag{14.16}$$

Now consider this consumer's dual problem of constrained cost minimization. Let U^0 be a fixed level of utility. Using (14.5), the first-order conditions are

$$\begin{aligned}
P_x - \mu\frac{\partial U}{\partial x} &= P_x - \mu A\alpha x^{\alpha-1}y^\beta = 0 \\
P_y - \mu\frac{\partial U}{\partial y} &= P_y - \mu A\beta x^\alpha y^{\beta-1} = 0 \\
U^0 - U(x, y) &= U^0 - Ax^\alpha y^\beta = 0.
\end{aligned} \tag{14.17}$$

[8]In general, the quantity demanded of each good will depend on the prices of both goods. In (14.14) the quantity demanded of a good depends only on its own price because of the simple nature of the utility function.

Proceeding as we did to obtain (14.13), the first two conditions in (14.17) imply that the optimal values x^0 and y^0 for the cost-minimization problem satisfy

$$\frac{\alpha}{\beta} \frac{y^0}{x^0} = \frac{P_x}{P_y}. \tag{14.18}$$

If we use (14.18) to substitute for y^0 in the third condition in (14.17), we obtain

$$U^0 = A(x^0)^{\alpha+\beta} \left(\frac{\beta}{\alpha} \frac{P_x}{P_y} \right)^{\beta}. \tag{14.19}$$

Solving (14.19) for x^0 we find

$$(x^0)^{\alpha+\beta} = \frac{U^0}{A} \left(\frac{\alpha}{\beta} \frac{P_y}{P_x} \right)^{\beta},$$

so that

$$x^0 = \left[\frac{U^0}{A} \left(\frac{\alpha}{\beta} \frac{P_y}{P_x} \right)^{\beta} \right]^{1/(\alpha+\beta)}, \tag{14.20a}$$

and by symmetry

$$y^0 = \left[\frac{U^0}{A} \left(\frac{\beta}{\alpha} \frac{P_x}{P_y} \right)^{\alpha} \right]^{1/(\alpha+\beta)}. \tag{14.20b}$$

These are *Hicksian* demand functions; they specify the cost-minimizing consumption bundle that will provide the utility U^0, given the prices and the utility function.

If the optimal value of the objective function in the maximization problem is used as the level of the constraint in the minimization problem, then

1. The optimal consumption bundles for the two problems will be identical.
2. The optimal value of the objective function for the second problem will be equal to the level of the constraint in the first problem.
3. The values of the Lagrange multipliers will be reciprocals of each other.

To show these properties, we let U^* in (14.15) be the optimal level of utility in the utility-maximization problem when income is fixed at I. Then we let U^0 be the level of the utility constraint in the cost-minimization problem. If we set U^* equal to U^0, we can show that the optimal quantity x^* of good X for the maximization problem is equal to the optimal quantity x^0 of good X for the minimization problem. Similarly, for good Y, the optimal quantities y^* and y^0 are equal to each other. Equations (14.20a and b) define the cost-minimizing quantities x^0 and y^0 as functions of the target utility U^0 and other parameters. Using (14.20a and b) to evaluate the functions x^0 (U^0) and y^0 (U^0) when $U^0 = U^*$, we have

$$x^0(U^*) = \left[\frac{U^*}{A} \left(\frac{\alpha}{\beta} \frac{P_y}{P_x} \right)^{\beta} \right]^{1/(\alpha+\beta)}$$

$$\tag{14.21}$$

$$y^0(U^*) = \left[\frac{U^*}{A} \left(\frac{\beta}{\alpha} \frac{P_x}{P_y} \right)^{\alpha} \right]^{1/(\alpha+\beta)}.$$

Now using (14.15) to substitute for U^* in the expression for $x^0(U^*)$ and gathering terms with common exponents, we have

$$x^0(U^*) = \left[\frac{A}{A} \left(\frac{I}{P_x} \frac{\alpha}{\alpha+\beta} \right)^\alpha \left(\frac{I}{P_y} \frac{\beta}{\alpha+\beta} \right)^\beta \left(\frac{\alpha}{\beta} \frac{P_y}{P_x} \right)^\beta \right]^{1/(\alpha+\beta)},$$

or

$$x^0(U^*) = \left[\left(\frac{I}{P_x} \frac{\alpha}{\alpha+\beta} \right)^{\alpha+\beta} \left(\frac{\beta}{P_y} \frac{P_y}{\beta} \right)^\beta \right]^{1/(\alpha+\beta)},$$

so that

$$x^0(U^*) = \frac{I}{P_x} \left(\frac{\alpha}{\alpha+\beta} \right) = x^*(I) \qquad (14.22)$$

upon using (14.14). By symmetry, we could show that the optimal quantities for good Y satisfy $y^0(U^*) = y^*(I)$.

We emphasize the distinction between the optimal choice *functions* and particular optimal *values* of those functions. For good X, for example, the optimal functions $x^* = x^*(I, P_x, \alpha, \beta)$ in (14.14) and

$$x^0 = \left[\frac{U^0}{A} \left(\frac{\alpha}{\beta} \frac{P_y}{P_x} \right)^\beta \right]^{1/(\alpha+\beta)}$$

in (14.20) are obviously different functions. When U^0 is set equal to U^* and U^* is the maximal level of utility for the maximization problem, given I and the other parameters, the utility-maximizing and cost-minimizing quantities of good X are the same. That is, $x^*(I) = x^0(U^*)$. Similarly, $y^*(I) = y^0(U^*)$.

The optimal level of expenditure in the cost-minimization problem is $E^0 = P_x x^0 + P_y y^0$. Using (14.22), we have

$$E^0 = P_x x^0 + P_y y^0 = P_x \left(\frac{I}{P_x} \frac{\alpha}{\alpha+\beta} \right) + P_y \left(\frac{I}{P_y} \frac{\beta}{\alpha+\beta} \right)$$

$$= I \left(\frac{\alpha}{\alpha+\beta} + \frac{\beta}{\alpha+\beta} \right) = I, \qquad (14.23)$$

so that the optimal value of the objective function in the cost-minimization problem is equal to the level of the constraint in the utility-maximization problem. This is the second duality property.

Using the first equations of (14.12) and (14.17) and the fact that $x^* = x^0$, we see that the optimal values of the Lagrangian multipliers satisfy

$$\lambda^* = \frac{\frac{\partial U}{\partial x}\big|_{x=x^*}}{P_x} = \frac{\frac{\partial U}{\partial x}\big|_{x=x^0}}{P_x} = \frac{1}{\mu^0}, \qquad (14.24)$$

which establishes the third duality property.

We have just demonstrated that the three duality properties for the consumer hold when we substitute the optimal value of the objective function in the maximization problem for the level of the constraint in the minimization problem. In Problem 14.1 at the end of this chapter you may show that the three duality properties hold if we proceed from the minimization problem to the maximization problem. That is, duality relationships are symmetric.

Sometimes it is easier to solve a given problem by first solving its dual and then inferring the solution to the original problem using duality properties. The foregoing analysis of a consumer is an example of this. Define the *primal* problem as that of *minimizing* the cost of obtaining the fixed level of utility U^0. The *dual* is then the constrained utility *maximization* problem. We know that the demand functions x^0 and y^0 for the primal problem are complicated exponential functions. The demand functions x^* and y^* for the dual problem are quite simple. Using the value function (14.15) to define U^* as a function of I (and the other parameters), we used (14.22) to generate the demand functions for the maximization problem from the demand functions for the minimization problem. We could have gone in the opposite direction using an expenditure function as the value function.

14.3   ROY'S IDENTITY

Roy's identity states that the optimal (utility-maximizing) quantity demanded of good X is equal to a ratio of partial derivatives of the consumer's value function. We will derive Roy's identity in general by applying the envelope theorem to the value function for a utility-maximization problem. We will then apply the result to the specific utility function used in the preceding section.

Let $U(x, y; \alpha)$ be a utility function in which x and y are the quantities demanded of goods X and Y and α is a vector of parameters. Let the budget constraint be $I - P_x x - P_y y \geq 0$, in which I is the consumer's income and P_x and P_y are the prices. The consumer's problem is

$$\text{Maximize}_{x, y} \quad U = U(x, y; \alpha)$$
$$\text{subject to} \quad g_0 - g(x, y; \alpha) = I - P_x x - P_y y \geq 0. \tag{14.25}$$

The Lagrangian for this problem is

$$\mathscr{L}(x, y, \lambda; \alpha) = U(x, y; \alpha) + \lambda(I - P_x x - P_y y). \tag{14.26}$$

From Chapter 13, the consumer's value function is the indirect utility function

$$V(\alpha) = U(x^*(\alpha), y^*(\alpha); \alpha). \tag{14.27}$$

Using the envelope theorem for the parameters I and P_x, we find

$$\frac{\partial V}{\partial I} = \frac{\partial U}{\partial I} + \lambda^* \frac{\partial}{\partial I}(I - P_x x - P_y y) = 0 + \lambda^* = \lambda^* \tag{14.28}$$

and

$$\frac{\partial V}{\partial P_x} = \frac{\partial U}{\partial P_x} + \lambda^* \frac{\partial}{\partial P_x}(I - P_x x - P_y y) = 0 + \lambda^*(-x^*) = -\lambda^* x^*, \tag{14.29}$$

recognizing that neither I nor P_x appears in the objective function U.

Combining (14.28) and (14.29) produces (14.30).

Roy's Identity

$$x^* = -\frac{\partial V / \partial P_x}{\partial V / \partial I},\qquad\qquad (14.30)$$

That is, the optimal quantity demanded of good X is the negative of the ratio of two partial derivatives of the value function. The numerator of this ratio is the marginal effect of the price of good X on the maximal utility the consumer can obtain, given prices and income; the denominator is the marginal utility of income. The demand for good X is positive because the marginal effect of P_x on V is negative and the marginal utility of income is positive. Of course, both marginal effects are measured at the point $x = x^*$ and $y = y^*$.

Now we can use Roy's identity to obtain the specific (Marshallian) demand functions for the consumer we analyzed in Section 14.2.2. That consumer's value function is (14.15), which we repeat for convenience as (14.31):

$$U^* = V(I, P_x, P_y, \alpha, \beta) = A \left(\frac{I}{\alpha + \beta} \right)^{\alpha + \beta} \left(\frac{\alpha}{P_x} \right)^{\alpha} \left(\frac{\beta}{P_y} \right)^{\beta}. \qquad (14.31)$$

Applying Roy's identity for x^* yields

$$x^* = -\frac{\frac{\partial V}{\partial P_x}}{\frac{\partial V}{\partial I}} = -\frac{A \left(\frac{I}{\alpha+\beta} \right)^{\alpha+\beta} \alpha \left(\frac{\alpha}{P_x} \right)^{\alpha-1} \left(\frac{-\alpha}{(P_x)^2} \right) \left(\frac{\beta}{P_y} \right)^{\beta}}{(\alpha + \beta) A \left(\frac{I}{\alpha+\beta} \right)^{\alpha+\beta-1} \left(\frac{1}{\alpha+\beta} \right) \left(\frac{\alpha}{P_x} \right)^{\alpha} \left(\frac{\beta}{P_y} \right)^{\beta}}$$

$$= -\frac{\alpha \left(\frac{\alpha}{P_x} \right)^{-1} \left(\frac{-\alpha}{(P_x)^2} \right)}{\left(\frac{I}{\alpha+\beta} \right)^{-1}} = \frac{\alpha \left(\frac{P_x}{\alpha} \right) \left(\frac{\alpha}{(P_x)^2} \right)}{\frac{\alpha+\beta}{I}} \qquad (14.32)$$

$$= \frac{\alpha / P_x}{(\alpha + \beta)/I} = \frac{I}{P_x} \left(\frac{\alpha}{\alpha + \beta} \right),$$

which is identical to the result we obtained for x^* in (14.14) by solving the consumer's constrained utility-maximization problem directly. Replacing x^* by y^* and $[\partial V / \partial P_x]$ by $[\partial V / \partial P_y]$ in the first line of (14.32) will produce the expression for y^* in (14.14).

14.4 SHEPHARD'S LEMMA

Shephard's lemma states that the partial derivative of an expenditure function with respect either to any price or to the level of the constraint is equal to the optimal value of the corresponding choice variable in the Lagrangian for the associated constrained minimization problem. An expenditure function is a value function for a constrained minimization problem. The arguments of a consumer's expenditure function are the prices of the consumption goods and the constrained level of utility. The corresponding choice variables are the quantities of the consumption goods and the Lagrange multiplier for the constraint on utility.

For a firm, the arguments are factor prices and the constrained level of output. In this case, the choice variables are the quantities of inputs and the Lagrange multiplier for the output constraint. Shephard's lemma states that the derivative of an expenditure function with respect to the price of good i (or factor i) is equal to the optimal value of good i (or factor i). Furthermore, the derivative of the expenditure function with respect to the level of the constraint is equal to the optimal value of the Lagrange multiplier for that constraint. Notice the reversal of the roles of prices and quantities between the last sentence and the preceding one. The level of the constraint is a quantity and the Lagrange multiplier is a (shadow) price.

Consider a consumer who minimizes expenditure subject to a constraint on utility. The Lagrangian is

$$\mathcal{L}(x, y, \lambda; P_x, P_y, U^0, H) = P_x x + P_y y + \mu[U^0 - U(x, y; H)] \tag{14.33}$$

in which H is a vector of parameters.

The value function is the expenditure function

$$E(P_x, P_y, U^0, H) = P_x x^0 + P_y y^0, \tag{14.34}$$

in which x^0 and y^0 are the optimal values of the choice variables. Differentiating the value function with respect to P_x, P_y, and U^0, we have (14.35) by the envelope theorem.

DEFINITION ## 14.2

Shephard's Lemma

$$\frac{\partial V}{\partial P_x}(P_x, P_y, U^0, H) = x^0(P_x, P_y, U^0, H)$$

$$\frac{\partial V}{\partial P_y}(P_x, P_y, U^0, H) = y^0(P_x, P_y, U^0, H) \tag{14.35}$$

$$\frac{\partial V}{\partial U^0}(P_x, P_y, U^0, H) = \mu^0(P_x, P_y, U^0, H),$$

---

in which $\mu^0(P_x, P_y, U^0, H)$ is the optimal value of λ.

The significance of Shephard's lemma is that we can obtain Hicksian (or income-compensated) demand functions by differentiating the expenditure function. For example, when the prices are P_x and P_y, the function $x^0 = x^0(P_x, P_y, U^0, H)$ specifies the cost-minimizing quantity of good X *conditional* on achieving a level of utility at least U^0.

For a simple illustration of Shephard's lemma, consider the problem of minimizing the expenditure required to obtain the utility level U^0 if the utility function is $U(x, y) = xy$. The Lagrangian is identical to (14.33) with $U(x, y; H)$ replaced by xy. It is easy to show that the optimal choice functions are

$$x^0(P_x, P_y, U^0) = \left(\frac{P_y U^0}{P_x}\right)^{1/2}$$

$$y^0(P_x, P_y, U^0) = \left(\frac{P_x U^0}{P_y}\right)^{1/2} \tag{14.36}$$

$$\mu^0(P_x, P_y, U^0) = \left(\frac{P_x P_y}{U^0}\right)^{1/2}.$$

Then the value function for this problem is the expenditure function

$$E(P_x, P_y, U^0) = 2(P_x P_y U^0)^{1/2}. \tag{14.37}$$

We can illustrate Shephard's lemma for the price of good X by differentiating (14.37) with respect to P_x. We have

$$\frac{\partial}{\partial P_x} E(P_x, P_y, U^0) = \left(\frac{1}{2}\right) 2(P_x P_y U^0)^{-1/2}(P_y U^0) = \left(\frac{P_y U^0}{P_x}\right)^{1/2} = x^0(P_x, P_y, U^0).$$

$$\tag{14.38}$$

Finally, we consider Shephard's lemma for the level U^0 of the constraint. Differentiating (14.37) with respect to U^0 yields

$$\frac{\partial}{\partial U^0} E(P_x, P_y, U^0) = (P_x P_y U^0)^{-1/2}(P_x P_y) = \left(\frac{P_x P_y}{U^0}\right)^{1/2} = \mu^0(P_x, P_y, U^0). \tag{14.39}$$

The partial derivative of the expenditure function with respect to the level of the utility constraint is the marginal cost of utility. We know that the optimal value μ^0 of the Lagrange multiplier for the constrained cost-minimization problem is also the marginal cost of utility. Shephard's lemma in (14.39) for the level of the constraint confirms the equivalence of the two approaches to marginal cost.

14.5   THE SLUTSKY EQUATION

The Slutsky equation resolves the effect of a change in the price of a good on the quantity demanded of that good into two parts, called the *(pure) substitution effect* and the *income effect*. We will derive the Slutsky equation for a two-good economy by using the envelope theorem and duality.

Let a consumer maximize utility by allocating a fixed income I between goods X and Y whose fixed prices are P_x and P_y. The optimal choice functions $x^* = x^*(P_x, P_y, I)$ and $y^* = y^*(P_x, P_y, I)$ that solve this constrained maximization problem are the Marshallian demand functions that we derived in Section 14.2. The Hicksian demand functions $x^0 = x^0(P_x, P_y, U^0)$ and $y^0 = y^0(P_x, P_y, U^0)$ discussed in Section 14.2 minimize the consumer's expenditure, subject to obtaining the level of utility U^0. The indirect utility function $V(I, P_x, P_y, \alpha)$ is the value function for the utility-maximization problem, and the expenditure function $E(U^0, P_x, P_y, \alpha)$ is the value function for the expenditure-minimization problem. Then by duality

$$V[E(U^0, P_x, P_y, \alpha), P_x, P_y, \alpha] = U^0$$
$$E[V(I, P_x, P_y, \alpha), P_x, P_y, \alpha] = I. \tag{14.40}$$

The first equation in (14.40) states that, given the prices and the utility function, the *maximum* utility obtainable when the available income is the *minimum* expenditure required to achieve the utility U^0 is U^0. The second equation in (14.40) states the analogous property of duality in the other direction. Given the prices and the utility function, the *minimum* expenditure required to achieve the *maximum* level of utility permitted by an income of I is I.

Duality also implies that if the income and the level of utility jointly satisfy (14.40), the optimal consumption bundle is the same for both problems. That is,[9]

$$x^0(P_x, P_y, U^0) = x^*(P_x, P_y, I)$$
$$y^0(P_x, P_y, U^0) = y^*(P_x, P_y, I). \tag{14.41}$$

Using the second equation in (14.40), we may rewrite the first equation of (14.41) as the identity

$$x^0(P_x, P_y, U^0) = x^*[P_x, P_y, E(U^0, P_x, P_y, \boldsymbol{\alpha})]. \tag{14.42}$$

Differentiating (14.42) with respect to P_x using the chain rule, we have

$$\frac{\partial x^0}{\partial P_x} = \frac{\partial x^*}{\partial P_x} + \frac{\partial x^*}{\partial E}\frac{\partial E^0}{\partial P_x} = \frac{\partial x^*}{\partial P_x} + \frac{\partial x^*}{\partial I} \cdot x^*$$

or

$$\frac{\partial x^*}{\partial P_x} = \frac{\partial x^0}{\partial P_x} - x^*\frac{\partial x^*}{\partial I}, \tag{14.43}$$

which is the Slutsky equation for good x.

The substitution of x^* for the derivative $\partial E^0/\partial P_x$ is justified by Shephard's lemma. More fundamentally, the substitution is an implication of the envelope theorem. $E(U^0, P_x, P_y, \boldsymbol{\alpha})$, or equivalently $E[V(I, P_x, P_y, \boldsymbol{\alpha}), P_x, P_y, \boldsymbol{\alpha}]$, is the value function for the constrained cost-minimization problem. The objective function is $f(x, y) = P_x x + P_y y$ and the constraint is $g(x, y; \boldsymbol{\alpha}) \geq g^0$, or $U^0 - U(x, y; \boldsymbol{\alpha}) \leq 0$. Since P_x does not appear in the constraint, the envelope theorem implies that $\partial E/\partial P_x = \partial f/\partial P_x = x^0$ which by duality is equal to x^*.

In (14.43) the term $\partial x^*/\partial P_x$ on the left is a partial derivative of the *Marshallian* demand function for good X. This derivative is the total effect of an increase in P_x on the quantity demanded of good X, with the price of good Y and the income held constant. The level of utility is *not* held constant; an increase in P_x, with P_y and I held constant, will, in general, force the consumer to a lower indifference curve. The two terms on the right of (14.43) resolve this total effect on x^* into the substitution and income effects. The quantity x^0 is a *Hicksian* quantity demanded, which requires that the level of utility be held constant. Hence, the derivative $\partial x^0/\partial P_x$ is the pure substitution effect of the change in P_x. We show in Figure 14.2 that the magnitude $(\partial x^0/\partial P_x)dP_x$ is the movement in the x-dimension along the indifference curve for the level of utility U^0 when P_x changes by the amount dP_x.

The product term $-x^*(\partial x^*/\partial I)$ is the income effect of the increase in P_x. If the consumer is behaving optimally before P_x changes, an increase of \$1 in P_x reduces income by $\$1(x^0) = \$x^0$. Then $-x^*(\partial x^*/\partial I)$ is the (marginal) income effect of P_x on the (Marshallian) quantity demanded of good X.

Now we can apply the Slutsky equation to the (Marshallian) demand function for good X that we obtained in (14.14), repeated here as

$$x^* = \frac{I}{P_x}\left(\frac{\alpha}{\alpha + \beta}\right). \tag{14.44}$$

---

[9]Remember that $x^0(P_x, P_y, U^0)$ and $x^*(P_x, P_y, I)$ are different functions that have the same value when U^0 and I satisfy (14.40). The same is true for the functions for y^0 and y^*.

FIGURE 14.2

Substitution Effect for Good *X* Following an Increase in P_x

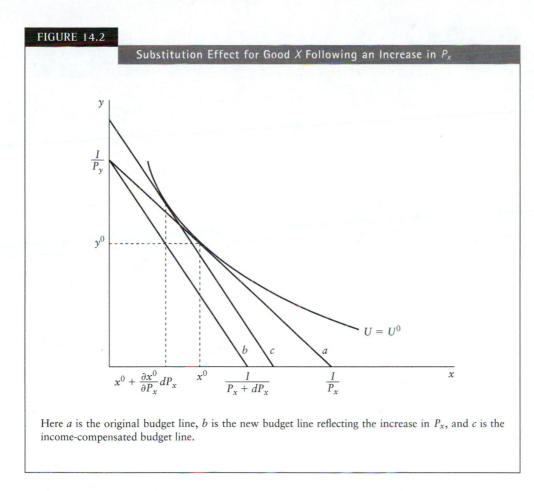

Here *a* is the original budget line, *b* is the new budget line reflecting the increase in P_x, and *c* is the income-compensated budget line.

Using (14.43) and (14.44), we have

$$\frac{\partial x^0}{\partial P_x} = \frac{\partial x^*}{\partial P_x} + x^* \frac{\partial x^*}{\partial I}$$

$$= -\frac{I}{(P_x)^2}\left(\frac{\alpha}{\alpha+\beta}\right) + \frac{I}{P_x}\left(\frac{\alpha}{\alpha+\beta}\right)\frac{1}{P_x}\left(\frac{\alpha}{\alpha+\beta}\right)$$

$$= -\frac{I}{(P_x)^2}\left(\frac{\alpha}{\alpha+\beta}\right)\left(1-\frac{\alpha}{\alpha+\beta}\right)$$

$$= -\left(\frac{\beta}{\alpha+\beta}\right)\frac{x^*}{P_x}, \tag{14.45}$$

which is the pure substitution effect on the demand for good *X* of a change in its price.

We have used the results in (14.45) to exhibit in Figure 14.3 the income and substitution effects on the demand for good *X* of a change in its price for a consumer whose utility function is $U(x, y) = Ax^\alpha y^\beta$.[10]

---

[10]In this section we have discussed the Slutsky equation for the effect of a change in the price of a good on the quantity demanded of *that* good. There are also Slutsky equations for cross-price effects to show the income and substitution effects of a change in the price of good *Y* on that demand for good *X*. We will use these cross-price effects when we discuss the Ramsey tax problem in Section 14.8.

FIGURE 14.3

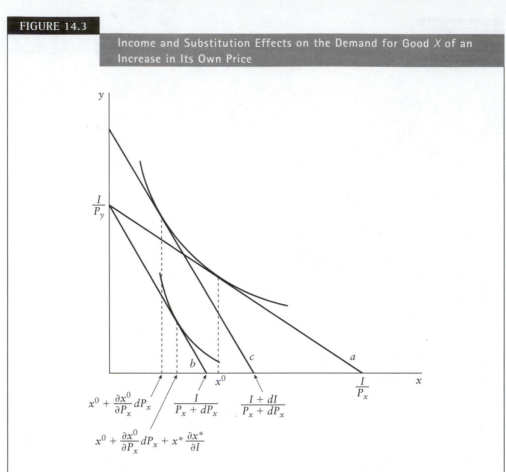

Here a is the original budget line, b is the new budget line reflecting the increase in P_x, and c is the income-compensated budget line.

14.6 COST FUNCTIONS FOR FIRMS: RECIPROCITY RELATIONS AND ENVELOPE CURVES

In this section we develop cost functions as value functions for a firm's constrained cost-minimization problem. To construct a cost function, a firm chooses quantities of inputs to minimize its expenditure on inputs subject to a constraint on the level of output. The resulting optimal choice functions determine the cost-minimizing input combination for the specified level of output. The value function for this problem has the input prices and the constrained level of output as its arguments. By allowing the level of output to vary, the value function specifies the minimum total cost as a function of the rate of output. We use a Cobb-Douglas production function to derive short-run cost functions, which we then use to construct long-run functions. Next we use the envelope theorem to show that the long-run average cost curve is an envelope of the short-run average cost curves. We conclude this section with reciprocity relations.

Let the firm have the following Cobb-Douglas production function:

$$Q = F(L, R, K, A, \alpha, \beta, \gamma) = AL^{\alpha} R^{\beta} K^{\gamma}, \qquad \textbf{(14.46)}$$

in which Q is the rate of output; L, R, and K are the quantities of labor, raw materials, and capital; and A, α, β, and γ are positive constants.[11] Finally, let P_L, P_R, and P_K be the factor prices, which are positive. We will now construct short- and long-run cost functions as value functions based on the production function in (14.46).

14.6.1 Short–Run Cost Functions

A short-run cost function is one in which the level of at least one input is fixed. Let the level of capital be fixed at K^0. To conserve notation, let $\mathbf{h} = (P_L, P_R, P_K, A, \alpha, \beta, \gamma)$ be the vector of factor prices and the parameters in the production function. For any output Q the Lagrangian for the firm's short-run cost-minimization problem is

$$\mathscr{L}(L, R, \mu; Q, K^0, \mathbf{h}) = P_L L + P_R R + P_K K^0 + \mu[Q - AL^\alpha R^\beta (K^0)^\gamma]. \quad \textbf{(14.47)}$$

Since the factor prices are positive, the constraint will be binding. Proceeding as we did in Chapter 9, we can show that the optimal choice functions for labor and raw materials are

$$L^0 = L^0(Q, K^0, \mathbf{h}) = \left[\frac{Q}{A} \left(\frac{P_R \alpha}{P_L \beta} \right)^\beta \left(\frac{1}{K^0} \right)^\gamma \right]^{1/(\alpha+\beta)}$$

$$R^0 = R^0(Q, K^0, \mathbf{h}) = \left[\frac{Q}{A} \left(\frac{P_L \beta}{P_R \alpha} \right)^\alpha \left(\frac{1}{K^0} \right)^\gamma \right]^{1/(\alpha+\beta)} . \qquad \textbf{(14.48)}$$

Clearly, the optimal level for each of the variable factors of labor and raw materials depends positively on the rate of output and the price of the other variable factor, and negatively on its own price and the fixed level of capital, just as we would expect. The price of capital does not appear in the functions for L^0 and R^0 because the price of a fixed factor is irrelevant to determining the optimal combination of the variable factors, given a constraint on output.[12]

The **short-run cost function** (SRC) when capital is fixed at K^0 is the value function for the Lagrangian (14.47); that is,

$$SRC(Q, K^0, \mathbf{h}) = P_L L^0(Q, K^0, \mathbf{h}) + P_R R^0(Q, K^0, \mathbf{h}) + P_K K^0, \qquad \textbf{(14.49)}$$

in which the optimal choice functions L^0 and R^0 are specified by (14.48). Let $\mu^0 = \mu^0(Q, K^0, L)$ be the optimal value of μ for the Lagrangian (14.47). Using the first-order conditions for the solution to (14.47), we can show that

$$\mu^0(Q, K^0, \mathbf{h}) = \frac{P_L}{\alpha A (L^0)^{\alpha-1}(R^0)^\beta (K^0)^\gamma} = \frac{P_R}{\beta A (L^0)^\alpha (R^0)^{\beta-1}(K^0)^\gamma} \qquad \textbf{(14.50)}$$

in which L^0 and R^0 are the optimal choice functions in (14.48).

---

[11]In (14.46) the symbol α is a single parameter, not a vector of parameters.

[12]The firm's fixed cost is $P_K K^0$. Therefore, the price of capital is relevant for determining the firm's *total* cost for producing the target level of output. Thus, the firm's capacity to produce a given Q profitably, even assuming that the variable costs with respect to labor and raw materials are minimized, does depend on P_K.

14.6.2 Long-Run Cost Functions

In the long run all inputs are variable. The long-run cost function is the value function of the Lagrangian

$$\mathscr{L}(L, R, K, \mu; Q, \mathbf{h}) = P_L L + P_R R + P_K K + \mu[Q - AL^\alpha R^\beta K^\gamma]. \qquad \textbf{(14.51)}$$

Note that (14.51) is the same as (14.47), except that K is now a choice variable rather than a parameter. Again using the techniques of Chapter 9, we can show that the solution to (14.51) consists of the optimal choice functions

$$L^* = L^*(Q, \mathbf{h}) = \left[\frac{Q}{A} \left(\frac{\alpha}{P_L} \right)^{\beta+\gamma} \left(\frac{P_R}{\beta} \right)^\beta \left(\frac{P_K}{\gamma} \right)^\gamma \right]^{1/(\alpha+\beta+\gamma)}$$

$$R^* = R^*(Q, \mathbf{h}) = \left[\frac{Q}{A} \left(\frac{\beta}{P_R} \right)^{\alpha+\gamma} \left(\frac{P_L}{\alpha} \right)^\alpha \left(\frac{P_K}{\gamma} \right)^\gamma \right]^{1/(\alpha+\beta+\gamma)}$$

$$K^* = K^*(Q, \mathbf{h}) = \left[\frac{Q}{A} \left(\frac{\gamma}{P_K} \right)^{\alpha+\beta} \left(\frac{P_L}{\alpha} \right)^\alpha \left(\frac{P_R}{\beta} \right)^\beta \right]^{1/(\alpha+\beta+\gamma)} \qquad \textbf{(14.52)}$$

$$\mu^* = \mu^*(Q, \mathbf{h}) = \frac{P_L}{\alpha A (L^*)^{\alpha-1} (R^*)^\beta (K^*)^\gamma}.$$

The Lagrange multiplier μ^* is the long-run marginal cost.[13] The **long-run cost function** (LRC) is the value function

$$\text{LRC}(Q, \mathbf{h}) = P_L L^* + P_R R^* + P_K K^*, \qquad \textbf{(14.53)}$$

in which L^*, R^*, and K^* are the optimal choice functions specified in (14.52).

We now assert (without proof) a relationship between the short-run and long-run cost functions that we will use in the next section on envelope curves: The long-run cost for any output Q is equal to the short-run cost for that Q when the level of capital is optimally adjusted to that Q. For any fixed level of capital K^0 the short-run cost function is the value function (14.49). The functions $L^0(Q; K^0, \mathbf{h})$ and $R^0(Q; K^0, \mathbf{h})$ in (14.49) specify the optimal levels of labor and raw materials to employ when capital is fixed at K^0 and the output is Q. In the long run all inputs, including capital, are optimally adjusted to the level of output. Therefore, $K^* = K^*(Q, \mathbf{h})$ in (14.52) is the value of K^0 that minimizes $\text{SRC}(Q; K^0, \mathbf{h})$ for a given output Q and set of parameters $\mathbf{h}$. Furthermore, we will have the following relationships between the short-run and the long-run optimal choice functions for L, R, and μ:

$$L^*(Q, \mathbf{h}) = L^0[Q; K^*(Q, \mathbf{h}), \mathbf{h}]$$
$$R^*(Q, \mathbf{h}) = R^0[Q; K^*(Q, \mathbf{h}), \mathbf{h}] \qquad \textbf{(14.54)}$$
$$\mu^*(Q, \mathbf{h}) = \mu^0[Q; K^*(Q, \mathbf{h}), \mathbf{h}].$$

The value functions for the short-run and long-run problems therefore satisfy

$$\text{LRC}(Q, \mathbf{h}) = \text{SRC}[Q; K^*(Q, \mathbf{h}), \mathbf{h}]. \qquad \textbf{(14.55)}$$

---

[13]Equivalently, $\mu^* = P_R/[\beta A(L^*)^\alpha (R^*)^{\beta-1}(K^*)^\gamma] = P_K/[\gamma A(L^*)^\alpha (R^*)^\beta (K^*)^{\gamma-1}]$.

14.6.3 The Long–Run Average Cost Curve as an Envelope Curve

We will use the envelope theorem to establish that a firm's long-run average cost curve is the envelope of its short-run average cost curves.

Each short-run average cost curve is associated with a fixed value for (at least) one of the firm's inputs. Let this input be capital and let K^0 and K', with $K^0 < K'$, be fixed levels for capital. We have drawn in Figure 14.4 the short-run average cost (SRAC) curves for the two fixed levels of capital and the long-run average cost (LRAC) curve.

We know from the preceding section that for any rate of output long-run cost is equal to short-run cost when the level of the fixed input is optimally adjusted to that rate of output. Let Q^0 be a fixed rate of output and let $K^0 = K^*(Q^0; \mathbf{h})$ be the optimal level of capital for Q^0. Then

$$\underset{K}{\text{Minimum}} \quad \text{SRC}(Q^0; K, \mathbf{h}) = \text{SRC}(Q^0; K^0, \mathbf{h}) = \text{LRC}(Q^0; \mathbf{h}). \qquad \textbf{(14.56)}$$

Note that the minimization in (14.56) is not a condition for being at the minimum point of a *given* short-run average cost curve. The minimization in (14.56) is with respect to K, not Q. To minimize with respect to K, we move vertically at $Q = Q^0$ from one short-run average cost curve to another until we find the lowest curve for that rate of output. In Figure 14.4 to minimize $\text{SRAC}(Q, K, \mathbf{h})$ with respect to K, holding Q fixed at Q^0, is to move from a curve like $\text{SRAC}(Q, K', L)$ to the curve $\text{SRAC}(Q, K^0, \mathbf{h})$. To minimize with respect to Q, on the other hand, is to move *along* a given curve (specified by the fixed level of K) until we find the minimum point of that curve.

FIGURE 14.4

Long–Run Average Cost Curve as an Envelope of Short–Run Average Cost Curves

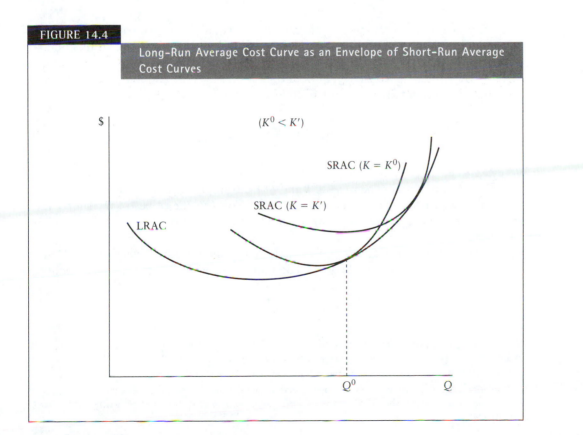

With K optimally adjusted to the output Q^0, the short-run and long-run costs are equal at Q^0. Therefore, the short-run and long-run average cost functions have the same height at Q^0.[14] To establish that the long-run average cost curve envelopes the short-run curves, we must show that the slopes of the two curves with respect to Q are equal at the point Q^0.

We will use a general relationship between marginal and average cost functions.[15] Let long-run average cost be $\text{LRAC}(Q, \mathbf{h})$. Then for any output Q

$$\text{LRAC}(Q, \mathbf{h}) = \frac{\text{LRC}(Q, \mathbf{h})}{Q}. \tag{14.57}$$

The slope of long-run average cost at any Q is

$$\frac{\partial}{\partial Q}\text{LRAC}(Q, \mathbf{h}) = \frac{\partial}{\partial Q}\left[\frac{\text{LRC}(Q, \mathbf{h})}{Q}\right] = \frac{Q\frac{\partial}{\partial Q}\text{LRC}(Q, \mathbf{h}) - \text{LRC}(Q, \mathbf{h})}{Q^2}$$
$$\tag{14.58}$$
$$= \frac{\frac{\partial}{\partial Q}\text{LRC}(Q, \mathbf{h})}{Q} - \frac{\text{LRAC}(Q, \mathbf{h})}{Q}.$$

Let $\text{LRMC}(Q, \mathbf{h})$ be long-run marginal cost, which is the derivative $\partial[\text{LRC}(Q, \mathbf{h})]/\partial Q$ in the last line of (14.58). Multiplying (14.58) by Q and solving for long-run marginal cost yields

$$\text{LRMC}(Q, \mathbf{h}) = \text{LRAC}(Q, \mathbf{h}) + Q\frac{\partial}{\partial Q}\text{LRAC}(Q, \mathbf{h}). \tag{14.59}$$

At any rate of output, long-run marginal cost is equal to long-run average cost plus the product of the rate of output and the *slope* of the long-run average cost function at that point.[16] The same relationship holds for short-run marginal and average cost. Let $\text{SRAC}(Q, K, \mathbf{h})$ and $\text{SRMC}(Q, K, \mathbf{h})$ be the short-run average and marginal cost functions, respectively, in which K is the fixed level of capital. Then

$$\text{SRMC}(Q, K, \mathbf{h}) = \text{SRAC}(Q, K, \mathbf{h}) + Q\frac{\partial}{\partial Q}\text{SRAC}(Q, K, \mathbf{h}). \tag{14.60}$$

We established in (14.56) that when $K = K^0$ and $Q = Q^0$, short-run and long-run average costs are equal. Then (14.59) and (14.60) imply that if short- and long-run marginal costs are also equal when $K = K^0$ and $Q = Q^0$, the slopes of the short- and long-run average cost functions (with respect to the rate of output) must be equal at that point.

To prove that short- and long-run marginal costs are equal when $K = K^0$ and $Q = Q^0$, we use the envelope theorem. The short-run cost function is a value function.

---

[14]This is true because average cost equals (total) cost divided by the rate of output. If short-run and long-run (total) costs are equal, then short- and long-run average costs must be equal at Q^0.

[15]This relationship holds for any marginal and average functions, not just cost functions.

[16]This relationship establishes the familiar property that a marginal cost function lies above (below) the associated average cost function when the latter is rising (falling). The two functions intersect at the minimum of average cost.

Value Functions and the Envelope Theorem: Duality and Other Applications

Using the envelope theorem, short-run marginal cost when $Q = Q^0$ and $K = K^0$ is

$$\left. SRMC(Q, K, \mathbf{h}) \right|_{Q=Q^0, K=K^0} = \left. \frac{\partial}{\partial Q} SRC(Q, K, \mathbf{h}) \right|_{Q=Q^0, K=K^0}$$

$$= \frac{\partial}{\partial Q}(P_L L^0 + P_R R^0 + P_K K^0) \qquad \textbf{(14.61)}$$

$$+ \mu^0 \frac{\partial}{\partial Q}[Q^0 - F(L^0, R^0, K^0)] = \mu^0.$$

We now show that long-run marginal cost at $Q = Q^0$ is also equal to μ^0. Recall that $K^0 = K^*(Q^0, \mathbf{h})$ is the optimal (long-run) level of capital when $Q = Q^0$. Using (14.56), long-run marginal cost at $Q = Q^0$ is

$$\left. LRMC(Q) \right|_{Q=Q^0} = \left. \frac{\partial}{\partial Q} LRC(Q, \mathbf{h}) \right|_{Q=Q^0} = \left. \frac{\partial}{\partial Q} SRC(Q, K, \mathbf{h}) \right|_{K=K^0, Q=Q^0}. \qquad \textbf{(14.62)}$$

The right-hand side of (14.62) requires total differentiation because K depends on Q in the long run. Using total differentiation, we find

$$\left. LRMC(Q) \right|_{Q=Q^0} = \left. \frac{\partial}{\partial Q} LRC(Q, \mathbf{h}) \right|_{Q=Q^0}$$

$$= \left[\frac{\partial}{\partial Q} SRC(Q, K, \mathbf{h}) + \frac{\partial}{\partial K} SRC(Q, K, \mathbf{h}) \frac{\partial K^*}{\partial Q}(Q, \mathbf{h}) \right]\Bigg|_{Q=Q^0, K=K^0}. $$

$$\textbf{(14.63)}$$

The first term in the brackets is equal to μ^0 by (14.61). The second term vanishes because (14.56) implies that the optimal level of capital K^0 for any output Q^0 satisfies the first-order condition

$$\left. \frac{\partial}{\partial K} SRC(Q, K, \mathbf{h}) \right|_{Q=Q^0, K=K^0} = 0. \qquad \textbf{(14.64)}$$

Then long-run marginal cost at Q^0 is

$$\left. LRMC(Q) \right|_{Q=Q^0} = \mu^0 + 0 \left. \frac{\partial K^*}{\partial Q}(Q, \mathbf{h}) \right|_{Q=Q^0} = \mu^0. \qquad \textbf{(14.65)}$$

Since short-run and long-run marginal costs at $Q = Q^0$ and $K = K^0$ are both equal to μ^0, (14.59) and (14.60) imply that the short-run and long-run average cost curves are tangential at $Q = Q^0$.

14.6.4 Reciprocity Relations

A **reciprocity relation** is an equality between certain kinds of partial derivatives of choice variables in a constrained optimization problem. Each choice variable is associated with a parameter, which is usually a price. For example, if the choice variable is the level of the labor input, the associated parameter is the wage rate. The reciprocity relation states that at the optimal solution of a constrained optimization problem, the partial derivative of any choice variable with respect to the parameter associated with a second choice variable is equal to the partial derivative of the second choice variable with respect to the parameter associated with the first variable.

Consider an example. Suppose that a firm has the production function $Q = F(L, K)$ and chooses quantities of labor and capital to minimize the cost of producing a fixed rate of output Q^0. The factor prices are P_L and P_K. Let the optimal choice functions be $L^* = L^*(P_L, P_K, Q^0, \mathbf{h})$ and $K^* = K^*(P_L, P_K, Q^0, \mathbf{h})$. The reciprocity relation states that the marginal effects on the optimal employment of each input with respect to a change in the price of the other input are equal. That is,

$$\frac{\partial K^*}{\partial P_L}(P_L, P_K, Q^0, \mathbf{h}) = \frac{\partial L^*}{\partial P_K}(P_L, P_K, Q^0, \mathbf{h}). \tag{14.66}$$

The optimal levels of capital and labor are, in general, sensitive to the values of all parameters, not just their own prices. The relation in (14.66) states that the marginal effect on the optimal level of capital of an increase in the price of *labor* is equal to the marginal effect on the optimal level of labor of an increase in the price of capital.

We can establish (14.66) by using the envelope theorem and Young's theorem. The Lagrangian for this cost-minimization problem is

$$\mathscr{L}(L, K, \lambda; Q^0, P_L, P_K, \mathbf{h}) = P_L L + P_K K + \mu[Q^0 - F(K, L; \mathbf{h})] \tag{14.67}$$

and the value function is the expenditure function

$$E(P_L, P_K, Q^0, \mathbf{h}) = P_L L^*(P_L, P_K, Q^0, \mathbf{h}) + P_K K^*(P_L, P_K, Q^0, \mathbf{h}). \tag{14.68}$$

The envelope theorem provides

$$\frac{\partial E}{\partial P_K} = K^* \quad \text{and} \quad \frac{\partial E}{\partial P_L} = L^*. \tag{14.69}$$

Differentiating each equation in (14.69) with respect to the price of the other factor and using Young's theorem, we have

$$\frac{\partial^2 E}{\partial P_K \partial P_L} = \frac{\partial K^*}{\partial P_L} \quad \text{and} \quad \frac{\partial^2 E}{\partial P_L \partial P_K} = \frac{\partial L^*}{\partial P_K}$$

so that

$$\frac{\partial K^*}{\partial P_L} = \frac{\partial L^*}{\partial P_K}. \tag{14.70}$$

Let us consider reciprocity relations in a broader context. In most constrained optimization problems a natural association exists between prices and quantities. Usually, if a choice variable is a quantity, its associated price will be one of the parameters, and conversely. The reciprocity relation in (14.70) equates two partial derivatives evaluated at the optimal solution. Each partial derivative is the marginal effect of the price of one good on the optimal quantity of a *different* good. The reciprocity relation differentiates quantities with respect to prices in the "reciprocal" manner shown in (14.70).

Another reciprocity relation is implied by the cost-minimization problem. One of the parameters is the constrained level of output Q, which is clearly a *quantity*. The associated choice variable is the Lagrangian multiplier, whose optimal value is μ^*. It is reasonable to consider μ^* as a price because μ^* is the marginal cost of an increase in Q, just as P_x is the marginal cost of an increase in K. Note that in this problem two of the choice variables (K and L) are quantities and their associated parameters are prices, while the third choice variable (μ) is a price whose associated parameter (Q) is a quantity.

The envelope theorem provides that

$$\frac{\partial E}{\partial Q^0} = \mu^* \quad \text{and} \quad \frac{\partial E}{\partial P_K} = K^*. \tag{14.71}$$

Using Young's theorem, we have

$$\frac{\partial^2 E}{\partial Q^0 \partial P_K} = \frac{\partial^2 E}{\partial P_K \partial Q^0}$$

so that

$$\frac{\partial \mu^*}{\partial P_K} = \frac{\partial K^*}{\partial Q^0}. \tag{14.72}$$

In words, the size of the vertical shift of the marginal cost curve, given an incremental increase in the price of capital, is equal to the increase in the optimal quantity of capital to employ, given an incremental increase in the constrained output.

Reciprocity relations are interesting as theoretical phenomena. These relations can also be useful in simplifying econometric studies by justifying *a priori* constraints (or assumptions) about properties of the equations or functions that the economist is trying to estimate.

Now we can apply reciprocity relations to the example of constrained cost minimization for the consumer we analyzed in Section 14.4. We repeat for convenience the optimal choice functions (14.36) as

$$x^0 = \left(\frac{P_y U^0}{P_x}\right)^{1/2}, \quad y^0 = \left(\frac{P_x U^0}{P_y}\right)^{1/2}, \quad \mu^0 = \left(\frac{P_x P_y}{U^0}\right)^{1/2}. \tag{14.73}$$

Differentiating each quantity demanded with respect to the price of the other good, we have

$$\frac{\partial x^0}{\partial P_y} = \frac{1}{2}\left(\frac{P_y U^0}{P_x}\right)^{-1/2}\left(\frac{U^0}{P_x}\right) = \frac{1}{2}\left(\frac{U^0}{P_y P_x}\right)^{1/2}$$

$$\frac{\partial y^0}{\partial P_x} = \frac{1}{2}\left(\frac{P_x U^0}{P_y}\right)^{-1/2}\left(\frac{U^0}{P_y}\right) = \frac{1}{2}\left(\frac{U^0}{P_x P_y}\right)^{1/2}, \tag{14.74}$$

which illustrates the reciprocity relation between consumption goods and their prices. We also have

$$\frac{\partial \mu^0}{\partial P_x} = \frac{1}{2}\left(\frac{P_x P_y}{U^0}\right)^{-1/2}\left(\frac{P_y}{U^0}\right) = \frac{1}{2}\left(\frac{P_y}{P_x U^0}\right)^{1/2}$$

$$\frac{\partial x^0}{\partial U^0} = \frac{1}{2}\left(\frac{P_y U^0}{P_x}\right)^{-1/2}\left(\frac{P_y}{P_x}\right) = \frac{1}{2}\left(\frac{P_y}{P_x U^0}\right)^{1/2}, \tag{14.75}$$

which is the reciprocity relation between good X and the constrained level of utility U^0.

14.7 TWO-PART TARIFFS

A **tariff** is a structure of prices under which a good is offered for sale.[17] The simplest tariff allows a consumer to purchase any number of units at a fixed price per unit. One

---

[17]A tax on imports is also called a tariff.

kind of tariff—a "quantity-discount" tariff—specifies that the price per unit of a good depends on the number of units purchased. We see an example of this kind of tariff when we purchase one can of soup for $1 or two cans for $1.50.

A pricing structure often used by monopolists is the **two-part tariff.** The consumer pays a fixed fee T for the *right* to purchase units at an incremental price per unit equal to P. Under this two-part tariff, the consumer must pay $T + Px$ to purchase x units. Thus, the *average* cost to purchase x units is $T/x + P$, and the *marginal* cost is P. If x is a rate of consumption per unit of time, then T is a fee per unit of time. For example, by paying annual dues of T at a country club, a member may play any number of rounds of golf during the following year by paying an additional cost of P per round. The annual dues serve as an "entry fee" required to participate in the monopolist's market.

We will use value functions and constrained maximization to analyze the optimal structure of a monopolist's two-part tariff. Briefly, the monopolist imposes the entry fee to capture the amount that consumers are willing to pay for the right to purchase his good at the incremental price P. The lower the value of P, the higher the entry fee the monopolist can charge. But for any given P, an excessively high value for T will cause consumers to abandon the monopolist's good, leaving him with zero profits.

Let there be a representative consumer who allocates a fixed income I between the monopolist's good and a composite good. Let x and y be the quantities of the monopolist's good and the composite good purchased. The price of the composite good is normalized at $1.[18]$ If the consumer purchases the monopolist's good, the entry fee has the effect of a lump-sum tax, reducing income from I to $I - T$. Let $V(I - T, P, \boldsymbol{\alpha})$ be the value function if the consumer purchases any of the monopolist's good. By definition, $V(I - T, P, \boldsymbol{\alpha})$ is the maximal utility obtainable by a consumer who allocates an income of $I - T$ between buying the monopolist's good at P per unit and the composite good at $1 per unit.

Let $U(x, y)$ be the utility function, and define $U^0 = U(0, I)$ as the utility obtained if the consumer allocates his entire income to the composite good. In choosing values for T and P, the monopolist must observe the constraint

$$V(I - T, P, \boldsymbol{\alpha}) \geq U^0; \qquad (14.76)$$

otherwise the consumer will abandon the monopolist's market.

Let the (Marshallian) demand function for the monopolist's good be $x = f(P)$. The monopolist's objective function is the profit function

$$Px + T - C(x), \qquad (14.77)$$

in which $C(x)$ is the cost function.

The monopolist's problem is to choose values for P and T to maximize (14.77) subject to (14.76). The Lagrangian is

$$\mathcal{L}(P, T, \lambda; \boldsymbol{\alpha}) = Px + T - C(x) + \lambda[V(I - T, P, \boldsymbol{\alpha}) - U^0]. \qquad (14.78)$$

---

[18]We can regard the *composite good* as income that the consumer allocates optimally among all other goods in the economy. The consumer solves a two-stage problem by first allocating income between the monopolist's good and the composite good, and then optimally distributing among all other goods in the economy the income allocated to the composite good.

The first-order conditions are[19]

$$\frac{\partial \mathcal{L}}{\partial P} = x + Pf'(P) - C'(x)f'(P) + \lambda\frac{\partial V}{\partial P} = 0$$

$$\frac{\partial \mathcal{L}}{\partial T} = 1 - \lambda\frac{\partial V}{\partial I} = 0 \qquad\qquad (14.79)$$

$$\frac{\partial \mathcal{L}}{\partial \lambda} = V(I - T, P, \boldsymbol{\alpha}) - U^0 = 0.$$

We can see that the first two conditions require the monopolist to produce the rate of output at which price equals marginal cost. Let $x^* = f(P^*)$ be the consumer's optimal quantity of the monopolist's good when the monopolist chooses the profit-maximizing values for P and T. The second condition in (14.79) requires $\lambda^* = 1/[\partial V/\partial I]$. Then by Roy's identity

$$x^* = -\frac{\partial V/\partial P}{\partial V/\partial I} = -\lambda^*\frac{\partial V}{\partial P}. \qquad\qquad (14.80)$$

Using (14.80) in the first condition of (14.79) establishes

$$P^* = \frac{\partial C(x^*)}{\partial x}, \qquad\qquad (14.81)$$

so that at the optimal rate of output price equals marginal cost. Given the optimal price, the third condition in (14.79) requires that the entry fee make the consumer indifferent between entering the monopolist's market or not.

A two-part tariff eliminates the allocative inefficiency of the monopolist because price and marginal cost are equal, just as they would be in perfect competition. If the monopolist used ordinary pricing, price would exceed marginal cost at the optimal output. A two-part tariff allows the monopolist to collect an entry fee that exceeds the profit he loses by expanding output to the point at which price equals marginal cost.

The second condition in (14.79) establishes Pareto optimality in the distribution of income between the consumer and the monopolist. The optimal value λ^* is the monopolist's marginal profit if the constraint on the consumer's utility is relaxed.[20] In this example the monopolist's utility function is his profit function. Then λ^* is the marginal utility for the monopolist of a one-unit reduction in the consumer's utility. Hence $1/\lambda^*$ is the additional utility the consumer can obtain if the monopolist forgoes one unit of profit.[21] The term $\partial V/\partial I$ is the consumer's marginal utility of income. Then the second condition in (14.79), $1/\lambda^* = \partial V/\partial I$, requires equality of the marginal utilities of profit for the monopolist and income for the consumer. Because the marginal utilities are equal, no transfer of income between the monopolist and the consumer can benefit one party without harming the other.

---

[19]The term $\partial V/\partial I$ is the partial derivative of the value function with respect to income. For the consumer who purchases the monopolist's good, income is $I - T$. Then $\partial V/\partial T = [\partial V/\partial I][\partial(I - T)/\partial T] = -\partial V/\partial I$. We assume a binding constraint and an interior solution.

[20]To relax the constraint in this problem would require reducing the value of U^0. The monopolist could then adjust P and T to drive $V(I - T, P, \boldsymbol{\alpha})$ down further, reducing the consumer's utility and transferring more wealth (utility) to himself.

[21]The consumer would gain $(1/\lambda^*)$ in utility if the constraint were tightened (U^0 increased) by enough to deprive the monopolist of one unit of profit.

The third condition in (14.79) specifies the optimal entry fee T^* if we know the consumer's value function. Consider a simple example. Let the consumer's utility function be $U(x, y) = xy + y$. The Lagrangian for the consumer if he purchases any of the monopolist's good is

$$\mathcal{L}(x, y, \mu) = xy + y + \mu(I - T - Px - y), \tag{14.82}$$

where the price of good Y is \$1. The first-order conditions for (14.82) are

$$\frac{\partial \mathcal{L}}{\partial x} = y - \mu P = 0, \qquad \frac{\partial \mathcal{L}}{\partial y} = x + 1 - \mu = 0, \qquad \frac{\partial \mathcal{L}}{\partial \mu} = I - T - Px - y = 0, \tag{14.83}$$

which yield the demand functions

$$x^* = \frac{I - T - P}{2P} \quad \text{and} \quad y^* = \frac{I - T + P}{2}. \tag{14.84}$$

Then the consumer's value function when he pays the entry fee is

$$
\begin{aligned}
V(I - T, P, \boldsymbol{\alpha}) = U(x^*, y^*) &= x^* y^* + y^* \\
&= \left(\frac{I - T - P}{2P} + 1 \right) \left(\frac{I - T + P}{2} \right) = \frac{(I - T + P)^2}{4P}.
\end{aligned} \tag{14.85}
$$

If the consumer avoids the monopolist's good, his utility is

$$U^0 = U(0, I) = I. \tag{14.86}$$

Substituting (14.85) into the third condition in (14.79), the optimal entry fee T^* is the solution to

$$\frac{(I - T + P^*)^2}{4P^*} - I = 0, \tag{14.87}$$

in which P^* is the value of the price at which the consumer's demand curve intersects the monopolist's marginal cost curve. That is, P^* and T^* solve (14.79).

14.8   THE RAMSEY TAX PROBLEM

We conclude this chapter with an application of the Lagrangian technique and the envelope theorem to a simplified version of the Ramsey tax problem.

The **Ramsey tax problem** is a constrained maximization problem in which the government must raise a fixed amount of revenue by imposing unit taxes on consumption goods in a way that will maximize consumers' aggregate utility. We will conduct our analysis in terms of a representative consumer. The government's objective function is the representative consumer's value function, which measures the maximum level of utility he can obtain given his income, the prices of the consumption goods, and the unit tax rates. The government's choice variables are the tax rates. Should the government tax all consumption goods at the same rate? Should "luxuries" be taxed more heavily than "necessities"? Should the tax rates be proportional to equilibrium quantities supplied and demanded?

We will consider a simple example.[22] The economy has two perfectly competitive industries that produce goods X and Y. Both industries have perfectly elastic (horizontal) supply functions. Thus, the long-run equilibrium prices P_x and P_y are unaffected by shifts in consumers' demand functions. Define the units of these goods so that $P_x = P_y = 1$.

The representative consumer supplies labor to the producers and purchases goods X and Y. Her only income is from the sale of labor. She allocates her limited time between labor and leisure, and then allocates the resulting income between goods X and Y. Her object is to maximize utility, which depends on the quantity of leisure and the quantities of consumption goods. Thus, the consumer solves a constrained maximization problem in which the parameters are the prices (inclusive of the tax rates) of the two consumption goods, the wage rate, and the endowment of time. There are two constraints. One prohibits the consumer from spending more on consumption than she earns by supplying labor. The second constraint is her endowment of time.

Let t_x and t_y be the unit tax rates. Then, because the supply functions for both goods are perfectly elastic and $P_x = P_y = 1$, the equilibrium prices of the goods, inclusive of the unit taxes, may be written as $Z_x = 1 + t_x$ and $Z_y = 1 + t_y$. Let w be the wage rate and J be the consumer's endowment of time. Finally, let $U(x, y, j)$ be the consumer's utility function, in which x and y are the quantities of goods X and Y consumed and j is the amount of time allocated to leisure. The consumer's Lagrangian is

$$\mathcal{L}(x, y, j, \mu, \eta; \mathbf{h}) = U(x, y, j) + \mu[w(J - j) - Z_x x - Z_y y] + \eta(J - j), \qquad \text{(14.88)}$$

in which the Lagrangian multipliers are μ for the budget constraint and η for the time constraint.

We assume that the consumer's optimal choices x^*, y^*, and j^* will make the budget constraint binding and the time constraint slack. Then μ^* will be positive and η^* will be zero. The consumer's value function, or indirect utility function, is

$$V(t_x, t_y, w, J) = U(x^*, y^*, j^*). \qquad \text{(14.89)}$$

Let R^0 be the amount of revenue that the government must collect from the representative consumer. The government seeks to

$$\begin{aligned} \underset{t_x, t_y}{\text{Maximize}} \quad & V(t_x, t_y, w, J) \\ \text{subject to} \quad & t_x x^* + t_y y^* = R^0. \end{aligned} \qquad \text{(14.90)}$$

The consumer's constrained maximization problem is incorporated in the government's constrained maximization problem because the government's objective function is the consumer's indirect utility function.

The Lagrangian for the government's problem is

$$L(t_x, t_y, \lambda; w, J, R^0) = V(t_x, t_y, w, J) + \lambda[t_x x^* + t_y y^* - R^0]. \qquad \text{(14.91)}$$

---

[22]Our purpose is not to present a complete and rigorous treatment of a problem, but to illustrate the use of the Lagrangian technique and the envelope theorem. For a more sophisticated analysis that relaxes several of the assumptions we make here, see Anthony B. Atkinson and Joseph E. Stiglitz, *Lectures on Public Economics*, McGraw-Hill, New York, 1980, pp. 370–382.

Assuming that the constraint will be binding, the necessary conditions are

$$
\frac{\partial \mathcal{L}}{\partial t_x} = \frac{\partial V}{\partial t_x} + \lambda \left(x^* + t_x \frac{\partial x^*}{\partial Z_x} + t_y \frac{\partial y^*}{\partial Z_x} \right) = 0
$$

$$
\frac{\partial \mathcal{L}}{\partial t_y} = \frac{\partial V}{\partial t_y} + \lambda \left(y^* + t_x \frac{\partial x^*}{\partial Z_y} + t_y \frac{\partial y^*}{\partial Z_y} \right) = 0 \qquad (14.92)
$$

$$
\frac{\partial \mathcal{L}}{\partial \lambda} = t_x x^* + t_y y^* - R^0 = 0
$$

where the price of good X is $Z_x = 1 + t_x$, and similarly for good Y.

The envelope theorem applied to the consumer's value function provides

$$
\frac{\partial V}{\partial t_x} = \mu^* \left(-x^* \frac{\partial Z_x}{\partial t_x} \right) = -\mu^* x^* \quad \text{and} \quad \frac{\partial V}{\partial t_y} = \mu^* \left(-y^* \frac{\partial Z_y}{\partial t_y} \right) = -\mu^* y^*. \qquad (14.93)
$$

Using (14.93), we may rewrite the first two conditions in (14.92) as

$$
t_x \frac{\partial x^*}{\partial Z_x} + t_y \frac{\partial y^*}{\partial Z_x} = \frac{-(\lambda^* - \mu^*)}{\lambda^*} \cdot x^*
$$

$$
t_x \frac{\partial x^*}{\partial Z_y} + t_y \frac{\partial y^*}{\partial Z_y} = \frac{-(\lambda^* + \mu^*)}{\lambda^*} \cdot y^*. \qquad (14.94)
$$

The partial derivatives in (14.94) can be expanded using the Slutsky equations:

$$
\frac{\partial x^*}{\partial Z_x} = S_{xx} - x^* \frac{\partial x^*}{\partial I}, \quad \frac{\partial y^*}{\partial Z_y} = S_{yy} - y^* \frac{\partial y^*}{\partial I},
$$

$$
\frac{\partial x^*}{\partial Z_y} = S_{xy} - y^* \frac{\partial x^*}{\partial I}, \quad \frac{\partial y^*}{\partial Z_x} = S_{yx} - x^* \frac{\partial y^*}{\partial I}. \qquad (14.95)
$$

In (14.95) S_{xx} is the (pure) substitution effect on the demand for good X of a change in its own price. The substitution effect on the compensated (Hicksian) demand for good X of a change in the price of good Y is S_{xy}. The terms S_{yy} and S_{yx} are defined analogously. The income effect on the demand for good X of a change in its own price is $-x^*[\partial x^*/\partial I]$ because an increase in the price of good X by \$1 reduces the consumer's income by x^*. The term $-y^*[\partial x^*/\partial I]$ is the income effect on the demand for good X of a change in the price of good Y. The other terms are defined analogously.

Substituting (14.94) into (14.95) and using the fact that the Slutsky derivatives are symmetric ($S_{xy} = S_{yx}$) due to Young's theorem, we have

$$
t_x S_{xx} + t_y S_{xy} = \theta x^* \quad \text{and} \quad t_x S_{yx} + t_y S_{yy} = \theta y^*, \qquad (14.96)
$$

in which θ is defined as

$$
\theta = -1 + t_x \frac{\partial x^*}{\partial I} + t_y \frac{\partial y^*}{\partial I} + \frac{\mu^*}{\lambda^*}. \qquad (14.97)
$$

The quantity $t_x S_{xx} + t_y S_{xy}$ in the first equation of (14.96) is the change in the income-compensated (Hicksian) quantity demanded of good X induced by imposing unit taxes on goods X and Y at rates t_x and t_y. The second equation has the same interpretation for good Y. Dividing the first equation in (14.96) by x^* and the second equation by y^* produces

$$
\theta = \frac{t_x S_{xx} + t_y S_{xy}}{x^*} = \frac{t_x S_{yx} + t_y S_{yy}}{y^*}. \qquad (14.98)
$$

This is the Ramsey tax result: the optimal unit tax rates equate the percentage changes in the (income-compensated) quantities demanded of the two goods. Note that the optimal tax rates need not be equal to each other, nor is it necessary to tax "luxuries" more heavily than "necessities."

Problems

14.1 Consider a consumer whose utility function is $U(x, y) = axy + by$, in which a and b are positive constants. Demonstrate that the solutions to this consumer's constrained utility-maximization problem and constrained cost-minimization problem satisfy the three duality properties specified in Section 14.2.1.

14.2 For the preceding problem specify the consumer's Marshallian and Hicksian demand functions.

14.3 Demonstrate Roy's identity for a consumer whose utility function is $U(x, y) = axy + by$. The consumer has a fixed income M, x and y are the quantities consumed of goods X and Y, and P_x and P_y are the prices of these goods.

14.4 Demonstrate Shephard's lemma for the consumer in the preceding problem. Let U^0 be the constrained level of utility.

14.5 Use the envelope theorem to demonstrate the Slutsky equation for the consumer of Problem 14.1.

14.6 Using the model of a monopolist's optimal two-part tariff in Section 14.7, find an expression to determine the optimal entry fee T in terms of the optimal price P and the consumer's income M if the consumer's utility function is

(a) $U(x, y) = Ax^\alpha y^\beta$, in which A, α, and β are positive constants

(b) $U(x, y) = axy + by$, in which a and b are positive constants.

14.7 Let the consumer's utility function be $U(x, y, j) = Ax^\alpha y^\beta j^\gamma$, in which x and y are the quantities of goods X and Y consumed, j is the quantity of leisure, and A, α, β, and γ are positive constants. The government wants to impose unit tax rates on goods X and Y so as to raise at least R^0 in revenue while maximizing the utility of a representative consumer. Determine the optimal unit tax rates in terms of the parameters.

15 Introduction to Dynamics: Theory

15.1   INTRODUCTION

So far in this book we have been concerned mostly with comparative static analysis: comparisons of old and new equilibrium values of endogenous variables when one or more exogenous variables change. But many of the examples used in economic theory classes also include a description of dynamics: how the endogenous variables adjust to their new equilibrium values. In addition, the validity of comparative static analysis depends on the character of the dynamic process of getting from one equilibrium to another—in particular, if the dynamic process is such that the new equilibrium will never be reached, the comparative static analysis is not very useful! This chapter introduces difference and differential equations as ways to analyze these dynamic processes. We will focus on three aspects of dynamic processes: identification of steady-state (equilibrium) values, analysis of whether those steady states are stable, and description of the time path of adjustment to new steady states, including how quickly the vicinity of the steady state is reached. An appendix to the chapter introduces dynamic programming with a simple example of a competitive firm choosing a time path of when to sell from its inventory.

One of the very first microeconomic models encountered in introductory economics courses is a simple demand-and-supply model. When consumers' incomes increase, this shifts the demand curve to the right and increases equilibrium price and output. This is the comparative static result. But classroom explanations also include a description of how price and output adjust to their new, higher levels. The higher incomes create excess demand, which leads firms to increase price, which leads to higher quantity supplied and lower quantity demanded, which reduces the excess demand. This process continues until the excess demand is completely eliminated. In this chapter we will see how to use difference equations to model this process. In particular, we will see under what conditions a market will arrive at new equilibrium values of price and output when an exogenous variable changes and under what conditions the market will be unstable and never get to the new equilibrium values. We will also investigate when excess demand in a market will be eliminated smoothly and when firms will overshoot in their adjustment, leading to prices and output that oscillate around equilibrium values before eventually reaching them.

Another example from introductory economics is the expenditure multiplier in macroeconomics. In the simplest model, an exogenous increase in aggregate demand (caused, for instance, by an increase in government purchases of goods and services) leads to more revenues for firms and more income for households as firms pay them for the factors of production needed to satisfy the higher aggregate demand. This in turn leads to higher consumption by households, which provides another round of increases in spending, revenue, and income, leading to yet another round of increased spending by consumers. The eventual increase in equilibrium output is, in the simple model, a multiple of the original exogenous increase in demand. More sophisticated models yield different increases in equilibrium output by changing the description of how various elements of aggregate demand change as the economy adjusts to the exogenous increase in demand, but the general description of the dynamic process that leads to the new equilibrium value of output is similar.

A third example is a model of economic growth. Most models of economic growth focus on investment in physical capital: current investment leads to a higher future capital stock, which leads to higher labor productivity and therefore more potential output. We will use differential equations to model this dynamic process and investigate the factors that influence the amount by which output grows in an economy and whether in response to an exogenous shock the economy adjusts smoothly to a new steady state.

We begin with discrete-time models, in which variables change their values in discrete time periods, for example, once a month or once a year. This is appropriate for many economic variables whose values do not in fact change continuously: the prices of most goods change only periodically; interest on bank account balances is typically added to the account balance once a month (though in many cases the interest accrues continuously); updates to the National Income and Product Accounts, including gross domestic product (GDP), are made quarterly. Some economic processes are truly discrete-time processes; many others are continuous, but, because data are only collected and published in discrete time intervals, discrete-time models are a good way to provide the theoretical underpinnings of empirical work. Discrete-time models lead to difference equations: the value of a variable in one time period is related to its value in the previous time period (as well as other variables, potentially). Most of our analysis is in the context of single-equation models, but we also describe how to analyze systems of difference equations. Appendix 15.1 describes the matrix algebra concepts of eigenvalues and eigenvectors needed to analyze systems of equations.

To model variables whose values change continuously, we use differential equations in which the rate of change through time of a variable is related to its value (as well as, potentially, other variables). Differential equations are similar in many respects to difference equations, but because difference equations are easier to work with initially, we will discuss differential equations after considering difference equations.

Appendix 15.2 is an introduction to dynamic programming. A dynamic programming problem consists of a sequence of periods over which an agent pursues an objective: for example, a firm maximizing profit. Two kinds of variables define each period: state variables and decision variables. The state variables for a given period are parameters in the agent's objective function for that period. The decision variables for that period are the arguments in that objective function. The essence of a dynamic programming problem is that the agent's decisions today affect the values of the state variables for future periods, and those state variables determine the maximal levels of the objective function that the agent can obtain in those future periods. The appendix focuses on the economic example of a firm choosing how much to sell (from a fixed inventory) in each time period, when the price in each period is a random variable and therefore unknown in advance.

15.2   DIFFERENCE EQUATIONS

After introducing notation and some basic concepts, we describe solutions to linear first-order difference equations. After an economic example, we introduce a useful graphical tool called a phase diagram. This leads into an introduction to nonlinear first-order difference equations. Then we characterize solutions to second-order linear difference equations and end our discussion of difference equations by discussing systems of linear difference equations.

15.2.1   Notation and the Concepts of Steady States and Stability

The notation x_t denotes the value that a variable x takes at a particular moment in time t. Thus for a discrete-time process a complete description of the values that x takes from an arbitrary starting time 0 to an ending time T is given by the sequence $\{x_t\} = x_0, x_1, \ldots, x_T$. Another way of depicting the sequence $\{x_t\}$ is with an equation that shows how the value of x at time t is related to the value of x at other time periods. This kind of equation is called a difference equation. A first-order difference equation is one in which the value of x at time t depends on the value of x in the previous time period (as well as other constants and variables) but not in any earlier time periods. For example,

$$x_t = ax_{t-1} + b \tag{15.1}$$

is a first-order difference equation. Since equation (15.1) is linear, it is a linear first-order difference equation. An example of a linear second-order difference equation is

$$x_t = ax_{t-1} + bx_{t-2} + c. \tag{15.2}$$

The process $\{x_t\}$ is said to **converge** if, given enough time, x eventually reaches a value which, in the absence of exogenous changes (in equation (15.1), this would mean changes to a or b), does not change in the future. For example, the process described by equation (15.1) converges if eventually x reaches a value x^* such that $x^* = ax^* + b$. Said another way, the limit of x_t as t gets large is x^*: $\lim_{t\to\infty} x_t = x^*$. If $\{x_t\}$ converges, then the difference equation that describes it is said to be **stable** and the value x^* is called the **steady-state value** of x. If there is no unique, finite, real limit to $\{x_t\}$, then the process is **divergent,** the difference equation is said to be **unstable,** and there is no steady state. It is possible for a process to have an unstable steady state: a value x^* such that $x^* = ax^* + b$ but that is reached only if the initial value x_0 happens to equal x^*. This steady state is unstable since if x_0 differs from x^* by even a little, the process will diverge away from x^*.

If the process described by equation (15.1) converges, then $x^* = ax^* + b$, which implies that the steady-state value of x is

$$x^* = \frac{b}{1-a}. \tag{15.3}$$

Clearly a cannot equal 1, so if in equation (15.1) $a = 1$, the process is divergent (except for the special case where $b = 0$). In fact, we will see later that equation (15.1) is stable only if $|a| < 1$.

The solution to a differential equation is a description of the values x takes in all time periods, as a function of time and the values of the parameters and exogenous variables in the equation. The solution also depends on the value x takes in one arbitrarily chosen time period, since the values of x in all other time periods are in general functions of each other. Usually the initial value x_0 is specified, but sometimes it is the terminal value x_T that is specified as part of the solution.

The notion of a steady-state value is essentially the same as the notion we have used throughout the book of an equilibrium value. The steady-state value is the value an endogenous variable takes as a function of exogenous parameters and variables. The difference is that we now explicitly recognize that x may not move to this new equilibrium value immediately. By modeling the adjustment process we can, in addition to finding the equilibrium value, investigate how long it takes to reach the new equilibrium in response to an exogenous change. We can also analyze what happens to the value of x during all the time periods in between when an exogenous change occurs and when the new steady state (equilibrium) is attained.

As an example, consider a simple Keynesian macroeconomic model for a closed economy. Output in time period t, Y_t, equals the sum of that period's consumption C_t, investment I_t, and government purchases G_t. Investment and government purchases are exogenous while consumption is a linear function of the previous time period's income: $C_t = c + a Y_{t-1}$ (so a is the marginal propensity to consume out of last period's income). Substituting in for consumption yields a difference equation in Y:

$$Y_t = b_t + a Y_{t-1} \tag{15.4}$$

where $b_t = c + I_t + G_t$. If investment and government purchases are assumed to be constant through time, then equation (15.4) takes the form of equation (15.1):

$$Y_t = a Y_{t-1} + b. \tag{15.5}$$

Thus the steady-state (equilibrium) level of output is $Y^* = b/(1-a)$. So if, say, government purchases increase marginally, then the steady-state level of output will increase by $\partial Y^*/\partial b = 1/(1-a)$. This is the simple multiplier formula familiar to most introductory macroeconomics students. But equation (15.5) allows us to investigate more fully the dynamics of how the economy adjusts to its new equilibrium. We now turn to the description of the full dynamics of a difference equation.

15.2.2 Linear First-Order Difference Equations: Solutions and Stability

In this section we derive the solutions to first-order difference equations, first by repeated substitution and then with an alternative technique that will be valuable when we turn to more complicated difference equations and to differential equations. We also discuss the conditions necessary for difference equations to be stable.

Repeated Substitution

The most intuitive way to solve a linear difference equation is through repeated substitution, also called repeated iteration. For the process described by equation (15.1),

$$x_t = a x_{t-1} + b = a(a x_{t-2} + b) + b = a(a(a x_{t-3} + b) + b) + b$$
$$= \cdots = a^t x_0 + (1 + a + a^2 + \cdots + a^{t-1})b = a^t x_0 + \frac{1 - a^t}{1 - a}b \tag{15.6}$$

where the final step is derived as follows: let $A = (1 + a + a^2 + \cdots + a^{t-1})$. Then

$$aA = (a + a^2 + \cdots + a^{t-1} + a^t) \quad \text{and}$$

$$A - aA = (1 + a + a^2 + \cdots + a^{t-1}) - (a + a^2 + \cdots + a^{t-1} + a^t) = 1 - a^t. \quad \textbf{(15.7)}$$

So $A(1 - a) = 1 - a^t$ and $A = (1 + a + a^2 + \cdots + a^{t-1}) = \dfrac{1 - a^t}{1 - a}$.

We can use equation (15.6) to show that $\{x_t\}$ converges if $|a| < 1$ (also for the uninteresting case of $b = 0$ and $a = 1$) and $\{x_t\}$ diverges if $|a| > 1$ (or if $a = -1$ or if $a = 1$ and $b \neq 0$). We begin by discussing the divergent cases. As t gets infinitely large, a^t becomes infinitely large and positive if $a > 1$, and a^t becomes infinitely large in absolute value if $a < -1$.[1] So if $b = 0$, equation (15.6) shows that x_t has no finite limit when $|a| > 1$. The sequence $\{x_t\}$ also diverges when $|a| > 1$ and $b \neq 0$ since $(1 - a^t)/(1 - a)$ becomes infinitely large in absolute value if either $a > 1$ or $a < -1$. If $a = 1$, then $x_t = x_{t-1} + b = (x_{t-2} + b) + b = \cdots = x_0 + tb$ which, as t gets infinitely large, tends to positive infinity if $b > 0$ and negative infinity if $b < 0$. (If $b = 0$, then $x_t = x_0$ for every time period, so the process converges since x always equals its initial value.) If $a = -1$, then there is no unique value for $\lim_{t \to \infty} x_t$ since a^t oscillates between negative and positive one. But if $|a| < 1$, the process converges since a^t approaches zero as t gets infinitely large. So, except for the uninteresting case of $a = 1$ and $b = 0$, $|a| < 1$ is the condition for a first-order linear difference equation to be stable.

The steady state of the first-order linear difference equation (15.1) was shown in equation (15.3). We can confirm that solution by taking the limit of equation (15.6): if $|a| < 1$, then $\lim_{t \to \infty} x_t = \lim_{t \to \infty}(a^t x_0 + b(1 - a^t)/(1 - a)) = 0 + b(1/(1 - a)) = b/(1 - a)$. So if the process converges, its steady-state value is $x^* = b/(1 - a)$.

We can derive another important result about first-order linear difference equations by substituting this steady-state value into equation (15.6):

$$x_t = a^t x_0 + \frac{1 - a^t}{1 - a} b = a^t x_0 + (1 - a^t) x^* = x^* - a^t(x^* - x_0). \quad \textbf{(15.8)}$$

So if $0 < a < 1$, x_t gradually and smoothly changes from its original value to its steady state value; that is, the process **converges monotonically**. But if $-1 < a < 0$, x_t **converges with oscillations** around the steady-state value, with the oscillations damping over time until the steady state is reached.[2] Figure 15.1 illustrates how x_t changes through time for several different values of a and b.

Figure 15.1 and equation (15.8) also indicate another important aspect of difference equations: how quickly the vicinity of the steady state is reached. From Figure 15.1 one might guess that values of a close to zero lead to rapid convergence, while values of a closer to 1 in absolute value lead to slower convergence. Equation (15.8) suggests why this is generally true. If a is close to zero, then so will be a^t, even for small values of t. So x_t will always be close to x^*, even when t is small. But if a is close to 1 in absolute value, a^t will also be close to 1 (or -1, if $a < 0$ and t is odd) for small values of t and it will take larger values of t to make a^t approach zero. So the closer a is to 0, the more quickly the process will tend to converge; the larger a is in absolute value (as long as it stays less than 1 in absolute value), the more slowly the process will tend to converge. The notation of equation (15.8) hides the fact, however, that the value of a affects the steady state value x^* as well as the value of a^t. So it is possible that a change

---

[1] If a is negative, a^t is positive when t is even and negative when t is odd.

[2] Equation (15.8) also shows what happens when the difference equation is unstable, though the notation is misleading since in that case there is no steady state. If $a > 1$ or $a < -1$, then x_t moves farther and farther away from its original value (oscillating if $a < -1$) and never reaches an equilibrium.

FIGURE 15.1

(a) x_t for $a = .7$, $b = 2$, $x_0 = 1$
(b) x_t for $a = .9$, $b = 2$, $x_0 = 1$
(c) x_t for $a = -.7$, $b = 2$, $x_0 = 1$
(d) x_t for $a = 1.1$, $b = 2$, $x_0 = 1$
(e) x_t for $a = -1.1$, $b = 2$, $x_0 = 1$

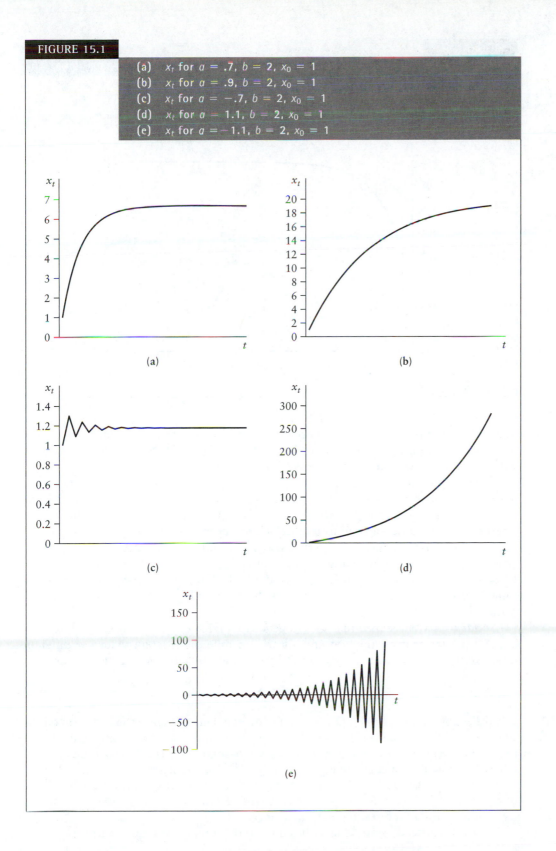

(a)

(b)

(c)

(d)

(e)

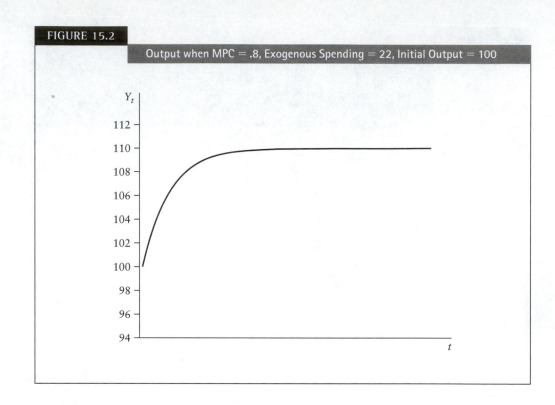

FIGURE 15.2

Output when MPC = .8, Exogenous Spending = 22, Initial Output = 100

in a will make a process converge more quickly by moving the steady-state value closer to the starting value. However, for a given difference between starting and steady-state values, a process will converge faster if a is close to zero.

Returning to the expenditure multiplier example, Figure 15.2 shows how, for a marginal propensity to consume of .8 and exogenous spending of 22, output adjusts from its initial value (set at 100) to its steady-state value. The initial value can be interpreted as the original equilibrium, which has been affected by an increase in government purchases of goods and services. For example, if investment equals 30, government purchases of goods and services equal 10, and the consumption function is $C_t = -20 + .8Y_{t-1}$, steady-state output would be 100. If government purchases rise (at time 0) to 12, then output will follow the path shown in Figure 15.2 to its new equilibrium value of 110.

So far we have assumed that the exogenous variable b is constant through time. It is easy to relax that assumption and allow b to take on different values at different times. In this case $x_t = ax_{t-1} + b_t$ which, by repeated substitution, has the solution

$$x_t = a^t x_0 + \sum_{s=0}^{t-1} a^s b_{t-s}. \tag{15.9}$$

Regardless of the values of b_t, $\{x_t\}$ will not converge if $|a| > 1$ since the first term on the right-hand side of equation (15.9) will get infinitely large. If $b_t = 0$ for all t and $|a| = 1$, then either $\{x_t\}$ will not converge because x_t will oscillate between positive and negative x_0 or else $\{x_t\}$ will be a constant equal to x_0. But if $|a| < 1$, then as t gets infinitely large, a^t will approach 0, so x_t will have a finite limit if the summation on the right-hand side of equation (15.9) does. So, as was the case when b was constant, $|a| < 1$ is the stability condition for all interesting first-order linear difference equations.

If $\{x_t\}$ has a steady state, it must be equal to the limit of equation (15.9):

$$x^* = \lim_{t \to \infty} \left(a^t x_0 + \sum_{s=0}^{t-1} a^s b_{t-s} \right) = \lim_{t \to \infty} \left(\sum_{s=0}^{t-1} a^s b_{t-s} \right). \tag{15.10}$$

A permanent marginal increase in b starting at time 0 will increase the steady-state level of x by $1/(1-a)$, just as in the case when b was constant. This can be shown as follows: $dx^* = \lim_{t\to\infty}(\sum_{s=0}^{t-1} a^s\, db_{t-s} + \sum_{s=0}^{t-1} sa^{s-1}b_{t-s}\, da)$ so, assuming that the changes in b are all equal, $da = 0$, and $|a| < 1$, $(dx^*/db)|_{a\,\text{constant}} = \lim_{t\to\infty}(\sum_{s=0}^{t-1} a^s)$. Thus, using equation (15.7), $(dx^*/db)|_{a\,\text{constant}} = \lim_{t\to\infty}((1-a^t)/(1-a)) = 1/(1-a)$.

An Alternative Approach

An alternative way of deriving the solution to a first-order difference equation will be helpful when analyzing higher-order difference equations as well as differential equations. Consider again the process described by equation (15.1). Its solution can be found by first finding the solution for the case of $b = 0$ and then considering the more general case.

If $b = 0$, then equation (15.1) can be rewritten as the **homogeneous equation**

$$x_t - ax_{t-1} = 0. \tag{15.11}$$

We begin by guessing that the solution to the homogeneous equation (15.11) takes the form $x_t = CA^t$; next we need to identify the constants C and A. If $x_t = CA^t$, then $x_{t-1} = CA^{t-1}$ so $x_t = Ax_{t-1}$. Using equation (15.11), $Ax_{t-1} = ax_{t-1}$ so $A = a$. The constant C remains to be determined.

The **general solution** to a first-order linear difference equation is the solution to the homogeneous equation (15.11) plus a constant that equals the steady-state value of the equation. This is clear for stable equations since the steady-state value must equal $\lim_{t\to\infty} x_t$; but if the equation is stable, then $|a| < 1$, which implies that $\lim_{t\to\infty} Ca^t = 0$. So to get $\lim_{t\to\infty} x_t$ to equal the steady-state value, the steady-state value must be added to Ca^t. We know the steady state from equation (15.3), so the general solution to equation (15.1) is

$$x_t = Ca^t + \frac{b}{1-a}. \tag{15.12}$$

To determine C, consider the initial value x_0. From equation (15.12), $x_0 = Ca^0 + b/(1-a)$, which implies that $C = x_0 - b/(1-a)$. This gives the **definite solution** to equation (15.1):

$$x_t = \left(x_0 - \frac{b}{1-a}\right)a^t + \frac{b}{1-a} = a^t x_0 + \left(\frac{b}{1-a}\right)(1-a^t), \tag{15.13}$$

equivalent to equation (15.6), which we derived in another way.

15.2.3 Demand–and–Supply Example

We now turn to another example familiar to introductory economics students: the adjustment to a new equilibrium in a linear demand-and-supply model. Let the demand function be

$$Q_t^d = a - bP_t \tag{15.14}$$

and the supply function be

$$Q_t^s = c + dP_t \tag{15.15}$$

where a, b, and d are positive constants; c might be either positive or negative, as long as the right-hand side of equation (15.15) is nonnegative. But assume that price does not adjust instantaneously when the market is out of equilibrium. Instead, price increases (decreases) proportionally to the amount of excess demand (supply) in the market in the previous period:

$$P_t - P_{t-1} = k(Q_{t-1}^d - Q_{t-1}^s) \tag{15.16}$$

where $k > 0$. Substituting equations (15.14) and (15.15) into equation (15.16) yields a first-order linear difference equation in P:

$$P_t = P_{t-1} + k((a - bP_{t-1}) - (c + dP_{t-1})) = (1 - k(b + d))P_{t-1} + k(a - c). \qquad \textbf{(15.17)}$$

Comparing this equation with the generic difference equation (15.1), it is clear that the dynamics of this system depend on the signs and magnitudes of $1 - k(b + d)$ and $k(a - c)$. Since k, b, and d are all assumed to be positive, $1 - k(b + d) < 1$, so the price process will converge as long as k is small enough so that $1 - k(b + d) > -1$; otherwise the market will never reach equilibrium unless P_0 happens to equal the equilibrium price and none of the constants ever change. Assuming that $1 - k(b + d) > -1$, equation (15.17) is stable and the steady-state price is

$$P^* = \frac{k(a - c)}{1 - (1 - (k(b + d)))} = \frac{k(a - c)}{k(b + d)} = \frac{(a - c)}{(b + d)}. \qquad \textbf{(15.18)}$$

This makes sense only if $a > c$; otherwise the steady-state price is negative, a situation shown in Figure 15.3. If $a < c$, the quantity intercept of the demand curve lies to the left of the quantity intercept of the supply curve, so the curves do not intersect at any nonnegative price.

Assuming that equation (15.17) is stable and that the steady-state price is positive, the definite solution to (15.17) is

$$\begin{aligned}
P_t &= (1 - k(b + d))^t P_0 + \frac{1 - (1 - k(b + d))^t}{k(b + d)} k(a - c) \\
&= (1 - k(b + d))^t P_0 + (1 - (1 - k(b + d))^t) \frac{(a - c)}{(b + d)}.
\end{aligned} \qquad \textbf{(15.19)}$$

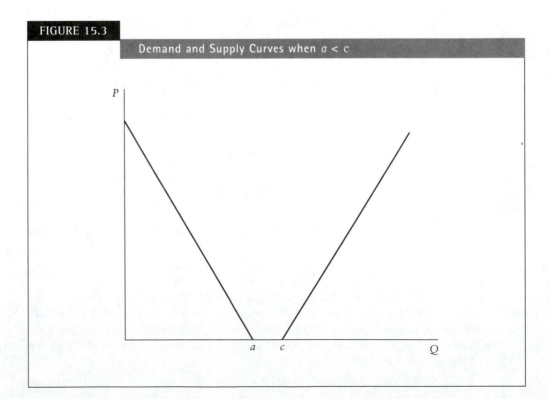

FIGURE 15.3

Demand and Supply Curves when $a < c$

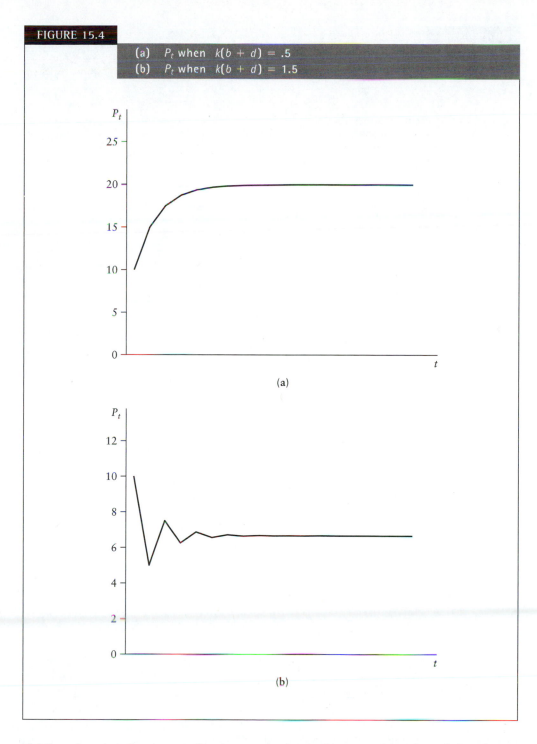

FIGURE 15.4

  (a) P_t when $k(b+d) = .5$
  (b) P_t when $k(b+d) = 1.5$

(a)

(b)

If $k(b+d) < 1$, price approaches its steady-state value smoothly; if $1 < k(b+d) < 2$, price oscillates around its steady-state value. Figure 15.4 shows the two cases: in panel (a) $k(b+d) = .5$ while in panel (b) $k(b+d) = 1.5$.[3]

Equation (15.19) can also be used to show how prices adjust to an exogenous change in the market, for example, a rightward shift of the demand curve (that is, an

---

[3]Panel (a) is based on $Q_t^d = 80 - 3P_t$, $Q_t^s = 30 + 2P_t$, and $k = .01$; in panel (b) all parameters are the same except $k = .03$.

FIGURE 15.5

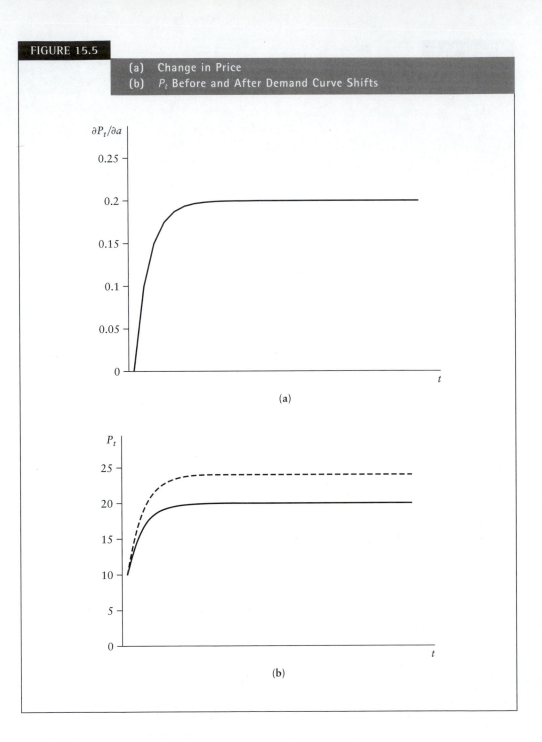

(a) Change in Price
(b) P_t Before and After Demand Curve Shifts

(a)

(b)

increase in a). The partial derivative of P_t with respect to a is

$$\frac{\partial P_t}{\partial a} = \frac{1 - (1 - k(b+d))^t}{b+d} \qquad \textbf{(15.20)}$$

which is positive for stable processes. Thus price is higher in every period. Figure 15.5 shows the dynamics of the price adjustment. Panel (a) shows the behavior of equation (15.20) for the case illustrated in panel (a) of Figure 15.4; panel (b) of Figure 15.5 shows the complete price process for the same case and, in the higher curve, for the same case except with a increased from 80 to 90 (so that the change is large enough to be clearly visible on the graph). As panel (a) shows, the change in price gets larger and larger through time, though at a slower and slower rate; panel (b) shows that this occurs as price gradually approaches its new, higher, steady-state value.

15.2.4 Phase Diagrams

Another useful way to depict a process $\{x_t\}$ is with a special graph called a phase diagram. For discrete-time processes like that described by equation (15.1), a phase diagram plots x_t against x_{t-1}. The difference equation is plotted and, in addition, a 45-degree line shows when x_t and x_{t-1} are equal. The value of x_t at which the graph of the difference equation crosses the 45-degree line is the steady-state value. For other values of x_t, the phase diagram can be used to figure out how the value of x will change in the next period. To illustrate, Figure 15.6 is the phase diagram for equation (15.1) for $0 < a < 1$ and positive b. Starting at the initial value x_0, the height of the graph of the difference equation shows the value of x_1. To see what the value will be in the following period, move horizontally from the difference equation graph to the 45-degree line. This locates the value of x_1 on the horizontal axis. The height of the difference equation graph above this point shows the value of x_2. Continue this process until the steady state is reached.

Figure 15.7 shows phase diagrams for other values of a. From them it is clear that a process will oscillate if $a < 0$ and will be monotonic if $a > 0$. Processes will converge if $|a| < 1$. Said another way, a process will converge monotonically if the graph of the difference equation is upward-sloping and intersects the 45-degree line from above. The process will oscillate but converge if the graph of the difference equation is downward-sloping but with slope algebraically greater than -1. These are important

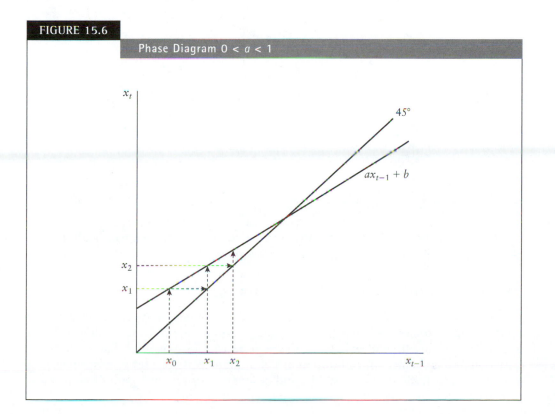

FIGURE 15.6

Phase Diagram $0 < a < 1$

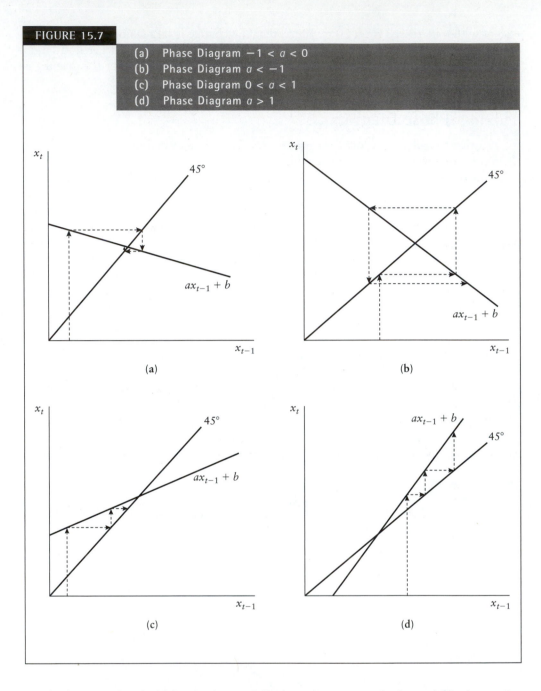

FIGURE 15.7

(a) Phase Diagram $-1 < a < 0$
(b) Phase Diagram $a < -1$
(c) Phase Diagram $0 < a < 1$
(d) Phase Diagram $a > 1$

results because they hold for nonlinear difference equations, which are difficult to solve explicitly, as well as the linear equations discussed here.

15.2.5 Nonlinear Difference Equations

So far we have restricted our attention to linear difference equations, but the same concepts apply to nonlinear equations. Although explicit solutions to nonlinear difference equations are beyond our scope, the main aspects of dynamics that we have focused on in this chapter—identification of steady-state (equilibrium) values, analysis of whether those steady states are stable, and description of the time path of adjustment to new steady states—can be analyzed using phase diagrams.

A nonlinear difference equation can be written as

$$x_t = f(x_{t-1}, t, a) \tag{15.21}$$

where a represents a parameter or a vector of parameters that influences the functional relationship between past and present values of x. If a steady-state value x^* exists, then

$$x^* = f(x^*, t, a) \tag{15.22}$$

for all values of t. Comparative statics of the steady state can be analyzed using equation (15.22). The dynamics of equation (15.21) can be characterized if enough is known about the function f to depict it graphically.

As with a linear difference equation, a phase diagram for a first-order nonlinear difference equation includes a graph of the difference equation, that is, equation (15.21), and a 45-degree line. Steady-state values are those values of x at which the graph of the difference equation crosses the 45-degree line. If the graph of the function f is upward-sloping as it intersects the 45-degree line, then the process is monotonic in the vicinity of that steady state. If the graph of f is downward-sloping at the intersection, then the process is oscillatory The steady state is unstable if the graph of f intersects the 45-degree line from below or if the graph of f is downward-sloping with slope algebraically less than -1. The steady state is stable if the graph of f is upward-sloping and intersects the 45-degree line from above or if it is downward-sloping with slope algebraically greater than -1 at the intersection.

Figure 15.8 shows three examples of phase diagrams for nonlinear first-order difference equations. In panel (a) there are two steady-state values, at 1 and 9. The first is stable, and the process will converge monotonically to this value as long as the initial value is less than 9. If the initial value of x equals 9, all future values will also equal 9. So 9 is a steady-state value, but it is unstable: any deviation from $x = 9$ will result in future values of x moving away from 9. So for any starting value less than 9, the process will converge monotonically to 1. For any starting value greater than 9, the process will be divergent because x gets larger and larger in every subsequent time period.

The process graphed in panel (b) also has two steady states, one stable and one unstable. But in this case the lower steady state is the unstable equilibrium and the upper value is stable. Since the graph of the difference equation is everywhere upward-sloping, the process will converge monotonically to the upper steady-state value as long as the starting value is greater than the lower steady-state value. If the starting value is lower than the lower steady-state value, the process does not converge.

Panel (c) shows a more complicated process that has three steady-state values. Because the graph of the difference equation slopes downward as it intersects the 45-degree line at the lowest steady state, the process will oscillate in the vicinity of that steady state, which is stable since the slope of the difference equation is less than one in absolute value. The middle steady state is unstable since the graph of the difference equation intersects the 45-degree line from below. The highest steady state is stable and the process will converge monotonically to it as long as the starting value is greater than the middle steady-state value. Unlike the processes shown in panels (a) and (b), this process will converge to a steady state regardless of the initial value. If the initial value is less than the middle steady-state value, then the process will converge to the lowest steady state (oscillating in the vicinity of the steady state). If the initial value is greater than the middle steady-state value, then the process will converge monotonically to the highest steady-state value.

An economic example of a nonlinear difference equation comes from a simple model of economic growth driven by investment in physical capital. Output Y comes from an aggregate production function $F(K, L)$ where K is the aggregate capital stock

FIGURE 15.8

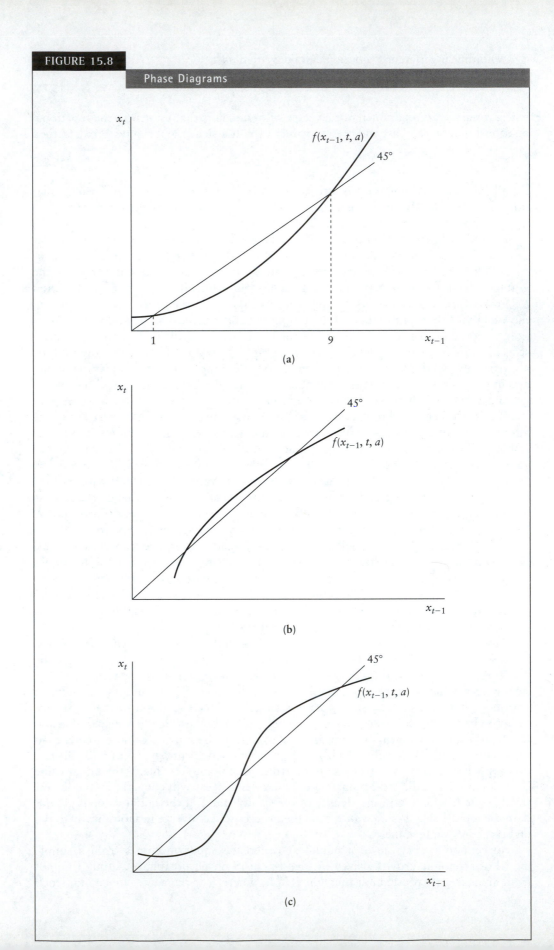

FIGURE 15.9

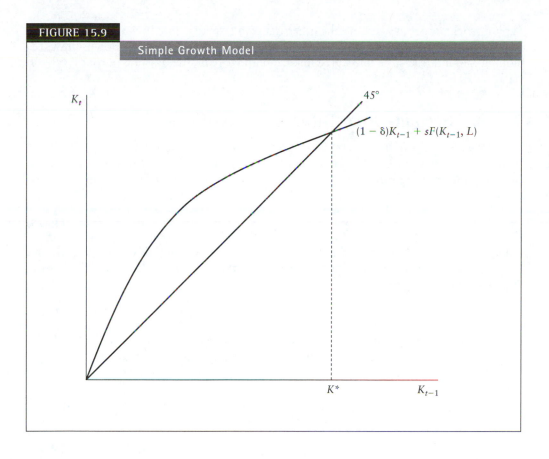

Simple Growth Model

and L is the aggregate labor supply:

$$Y_t = F(K_t, L_t, t) \tag{15.23}$$

where the argument t allows for technological change (the production function changes through time). Assuming no technological change and a constant labor supply, growth in output is driven entirely by changes in the capital stock, that is, by investment in physical capital. Investment must equal savings, so assume that savings is a constant percentage of income, which equals output. Since existing capital depreciates and new capital takes time to produce, the capital stock in period t equals the capital stock in the previous period minus depreciation plus the previous period's investment:

$$K_t = K_{t-1}(1 - \delta) + I_{t-1} = K_{t-1}(1 - \delta) + s Y_{t-1} \tag{15.24}$$

where δ is the (constant) depreciation rate and s is the (constant) marginal propensity to save. Substituting equation (15.23) into equation (15.24) and incorporating the assumptions of constant labor force and no technological change yields

$$K_t = K_{t-1}(1 - \delta) + s F(K_{t-1}, L) \tag{15.25}$$

which, assuming F is nonlinear, is a nonlinear first-order difference equation. If F is a Cobb-Douglas production function $F(K_t, L) = AK_t^\beta L^{1-\beta}$, then the phase diagram of equation (15.25) looks like Figure 15.9.[4] There are two steady-state values of the

---

[4]The general shape of the curve in Figure 15.9 can be confirmed by taking the first and second derivatives of (15.25) with respect to K_{t-1}.

capital stock: one is at zero and the other is a stable steady state to which the economy will converge monotonically, since the graph of the difference equation is upward-sloping and intersects the 45-degree line from above. The steady-state output level will be $Y^* = F(K^*, L)$.

15.2.6 Second–Order Difference Equations

Although we have focused on first-order difference equations so far, we introduced second-order equations at the beginning of the chapter. Equation (15.2), a general form of a second-order linear difference equation, is reprinted here and renumbered for convenience:

$$x_t = ax_{t-1} + bx_{t-2} + c. \tag{15.26}$$

Analyzing second-order linear difference equations is more complicated than analyzing first-order equations, but the general procedure is the same: first, determine the steady-state value(s), if they exist; then solve the homogeneous equation

$$x_t - ax_{t-1} - bx_{t-2} = 0; \tag{15.27}$$

then combine the two results to yield the general solution to equation (15.26).

If a steady-state solution to equation (15.26) exists, then it must satisfy

$$x^* = ax^* + bx^* + c, \tag{15.28}$$

so

$$x^* = \frac{c}{1 - (a + b)}. \tag{15.29}$$

This implies that $a + b$ cannot equal 1, a condition analogous to $a \neq 1$ in the case of the first-order linear difference equation (15.1).

To solve the homogeneous second-order equation (15.27), we conjecture that the solution will be of the form

$$x_t = CA^t \tag{15.30}$$

where C and A are constants to be determined. If this conjecture is correct, then (substituting into (15.27))

$$CA^t - aCA^{t-1} - bCA^{t-2} = 0. \tag{15.31}$$

Clearly any value of C is consistent with equation (15.31) so we focus on determining the constant A:

$$CA^{t-2}(A^2 - aA - b) = 0$$
$$(A^2 - aA - b) = 0$$
$$A = \frac{a \pm \sqrt{a^2 + 4b}}{2}. \tag{15.32}$$

There are three possibilities: $a^2 + 4b > 0$, in which case equation (15.31) has two distinct real roots; $a^2 + 4b = 0$, in which case equation (15.31) has a single, real root; and $a^2 + 4b < 0$, in which case equation (15.31) has roots that are complex numbers (that is, numbers that have both real and imaginary—functions of $i = \sqrt{-1}$—parts).

Distinct Real Roots

If $a^2 + 4b > 0$, then there are two possible values of A: $A_1 = a/2 + \sqrt{(a/2)^2 + b}$ and $A_2 = a/2 - \sqrt{(a/2)^2 + b}$. Any linear combination of these two values will yield a solution to the homogeneous equation (15.27):

$$x_t = C_1 A_1^t + C_2 A_2^t \tag{15.33}$$

where the two terms on the right-hand side must be linearly independent.

The general solution adds the steady-state value of x to the right-hand side of equation (15.33):

$$x_t = C_1 A_1^t + C_2 A_2^t + x^* = C_1 A_1^t + C_2 A_2^t + \frac{c}{1 - (a + b)}. \tag{15.34}$$

The constants C_1 and C_2 can be determined by specifying two periods' values of x, typically the initial value x_0 and the next period's value x_1. From equation (15.34),

$$x_0 = C_1 + C_2 + \frac{c}{1 - (a + b)}$$

and $\tag{15.35}$

$$x_1 = C_1 A_1 + C_2 A_2 + \frac{c}{1 - (a + b)},$$

from which C_1 and C_2 can be solved as functions of a, b, c, and the two initial values x_0 and x_1.

The steady-state value will be stable as long as, regardless of initial values, x_t approaches the steady-state value as t gets large. That is, there will be a stable steady state as long as

$$\lim_{t \to \infty} x_t = x^* = \frac{c}{1 - (a + b)}. \tag{15.36}$$

It is clear from equation (15.34) that a sufficient condition for a stable steady state is that both $|A_1| < 1$ and $|A_2| < 1$.

As an example, consider the difference equation

$$x_t = .6x_{t-1} + .16x_{t-2} + 48 \tag{15.37}$$

which has a steady-state value of $48/(1 - (.6 + .16)) = 200$. The two roots of the related homogeneous difference equation are $A_1 = (.6/2) + \sqrt{(.6/2)^2 + .16} = .3 + \sqrt{.25} = .8$ and $A_2 = .3 - \sqrt{.25} = -.2$. If the initial values of the process are $x_0 = 100$ and $x_1 = 110$, then

$$100 = C_1 + C_2 + 200$$

and $\tag{15.38}$

$$110 = .8C_1 - .2C_2 + 200$$

which can be solved to show that $C_1 = -110$ and $C_2 = 10$. Thus the definite solution to the second-order linear difference equation (15.37), for the two initial values as given, is

$$x_t = (-110)(.8)^t + 10(-.2)^t + 200. \tag{15.39}$$

The steady state of this process is stable since $|.8| < 1$ and $|-.2| < 1$.

Single Real Root

If $a^2 + 4b = 0$, then from equation (15.32) $A = a/2$ is the only root of equation (15.31). This suggests that the solution to the homogeneous equation (15.27) might be $x_t = A^t$. This (and anything proportional to it) is, in fact, a solution to (15.27); but so too is $x_t = tA^t$ (and anything proportional to it). Any linear combination of these two values will solve (15.27):

$$x_t = (C_1 + C_2 t)\left(\frac{a}{2}\right)^t. \tag{15.40}$$

The general solution to the difference equation (15.26) is obtained by adding the steady-state value of x to the right-hand side of equation (15.40):

$$x_t = (C_1 + C_2 t)\left(\frac{a}{2}\right)^t + \frac{c}{1 - (a + b)}. \tag{15.41}$$

The process converges to a stable steady state as long as $|a/2| < 1$ since then

$$\lim_{t \to \infty} x_t = \frac{c}{1 - (a + b)}.$$

As when there are two distinct real roots, the constants C_1 and C_2 can be determined by specifying the values of x_0 and x_1. From equation (15.41),

$$x_0 = C_1 + \frac{c}{1 - (a + b)}$$

and

$$x_1 = (C_1 + C_2)\left(\frac{a}{2}\right) + \frac{c}{1 - (a + b)}. \tag{15.42}$$

From equations (15.42), C_1 and C_2 can be solved for as functions of a, b, c, x_0, and x_1.

Complex Roots

When $a^2 + 4b < 0$, equation (15.31) has complex roots. The derivation of the solution to the difference equation (15.26) in this case is beyond the scope of this book,[5] but for completeness we present the solution:

$$x_t = C_1(\sqrt{-b})^t \cos\left(\frac{at}{2\sqrt{-b}}\right) + C_2(\sqrt{-b})^t \sin\left(\frac{at}{2\sqrt{-b}}\right) + \frac{c}{1 - (a + b)} \tag{15.43}$$

where C_1 and C_2 are constants that can be determined by specifying the values of x_0 and x_1. There will be a stable steady state as long as $-b < 1$ or, since when there are complex roots $a^2 + 4b < 0$, $a^2/4 < 1$, which in turn implies $a/2 < 1$, a condition similar to the stability condition for a single real root. Because of the nature of the sine and cosine functions, the process described by equation (15.43) is oscillatory.

15.2.7   Systems of Difference Equations

When the values of two or more variables are related because they each depend on past values of themselves and each other, their relationship can be described by a system

---

[5]For details see *Mathematics for Economic Analysis*, Knut Sydsaeter and Peter J. Hammond (Prentice-Hall, 1995), Chapter 20.

of difference equations. For example, consider the two-variable system of linear first-order difference equations

$$x_t = a_{11}x_{t-1} + a_{12}y_{t-1} + b_1$$
$$y_t = a_{21}x_{t-1} + a_{22}y_{t-1} + b_2$$

(15.44)

which can also be written in matrix notation as

$$\mathbf{z}_t = A\mathbf{z}_{t-1} + \mathbf{b}$$

(15.45)

where $\mathbf{z}_t = \begin{pmatrix} x_t \\ y_t \end{pmatrix}$, $A = \begin{pmatrix} a_{11} & a_{12} \\ a_{21} & a_{22} \end{pmatrix}$, and $\mathbf{b} = \begin{pmatrix} b_1 \\ b_2 \end{pmatrix}$.

The technique for solving systems of linear difference equations is to transform the system into an equivalent system of two univariate difference equations. This is called diagonalizing the system since it transforms the system into an equivalent system that has a diagonal matrix:

$$\mathbf{u}_t = \Lambda\mathbf{u}_{t-1} + \mathbf{c}$$

(15.46)

where $\mathbf{u}_t$ is a 2×1 vector of linear combinations of x_t and y_t, $\mathbf{c}$ is a 2×1 linear function of b_1 and b_2, and $\Lambda = \begin{pmatrix} \lambda_1 & 0 \\ 0 & \lambda_2 \end{pmatrix}$ is a diagonal matrix. Once the system has been diagonalized, each univariate difference equation can be solved using the methods already covered in this chapter. The resulting solutions can then be transformed back into solutions for the original variables x and y.

The transformation of system (15.45) into the diagonal system (15.46) uses **eigenvalues** (also called **characteristic values**) and **eigenvectors** (also called **characteristic vectors**). These matrix algebra concepts are covered in an appendix to this chapter, but for linear two-equation systems, it is fairly straightforward to analyze systems of difference equations.

As shown in the appendix, for the system (15.45) there are two eigenvectors:

$$\lambda_1, \lambda_2 = \frac{a_{11} + a_{22}}{2} \pm \frac{\sqrt{(a_{11} + a_{22})^2 - 4(a_{11}a_{22} - a_{12}a_{21})}}{2}.$$

(15.47)

Both of the linear first-order difference equations in system (15.46) will have stable steady states if both eigenvalues are real (that is, if $(a_{11} + a_{22})^2 - 4(a_{11}a_{22} - a_{12}a_{21}) \geq 0$) and less than one in absolute value. In this case, system (15.46), and by extension system (15.45) since the steady-state values of (15.45) are simply linear functions of the steady-state values of (15.46), is said to be stable. From equation (15.45) it is clear that the steady-state values of x and y, if they exist, satisfy

$$\mathbf{z}^* = A\mathbf{z}^* + \mathbf{b}$$

(15.48)

from which the steady-state values can be calculated using matrix inversion, Cramer's rule, or by writing out the two equations separately and solving by substitution.

The signs of the eigenvalues determine whether the two difference equations in (15.46) are monotonic or oscillatory. But recall that these are difference equations in the transformed variables; the signs of the eigenvalues do not necessarily indicate whether the original variables x and y approach their steady-state values monotonically. The complete solutions for x and y can be found using matrix algebra, as shown in the appendix. But recently many economists have instead used simulations, which for linear equations with known constants can be done quite easily with standard spreadsheet computer programs. Effects of changes in various parameters can also easily be investigated with simulations, and many times simulations can be used to analyze nonlinear systems of equations as well.

So a standard procedure is to compute eigenvalues to show that the system of equations is stable, calculate the steady-state values of x and y, and simulate the system of dynamic processes to show the paths from arbitrary initial values to the steady states of the two variables.

15.3   DIFFERENTIAL EQUATIONS

We now turn to continuous-time models. Instead of relating the current and past values of a variable, we now relate the current value of a variable to its rate of change through time. That is, we relate x_t to dx_t/dt. A common notation for the time derivative of a variable is $\dot{x}_t \equiv dx_t/dt$. The usual way of writing a first-order differential equation is in the general form

$$\dot{x}_t = f(x_t, t, a) \tag{15.49}$$

where a represents a parameter or a vector of parameters that influences the functional relationship between x and its rate of change. A first-order linear differential equation is

$$\dot{x}_t = a_t x_t + b_t \tag{15.50}$$

where a and b are parameters whose value may be different in different time periods. If the parameters are constant through time, then the first-order differential equation is

$$\dot{x}_t = a x_t + b. \tag{15.51}$$

A second-order differential equation is

$$\ddot{x}_t = a_t \dot{x}_t + b x_t + c_t \tag{15.52}$$

where $\ddot{x}_t \equiv d^2 x_t/dt^2$ is the second derivative of x with respect to time. In this chapter we will restrict our attention to differential equations with constant coefficients; the more general case requires integration.[6]

An economic example of a first-order differential equation is a price-adjustment model, similar to that represented by Section 15.2.3 but in continuous time. As in the earlier section, it is assumed that prices adjust proportionally to excess demand. But since time is continuous, the price adjustment equation relates the rate of change of prices to the current levels of demand and supply:

$$\dot{P}_t = k(Q_t^d - Q_t^s). \tag{15.53}$$

Assuming linear demand and supply functions as in Section 15.2.3 yields

$$\dot{P}_t = k((a - bP_t) - (c + dP_t)) = k(a - c) - k(b + d)P_t \tag{15.54}$$

or, rearranging into a form similar to equation (15.50),

$$\dot{P}_t = -k(b + d)P_t + k(a - c). \tag{15.55}$$

As was the case with difference equations, our interest is in determining whether there is a steady state and, if there is, its value; describing whether the steady state is

---

[6]Details can be found in texts such as Sydsaeter and Hammond, *op. cit.*

stable; and describing the dynamic path that the variable takes in getting to its steady-state value.

15.3.1 Solutions to First-Order Linear Differential Equations

The solution to the differential equation (15.49) shows the value that x takes as a function of time. As in the discrete-time case, a steady state implies that x does not change (unless there is a new shock—a change in one of the exogenous variables or parameters). The steady state is thus reached when $\dot{x}_t = 0$. Another way to characterize the steady state is to say it is the limit of x as t gets large: $\lim_{t \to \infty} x_t$. For a linear first-order differential equation with constant parameters, equation (15.51) immediately shows that if a steady state exists, its value is

$$x^* = -\frac{b}{a} \tag{15.56}$$

where obviously a cannot equal zero.

To get a complete solution for a differential equation, we employ the same strategy as in the section, An Alternative Approach, on page 367. First we look for a solution to the homogeneous equation

$$\dot{x}_t = ax_t \tag{15.57}$$

and then we will find a general solution. Looking at equation (15.57), it is clear that we need to find a function such that its time derivative is proportional to its value. This suggests an exponential function; recall that if $y = e^{ax}$, then $dy/dx = ay$. So a likely solution to equation (15.57) is

$$x_t = Ce^{At} \tag{15.58}$$

where C and A are constants to be determined. Taking the derivative with respect to time of equation (15.58),

$$\dot{x}_t = CAe^{At} = Ax_t \tag{15.59}$$

so, comparing to equation (15.57), $A = a$. The constant C will be determined by an initial condition.

As was the case for difference equations, the general solution to a nonhomogeneous first-order linear differential equation is found by adding the steady-state value to the solution for the homogeneous case. So the general solution to equation (15.51) is, using the result that $A = a$,

$$x_t = Ce^{at} - \frac{b}{a}. \tag{15.60}$$

The definite solution specifies the value of C by specifying the initial value of x, x_0. From equation (15.60),

$$x_0 = Ce^{a(0)} - \frac{b}{a} = C - \frac{b}{a} \tag{15.61}$$

from which it is clear that $C = x_0 + b/a$. So the definite solution to (15.51) is

$$x_t = \left(x_0 + \frac{b}{a}\right) e^{at} - \frac{b}{a}. \tag{15.62}$$

The reader can confirm that the derivative of (15.62) does indeed satisfy (15.51).

By taking the limit of equation (15.62) as t gets infinitely large, we can investigate the conditions under which there is a steady state. Clearly a cannot equal 0. Assuming $a \neq 0$, if the initial value of x equals $-b/a$, then x will equal that value in all future time periods as well:

$$x_t = \left(-\frac{b}{a} + \frac{b}{a}\right) e^{at} - \frac{b}{a} = -\frac{b}{a}. \tag{15.63}$$

If $x_0 \neq -b/a$, then

$$\lim_{t \to \infty} x_t = \left(x_0 + \frac{b}{a}\right) \lim_{t \to \infty} e^{at} - \frac{b}{a} \tag{15.64}$$

which is finite only if $a < 0$. If $a > 0$, then $\lim_{t \to \infty} x_t$ gets infinitely positive if $x_0 > -b/a$ and infinitely negative if $x_0 < -b/a$. If $a < 0$, then $\lim_{t \to \infty} e^{at} = 0$ so $\lim_{t \to \infty} x_t = -b/a$. So there is a steady state only if either $x_0 = -b/a$ or $a < 0$; in either case the steady state is $x^* = -b/a$.

The stability of the steady state can be seen by rewriting equation (15.51):

$$\dot{x}_t = ax_t + b = ax_t - a\left(-\frac{b}{a}\right) = a(x_t - x^*). \tag{15.65}$$

If a is negative, then when x_t is above its steady-state value, $\dot{x}_t < 0$; when x_t is below its steady-state value, $\dot{x}_t > 0$. Said another way, if x starts at anything other than its steady-state value, it will monotonically[7] approach the steady state through time. If a is positive, however, any time x deviates from the steady state it will continue to diverge in the future. That is, the steady state is unstable and can only be reached if the initial value equals the steady-state value.

Equation (15.62) can be rewritten to show when convergence will be rapid and when it will be slow. Since the steady-state value of x is $x^* = -b/a$,

$$x_t = (x_0 - x^*) e^{at} + x^* = x_0 e^{at} + x^*(1 - e^{at}). \tag{15.66}$$

So if a is large and negative, e^{at} is close to 0 even for small values of t, and x_t will always be close to x^*. But the closer a is to 0, the closer e^{at} will be to 1, and it will take large values of t to move x_t away from its starting point. As was the case with difference equations, changes in a will also affect the steady-state values, so in unusual circumstances changing a toward 0 might speed up convergence to the vicinity of the steady state because the steady state will move very close to the starting value. Generally speaking, though, the more negative is a, the faster will be convergence.

Summarizing, if $a < 0$, then there is a stable steady state at $x^* = -b/a$ to which the process monotonically converges. If $a > 0$, then there is an unstable steady state at $x^* = -b/a$. If $a = 0$, there is no steady state (except for the trivial case of $a = b = 0$ in which case $x_t = x_0$ for all t). The more negative is a, the faster the process will converge

---

[7]Unlike difference equations, single differential equations never have oscillatory dynamics. There will never be overshooting of the steady state since x changes continuously. As soon as the value of x gets to the steady state, $\dot{x} = 0$.

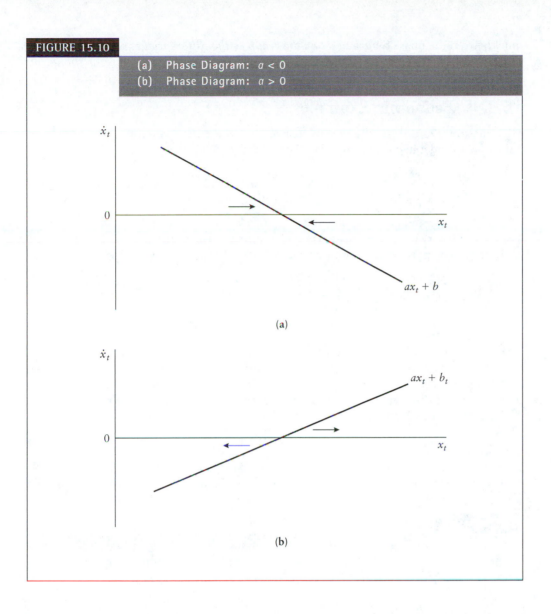

FIGURE 15.10

(a)   Phase Diagram: $a < 0$
(b)   Phase Diagram: $a > 0$

$ax_t + b$

(a)

$ax_t + b_t$

(b)

to the vicinity of its steady state, for a given difference between starting and steady-state values.

15.3.2   Phase Diagrams

The cases from the previous section are illustrated using phase diagrams in Figure 15.10. For differential equations, phase diagrams are graphs showing the relationship between $\dot{x}_t$ and x_t. Arrows show in which direction x will move from its current value. Clearly if $\dot{x}_t > 0$, then x will move to the right in the phase diagram; if $\dot{x}_t < 0$, then x will move to the left. Steady states exist where the graph crosses the horizontal axis, since then $\dot{x}_t = 0$. Panel (a) is for the case of $a < 0$ and shows the stability of the steady state. Panel (b) is for $a > 0$ and shows that the steady state is unstable since the arrows all point away from x^*.

Phase diagrams like Figure 15.10 can be used to analyze nonlinear differential equations, which can be very difficult to solve explicitly. Nonlinear differential equations may have several steady states. The stability of each steady state can be seen quickly from the phase diagram: *when the graph of the differential equation crosses the horizontal axis from below, it corresponds to an unstable steady state; if the graph crosses the horizontal axis from above, the corresponding steady state is stable.*

15.3.3 Economic Examples

Equation (15.55) is the linear first-order differential equation for the price-adjustment model in continuous time. The steady-state price is

$$P^* = -\frac{k(a-c)}{-k(b+d)} = \frac{a-c}{b+d} \tag{15.67}$$

which is positive and finite as long as $a > c$, since as in Section 15.2.3 a, b, and d are all assumed to be positive. The steady state is stable since the coefficient on P_t, $-k(b+d)$, is negative. The particular solution to the differential equation (15.55) is

$$P_t = \left(P_0 + \frac{k(a-c)}{-k(b+d)}\right) e^{-k(b+d)t} - \frac{k(a-c)}{-k(b+d)} = \left(P_0 - \frac{a-c}{b+d}\right) e^{-k(b+d)t} + \frac{a-c}{b+d}. \tag{15.68}$$

The phase diagram for this example will look like Figure 15.10, Panel (a).

An economic example of a nonlinear differential equation is a continuous-time, Cobb-Douglas growth model similar to that in Section 15.2.5. The only difference from that example is that depreciation plus gross investment now equals the time derivative of the capital stock instead of the discrete change $K_t - K_{t-1}$:

$$\dot{K}_t = -\delta K_t + I_t = -\delta K_t + s Y_t = -\delta K_t + s A K_t^\beta L^{1-\beta}. \tag{15.69}$$

Taking first and second derivatives of (15.69) with respect to K_t confirms that the graph of this differential equation has the shape shown in Figure 15.11. There is an unstable

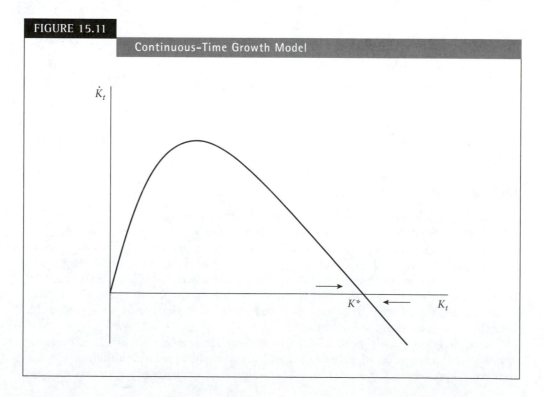

FIGURE 15.11

Continuous-Time Growth Model

steady state at $K = 0$ and a stable steady state (since the graph of the differential equation (15.69) crosses the horizontal axis from above) at some positive value K^*. This steady-state value can be found be setting the right-hand side of equation (15.69) equal to zero and solving for K:

$$0 = -\delta K + s A K^\beta L^{1-\beta}$$

$$\delta K = s A K^\beta L^{1-\beta}$$

$$\frac{K}{K^\beta} = \frac{s A}{\delta} L^{1-\beta}$$

$$K^{1-\beta} = \frac{s A}{\delta} L^{1-\beta} \tag{15.70}$$

$$K^* = \left(\frac{s A}{\delta}\right)^{\frac{1}{1-\beta}} L.$$

15.3.4 Using Phase Diagrams to Analyze Systems of Two Differential Equations

Systems of differential equations can be difficult to solve explicitly, but two-equation systems can often be analyzed easily with phase diagrams. Consider first the system of first-order linear differential equations

$$\dot{x}_t = a_{11} x_t + a_{12} y_t + c_1$$
$$\dot{y}_t = a_{21} x_t + a_{22} y_t + c_2. \tag{15.71}$$

A phase diagram of this system shows level curves of these two equations. In particular, two equations are graphed with x and y on the axes: all the combinations of x_t and y_t for which $\dot{x}_t = 0$, and all the combinations of x_t and y_t for which $\dot{y}_t = 0$. The graphs of these two equations divide the graph into four sections; arrows are drawn in each section showing the directions of motion of x and y.

Figure 15.12 is the phase diagram for an example with all coefficients positive. In this case, the $\dot{x}_t = 0$ and $\dot{y}_t = 0$ lines are both downward-sloping. (For example, the equation of the $\dot{x}_t = 0$ line is $x_t = -(c_1/a_{11}) - (a_{12}/a_{11})y_t$.) In the example shown, the two lines intersect; this corresponds to the steady state of the system since if x and y reach the values that correspond to the intersection point S, both $\dot{x}_t = 0$ and $\dot{y}_t = 0$ so x and y will stay at those values in future time periods.

The arrows of motion are derived as follows. Consider first the $\dot{x}_t = 0$ line. Starting with any combination of x and y for which $\dot{x}_t = 0$, any increase in either x or y will make $\dot{x}_t$ positive since $a_{11} > 0$ and $a_{12} > 0$. Similarly for any point lying below and to the left of the $\dot{x}_t = 0$ line, $\dot{x}_t$ will be negative. Using similar reasoning, for any point lying above and to the right of the $\dot{y}_t = 0$ line, $\dot{y}_t$ will be positive since $a_{21} > 0$ and $a_{22} > 0$; for any point lying below and to the left of the $\dot{y}_t = 0$ line, $\dot{y}_t$ will be negative. The arrows of motion show the four different cases, depending on which section of the diagram corresponds to the current values of x and y. Vertical arrows show the direction of change for x, and horizontal arrows the direction of change for y. Using the arrows of motion, it is clear that the steady state shown in the diagram is globally unstable: starting from any point other than the steady state, both x and y will move further away from their steady-state values.[8]

---

[8]The steady state is not always globally unstable when all coefficients are positive; the relative magnitudes of the coefficients matter as well. If the $\dot{y}_t = 0$ line is flatter than is the $\dot{x}_t = 0$ line, there can be a saddlepath.

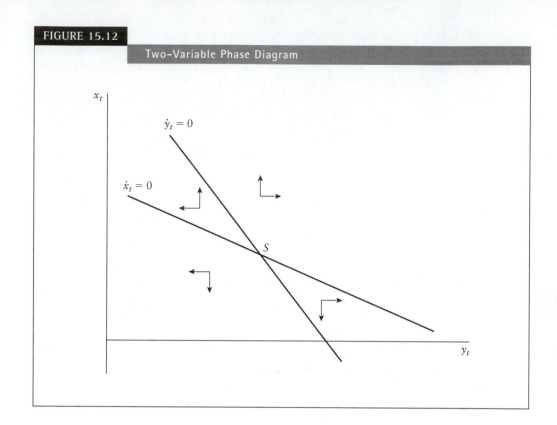

FIGURE 15.12

Two–Variable Phase Diagram

The dynamics of the system depend on the individual signs of $a_{11}, a_{12}, a_{21},$ and a_{22}. (The constants c_1 and c_2 affect the steady state—where the two lines in the phase diagram intersect—but not the dynamics of the system.) The possibilities are explored in an end-of-chapter problem; here we focus on two cases. In the first, the steady state is globally stable: it will be reached regardless of the initial values of x and y. The second case is called a **saddlepath-stable steady state**: for most initial conditions, the steady state will never be reached; but there is a **saddlepath** along which the steady state can be reached, so if the initial conditions are on that saddlepath the steady state will be achieved.

Figure 15.13 shows a globally stable case. In the example, a_{11} is negative and a_{12} is positive, so the $\dot{x}_t = 0$ line is upward-sloping. For any point above and to the left of the $\dot{x}_t = 0$ line $\dot{x}_t < 0$, while for any point below and to the right of the $\dot{x}_t = 0$ line $\dot{x}_t > 0$. Meanwhile, a_{21} and a_{22} are negative so the $\dot{y}_t = 0$ line is downward-sloping. For any point above and to the right of the $\dot{y}_t = 0$ line $\dot{y}_t < 0$, while for any point below and to the left of the $\dot{y}_t = 0$ line $\dot{y}_t > 0$. The arrows of motion indicate that, regardless of the starting point, both x and y will move toward their steady-state values. The relative magnitudes of the coefficients determine exactly what the time path of x and y is. One possibility is shown in the figure.

Figure 15.14 shows a saddlepath-stable steady state. As in the previous example, a_{11} is negative and a_{12} is positive, so the dynamics of x_t are the same as in Figure 15.13. Even though the $\dot{y}_t = 0$ line looks the same as in Figure 15.13, in this case a_{21} and a_{22} are both positive. So now for any point above and to the right of the $\dot{y}_t = 0$ line $\dot{y}_t > 0$, while for any point below and to the left of the $\dot{y}_t = 0$ line $\dot{y}_t < 0$. This yields arrows of motion as shown in Figure 15.14. From most initial conditions, the arrows of motion will take x and y away from the steady state. But for initial conditions lying on a particular path, x and y will follow that path to the steady state. This is called the saddlepath, and it is shown in Figure 15.14.

FIGURE 15.13

Global Stability

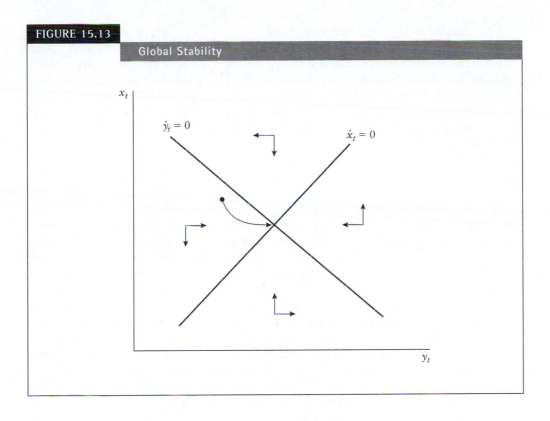

FIGURE 15.14

Saddlepath Stability

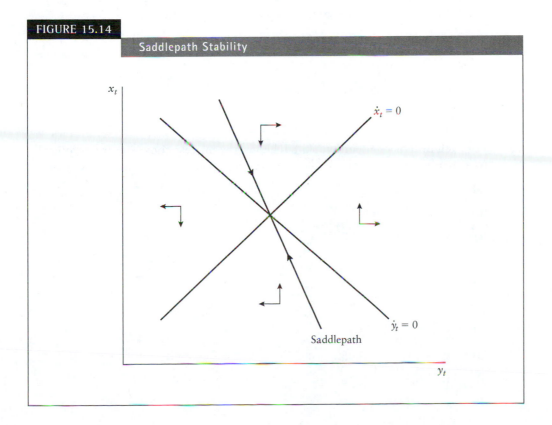

Phase diagrams like those shown in Figures 15.12–15.14 can be used for nonlinear as well as linear differential equations. As long as the $\dot{x}_t = 0$ and $\dot{y}_t = 0$ curves can be drawn, assumptions about the derivatives of the differential equations usually are sufficient to figure out the arrows of motion for the two variables. From the arrows of motion, it is often possible to deduce whether steady states are globally stable or globally unstable or whether there is saddlepath stability. This makes it possible to analyze systems of equations that would be very difficult or impossible to solve explicitly.

SUMMARY

In this chapter we have introduced dynamic analysis by discussing difference and difference equations. We have focused our attention on three aspects of dynamic processes: the identification of steady states, the analysis of stability of steady states, and the description of the time path a process takes from a starting point to a steady state, including how long it takes for the process to reach the vicinity of the steady state. Analyzing the dynamics of economic models adds richness to our understanding of economic processes. In addition, confirming that an economic model is convergent justifies the comparative static analysis we have done throughout the book. Comparative statics analyzes how equilibria change in response to changes in exogenous variables; but if the model is not convergent, the new equilibrium will never be reached. So for comparative static analysis to make sense, the processes being modeled must be convergent.

Although we have covered only the basics of difference and differential equations, the basics are enough to understand many economic applications, some of which are explored in the next chapter. Appendix 15.2 introduces a simple version of another technique economists use to study dynamics: dynamic optimization. The rest of the book explores game theory models; the final two chapters investigate dynamic games, which are yet another type of dynamic mathematical economic model.

Problems

15.1 For each difference equation, state whether the process converges monotonically to a steady state, oscillates while converging to a steady state, or is divergent.

(a) $x_t = 7.5x_{t-1} + 28$

(b) $x_t = .75x_{t-1} + 28$

(c) $x_t = -.75x_{t-1} + 28$

(d) $x_t = -.75x_{t-1}$

(e) $x_t = -.75x_{t-1} - 28$

(f) $x_t = -7.5x_{t-1} - 28$

(g) $x_t = 1.4x_{t-1} - 0.13x_{t-2} + 55$

(h) $x_t = 0.8x_{t-1} + 0.09x_{t-2} + 55$

(i) $x_t = 0.8x_{t-1} - 0.16x_{t-2} - 18$

(j) $x_t = -12 + x_{t-1}^2$

15.2 For each stable difference equation in Problem 15.1, compute the steady state.

15.3 For each stable first-order linear difference equation in Problem 15.1, find the definite solution if $x_0 = 10$ and calculate the values of x_1, x_2, and x_5.

15.4 For each first-order difference equation in Problem 15.1, draw a phase diagram to illustrate the process, starting from $x_0 = 10$.

15.5 For the following system of two first-order linear difference equations, find the eigenvalues of the system and state whether the system is stable.

$$x_t = 0.2x_{t-1} + 0.1y_{t-1} + 10$$
$$y_t = 0.6x_{t-1} + 0.1y_{t-1} + 20$$

15.6 For the following differential equations, draw a phase diagram, identify every steady state, and, for each steady state, say whether it is stable or unstable.

(a) $\dot{x}_t = 2.5x_t + 40$

(b) $\dot{x}_t = .25x_t + 40$

(c) $\dot{x}_t = -.25x_t + 40$

(d) $\dot{x}_t = -2.5x_t + 40$

(e) $\dot{x}_t = x_t^3 + 3x_t^2 - x - 3$

(f) $\dot{x}_t = 10 - \sqrt{x_t}$

15.7 For each of the following cases of coefficients in two-equation systems of first-order linear differential equations, draw a phase diagram with arrows of motion and label any steady states as globally stable, globally unstable, or stable along a saddlepath. For cases where the $\dot{x}_t = 0$ and $\dot{y}_t = 0$ curves have slopes of the same sign, assume that $a_{12}/a_{11} > a_{22}/a_{21}$.

(a) $a_{11} > 0, a_{12} > 0, a_{21} > 0, a_{22} > 0$

(b) $a_{11} > 0, a_{12} > 0, a_{21} < 0, a_{22} > 0$

(c) $a_{11} < 0, a_{12} < 0, a_{21} < 0, a_{22} < 0$

(d) $a_{11} < 0, a_{12} < 0, a_{21} > 0, a_{22} < 0$

(e) $a_{11} < 0, a_{12} < 0, a_{21} < 0, a_{22} > 0$

(f) $a_{11} < 0, a_{12} > 0, a_{21} > 0, a_{22} < 0$

(g) $a_{11} > 0, a_{12} < 0, a_{21} < 0, a_{22} > 0$

(h) $a_{11} > 0, a_{12} = 0, a_{21} < 0, a_{22} > 0$

(i) $a_{11} < 0, a_{12} > 0, a_{21} > 0, a_{22} = 0$

(j) $a_{11} > 0, a_{12} > 0, a_{21} < 0, a_{22} < 0$

Appendix 15.1:
Eigenvalues and Eigenvectors

15A1.1   INTRODUCTION

In Section 15.2.7 we introduced the concepts of eigenvalues and eigenvectors, also called *characteristic values* and *characteristic vectors,* saying they were useful for analyzing systems of linear difference equations. Eigenvalues and eigenvectors can be used to characterize any square matrix so, in addition to systems of difference and differential equations, they are useful in many other mathematical economics applications as well. Some examples are statistical and econometric theory, input-output analysis, and analysis of labor market transition.

In this appendix we will briefly describe eigenvalues and eigenvectors in general, including how they are used to diagonalize systems of linear equations. Then we will go thoroughly through the details of finding eigenvalues and eigenvectors for the special case of systems of two equations and how to use them to analyze a system of two difference equations. We use that special case as a basis for our explanation of how to find eigenvalues and eigenvectors when there are more than two equations and how to use them to analyze systems of more than two difference equations.

15A1.2   NOTATION AND DEFINITIONS

Consider a matrix A with n columns and n rows. Suppose there is a scalar λ and a nonzero vector $\mathbf{x}$ that solves the matrix equation

$$A\mathbf{x} = \lambda\mathbf{x}. \tag{15A1.1}$$

Then λ is an eigenvalue of the matrix A and $\mathbf{x}$ is the associated eigenvector. Eigenvectors are unique only up to a multiplicative constant: if $A\mathbf{x} = \lambda\mathbf{x}$ then, for any nonzero constant c, $A(c\mathbf{x}) = \lambda(c\mathbf{x})$, so $c\mathbf{x}$ is also an eigenvector associated with λ.

Rewriting equation (15A1.1) as

$$(A - \lambda I)\mathbf{x} = 0, \tag{15A1.2}$$

where I is an $n \times n$ identity matrix and $\mathbf{0}$ is in boldface to emphasize that it is an $n \times 1$ column vector of zeros, yields another useful interpretation of eigenvalues: an eigenvalue is any scalar that, when subtracted from each and every diagonal element of A, results in a *singular matrix*. Singular matrices, which we defined in Chapter 3, do not have inverses. If an inverse of the matrix $A - \lambda I$ existed, then using it to premultiply the left-hand side of equation (15A1.2) would yield $I\mathbf{x}$. But equation (15A1.2) shows that $(A - \lambda I)^{-1}(A - \lambda I)\mathbf{x} = (A - \lambda I)^{-1}0 = 0$, not $I\mathbf{x}$. So the inverse does not exist and $(A - \lambda I)$ is singular. Further, as we described in Chapter 3, a matrix is not invertible when its determinant equals zero. So there is a nonzero solution for $\mathbf{x}$ only if

$$|A - \lambda I| = 0. \tag{15A1.3}$$

Equation (15A1.3) is called the **characteristic equation** for the matrix A. We will use it when we solve for the eigenvalues of a matrix. It turns out that there are n eigenvalues (that is, the number of eigenvalues equals the number of rows in A), though sometimes two or more of the eigenvalues equal each other. Associated with the n eigenvalues are n eigenvectors.

15A1.3   DIAGONALIZING SYSTEMS OF EQUATIONS

For the purposes of analyzing systems of difference equations, the most important use of eigenvalues and eigenvectors is in diagonalizing systems of equations. That is, they are used to convert a matrix equation like

$$A\mathbf{y} = \mathbf{b} \qquad\qquad (15A1.4)$$

into an equivalent matrix equation

$$D\mathbf{z} = \mathbf{d} \qquad\qquad (15A1.5)$$

where D is a diagonal matrix, the vector $\mathbf{z}$ is a linear transformation of $\mathbf{y}$, and $\mathbf{d}$ is a linear transformation of $\mathbf{b}$.

Let X be an $n \times n$ matrix comprised of the n eigenvectors of A, one in each column of X. Based on equation (15A1.1), each column of AX equals $\lambda_i \mathbf{x}_i$, where the subscripts indicate the eigenvalue and associated eigenvector corresponding to column i of X. Written differently,

$$AX = XD \qquad\qquad (15A1.6)$$

where D is the $n \times n$ diagonal matrix with the eigenvalues of A on the diagonal. (Readers can confirm this by writing out the rows and columns of each matrix.) If all the eigenvectors are linearly independent,[1] then

$$X^{-1}AX = X^{-1}XD = ID = D. \qquad\qquad (15A1.7)$$

We now show how to use the eigenvalues and eigenvectors of A to diagonalize the system of equations (15A1.4). From equation (15A1.7),

$$XX^{-1}AX = XD$$
$$IAXX^{-1} = XDX^{-1}$$
$$AI = XDX^{-1} \qquad\qquad (15A1.8)$$
$$A = XDX^{-1}.$$

Substituting this into the matrix equation (15A1.4), we get

$$A\mathbf{y} = \mathbf{b}$$
$$XDX^{-1}\mathbf{y} = \mathbf{b}$$
$$DX^{-1}\mathbf{y} = X^{-1}\mathbf{b} \qquad\qquad (15A1.9)$$
$$D\mathbf{z} = \mathbf{d}$$

---

[1] As we mentioned in Chapter 3, a matrix is invertible if its columns are all linearly independent.

Introduction to Dynamics: Theory

where $\mathbf{z} = X^{-1}\mathbf{y}$ and $\mathbf{d} = X^{-1}\mathbf{b}$. So, by using the eigenvectors, we have turned the system of equations (15A1.4) into the diagonal system of equations (15A1.5), and the diagonal matrix D has the eigenvalues of A on the diagonal.

Putting this in the context of systems of difference equations, let $\mathbf{z}_t$ be a vector of the values of a set of variables at time t, let A be a square matrix of constants, and let $\mathbf{b}$ be a vector of constants. Then a system of difference equations can be represented by the matrix equation

$$\mathbf{z}_t = A\mathbf{z}_{t-1} + \mathbf{b}. \tag{15A1.10}$$

If X is the matrix of eigenvectors of A and D is the diagonal matrix of eigenvalues, then we can diagonalize the difference equations as follows:

$$X^{-1}\mathbf{z}_t = X^{-1}A\mathbf{z}_{t-1} + X^{-1}\mathbf{b}$$
$$X^{-1}\mathbf{z}_t = X^{-1}A(XX^{-1})\mathbf{z}_{t-1} + X^{-1}\mathbf{b}$$
$$X^{-1}\mathbf{z}_t = (X^{-1}AX)X^{-1}\mathbf{z}_{t-1} + X^{-1}\mathbf{b} \tag{15A1.11}$$
$$X^{-1}\mathbf{z}_t = DX^{-1}\mathbf{z}_{t-1} + X^{-1}\mathbf{b}$$
$$\mathbf{u}_t = D\mathbf{u}_{t-1} + \mathbf{d}$$

where $\mathbf{u}_t = X^{-1}\mathbf{z}_t$ and $\mathbf{d} = X^{-1}\mathbf{b}$. Except for notation, the last line of (15A1.11) is identical to equation (15.46).

Once the resulting univariate difference equations for each element of $\mathbf{u}$ are analyzed (for example, finding the steady states, determining if the process is monotonic or oscillatory, or finding definite solutions), we can transform the results back into the original variables using the definition of $\mathbf{u}$: since $\mathbf{u}_t = X^{-1}\mathbf{z}_t$, $\mathbf{z}_t = X\mathbf{u}_t$.

15A1.4 FINDING EIGENVALUES AND EIGENVECTORS FOR SYSTEMS OF TWO EQUATIONS

Equation (15A1.3) can be used to solve for eigenvalues and eigenvectors. If A has two rows and two columns, then equation (15A1.3) can be written as

$$\begin{vmatrix} a_{11} - \lambda & a_{12} \\ a_{21} & a_{22} - \lambda \end{vmatrix} = 0 \tag{15A1.12}$$

or, writing out the determinant,

$$(a_{11} - \lambda)(a_{22} - \lambda) - a_{12}a_{21} = 0$$
$$\lambda^2 - (a_{11} + a_{22})\lambda + (a_{11}a_{22} - a_{12}a_{21}) = 0. \tag{15A1.13}$$

This **characteristic polynomial** is a second-degree polynomial in λ. The eigenvalues are the roots of this equation, found by solving the polynomial:

$$\lambda_1, \lambda_2 = \frac{-(-(a_{11} + a_{22})) \pm \sqrt{(-(a_{11} + a_{22}))^2 - 4(1)(a_{11}a_{22} - a_{12}a_{21})}}{2(1)} \tag{15A1.14}$$
$$= \frac{a_{11} + a_{22}}{2} \pm \frac{\sqrt{(a_{11} + a_{22})^2 - 4(a_{11}a_{22} - a_{12}a_{21})}}{2}.$$

This is identical to equation (15.47).

As described in Chapter 15, eigenvalues are used to characterize the dynamics of systems of difference equations. To find steady states or definite solutions to systems of difference equations, we will also need the eigenvectors since, as shown in equation (15A1.9), they are used to transform the system of equations into a diagonal system. We can find the eigenvector that corresponds to each eigenvalue using equation (15A1.2). For our 2×2 case, equation (15A1.2) can be written as the system of equations

$$(a_{11} - \lambda)\, x_1 + a_{12} x_2 = 0$$
$$a_{21} x_1 + (a_{22} - \lambda)\, x_2 = 0. \tag{15A1.15}$$

Each eigenvector can be found by substituting one of the eigenvalues into (15A1.15) and then solving for x_1 and x_2.

15A1.5 A WORKED-OUT EXAMPLE OF A TWO-EQUATION SYSTEM

Consider the system of difference equations

$$x_t = .1x_{t-1} + .4y_{t-1} + 10$$
$$y_t = .2x_{t-1} + .3y_{t-1} + 20 \tag{15A1.16}$$

or, written in matrix notation, with $z_t = \begin{pmatrix} x_t \\ y_t \end{pmatrix}$,

$$z_t = \begin{pmatrix} .1 & .4 \\ .2 & .3 \end{pmatrix} z_{t-1} + \begin{pmatrix} 10 \\ 20 \end{pmatrix}. \tag{15A1.17}$$

The two eigenvalues of this system are

$$\lambda_1, \lambda_2 = \frac{.1 + .3}{2} \pm \frac{\sqrt{(.1 + .3)^2 - 4((.1)(.3) - (.4)(.2))}}{2}$$
$$= .2 \pm \frac{\sqrt{.16 - 4(.3 - .8)}}{2} \tag{15A1.18}$$
$$= .2 \pm \frac{\sqrt{.16 + .20}}{2} = .2 \pm \frac{.6}{2} = .2 \pm .3$$

or .5 and $-.1$. The associated eigenvectors can be found by substituting each eigenvalue into the system of equations

$$(.1 - \lambda)w_1 + .4w_2 = 0$$
$$.2w_1 + (.3 - \lambda)w_2 = 0 \tag{15A1.19}$$

and solving for w_1 and w_2. (We have used w as a symbol for elements of the eigenvector to avoid confusion with the variable x in the original system of difference equations (15A1.16)). If $\lambda = .5$, then equation (15A1.19) becomes

$$-.4w_1 + .4w_2 = 0$$
$$.2w_1 - .2w_2 = 0 \tag{15A1.20}$$

Introduction to Dynamics: Theory

which has as a solution $w_1 = w_2$. So the eigenvector associated with the eigenvalue .5 is $\binom{1}{1}$.[2] If $\lambda = -.1$, then equation (15A1.19) becomes

$$.2w_1 + .4w_2 = 0$$
$$.2w_1 + .4w_2 = 0$$

(15A1.21)

which has as a solution $w_1 = -2w_2$. So the eigenvector associated with the eigenvalue $-.1$ is $\binom{-2}{1}$.

The system of difference equations (15A1.16) can now be diagonalized using the eigenvalues and eigenvectors. The matrix of eigenvectors is

$$W = \begin{pmatrix} 1 & -2 \\ 1 & 1 \end{pmatrix}$$

(15A1.22)

and readers can confirm that its inverse is

$$W^{-1} = \begin{pmatrix} 1/3 & 2/3 \\ -1/3 & 1/3 \end{pmatrix}.$$

(15A1.23)

The diagonal matrix of eigenvalues is

$$D = \begin{pmatrix} .5 & 0 \\ 0 & -.1 \end{pmatrix}.$$

(15A1.24)

So the diagonalized system of difference equations is

$$\begin{pmatrix} 1/3 & 2/3 \\ -1/3 & 1/3 \end{pmatrix}\begin{pmatrix} x_t \\ y_t \end{pmatrix} = \begin{pmatrix} .5 & 0 \\ 0 & -.1 \end{pmatrix}\begin{pmatrix} 1/3 & 2/3 \\ -1/3 & 1/3 \end{pmatrix}\begin{pmatrix} x_{t-1} \\ y_{t-1} \end{pmatrix} + \begin{pmatrix} 1/3 & 2/3 \\ -1/3 & 1/3 \end{pmatrix}\begin{pmatrix} 10 \\ 20 \end{pmatrix}$$

$$\begin{pmatrix} \dfrac{x_t + 2y_t}{3} \\ \dfrac{-x_t + y_t}{3} \end{pmatrix} = \begin{pmatrix} .5 & 0 \\ 0 & -.1 \end{pmatrix}\begin{pmatrix} \dfrac{x_{t-1} + 2y_{t-1}}{3} \\ \dfrac{-x_{t-1} + y_{t-1}}{3} \end{pmatrix} + \begin{pmatrix} \dfrac{10 + 40}{3} \\ \dfrac{-10 + 20}{3} \end{pmatrix}.$$

(15A1.25)

Defining

$$u_{1,t} = \frac{x_t + 2y_t}{3} \quad \text{and} \quad u_{2,t} = \frac{-x_t + y_t}{3},$$

(15A1.26)

we can rewrite (15A1.25) as

$$u_{1,t} = .5u_{1,t-1} + 50/3$$
$$u_{2,t} = -.1u_{2,t-1} + 10/3$$

(15A1.27)

each of which can be analyzed using the methods of Chapter 15. The steady-state values of $u_{1,t}$ and $u_{2,t}$ are 100/3 and 100/33. Both are stable steady states, with $u_{1,t}$ approaching its steady state monotonically while $u_{2,t}$ oscillates. Definite solutions for $u_{1,t}$ and $u_{2,t}$ can be obtained using equation (15.13).

---

[2]Because eigenvectors are unique only up to a multiplicative constant, the equations used to solve for eigenvectors will always be redundant, and elements of the eigenvector can only be solved for as functions of the other elements. Typically one of the elements is normalized to equal 1 and the rest of the elements are solved for relative to the chosen element. Also, remember that the solution $w_1 = w_2 = 0$ is not used, since eigenvectors are defined to be nonzero solutions to equation (15A1.3).

Once steady-state values or definite solutions for $u_{1,t}$ and $u_{2,t}$ are obtained, (15A1.26) can be used to transform them back into corresponding solutions for x_t and y_t. Rewriting (15A1.26),

$$3u_{1,t} = x_t + 2y_t$$
$$3u_{2,t} = -x_t + y_t \qquad\qquad \textbf{(15A1.28)}$$

or

$$\begin{pmatrix} 1 & 2 \\ -1 & 1 \end{pmatrix} \begin{pmatrix} x_t \\ y_t \end{pmatrix} = \begin{pmatrix} 3u_{1,t} \\ 3u_{2,t} \end{pmatrix}, \qquad\qquad \textbf{(15A1.29)}$$

so

$$x_t = \frac{3u_{1,t} - 6u_{2,t}}{3} = u_{1,t} - 2u_{2,t} \quad \text{and} \quad y_t = \frac{3u_{2,t} + 3u_{1,t}}{3} = u_{1,t} + u_{2,t}. \qquad \textbf{(15A1.30)}$$

Substituting in the steady-state values of $u_{1,t}$ and $u_{2,t}$, the steady-state values of x and y are $x^* = 100/3 - 200/33 = 300/11$ and $y^* = 100/3 + 100/33 = 400/11$. Readers can confirm, by substituting these values into (15A1.16), that these are indeed the steady-state values.

15A1.6   SYSTEMS WITH MORE THAN TWO EQUATIONS

The methods described in Section 15A1.4 can be extended readily to systems with more than two equations, though finding eigenvalues and eigenvectors can be quite cumbersome without computer programs. Consider first a three-equation system. If a matrix A has three rows and three columns, then equation (15A1.3) can be written as

$$\begin{vmatrix} a_{11} - \lambda & a_{12} & a_{13} \\ a_{21} & a_{22} - \lambda & a_{23} \\ a_{31} & a_{32} & a_{33} - \lambda \end{vmatrix} = 0 \qquad\qquad \textbf{(15A1.31)}$$

or, writing out the determinant,

$$\begin{aligned} &(a_{11} - \lambda)\left((a_{22} - \lambda)(a_{33} - \lambda) - a_{32}a_{23}\right) \\ &\quad - a_{21}\left(a_{12}(a_{33} - \lambda) - a_{32}a_{13}\right) \qquad\qquad \textbf{(15A1.32)} \\ &\quad + a_{31}\left(a_{12}a_{23} - a_{13}(a_{22} - \lambda)\right) = 0 \end{aligned}$$

which is a third-degree polynomial in λ. A fundamental theorem of algebra states that an nth-degree polynomial has n roots, so there are three eigenvalues of A. If the polynomial is easily factored, finding the three eigenvalues is often straightforward. But if factoring the polynomial is impossible, finding the three eigenvalues without the help of a computer program is burdensome.

Once the eigenvalues are discovered, finding the associated eigenvectors is relatively easy. For a 3×3 matrix, we can write equation (15A1.2) as

$$\begin{aligned} (a_{11} - \lambda)x_1 + a_{12}x_2 + a_{13}x_3 &= 0 \\ a_{21}x_1 + (a_{22} - \lambda)x_2 + a_{23}x_3 &= 0 \qquad\qquad \textbf{(15A1.33)} \\ a_{31}x_1 + a_{32}x_2 + (a_{33} - \lambda)x_3 &= 0 \end{aligned}$$

where the x_i are elements of the eigenvector. We then substitute one of the eigenvalues into (15A1.33) and solve the system for the (nonzero) x_i. Repeat the process for the other eigenvalues.

We can use the eigenvalues to form the diagonal matrix

$$D = \begin{pmatrix} \lambda_1 & 0 & 0 \\ 0 & \lambda_2 & 0 \\ 0 & 0 & \lambda_3 \end{pmatrix} \tag{15A1.34}$$

and we can place the eigenvectors into the columns of the matrix X. If x_{ij} is the jth element of the ith eigenvector, then

$$X = \begin{pmatrix} x_{11} & x_{21} & x_{31} \\ x_{12} & x_{22} & x_{32} \\ x_{13} & x_{23} & x_{33} \end{pmatrix}. \tag{15A1.35}$$

Finally, we can use the matrices D and X to diagonalize a system of three difference equations as shown in (15A1.11).

For systems of many difference equations, the procedure is the same. The eigenvalues are found from equation (15A1.3), which results in an nth-degree polynomial in λ:

$$|A - \lambda I| = \begin{vmatrix} a_{11} - \lambda & a_{12} & \cdots & a_{1n} \\ a_{21} & a_{22} - \lambda & \cdots & a_{2n} \\ \vdots & \vdots & \ddots & \vdots \\ a_{n1} & a_{n2} & \cdots & a_{nn} - \lambda \end{vmatrix} = 0. \tag{15A1.36}$$

This has n roots (some of which may be equal, and some of which may be complex numbers), which are the eigenvalues of the matrix A. We can find the eigenvector associated with each eigenvalue by substituting the eigenvalue into the following system of equations and then solving the system for the x_i, which are the elements of the eigenvector:

$$(a_{11} - \lambda) x_1 + a_{12} x_2 + \cdots + a_{1n} x_n = 0$$
$$a_{21} x_1 + (a_{22} - \lambda) x_2 + \cdots a_{2n} x_n = 0$$
$$\vdots \tag{15A1.37}$$
$$a_{n1} x_1 + a_{n2} x_2 + \cdots + (a_{nn} - \lambda) x_n = 0.$$

Then we can form the diagonal matrix D by placing the eigenvalues as the diagonal elements, and we can form the matrix X by placing the eigenvector associated with the ith eigenvalue into the ith column of X. Following the steps shown in (15A1.11), we can diagonalize the system of difference equations and then use the techniques of Chapter 15 to analyze the resulting univariate difference equations.

Appendix 15.2: Dynamic Optimization[1]

15A2.1 A RUDIMENTARY INTRODUCTION TO DYNAMIC PROGRAMMING

A dynamic programming problem consists of a sequence of periods over which the firm (or any agent) pursues an objective, such as maximizing profit. Two kinds of variables define each period: state variables and choice variables. The state variables for a given period are parameters in the firm's objective function for that period. The choice variables for that period are the arguments in that objective function. The essence of a dynamic programming problem is that the firm's choices today affect the values of the state variables for future periods, and those state variables determine the maximal levels of profit that the firm can obtain in those future periods.

The number of periods in the problem can be either finite or infinite. We will restrict our attention to problems that contain a finite number of periods.

To distinguish among the firm's objective, the state variables, and the choice variables, consider the following example of a dynamic programming problem for a monopolist.

The monopolist produces and sells a product today and tomorrow. The product is perishable; the monopolist can sell on each day only the units produced on that day. The demand functions for each day are linear. The state variables for a given day are the slope and the intercept of the demand function for that day. The choice variables for that day are the price and the quantity that the monopolist chooses for that day. Her objective is to maximize the sum of her profits over the two days.

Suppose that the demand functions for each day are fixed (although not necessarily identical). The monopolist's optimal choices in this case are obvious. She should produce and sell on each day the quantity for which marginal revenue and marginal cost for that day are equal, and set the price for each day according to the demand function for that day. This problem is not dynamic because the monopolist's choices for price and quantity today do not affect the demand function or the cost function for tomorrow.

Suppose instead that the demand function for tomorrow depends on the price that the monopolist sets for today. For example, a low price today could shift the demand function for tomorrow rightward by attracting more consumers into the market tomorrow. Or, a *high* price today could shift the demand function for tomorrow rightward by persuading consumers that the monopolist's product is a "status good." This problem *is* dynamic; the price that the monopolist sets today determines the *position* of the demand function for tomorrow. The maximal profit that she can obtain tomorrow

[1]Our presentation of dynamic programming draws substantively on the work of Edward Zabel, particularly his paper "A Dynamic Model of the Competitive Firm" (*International Economic Review*, Vol. 8, No. 2, pp. 194–208). This paper is the first of his several papers using dynamic programming to analyze the optimal behavior of competitive and monopolistic firms.

depends on the position of the demand function on that day. Therefore, to maximize the sum of her profits over the two days she must consider the consequence for tomorrow of the price that she chooses for today.

In Chapter 1 we stated that a mathematical entity can be an endogenous variable in one context of a problem and a parameter in a different context of the same problem. We have a clear example of that here. From the perspective of today, the state variables for the problem tomorrow are variables because the position of the demand function for tomorrow depends on the monopolist's choice for the price today. Once tomorrow arrives, however, the state for that day is fixed. The mathematical entities that were variables from the perspective of today become parameters tomorrow.

We will use dynamic programming to analyze a simple problem in dynamic optimization for a perfectly competitive firm. In Section 15A2.2 we present a verbal description of the general structure of a dynamic programming problem. In Section 15A2.3 we define mathematically a dynamic optimization problem for a competitive firm that has T periods in which to sell a fixed inventory when the prices in future periods are uncertain. In Section 15A2.4 we analyze that problem when the firm has only two periods during which to sell its inventory. We extend the analysis to three periods in Section 15A2.5. In Section 15A2.6 we describe some more complicated problems that economists have examined using dynamic programming. Section 15A2.7 contains a brief conclusion.

15A2.2   THE STRUCTURE OF A DYNAMIC PROGRAMMING PROBLEM THAT HAS A FINITE NUMBER, T, OF PERIODS

In a dynamic programming problem the optimal values for the choice variables in a given period depend both on the values of the state variables for that period and on the number of periods remaining in the problem. For that reason it is conventional to number the periods in reverse chronological order. If a firm has T periods in which to act, Period T is the first period. Period T is followed by period $T - 1$. The final period in the problem is period 1. The solution of a dynamic programming problem is a set of optimal choice functions, one for each period in the problem. Define the current period as Period t, and let t be any of the integers 1, T. The optimal choice function for Period t specifies the optimal values of the choice variables for Period t, based on the values of the state variables in that period, and on the fact that the firm has $t - 1$ additional periods in which to make choices.

The strategy for solving a dynamic programming problem is to divide the t periods into the current period and the remaining $t - 1$ periods, and consider those $t - 1$ periods *as if they were a single period*. We then construct a value function and an objective function for each period. The *value function* for period t is the *maximal* present value of the sequence of profits that the firm can obtain when t periods remain in the problem. The arguments of the value function for period t are the state variables for that period. The *objective function* for period t is the sum of the profit in that period and the expected present value of the value function for period $t - 1$. The arguments of the objective function for any period are the state variables and the choice variables for that period.

The objective function for period t is connected (dynamically) to the value function for period $t - 1$ because the firm's choices in period t affect one or more of the state variables for the value function in period $t - 1$. Using this fact, we can obtain the optimal policy for period t by differentiating the objective function for that period with respect to the choice variable for that period. This procedure works regardless of the total number, T, of periods in the problem, and regardless of the number, t, of periods (including the current period) that remain in the problem. By partitioning the t periods into the current period and all future periods, we convert a problem that contains t periods into an equivalent problem that contains only two periods. This technique, known as the *optimality principle*, makes dynamic programming a powerful tool.

15A2.3 A SIMPLE APPLICATION OF DYNAMIC PROGRAMMING

15A2.3.1 The Problem

We will determine the optimal behavior for a perfectly competitive firm that has a fixed level of inventory that it can sell over t periods. The firm cannot produce additional units for its inventory. The firm knows the price at which it can sell units in the present period before choosing the quantity that it will sell. The prices at which it can sell units in future periods are unknown; the firm knows only the probability distributions of those prices.

In any period the firm must choose the portion of its inventory to sell at the known current price, and the portion of inventory to retain for sales in future periods, when prices might be higher than the current price. The firm incurs a holding cost for units stored from one period to the next.

In any period t the state variables are the level of inventory with which the firm enters the period, the price at which the firm can sell units in the current period, and the number of periods that remain during which the firm can sell its inventory. We assume that any inventory that remains unsold at the end of the final period has no value. The firm's objective is to maximize the present expected value of its sales revenue, net of holding costs.

The solution to the problem is a set of choice functions, one for each period. The choice function for period t specifies the number of units that the firm should sell in period t, given the level of inventory with which the firm enters that period and given the price in that period.

15A2.3.2 The Notation for the State Variables and the Choice Variables

t = The number of periods that remain in the problem, including the current period.

X_t = The level of inventory with which the firm enters period t. X_t is a *state variable*.

S_t = The quantity that the firm chooses to sell during period t. S_t is a *choice variable*. We impose the constraint $S_t \leq X_t$ because the firm cannot increase the level of its inventory.

P_t = The price at which the firm can sell units during period t. P_t is a *state variable*; its value is an outcome of a random variable.

The level of inventory and the quantity of sales in one period are related to the level of inventory in the following period by

$$X_{t-1} = X_t - S_t. \qquad \text{(15A2.1)}$$

If the firm enters period t with X_t units in inventory, and sells S_t of those units during period t, then the firm will enter period $t - 1$ with $X_{t-1} = X_t - S_t$ units in inventory.

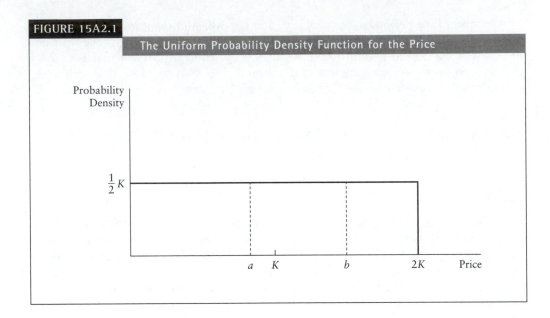

FIGURE 15A2.1

The Uniform Probability Density Function for the Price

15A2.3.3 Prices

The price is a random variable that is uniformly distributed over the interval from 0 to $2K$, in which K is a positive constant. Each day there is a new draw from this density function; draws on successive days are independent of each other.

Figure 15A2.1 is a graph of the probability density function for the price. The interval from 0 to $2K$ (including the end points) contains the possible values for the price. The probability density of any particular value for the price is measured on the vertical axis. The horizontal line with the vertical intercept $1/2K$ is the probability density function. The height of the density function for any given price is the probability (density) that the price will take that value. Since all values of a uniformly distributed random variable are equally likely, the probability density function is horizontal. The height of the density function is $1/2K$ because the area under the density function and over the interval of permissible values for the price must equal 1.

Since the price is a continuous variable we make a distinction between a *probability* and a *probability density*. "Probability" is the likelihood that the price will fall within a specified subinterval. This likelihood is the area between the horizontal axis and the probability density function over the subinterval. Since a specific value on the horizontal axis is an interval that has zero width, the probability that the price will take that exact value is zero. We can, however, consider a small subinterval that contains the specific value under consideration. As the width of the subinterval approaches zero (while still containing the specific value), the limit of the probability that the price will fall within the subinterval approaches the probability density associated with that specific value.

The probability that the price will take a value within the subinterval between points a and b on the horizontal axis in Figure 15A2.1 is $[1/2K] [b - a]$. Suppose that the subinterval is equal to the proportion, λ, of the entire interval. That is, $[b - a] = \lambda[2K - 0] = \lambda 2K$. Then the probability that the price will take a value within the subinterval from a to b is $[1/2K][\lambda 2K]$, or λ. In general, under a uniform probability distribution, the probability that the price will take a value within any subinterval is equal to the proportion of that subinterval to the entire interval. We will use this fact frequently.

Finally, the expected value of the price is equal to the midpoint, K, of the interval of its permissible values.

15A2.3.4 Holding Costs

The firm incurs a constant marginal holding cost, h, to carry its inventory from one period to the next. If X_t is the level of the inventory at the beginning of period t, and if the firm sells S_t units during that period, then $X_t - S_t$ is the level of inventory carried into period $t - 1$. The consequent holding cost is $h(X_t - S_t)$.

The holding cost to carry inventory from period t to period $t - 1$ is incurred at the beginning of period $t - 1$. Let r be the bank's rate of interest, and define $\delta = 1/[1 + r]$, so that δ is the one-period discount factor. Then a cost of \$1 incurred at the beginning of period $t - 1$ has a present value (at the beginning of period t) equal to $[1/[1 + r]]\$1 = \delta\$1$.

We specify that the marginal holding cost is less than the expected value of the price. That is, $h < K$. Without this condition, the optimal choice for the firm would be to sell its entire inventory immediately, regardless of the current price and the number of periods remaining in the horizon.

15A2.3.5 The Value Functions and the Objective Functions

Each period has a value function that specifies the *maximal* level of profit that the firm can obtain by choosing values for the choice variables *optimally* both in that period *and* throughout all the remaining periods in the problem. The arguments of the value function for a given period are the state variables for that period. The value function for period t is

$$V_t = V_t(X_t, P_t). \tag{15A2.2}$$

The definition of a value function subsumes optimal choices of values for the choice variables throughout the horizon.

Each period has an objective function defined as the sum of the firm's profit for that period and the expected present value of the value function for the following period. Let $R_t(S_t; X_t, P_t)$ be the objective function for period t. Then

$$R_t(S_t; X_t, P_t) = P_t S_t - \delta h(X_t - S_t) + \delta E[V_{t-1}(X_t - S_t, \tilde{P}_{t-1})], \tag{15A2.3}$$

in which the semicolon in $R_t(S_t; X_t, P_t)$ separates the choice variable, S_t, from the state variables, X_t and P_t. In the final term on the right side of equation (15A2.3) the first argument of V_{t-1} is $X_t - S_t$, because the level of inventory with which period $t - 1$ begins is equal to the inventory with which the firm began period t minus the quantity sold in that period. The expression $E[V_{t-1}(X_t - S_t, \tilde{P}_{t-1})]$ is the expected value[2] of $V_{t-1}(X_t - S_t, \tilde{P}_{t-1})$ with respect to the random variable $\tilde{P}_{t-1}$, which is the price for period $t - 1$. This expected value is calculated from the perspective of period t, which precedes period $t - 1$. In period t the firm knows the price for that period only; the prices for all subsequent periods are random variables from the perspective of period t.

---

[2]The expected value of a random variable is the sum of the values that the variable can take, with each value multiplied by the probability that the variable will take that value. Hence, the expected value of a random variable is a probability-weighted sum of the values that the variable can take. The expected value of $\tilde{P}_{t-1}$ is K, as we established in Section 15A2.3.3. In equation (15A2.3) we take the expected value of the *function* $V_{t-1}(X_t - S_t; \tilde{P}_{t-1})$, which depends on the random variable $\tilde{P}_{t-1}$.

The objective function in equation (15A2.3) for period t is the firm's revenue from sales in the current period, minus the present value of the holding cost to carry the unsold inventory into period $t - 1$, plus the *maximal* expected level of profits that the firm can obtain over the remaining $t - 1$ periods. Given the values of the state variables X_t and P_t, the value chosen for S_t determines the revenue for period t, the holding cost that the firm will incur at the beginning of period $t - 1$, and the level of inventory with which the firm will enter period $t - 1$. The value function V_{t-1} subsumes optimal behavior from period $t - 1$ onward. Since the objective function for period t contains the value function for period $t - 1$, the firm needs to take all future periods into account when choosing the quantity to sell in period t.

15A2.3.6   The Optimal Choice Functions

The firm's objective is to allocate a fixed initial level of inventory to sales over t periods so as to maximize the expected present value of the sequence of profits. In any given period, the firm chooses the quantity that it will sell during that period, given the current level of its inventory and the current value of the price. The firm defers its choices on the quantities to sell in all future periods until it observes the prices in those periods.

Notice that the choice variable for period t, S_t, appears in all three terms on the right side of equation (15A2.3). Define $S_t^* = S_t^*(X_t, P_t)$ as the optimal quantity to sell in period t. If the value function, V_{t-1}, on the right side of (15A2.3) is differentiable with respect to S_t, then S_t^* is the solution to

$$d[R_t(S_t^*; X_t, P_t)]/dS_t = 0, \quad \text{or}$$

$$P_t + \delta h + \delta d[E[V_{t-1}(X_t - S_t^*, \tilde{P}_{t-1})]]/dS_t = 0.$$

(15A2.4)

The appearance of S_t^* for S_t in the functions R_t and V_{t-1} indicates that we evaluate the derivatives at the point $S_t = S_t^*$.

The optimality condition in equation (15A2.4) requires the firm to choose the value for S_t^* that will equate marginal revenue to marginal cost, with both evaluated for their effect on the present value of the *sequence* of profits. Selling one more unit in period t will increase the present value of profits by P_t, the extra revenue from that unit, plus δh, the discounted marginal holding cost that the firm will no longer incur. On the other hand, selling one more unit in period t will reduce by one unit the level of inventory with which the firm will begin period $t - 1$. The marginal value of inventory on the maximal expected profit that the firm can obtain over $t - 1$ periods is the derivative of the expected value of the value function for period $t - 1$ with respect to the level of sales in period t. We must discount this marginal value because period $t - 1$ begins one period after period t.

15A2.4   SOLVING THE PROBLEM WHEN THERE ARE TWO PERIODS (T = 2)

The standard procedure for solving a dynamic programming problem is to work backward from the final period in which the firm can act. Accordingly, we will first determine the optimal policy for the final period and then use that result to determine the optimal policy for period 2, which is the first period in a two-period horizon.

15A2.4.1 Determining $S_1^*(X_1; P_1)$

Let X_1 be the firm's level of inventory at the beginning of the final period of the horizon, and let P_1 be the price for that period. The firm's objective function when only one period remains is

$$R_1(S_1; X_1, P_1) = P_1 S_1. \tag{15A2.5}$$

Since period 1 is the firm's final opportunity to sell its inventory, the obvious choice is to sell the entire inventory regardless of the price. Then the optimal policy for period 1 is

$$S_1^*(X_1; P_1) = X_1. \tag{15A2.6}$$

The firm's supply function in the final period is perfectly inelastic at the level of its inventory.

15A2.4.2 The Value Function for Period 1

The value function for period 1 specifies the maximal present value of the firm's profits when only one period remains. As is the case for any value function, we obtain the value function for period 1 by substituting the optimal policy for period 1 into the objective function for that period. The result is

$$V_1(X_1, P_1) = P_1 S_1^*(X_1; P_1) = P_1 X_1. \tag{15A2.7}$$

To proceed to the problem in which the firm has two periods remaining, we need the marginal expected value of the value function for period 1 with respect to the level of inventory in that period. From the perspective of period 2 the price for period 1 is a random variable. Then the expected value of $V_1(X_1, \tilde{P}_1)$ is

$$
\begin{aligned}
E[V_1(X_1, \tilde{P}_1)] &= E[\tilde{P}_1 X_1] \\
&= E[\tilde{P}_1] X_1 \\
&= K X_1.
\end{aligned} \tag{15A2.8}
$$

Differentiating equation (15A2.8) with respect to X_1 yields

$$d\{E[V_1(X_1, \tilde{P}_1)]\}/dX_1 = K. \tag{15A2.9}$$

The marginal effect on the expected value of profits in period 1 of an increase in the level of inventory for that period is equal to the expected value of the price for that period. This is understandable because the optimal policy in period 1 requires the firm to sell the entire inventory, X_1, with which it enters that in period.

15A2.4.3 Determining $S_2^*(X_2, P_2)$

Period 2 is the first period of a two-period horizon. We want to choose the level of sales in period 2 that will maximize the value of the objective function for that period. Recalling that the objective function for period 2 includes the discounted expected value of the value function for period 1, we have

$$R_2(S_2; X_2, P_2) = P_2 S_2 - \delta h(X_2 - S_2) + \delta E[V_1(X_2 - S_2, \tilde{P}_1)]. \tag{15A2.10}$$

Using equation (15A2.8) to evaluate the final term on the right side of equation (15A2.10), we can rewrite the objective function for period 2 as

$$R_2(S_2; X_2, P_2) = P_2 S_2 - \delta h(X_2 - S_2) + \delta K(X_2 - S_2). \qquad \textbf{(15A2.11)}$$

Differentiating $R_2(S_2; X_2, P_2)$ with respect to S_2 yields

$$d\{R_2(S_2^*; X_2, P_2)\}/dS_2 = P_2 + \delta h - \delta K. \qquad \textbf{(15A2.12)}$$

The derivative in equation (15A2.12) is a constant with respect to S_2; therefore, the optimal quantity to sell in period 2 is a corner solution. If the price in period 2 is at least as high as $\delta(K - h)$, the optimal choice in period 2 is to sell the entire inventory immediately. Alternatively, if the price is less than $\delta(K - h)$, the firm should sell nothing in period 2, and carry its entire inventory into period 1.

We can write the optimal sales policy for period 2 as

$$\begin{aligned} S_2^*(X_2, P_2) &= X_2 \quad \text{if } P_2 \geq \delta(K - h), \quad \text{and} \\ &= 0 \quad \text{if } P_2 < \delta(K - h). \end{aligned} \qquad \textbf{(15A2.13)}$$

We assume that if the firm is indifferent between selling its entire inventory in period 2 and selling nothing in that period, then the firm sells its entire inventory in period 2.

Define the firm's *reservation price* for period 2 as

$$P_2^R = \delta(K - h). \qquad \textbf{(15A2.14)}$$

A reservation price is a threshold below which the firm will not enter the market. We see from equation (15A2.13) that the firm will not enter the market in period 2 unless the price is at least as high as P_2^R. The form of $S_2^*(X_2; P_2)$ in equation (15A2.13) indicates that the firm's supply curve for period 2 coincides with the vertical axis between zero and the reservation price $\delta(K - h)$. For prices equal to or greater than $\delta(K - h)$ the firm's supply curve is perfectly inelastic at the quantity X_2.

The reservation price for period 2 depends positively on both the discount factor, δ, and the expected value of the price in period 1, K, and negatively on the marginal holding cost, h. An increase in δ increases the present value of future profits. Consequently, higher values for δ increase the threshold that the current price must reach if the firm is to sell any of its inventory in the current period. Similarly, an increase in the expected value, K, of the price for the next period increases the reservation price for period 2. An increase in the marginal holding cost, h, increases the cost of obtaining future profits. Consequently, larger values for h reduce the reservation price for period 2.

In equation (15A2.12) we differentiated R_2 with respect to S_2. Selling one more unit in period 2 means carrying one less unit into the inventory for period 1. Therefore, the derivative in equation (15A2.12) is equal to the marginal revenue (from sales) in period 2, plus the discounted marginal holding cost, minus the discounted expected marginal profit with respect to the level of inventory in period 1. Since the firm is a perfect competitor, the firm's marginal revenue for sales in period 2 is equal to the price, P_2. The discounted marginal holding cost is δh. This term appears with a plus sign in equation (15A2.12) because selling one more unit in period 2 allows the firm to save the marginal holding cost. From the perspective of period 2, the maximal expected profit for the final period is $E[V_1(X_2 - S_2, \tilde{P}_1)]$. Consequently, the (discounted) marginal effect of an increase in sales in period 2 on the expected maximal profit for the final period is (using the chain rule and the fact that $X_1 = X_2 - S_2$)

$$\delta d\{E[V_1(X_1, \tilde{P}_1)]\}/dS_2 = \delta d\{[E[V_1(X_1, \tilde{P}_1)]/dX_1\}[dX_1/dS_2]$$
$$= \delta K(-1) \tag{15A2.15}$$
$$= -\delta K.$$

15A2.5 SOLVING THE PROBLEM WHEN THERE ARE THREE PERIODS (T = 3)

15A2.5.1 The Value Function for Period 2

The first step in solving the problem for three periods is to obtain the value function, $V_2(X_2, P_2)$, for period 2. Using the value function for period 1 that we obtained in equation (15A2.7), the objective function for period 2 specified in equation (15A2.10), and the optimal policy for period 2 specified in equation (15A2.13), the value function for period 2 is

$$V_2(X_2, P_2) = R_2(S_2^*; X_2, P_2)$$
$$= P_2 S_2^*(X_2; P_2) - \delta h[X_2 - S_2^*(X_2; P_2)] \tag{15A2.16}$$
$$+ \delta E[V_1(X_2 - S_2^*(X_2; P_2), P_1)].$$

In the final step above we substituted $S_2^*(X_2; P_2)$ for S_2^* to recognize that the optimal level of sales in period 2 is a function of the parameters for that period, X_2 and P_2.

Next, we obtain the expected value of V_2 with respect to $\tilde{P}_2$. The firm calculates this expected value from the perspective of period 3, when it is deciding how much to sell in that period, and consequently how much inventory to bequeath to period 2. This expected value is

$$E[V_2(X_2, \tilde{P}_2)] = E[R_2(S_2^*(X_2; \tilde{P}_2); X_2, \tilde{P}_2]$$
$$= E[P_2 S_2^*(X_2; \tilde{P}_2)] - \delta h[X_2 - S_2^*(X_2; \tilde{P}_2)] \tag{15A2.17}$$
$$+ \delta E[V_1(X_2 - S_2^*(X_2; \tilde{P}_2), P_1)].$$

The calculation of the expected value in equation (15A2.17) is complicated because $S_2^*(X_2, \tilde{P}_2)$ depends on the random variable $\tilde{P}_2$.

From equation (15A2.13) we know that the firm should sell its entire inventory in period 2 if $P_2 \geq \delta(K - h)$, and sell nothing in that period if $P_2 < \delta(K - h)$. Consequently, we can calculate the expected value of $V_2(X_2, \tilde{P}_2)$ by considering separately the two cases determined by the boundary $\delta(K - h)$ in Figure 15A2.2.[3]

Suppose that P_2 takes a value in the upper subinterval $[\delta(K - h), 2K]$. In this event the firm sells its entire inventory of X_2 immediately, leaving no inventory to sell in the final period. Consequently, the only revenue that the firm obtains is from sales in period 2. The expected value of this revenue is equal to X_2 multiplied by the probability-weighted sum of the values that the price can take over the subinterval $[\delta(K - h), 2K]$. This expected value is equal to[4]

$$X_2(4K^2 - \delta^2 K^2 + 2\delta^2 Kh - \delta^2 h^2)(1/4K). \tag{15A2.18}$$

---

[3] Remember that the prices are identically distributed for all periods.

[4] Obtaining this result requires a simple application of integral calculus. The expected value of the price over the subinterval $[\delta(k - h), 2K]$ is $(1/2K)$ times the value of the integral of $P\,dP$ from $P = \delta(k - h)$, to $P = 2K$.

Introduction to Dynamics: Theory

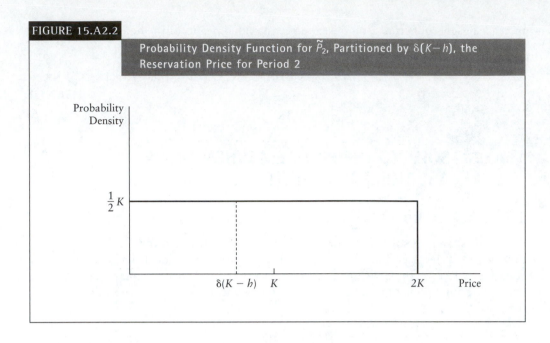

FIGURE 15.A2.2

Probability Density Function for $\tilde{P}_2$, Partitioned by $\delta(K-h)$, the Reservation Price for Period 2

Probability Density

$\frac{1}{2}K$

$\delta(K-h)$ K $2K$ Price

Suppose instead that P_2 takes a value in the lower subinterval $[0, \delta(K - h)]$. In this event the firm sells none of its inventory in period 2. The probability of this event is equal to the proportion of the entire interval occupied by the subinterval, namely $[\delta(K - h) - 0]/([2K - 0] = \delta(K - h)/2K$. The expected value of the price in period 1 is K. The expected value of the firm's revenue in period 1 is equal to the probability that the firm will defer sales until that period, multiplied by the expected value of the price in that period, multiplied by the level of inventory carried into that period. Consequently, the expected value of revenue from period 1 is

$$X_2\delta[(K - h)/2K]K = X_2\delta(K - h)/2. \tag{15A2.19}$$

The expected value of the value function for period 2 is the sum of equation (15A2.18) and equation (15A2.19). After simplification we have

$$\begin{aligned} E[V_2(X_2, \tilde{P}_2)] &= X_2(1/4K)(4K^2 - \delta^2 K^2 + 2\delta^2 Kh - \delta^2 h^2) + X_2\delta(K - h)/2 \\ &= X_2[K^2(4 - \delta^2) + 2\delta K(K - h) + \delta^2 h(2K - h)](1/4K), \end{aligned} \tag{15A2.20}$$

which is positive because $K > h$ and $\delta < 1$.

For convenience, define

$$\hat{P_2} = [K^2(4 - \delta^2) + 2\delta K(K - h) + \delta^2 h(2K - h)](1/4K). \tag{15A2.21}$$

Then

$$E[V_2(X_2, \tilde{P}_2)] = \hat{P_2}X_2. \tag{15A2.22}$$

Notice that $\hat{P_2}$ is a constant that depends only on the fixed parameters δ, h, and K. That is, $\hat{P_2}$ is not a state variable.

15A2.5.2   A Digression for Comparative Statics

From equations (15A2.21) and (15A2.22) it is immediate that

$$d\{E[V_2(X_2, \tilde{P}_2)]\}/dX_2 = d\{\hat{P_2}X_2\}/dX_2 = \hat{P_2} > 0. \tag{15A2.23}$$

The marginal effect of inventory on the maximal expected present value of profits over two periods is the positive constant $\hat{P_2}$. Obviously, a larger inventory at the beginning of period 2 means a larger expected profit over the two-period horizon.

We also have

$$d\{E[V_2(X_2, \tilde{P}_2)]\}/dh = X_2 d[\hat{P_2}]/dh$$
$$= X_2(1/2K)\,\delta[-\delta h + K(\delta - 1)] < 0. \qquad \textbf{(15A2.24)}$$

The marginal effect on the maximal expected present value of profits over the two-period horizon of an increase in the holding cost is negative, as we would expect.

Finally, the marginal effect of an increase in the expected value of the price is (after several simplifying steps)

$$d\{E[V_2(X_2, \tilde{P}_2)]\}/dK = X_2 d[P_2^R]/dK$$
$$= X_2(1/16\,K^2)[16\,K^2 + 8\,\delta K(K - h) + 4\,\delta h(2 + \delta h)] > 0,$$

$$\textbf{(15A2.25)}$$

because $K > h$. An increase in the expected value of the price increases the expected value of profits for the two-period horizon.

15A2.5.3 Determining $S_3^*(X_3, P_3)$

The objective function for period 3 is

$$R_3(S_3; X_3, P_3) = P_3 S_3 - \delta h(X_3 - S_3) + \delta E[V_2(X_2, \tilde{P}_2)]$$
$$= P_3 S_3 - \delta h(X_3 - S_3) + \delta P_2^R X_2 \qquad \textbf{(15A2.26)}$$
$$= P_3 S_3 - \delta h(X_3 - S_3) + \delta P_2^R(X_3 - S_3),$$

using equation (15A2.22) and the fact that $X_2 = X_3 - S_3$.

The derivative of R_3 with respect to S_3 is:

$$d\{R_3(S_3; X_3, P_3)\}/dS_3 = P_3 + \delta h - \delta P_2^R, \qquad \textbf{(15A2.27)}$$

which is a constant with respect to S_3. Consequently, the optimal quantity to sell in period 3 is a corner solution that is analogous to the optimal quantity to sell in period 2. Specifically,

$$S_3^*(X_3; P_3) = X_3 \quad \text{if } P_3 \geq \delta(P_2^R - h), \quad \text{and}$$
$$S_3^*(X_3; P_3) = 0 \quad \text{if } P_3 < \delta(P_2^R - h). \qquad \textbf{(15A2.28)}$$

Define the firm's reservation price for period 3 as

$$P_3^R = \delta(P_2^R - h). \qquad \textbf{(15A2.29)}$$

As is the case for period 2, the firm's supply function for period 3 is perfectly inelastic above a reservation price.

The reservation prices increase as the number of periods that remain increases. Let P_1^R be the reservation price for period 1. We know that is $P_1^R = 0$; no matter what the price in period 1, the optimal choice is to sell the entire inventory. The reservation price for period 2 is $P_2^R = \delta[K - h]$, and the reservation price for period 3 is $P_3^R = \delta[P_2^R - h]$.

Subtracting P_2^R from P_3^R, we have (after simplification)

$$P_3^R - P_2^R = (1/4K)[K^2(4 + \delta^2) + 2\delta K(K - h) + \delta^2 h(2K - h)] > 0, \qquad \textbf{(15A2.30)}$$

because $K > h$ and $\delta < 1$. Then

$$P_3^R > P_2^R > P_1^R = 0. \qquad \textbf{(15A2.31)}$$

The reservation prices increase as the length of the horizon increases from one period to three periods.[5] This is reasonable. The more opportunities that the firm has to wait for a better price, the higher the current price must be if the firm is to enter the market in the current period.

15A2.6   A BRIEF DESCRIPTION OF MORE COMPLICATED PROBLEMS[6]

Economists have applied dynamic programming to more complicated problems than the problem that we analyzed in this appendix. For example, we could allow the marginal holding cost to increase as the quantity carried into the next period increases. We could also allow the firm to produce additional units of its product in any period. In this case the firm would have *two* choice variables. Each period the firm would choose both the quantity to produce (which could be zero) and the quantity to sell (which could not exceed the level of inventory with which the firm entered that period plus whatever the firm produces in that period).

In this appendix we assumed that the firm knows the probability distribution for the prices. We could instead suppose that the firm does not know the distribution but adjusts its beliefs about it based on the observation of prices over time. We could also extend this analysis to the case of a monopolist who is uncertain of the demand function for her product and who experiments by changing the price from one period to the next. In this case, the firm's choice variables in each period are the quantity to produce and the price to charge. The state variables for each period are the level of inventory with which she enters that period and her accumulated knowledge about the demand function obtained from her observations of the quantities that she was able to sell at various prices in earlier periods.

15A2.7   CONCLUSION

In this appendix we introduced the mathematical technique of dynamic programming to analyze problems of dynamic optimization. An optimization problem is *dynamic* if the objective function is defined over a sequence of periods and if the agent's choices in one period affect the values of the state variables in subsequent periods.

To obtain the agent's optimal choices in a dynamic programming problem we use the *optimality principle*. This principle simplifies the problem by enabling us to collapse t periods into two periods. The current period[7] is the first of these periods. The

---

[5]We assert, without proof, that the reservation prices continue to increase as the horizon increases beyond three periods. To prove this assertion one would use mathematical induction to show that if P_t^R is the reservation price when the firm has t more periods during which to sell its inventory, and P_{t+1}^R is the reservation price when there are $t + 1$ more periods, then for any integral value of t, $P_{t+1}^R > P_t^R$.

[6]For examples of these extensions of dynamic programming see the references at the end of this appendix.

[7]Recall that we number the periods in reverse chronological order.

second period consists of the remaining $t - 1$ periods collapsed into a single period. The agent's choices in the current period affect the value of the objective function for the original t periods directly through the current period and indirectly by affecting the state variables for the following period. Then for any value of $t \geq 2$, we obtain the agent's optimal choices for the current period by equating the direct and the indirect marginal values of those choices.

The mathematical technique of dynamic programming and the *optimality principle* enable economists to conduct analyses that would otherwise be impracticably difficult, if not impossible. For an analogy, the reader should recall our use of matrix algebra in several of the earlier chapters to conduct comparative static analyses that would be impossible without matrix algebra.

REFERENCES

Elmaghraby, W., and P. Keskinocak, "Dynamic pricing in the presence of inventory considerations: Research overview, current practices, and future directions," *Management Science*, 49(10), pp. 1287–1309, October, 2003.

Fisher, M., K. Ramdas, and Y. S. Zheng, "Ending inventory valuation in multiperiod production scheduling," *Management Science*, 47(5), pp. 679–692, May, 2001.

Hsu, V. N., "Dynamic economic lot size with perishable inventory," *Management Science*, 46(8), pp. 1159–1169, August, 2000.

Petruzzi, N. C., and M. Dada, "Dynamic pricing and inventory control with learning," *Naval Research Logistics*, 49(3), pp. 303–325, April, 2002.

Zabel, Edward, "A Dynamic Model of a Competitive Firm," *International Economic Review*, Vol. 8, No. 2, June 1967, pp. 194–208.

_____, "The Competitive Firm and Price Expectations," *loc. cit.*, Vol. 10, No. 3, October 1969, pp. 467–478.

_____, "The Competitive Firm, Uncertainty, and Capital Accumulation," *loc. cit.*, Vol. 14, No. 3, October 1973, pp. 765–779.

16 Difference and Differential Equations: Applications

16.1 INTRODUCTION

In this chapter we present several economic applications of difference and differential equations. Many applications can be analyzed using both discrete- or continuous-time models, so we discuss both difference and differential equations models in several sections. Readers may find it helpful to compare the notation used in each application with the generic notation of Chapter 15, for example, the linear first-order difference and differential equations (15.1) and (15.50).

We begin with a generic partial-adjustment model; such models are used frequently in both micro- and macroeconomics. We therefore discuss the dynamics of four versions of the model: both linear and nonlinear, using both discrete and continuous time. Specific economic applications of partial-adjustment models are explored in some end-of-chapter problems as well as in some of the later applications in this chapter.

In Chapter 15 we used as an example the dynamics of adjustment to market equilibrium in a demand-and-supply model. (That was another example of a partial-adjustment model.) In that example, prices (partially) adjusted when the market was out of equilibrium. This kind of market adjustment is called Walrasian adjustment, after the French economist Leon Walras (1834–1910). Another possible adjustment is Marshallian adjustment, named after the British economist Alfred Marshall (1842–1924); Marshall assumed that when markets were out of equilibrium, firms would see that price was not equal to marginal cost and would therefore adjust quantity in order to increase profits. In Section 16.3 we explore Marshallian dynamics in both discrete and continuous time. The analysis of the dynamics of market adjustment is often referred to as stability analysis, since a key issue is whether the dynamic process has a stable steady state. If not, the comparative statistic analyses we have done in previous chapters are invalid since the market will never reach a new equilibrium when a parameter changes. In the next application we investigate yet another market adjustment mechanism. This linear model, in discrete time, is often called a cobweb model because a graph showing the dynamic path of prices and quantities in the market resembles a cobweb.

We then turn to two oligopoly applications. In the first, a Cournot model, the number of firms adjusts through time in response to profit opportunities in the market. We include both discrete-time and continuous-time versions. Section 16.6 explores the dynamics of a Cournot duopoly with partial adjustment. This is an example of a two-variable system of equations, and we discuss both difference and differential equations systems.

The final microeconomic application is a simple model of a fishery. This nonlinear, continuous-time model investigates the impact of human harvesting of fish on the dynamics of the fish population. Similar models can be used for other renewable resources, for example, models of forestry and hunting.

The chapter ends with four macroeconomic applications. The first is a partial-adjustment model based on a linear IS–LM model; the dynamics are provided by a Fed reaction function in which monetary policy reacts to differences between the equilibrium interest rate and a target interest rate. In the application we use discrete time and therefore difference equations; problems at the end of the chapter analyze the model using continuous time. Section 16.9 explores a simple dynamic IS–LM model in continuous time; it is thus an example of a two-equation system of differential equations. The third macroeconomic application is a discrete-time model of an expectations-augmented Phillips curve; this is an example of a second-order difference equation. The final section of the chapter discusses a model of economic growth. We analyze a fairly simple version of a Solow growth model, named after the American economist Robert Solow (1924–). This nonlinear differential equation model can be extended in many ways, some of which are explored in the problems that conclude the chapter.

16.2   PARTIAL–ADJUSTMENT MODELS

Consider an economic agent with a choice variable x. The agent has a target level $\hat{x}$ for the choice variable, for example, the value that optimizes some objective function, but at any particular time x may not equal its target level. This is often the case because the agent must choose a value for x before all relevant information is known. Furthermore, it may be costly to adjust x quickly, so the agent prefers gradual changes in x even though this implies choices that differ from the optimal levels at particular times. One example of when this model would be appropriate is when firms have a target level of inventories but, because demand differs from what they anticipated, their actual inventories do not always match their planned inventories. Another example is when workers supply labor based on their expectations of the real wage, but their expectations take time to adjust to changes in the aggregate price level. This worker-misperception model is often used to justify an upward-sloping aggregate supply curve.

To formalize, at time t the value of the choice variable is x_t while the target level is $\hat{x}$. Because of the interest in gradual changes, the agent changes the value of x based on the difference $(x_t - \hat{x})$. We will investigate four versions of this adjustment model:

$$x_{t+1} = x_t + \gamma(\hat{x} - x_t) \qquad 0 < \gamma \leq 1 \qquad \text{(discrete linear)} \qquad \textbf{(16.1)}$$

$$x_{t+1} = x_t + f(\hat{x} - x_t) \qquad f' \geq 0,\, f(0) = 0 \quad \text{(discrete nonlinear)} \qquad \textbf{(16.2)}$$

$$\dot{x}_t = \delta(\hat{x} - x_t) \qquad\qquad \delta > 0 \qquad\qquad \text{(continuous linear)} \qquad \textbf{(16.3)}$$

$$\dot{x}_t = g(\hat{x} - x_t) \qquad\qquad g' \geq 0,\, g(0) = 0 \quad \text{(continuous nonlinear)} \qquad \textbf{(16.4)}$$

16.2.1   Discrete Linear Case

For a linear partial-adjustment model in discrete time, equation (16.1) is the difference equation to be analyzed. Rewritten to conform to the notation in Chapter 15, the first-order difference equation is

$$x_t = (1 - \gamma) x_{t-1} + \gamma \hat{x}. \tag{16.5}$$

This has a stable steady state: the assumption that $0 < \gamma \leq 1$ ensures that $0 \leq 1 - \gamma < 1$ so x_t will approach a stable steady state monotonically. Note that if $\gamma = 1$ x_t always equals the target level, so there are no interesting dynamics. If $\gamma < 1$ the agent partially adjusts toward the target level, which yields a gradual and monotonic approach to the stable steady state.

The steady-state value of x is

$$x^* = \frac{\gamma \hat{x}}{1 - (1 - \gamma)} = \hat{x} \tag{16.6}$$

and the definite solution to (16.5) is

$$x_t = (1 - \gamma)^t x_0 + \hat{x}(1 - (1 - \gamma)^t). \tag{16.7}$$

The closer γ is to one, the faster is the adjustment process; for small values of γ, the choice variable is changed by small amounts each period. It takes more periods to reach the steady state, for a given difference between x_0 and the steady-state value.

Now suppose that the target level of x, $\hat{x}$, depends on some parameter α. If α changes, so does the steady-state value of x. But α does not affect the stability of the steady state, which continues to depend only on γ. Nor does α affect the speed of adjustment, except by changing the steady-state value and therefore the difference between the initial and steady-state values. So a permanent change in the parameter starting at an arbitrary time changes the steady-state value and x changes smoothly and monotonically to the new steady state, with the amount of time needed to reach the new steady state depending on γ and how far apart are the new steady state and the starting value of x.

An alternative modeling approach would be to think of this as a special case of a time-varying exogenous term in the difference equation. That is, the difference equation might be modeled as

$$x_t = (1 - \gamma)x_{t-1} + \gamma \hat{x}_t. \tag{16.8}$$

This is probably a realistic model of many economic processes, where the equilibrium value of a variable changes periodically in response to changes in economic conditions, but economic agents do not immediately and fully change to the new equilibrium values. As long as $0 < \gamma < 1$, the variable will always adjust toward its target (equilibrium) level, whatever that level may be at the time. But if the target level changes frequently, or if the speed of adjustment is low, x may not reach its new steady state before the target level changes again and creates a new steady state. So the variable is constantly adjusting (partially) toward a moving target.

16.2.2 Discrete Nonlinear Case

We now generalize the partial-adjustment model so that the adjustment is a (possibly nonlinear) function of the difference between the target and past levels of the variable x. A reasonable functional form in many economic contexts would be one similar

FIGURE 16.1

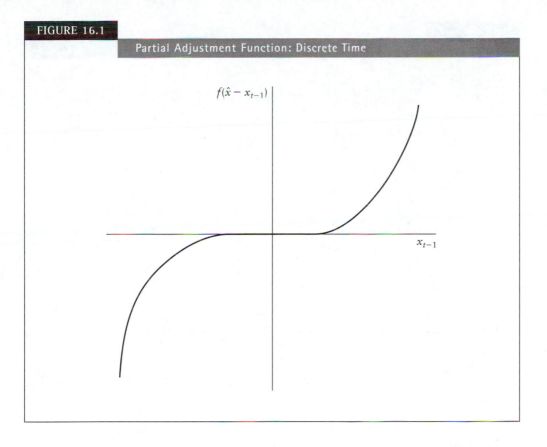

to that graphed in Figure 16.1. As long as the previous level of x is in the vicinity of the target level, no adjustment is made since there may be fixed costs to adjustment. The further away from the target is the previous level of x, the greater will be the adjustment. Rewriting equation (16.2) as a difference equation in x_t rather than x_{t+1} yields

$$x_t = x_{t-1} + f(\hat{x} - x_{t-1}) \tag{16.9}$$

from which it is clear that a steady-state value is $\hat{x}$ since $f(0) = 0$:

$$x^* = x^* + f(\hat{x} - x^*) \Rightarrow f(\hat{x} - x^*) = 0 \Rightarrow x^* = \hat{x}. \tag{16.10}$$

For the functional form shown in Figure 16.1, however, there are multiple steady states: as long as x_{t-1} is in the vicinity of $\hat{x}$, the function f will equal zero and x will be in a steady state since its value will not change.

To sketch the phase diagram of equation (16.9), take the derivative with respect to x_{t-1}:

$$\frac{\partial x_t}{\partial x_{t-1}} = 1 - f' \tag{16.11}$$

which, based on the assumptions about f', is less than or equal to one. The stability of the steady state and whether or not the process oscillates depend on the magnitude of f'. If $f' = 0$, which for the functional form shown in Figure 16.1 is true when x_{t-1} is in

FIGURE 16.2

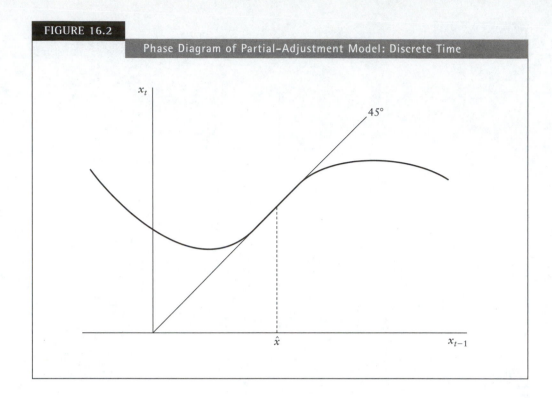

Phase Diagram of Partial-Adjustment Model: Discrete Time

the vicinity of the target level $\hat{x}$, the steady state is not stable in the sense that if x is at the steady state and then changed slightly, it will remain at its new level rather than returning to its old level. The new level will also be a steady state, unstable in the same sense as long as x is in the range for which $f' = 0$. Outside of that range, $f' > 0$ so the difference equation is either downward-sloping (if $f' > 1$), upward-sloping with slope less than one (if $0 < f' < 1$), or horizontal (if $f' = 1$). The phase diagram corresponding to the functional form shown in Figure 16.1 is shown in Figure 16.2. There is a region of steady-state values of x: all values of x_{t-1} for which $f(\hat{x} - x_{t-1}) = 0$. According to Figure 16.2, x will move toward this region of steady-state values for all starting points except potentially ones very far away from $\hat{x}$. Depending on the starting point, there might be some oscillations, but for initial values near $\hat{x}$ the process will be monotonic.

Although the adjustment function shown in Figure 16.1 seems economically reasonable, the results based on this function cannot be generalized. It is at least as reasonable to suppose that there is only one value of x_{t-1} for which $f(\hat{x} - x_{t-1}) = 0$, in which case there is a unique steady state. And it is certainly easy to find functions for which $f'(\hat{x} - x_t) > 2$ for values of x_t near $\hat{x}$, implying an unstable steady state, though it would be more difficult to find a reasonable economic justification for such an adjustment function.

16.2.3   Continuous Linear Case

A continuous-time version of the partial-adjustment model is given by equation (16.3), which can be rewritten as

$$\dot{x}_t = -\delta x_t + \delta \hat{x} \tag{16.12}$$

which clearly has a stable steady state since $-\delta < 0$. Since the differential equation is linear, the approach to the steady state will be monotonic. The steady-state value is

$$x^* = -\frac{\delta \hat{x}}{(-\delta)} = \hat{x} \qquad (16.13)$$

so the dynamic process converges toward the target value of x, with the speed of adjustment being greater the larger is δ. The definite solution to the differential equation (16.12) is

$$x_t = \left(x_0 + \frac{\delta \hat{x}}{(-\delta)} \right) e^{-\delta t} - \frac{\delta \hat{x}}{(-\delta)} = (x_0 - \hat{x})e^{-\delta t} + \hat{x} = x_0 e^{-\delta t} + \hat{x}(1 - e^{-\delta t}). \qquad (16.14)$$

Using the linear partial-adjustment model in continuous time is therefore very convenient since a stable steady state and a monotonic process are guaranteed.

16.2.4   Continuous Nonlinear Case

A unique and stable steady state and a monotonic process are guaranteed even for the more general adjustment function shown in equation (16.4), as long as $g'(\hat{x} - x_t) > 0$ everywhere. This is because the differential equation (16.4) is everywhere downward-sloping: $\partial \dot{x}_t / \partial x_t = -g' < 0$. Thus, as shown in Figure 16.3, the graph of the differential equation crosses the horizontal axis in only one place, where $g(\hat{x} - x_t) = 0 \Rightarrow x_t = \hat{x}$. If, similar to the discrete-time adjustment function shown in Figure 16.1, there is a range of values near $\hat{x}$ for which $g' = 0$, then all of those values are steady states and the dynamic

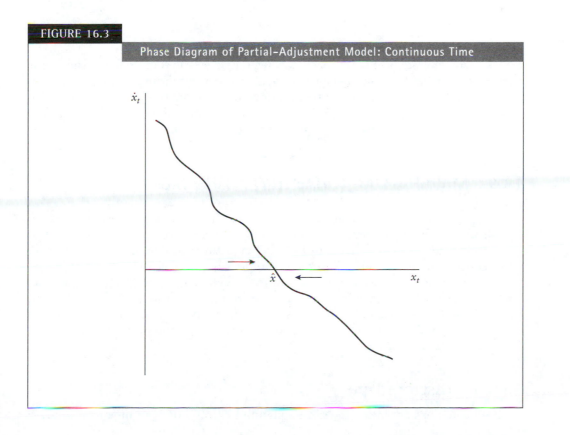

FIGURE 16.3

Phase Diagram of Partial-Adjustment Model: Continuous Time

FIGURE 16.4

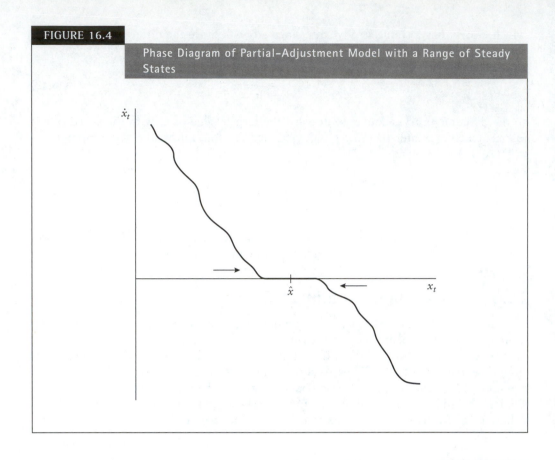

process will monotonically approach the region of steady-state values. The phase diagram for this case is shown in Figure 16.4. As was true for the linear case, the nonlinear partial-adjustment model in continuous time is considerably simpler to analyze than its counterpart in discrete time. This is one reason many economists choose to use differential-equation models rather than difference-equation models.

16.3 MARSHALLIAN (QUANTITY) ADJUSTMENT

In Chapter 15 we used as an example a Walrasian model of markets adjusting to equilibrium. In the Walrasian model, prices adjust when market supply does not equal market demand. A different model of market adjustment is due to Alfred Marshall, who posited that firms would adjust their quantity supplied when they were not maximizing profits because price did not equal the marginal cost of production. A mathematical model of Marshallian dynamics consists of three equations: a market demand curve, a marginal cost curve for the industry (in competitive markets this is the market supply curve), and an equation showing how market quantity adjusts when price does not equal marginal cost.

16.3.1 Discrete Time

If all equations are linear, the three-equation system in discrete time is as follows:

$$P_t = a - bQ_t \tag{16.15}$$

$$MC_t = c + dQ_t \tag{16.16}$$

$$Q_{t+1} = Q_t + \gamma(P_t - MC_t) \tag{16.17}$$

where all parameters are assumed to be positive. Substituting equations (16.15) and (16.16) into (16.17) yields a first-order linear difference equation:

$$Q_{t+1} = Q_t + \gamma(a - bQ_t - c - dQ_t) = (1 - \gamma(b + d))Q_t + \gamma(a - c) \tag{16.18}$$

or, written as a difference equation for Q_t rather than Q_{t+1},

$$Q_t = (1 - \gamma(b + d))Q_{t-1} + \gamma(a - c). \tag{16.19}$$

The steady state for this process equals

$$Q^* = \frac{\gamma(a - c)}{1 - (1 - \gamma(b + d))} = \frac{a - c}{b + d}. \tag{16.20}$$

(Note that this is the same as the steady state from the Walrasian dynamics model of Chapter 15. The reader will have to be careful when confirming this because of differences in notation: in Chapter 15 a, b, c, and d are parameters of the demand and supply functions whereas here they are parameters of the *inverse* demand and supply functions.) Stability and dynamics of the system depend on the sign and magnitude of $1 - \gamma(b + d)$. The assumption that all parameters of the three equations (16.15)–(16.17) are positive implies that $1 - \gamma(b + d) < 1$, but stability also requires further assumptions to ensure that $1 - \gamma(b + d) > -1$. The extra condition required is that $\gamma(b + d) < 2$ or $\gamma < 2/(b + d)$. In order for the process to be monotonic, $1 - \gamma(b + d)$ has to exceed zero, so $\gamma < 1/(b + d)$. If $1/(b + d) < \gamma < 2/(b + d)$, then the process oscillates around the steady-state value as it approaches the steady state.

The definite solution for Q_t is

$$Q_t = (1 - \gamma(b + d))^t Q_0 + \left(\frac{a - c}{b + d}\right)(1 - (1 - \gamma(b + d))^t). \tag{16.21}$$

How quickly the process approaches the steady state (assuming it exists) depends not only on the adjustment parameter γ but also on how steep the demand and marginal cost curves are: the process will approach the steady state faster the closer $1 - \gamma(b + d)$ is to zero. Faster convergence with steeper demand and marginal cost curves makes some sense since with steep demand and marginal cost curves, quantity will never be far from the equilibrium value. If the demand and marginal cost curves are too steep, however, the process will oscillate; if $\gamma(b + d) > 2$ the process will oscillate explosively and there will be no steady state.

In comparison, the reader can confirm that with the Walrasian (price) dynamics discussed in Chapter 15, convergence to a steady state happens faster with flat demand and supply curves, though if the curves are too flat the process may oscillate explosively.

16.3.2   Continuous Time

Marshallian dynamics can also be modeled in continuous time. The three-equation system consists of the same (inverse) demand and marginal cost functions, equations

(16.15) and (16.16), but the adjustment function (16.17) is replaced by

$$\dot{Q}_t = \delta(P_t - MC_t).\qquad(16.22)$$

Substituting in equations (16.15) and (16.16) yields the first-order differential equation

$$\dot{Q}_t = \delta(a - bQ_t - c - dQ_t) = -\delta(b+d)Q_t + \delta(a-c).\qquad(16.23)$$

Since b, d, and δ are all positive, there is a stable steady state, which equals

$$Q^* = \frac{\delta(a-c)}{-(-\delta(b+d))} = \frac{a-c}{b+d}\qquad(16.24)$$

and the process will monotonically approach this steady state from any starting point.

The necessary conditions for stability and monotonicity of a linear continuous-time dynamic process are simpler than the corresponding conditions for a linear discrete-time process because for the continuous-time process there is no chance of overshooting the steady state. Oscillations are possible with continuous processes only when processes are interacting in a system of differential equations.

The definite solution of the differential equation (16.23) is

$$Q_t = \left(Q_0 - \frac{a-c}{b+d}\right)e^{-\delta(b+d)t} + \frac{a-c}{b+d} = Q_0 e^{-\delta(b+d)t} + \left(\frac{a-c}{b+d}\right)\left(1 - e^{-\delta(b+d)t}\right).$$

$$(16.25)$$

Since there is no possibility of overshooting the steady state, convergence happens unambiguously faster when the demand and marginal cost curves are steep. In addition, of course, convergence happens faster when the adjustment parameter δ is large.

16.4   COBWEB MODEL

The cobweb model of market adjustment is often motivated by considering agricultural markets, though it is also appropriate for some other kinds of industries. The key is that production decisions have to be made before firms know how big market demand will be.[1] A linear version of the model consists of a market demand function, where quantity demanded depends on current price, and a market supply function where quantity supplied depends on last period's price.[2] As described later, the cobweb model works only with discrete time, so the model consists of the following two equations:

$$Q_t = a - bP_t \qquad \text{(demand)}\qquad(16.26)$$

$$Q_t = c + dP_{t-1} \quad \text{(supply)}\qquad(16.27)$$

---

[1]Another important aspect of agriculture is the uncertainty about how much output will result from particular production decisions (how much to plant and fertilize, for example). The cobweb model does not include this aspect.

[2]A slightly more sophisticated justification, which leads to the same mathematical specification, is that firms base their production decisions on the expected market price, but firms have a simple, adaptive expectations process so the expected price is last period's price. An end-of-chapter problem explores a different expectations process, similar to partial-adjustment models.

where all parameters are assumed to be nonnegative and $a > c$. Setting demand equal to supply and rearranging yields the first-order difference equation

$$P_t = \frac{(a-c) - dP_{t-1}}{b} = -\frac{d}{b}P_{t-1} + \frac{a-c}{b}. \tag{16.28}$$

The steady state will equal

$$P^* = \frac{(a-c)/b}{1 - (-d/b)} = \frac{a-c}{b+d} \tag{16.29}$$

so a steady state will exist as long as the demand and supply curves are not both horizontal (b and d equaling zero). Note that the equilibrium price, and therefore the equilibrium quantity as well, of the cobweb model is the same as in a "normal" demand-and-supply model where both demand and supply are functions of current price.

If b and d are both positive (so the demand curve slopes down while the supply curve slopes up), $-d/b < 0$ so price will oscillate below and above the steady-state price. The steady state will be stable as long as $d/b < 1$, which implies that the demand curve is flatter than the supply curve (recall that when we graph these functions we actually graph their inverses, so for example a large value of d implies a flat supply curve).

The definite solution to the difference equation (16.28) is

$$P_t = \left(-\frac{d}{b}\right)^t P_0 + \left(\frac{a-c}{b+d}\right)\left(1 - \left(-\frac{d}{b}\right)^t\right). \tag{16.30}$$

Figure 16.5 shows the dynamic path of market price and quantity starting from an initial disequilibrium condition where quantity supplied is below its equilibrium value. Price will rise enough to reduce quantity demanded to this low level of quantity, so the initial price is higher than the equilibrium level, as shown in the figure. In period 1, firms will react to the high price and supply quantity Q_1. But that will drive period 1's price down to P_1, which is below the equilibrium value. In period 2 firms reduce quantity in

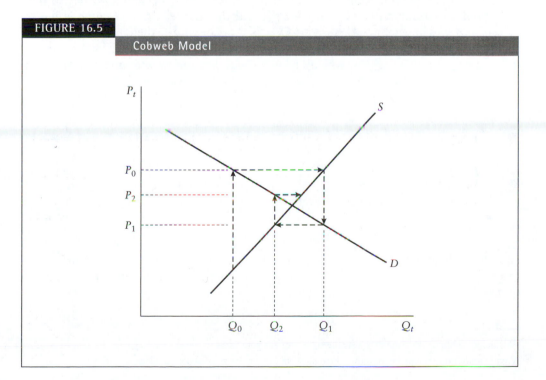

FIGURE 16.5

Cobweb Model

response to the low price in period 1, which drives the price in period 2 back up over its equilibrium value. As long as the demand curve is flatter than the supply curve, however, these oscillations are damped as time goes on: the difference between a period's price and the equilibrium price gets smaller and smaller through time, until finally the equilibrium is reached as a stable steady state. Figure 16.5 is the diagram that gives the cobweb model its name because, at least to some observers, the dynamic path of price and quantities resembles a cobweb.

The oscillations of the cobweb model cannot happen in a continuous-time model since linear differential equations cannot overshoot their steady states (unless they are part of a system of linear differential equations). The oscillations of the cobweb model occur because firms are always reacting to old prices, and "old" is discretely different than "current." In a continuous-time model, "old" and "current" are arbitrarily close together, so the price that influences supply decisions will be the same as the price that influences demand, and equilibrium will be reached instantaneously.

16.5   NUMBER OF FIRMS IN AN OLIGOPOLY

In Chapter 2 we investigated a short-run Cournot model of an oligopoly, including the effect of the number of firms on equilibrium output and price. In this section we make the number of firms endogenous and model the dynamic process of the number of firms in the oligopoly, assuming there are no barriers to entry or exit of firms.[3] Since this is a Cournot model, each firm takes other firms' outputs as exogenous but takes into account the effect its own output decision has on market price. In the long run, more firms will enter the industry if there are positive economic profits, while firms will leave the industry if there are negative economic profits. The dynamics of the number of firms, then, includes an adjustment equation that depends on $\Pi_{i,t}$, the profit of the (typical) i^{th} firm at time t.

If firms operate with constant marginal cost, fixed costs of f, and no taxes, so that their total cost function is $TC_{i,t} = f + cq_{i,t}$, and the market demand curve is linear, $P_t = a - bQ_t$, where market quantity is $Q_t = \sum_{i=1}^{n_t} q_{i,t} = n_t q_{i,t}$, then following the procedure of Section 2.6 of Chapter 2 the equilibrium quantity for each firm at time t is $q_{i,t}^* = (1/(n_t + 1))((a - c)/b)$, which makes the equilibrium market quantity at time t $Q_t^* = (n_t/(n_t + 1))((a - c)/b)$ and equilibrium market price at time t $P_t^* = a/(n_t + 1) + (n_t/(n_t + 1))c$. This makes the profit at time t of the typical firm

$$
\begin{aligned}
\Pi_{i,t}^* = P_t^* q_{i,t}^* - f - c q_{i,t}^* &= \left(\frac{a + n_t c}{n_t + 1} \right) \left(\frac{1}{n_t + 1} \right) \left(\frac{a - c}{b} \right) - f - c \left(\frac{1}{n_t + 1} \right) \left(\frac{a - c}{b} \right) \\
&= \left(\frac{1}{n_t + 1} \right) \left(\frac{a - c}{b} \right) \left(\left(\frac{a + n_t c}{n_t + 1} \right) - c \right) - f = \left(\frac{1}{n_t + 1} \right) \left(\frac{a - c}{b} \right) \left(\frac{a - c}{n_t + 1} \right) - f \\
&= \left(\frac{(a - c)^2}{b} \right) \left(\frac{1}{n_t + 1} \right)^2 - f
\end{aligned}
$$

(16.31)

and the industry will reach long-run equilibrium, when the number of firms has reached a steady state, when $\Pi_{i,t}^* = 0$, which implies the long-run equilibrium number of firms[4] is

---

[3]This model, without the dynamics, was the subject of Problem 2.12.

[4]We are ignoring the complication that the number of firms must be a whole number.

$$n^* = \frac{a-c}{\sqrt{bf}} - 1. \qquad \text{(16.32)}$$

16.5.1 Discrete Time

In a discrete-time model, the number of firms n in the oligopoly at time t will equal the number of firms from the previous period plus an adjustment based on the profit of the typical firm:

$$n_t = n_{t-1} + \gamma \Pi_{i,t-1}^* \qquad \text{(16.33)}$$

where $\gamma > 0$ measures the number of firms that enter (leave) the industry for each unit of positive (negative) economic profits. Given equation (16.31), equation (16.33) is a nonlinear first-order difference equation.

The steady state is given by equation (16.32); it is a steady state because with this number of firms, each firm's economic profit is zero and so $n_t = n_{t-1}$. Even though the difference equation is nonlinear, equation (16.32) shows that there is only one positive steady state; the number of firms cannot be negative so the negative square root solution to equation (16.32) is irrelevant.[5] To investigate the stability of this steady state, we could sketch the phase diagram of the difference equation (16.33). The slope of the difference equation is

$$\frac{\partial n_t}{\partial n_{t-1}} = 1 + \gamma \frac{\partial \Pi_{i,t-1}^*}{\partial n_{t-1}} = 1 - 2\gamma \left(\frac{(a-c)^2}{b} \left(\frac{1}{n_{t-1}+1} \right)^3 \right). \qquad \text{(16.34)}$$

The steady state will be stable if, evaluated at the steady state, this slope is less than one in absolute value. Given that b and γ are both assumed to be positive, equation (16.34) shows that the slope is always algebraically less than one. But depending on the relative magnitude of various parameters, the slope might be algebraically less than negative one, in which case the steady state would be unstable.

To see the conditions under which the steady state will be stable, evaluate equation (16.34) at the steady state value of n:

$$\left. \frac{\partial n_t}{\partial n_{t-1}} \right|_{n_{t-1}=n^*} = 1 - 2\gamma \left(\frac{(a-c)^2}{b} \left(\frac{1}{n^*+1} \right)^3 \right) = 1 - 2\gamma \left(\frac{(a-c)^2}{b} \left(\frac{\sqrt{bf}}{(a-c)} \right)^3 \right)$$

$$= 1 - 2\gamma \left(\frac{f\sqrt{bf}}{a-c} \right). \qquad \text{(16.35)}$$

The steady state will be stable as long as this is algebraically greater than -1 (remember we already know that it is less than 1). So the condition for a stable steady state is

$$2 - 2\gamma \frac{f\sqrt{bf}}{a-c} > 0 \Rightarrow 2 \left(1 - \gamma \frac{f\sqrt{bf}}{a-c} \right) > 0 \Rightarrow \gamma \frac{f\sqrt{bf}}{a-c} < 1 \Rightarrow \gamma < \frac{a-c}{f\sqrt{bf}}. \qquad \text{(16.36)}$$

---

[5]Mathematically, even with the positive square root of bf the equilibrium number of firms might be negative. But in that case the fixed costs are too costly for any profits to be made, even if there are only two firms (the minimal number for this model to make sense economically). So we assume that the steady-state number of firms is at least two.

The speed of adjustment must be sufficiently slow: if too many firms enter and exit when profits become slightly positive or slightly negative, then the number of firms oscillates explosively around the steady state, making the steady state unstable. As long as a small enough number of firms enter or exit in response to the small positive or negative profits associated with the steady-state (zero-profit) number of firms, the steady state is stable.

The number of firms will approach a stable steady state monotonically as long as the slope of the difference equation, evaluated at n^*, is positive (and less than one, which we already know is the case). So the condition for monotonicity is

$$1 - 2\gamma \frac{f\sqrt{bf}}{a-c} > 0 \Rightarrow \gamma \frac{f\sqrt{bf}}{a-c} < \frac{1}{2} \Rightarrow \gamma < \frac{1}{2}\left(\frac{a-c}{f\sqrt{bf}}\right). \qquad \textbf{(16.37)}$$

So firms must enter and exit half as fast as what is necessary for stability. This makes sense since the more slowly firms enter and exit, the smaller is the chance that the equilibrium number of firms will be overshot during the adjustment process.

16.5.2 Continuous Time

As is the case in other applications in this chapter, analysis of the continuous-time version of this model is more straightforward than is the analysis of the discrete-time version. This is because with continuous time the steady state cannot be overshot, so there are no oscillatory stable processes. Furthermore, in this case there is only one positive steady state, as shown by equation (16.32), so even though it turns out that we are dealing with a nonlinear difference equation, we need only consider the properties of this equation in the vicinity of the steady state.

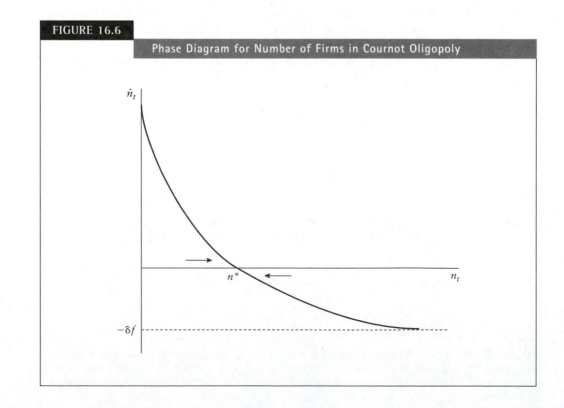

FIGURE 16.6

Phase Diagram for Number of Firms in Cournot Oligopoly

When time is modeled as a continuous variable, the equation showing how the number of firms responds to profits is

$$\dot{n}_t = \delta \Pi_{i,t}^* = \delta \left(\frac{(a-c)^2}{b} \left(\frac{1}{n_t+1} \right)^2 - f \right), \quad \delta > 0. \tag{16.38}$$

The slope of this nonlinear, first-order differential equation is

$$\frac{\partial \dot{n}_t}{\partial n_t} = -2\delta \left(\frac{(a-c)^2}{b} \left(\frac{1}{n_t+1} \right)^3 \right) < 0 \tag{16.39}$$

so the number of firms monotonically approaches a stable steady state. It can be confirmed easily that the second derivative of the differential equation is always positive and that as the number of firms gets infinitely large, $\dot{n}_t \to -\delta f$. So the phase diagram of the differential equation is as shown in Figure 16.6, with one stable steady state.

16.6 DYNAMICS OF COURNOT DUOPOLY

In this section we modify the Cournot model so that it takes some time for each firm to adjust to its optimal output level. Using a partial-adjustment model, we assume that each firm will change its output in the direction of the difference between the output level described by its Cournot reaction function and its current output level. This gives us a two-equation system of either difference or differential equations, depending on whether we treat time as discrete or continuous.

16.6.1 Discrete Time

When time is treated as discrete, we must make some assumption about whether current information is used to help determine each firm's choice of output or whether the firms use only past information when making these decisions. We will use a partial-adjustment, error-correction model where each firm changes its output based on how different last period's output was from the output indicated by its Cournot reaction function. This assumes that each firm does not know the other firm's current output when it makes its own output decision, so it reacts instead to its competitor's output in the previous period. But rather than just acting as if the competitor's output in the current period will be the same as in the last period, each firm adjusts its own output based on how its output last period compared to the optimal output as defined by the Cournot reaction function. This leads to the following adjustment equations for the two firms:

$$q_{1,t} = q_{1,t-1} + \gamma(R_1(q_{2,t-1}) - q_{1,t-1}) \quad \text{and} \quad q_{2,t} = q_{2,t-1} + \gamma(R_2(q_{1,t-1}) - q_{2,t-1})$$

$$\tag{16.40}$$

where $0 < \gamma \leq 1$; if $\gamma = 1$, then each firm acts as if its competitor's output will be the same as in the previous period. With linear demand, no fixed costs, constant marginal cost, and identical firms, these equations can be rewritten as

$$q_{1,t} = q_{1,t-1} + \gamma \left(\frac{a-c}{2b} - \frac{1}{2} q_{2,t-1} - q_{1,t-1} \right) = (1-\gamma) q_{1,t-1} - \frac{\gamma}{2} q_{2,t-1} + \frac{\gamma(a-c)}{2b}$$

$$\tag{16.41}$$

and

$$q_{2,t} = q_{2,t-1} + \gamma\left(\frac{a-c}{2b} - \frac{1}{2}q_{1,t-1} - q_{2,t-1}\right) = (1-\gamma)\,q_{2,t-1} - \frac{\gamma}{2}q_{1,t-1} + \frac{\gamma(a-c)}{2b}.$$

(16.42)

Written in matrix notation, this system of two first-order difference equations is

$$\begin{pmatrix} q_{1,t} \\ q_{2,t} \end{pmatrix} = \begin{pmatrix} 1-\gamma & -\dfrac{\gamma}{2} \\ -\dfrac{\gamma}{2} & 1-\gamma \end{pmatrix} \begin{pmatrix} q_{1,t-1} \\ q_{2,t-1} \end{pmatrix} + \begin{pmatrix} \dfrac{\gamma(a-c)}{2b} \\ \dfrac{\gamma(a-c)}{2b} \end{pmatrix}$$

(16.43)

so the characteristic equation of the system (see Appendix 15.1) is

$$\lambda^2 - ((1-\gamma) + (1-\gamma))\lambda + \left((1-\lambda)^2 - \left(-\frac{\gamma}{2}\right)^2\right) = 0.$$

(16.44)

The eigenvalues for this system are

$$\lambda_1, \lambda_2 = \frac{2(1-\gamma)}{2} \pm \frac{\sqrt{(2(1-\gamma))^2 - 4((1-\gamma)(1-\gamma) - (-\gamma/2)(-\gamma/2))}}{2}$$

$$= (1-\gamma) \pm \frac{\sqrt{4(1-\gamma)^2 - 4(1-\gamma)^2 + \gamma^2}}{2} = (1-\gamma) \pm \frac{\gamma}{2}.$$

(16.45)

Since $0 < \gamma \le 1$, one eigenvalue lies between $\frac{1}{2}$ and 1 and the other lies between $-\frac{1}{2}$ and 1. Both are less than one in absolute value, so the system is stable.

The steady state of the system can be computed various ways; one way is to substitute the steady-state values q_1^* and q_2^* into equations (16.41) and (16.42) and solve simultaneously. We can also use the fact that the firms are identical, and therefore q_1^* must equal q_2^*, to use just one equation and symmetry:

$$q^* = (1-\gamma)\,q^* - \frac{\gamma}{2}q^* + \frac{\gamma(a-c)}{2b}$$

$$\frac{3\gamma}{2}q^* = \frac{\gamma(a-c)}{2b}$$

$$q^* = \left(\frac{2}{3\gamma}\right)\left(\frac{\gamma(a-c)}{2b}\right) = \frac{a-c}{3b}.$$

(16.46)

So there is a single steady state that is stable and equal to the usual Cournot equilibrium. We could solve for the complete dynamic path from arbitrary starting values to this steady state by diagonalizing system (16.43) using the eigenvalues (16.45) or by simulation. The same techniques could be used to investigate the effect of a change in one of the parameters of the problem, for example, a shift of the market demand curve.

16.6.2 Continuous Time

When time is treated as continuous, the adjustment equations for the two firms are

$$\dot{q}_{1,t} = \delta(R_1(q_{2,t}) - q_{1,t}) \quad \text{and} \quad \dot{q}_{2,t} = \delta(R_2(q_{1,t}) - q_{2,t})$$

(16.47)

where $\delta > 0$. With linear demand, no fixed costs, constant marginal cost, and identical firms, the equations can be rewritten as

$$\dot{q}_{1,t} = \delta\left(\frac{a-c}{2b} - \frac{1}{2}q_{2,t} - q_{1,t}\right) \quad \text{and} \quad \dot{q}_{2,t} = \delta\left(\frac{a-c}{2b} - \frac{1}{2}q_{1,t} - q_{2,t}\right) \quad \textbf{(16.48)}$$

or

$$\dot{q}_{1,t} = -\delta q_{1,t} - \frac{\delta}{2}q_{2,t} + \delta\left(\frac{a-c}{2b}\right) \quad \textbf{(16.49)}$$

and

$$\dot{q}_{2,t} = -\delta q_{2,t} - \frac{\delta}{2}q_{1,t} + \delta\left(\frac{a-c}{2b}\right). \quad \textbf{(16.50)}$$

Thus the dynamics of the Cournot duopoly model are described by a two-equation system of linear, first-order differential equations.

The phase diagram of this system, shown in Figure 16.7, shows that the system has a globally stable steady state. With $q_{2,t}$ measured on the vertical axis and $q_{1,t}$ measured on the horizontal axis, the graph of the level curve of the differential equation (16.49) is downward-sloping with slope equal to -2: if $\dot{q}_{1,t} = 0$, then

$$q_{2,t} = \frac{2}{\delta}\left(-\delta q_{1,t} + \delta\left(\frac{a-c}{2b}\right)\right) = -2q_{1,t} + \frac{a-c}{b} \quad \textbf{(16.51)}$$

while the level curve of equation (16.50) is downward sloping with slope $-1/2$: if $\dot{q}_{2,t} = 0$,

$$q_{2,t} = \frac{1}{\delta}\left(-\frac{\delta}{2}q_{1,t} + \delta\left(\frac{a-c}{2b}\right)\right) = -\frac{1}{2}q_{1,t} + \frac{a-c}{2b}. \quad \textbf{(16.52)}$$

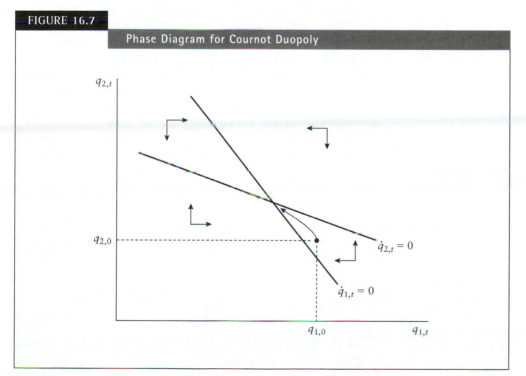

FIGURE 16.7

Phase Diagram for Cournot Duopoly

Difference and Differential Equations: Applications

Equation (16.49) shows that if, from any point on the level curve of equation (16.49) (that is, any values of $q_{1,t}$ and $q_{2,t}$ that satisfy (16.51)), either $q_{1,t}$ or $q_{2,t}$ increases, $\dot{q}_{1,t}$ will become negative. Thus for all points to the right and above the level curve, Figure 16.7 shows arrows of motion pointing to the left, while for all points to the left and below the level curve, arrows of motion point to the right. The directions of the vertical arrows of motion are determined by equation (16.50), which indicates that any point above or to the right of the level curve of equation (16.50) results in $\dot{q}_{2,t} < 0$, so the arrows of motion point down. For all points to the left and below this level curve, the arrows of motion point up. So, regardless of the starting point, the dynamic path of the variables will always move toward the steady-state equilibrium. A possible transition path is shown in Figure 16.7.

The steady state of this system is found by setting both $\dot{q}_{1,t}$ and $\dot{q}_{2,t}$ equal to zero and solving equations (16.51) and (16.52) simultaneously. Since these equations are mathematically equivalent to the two Cournot reaction functions, the steady state is the Cournot equilibrium: $q_1^* = q_2^* = (a - c)/3b$.

Therefore with time treated continuously and partial adjustment, the usual Cournot result holds as a globally stable steady state. How quickly this steady state is reached when there is a shock to the system (for example, a shift of the market demand curve) depends on how large the shock is and also on the size of the adjustment parameter δ.

16.7 FISHERIES

In this section we explore one of the classic problems of natural resource economics: the harvesting of a renewable resource. Our example is a fishery, with the amount of fish measured in biomass. This is appropriate for most fish harvested for food. For recreational fishing the number and age distribution of individual fish can be important; these concerns are often important for other kinds of renewable resources as well, for example, forests. But the focus on biomass is reasonable for many renewable resources in addition to fish. We treat time as continuous, which is typical in the literature.

Although the management of fisheries and other renewable natural resources typically involves many policy variables (for example, hatcheries that affect the growth rate of the population, regulations on harvesting technology, and concern for impacts on the larger ecosystem), we use a simple model where the only management decision is the amount of biomass harvested. This results in a model in which three variables interact: the current stock in biomass of fish, x_t, the net natural increase of fish population (in biomass), $F(x_t)$, and the amount of biomass harvested, h_t, which is the choice variable for fishery management. The fish population grows according to the first-order differential equation

$$\dot{x}_t = F(x_t) - h_t. \tag{16.53}$$

We will explore two models: a proportional growth model where

$$F(x_t) = \lambda x_t, \quad \lambda > 0 \tag{16.54}$$

and the classic logistic case:

$$F(x_t) = r x_t \left(1 - \frac{x_t}{k}\right), \tag{16.55}$$

where r and k are arbitrary positive constants. For each model, we compare the biological equilibrium, where there is no harvesting, to the equilibrium when there is a constant harvest ($h_t = h \; \forall t$). To investigate the economics of the harvest decision, we model the harvest as a function of population size and effort (number of fishing boats, for example). To keep things reasonably simple, we assume that effort is constant through time and that the harvest will be proportional to both population size and effort:

$$h_t = ax_t E \tag{16.56}$$

where a is a positive constant.

We treat price per unit of biomass and cost of effort as exogenous and investigate the economics of the decision of how much effort to undertake, consistent with a steady-state population size. With our assumptions, profit in the steady state will be

$$\Pi = Pax^* E - cE \tag{16.57}$$

where P is the constant price per unit of biomass, c is the constant cost per unit of effort, and the constant level of effort is consistent with the steady-state population size. We compare the maximum sustainable yield (the largest harvest consistent with a constant population of fish) to the results of open-access fishing (in which case positive profit leads to more and more effort until the profit from the fishery is driven to zero) and the optimal level of fishing effort—the level that results in the largest profit from the fishery (this assumes there are no externalities associated with fishing).

16.7.1 Proportional Growth

The assumption that the net natural increase of biomass is a constant proportional growth is a reasonable assumption for fisheries with very small populations or for some specific species of fish. It is not realistic, however, for commercial fisheries of most species. It does have the advantage of simplicity, however.

With proportional natural growth and a constant harvest rate, the differential equation showing the dynamics of population size is

$$\dot{x}_t = \lambda x_t - h. \tag{16.58}$$

This equation only applies if $x_t > 0$; once $x_t = 0$ no harvest is possible since there is no population left, and $\dot{x}_t = 0$ from that time forward. The biological equilibrium is found when $h = 0$. Equation (16.58) makes it clear that there is no interesting biological equilibrium: the only steady state is when $x_t = 0 \; \forall t$, and this steady state is unstable because $\lambda > 0$. So unless there is no population to begin with, the population will naturally increase through time forever. For completeness, the solution to the differential equation (16.58) when $h = 0$ is

$$x_t = \lambda^t x_0. \tag{16.59}$$

When $h > 0$ the steady state of equation (16.58) is

$$x^* = \frac{h}{\lambda} \tag{16.60}$$

which is unstable since $\lambda > 0$. The phase diagram of equation (16.58) is shown in Figure 16.8, for the cases of $h = 0$ and $h > 0$. For an initial population of x_0, the only harvest rate consistent with a steady state is $h = \lambda x_0$, so that (from equation (16.60)) $x^* = x_0$. For any other harvest rate, the population will be driven to zero, if the harvest

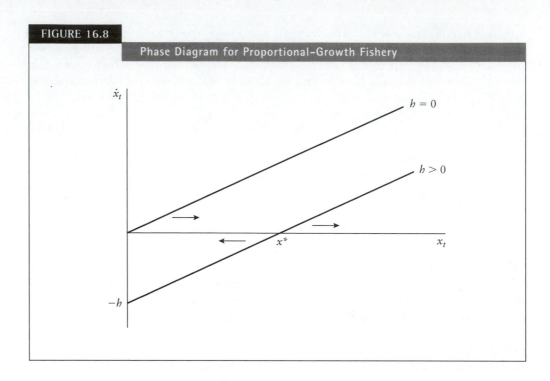

FIGURE 16.8

Phase Diagram for Proportional-Growth Fishery

rate is too large relative to the original population, or else biomass will grow forever, if the harvest rate is too small relative to the starting population.

If the harvest is proportional to both population size and effort, $h = axE$, the differential equation for population is

$$\dot{x}_t = \lambda x_t - ax_t E \tag{16.61}$$

and so the steady state can be achieved only when $E = \lambda/a$. This level of effort will lead to a steady state for any initial population size: if $E = \lambda/a$, then $\dot{x}_t$ will always equal 0 and the population will always stay at its initial size, just as when the harvest rate was chosen directly.

For the proportional growth case and a constant harvest rate (and no uncertainty), fishery management consists solely of choosing (and enforcing) the sustainable harvest rate $h = \lambda x_0$. Any other harvest rate leads to instability so managers have no room for error. Because there is no room for choice of the harvest rate, there is no interesting economic equilibrium. In the steady state, profits from the fishery equal

$$\Pi = Pax^*E^* - cE^* = Px^*\lambda - c\frac{\lambda}{a}. \tag{16.62}$$

Profits may be positive, zero, or negative, but since only one level of effort is consistent with the steady state, a managed fishery will maintain that level of effort regardless of its profitability.[6] In an open-access fishery, positive profit will encourage more effort, but this is inconsistent with the steady state. So, in the absence of any management or oversight, open access will tend to destabilize the fish population and drive it toward

---

[6]Presumably, if profits are negative the fishery will go out of business. With proportional biomass growth, this will lead to a growing fish population and eventually fishing will become profitable.

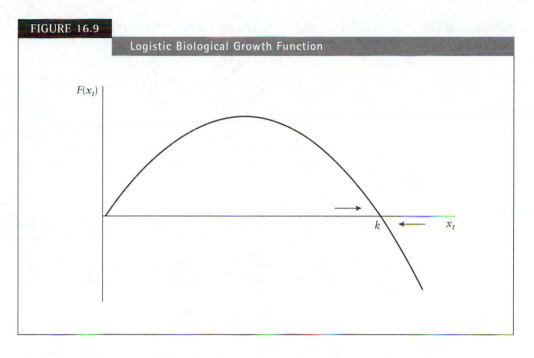

FIGURE 16.9

Logistic Biological Growth Function

$F(x_t)$

k

x_t

zero since, if $E > \lambda/a$, equation (16.61) shows that $\dot{x}_t < 0 \; \forall t$. This is the extreme version of the tragedy of the commons.[7]

In reality, of course, the harvest is not restricted to be constant through time. The formal analysis of this case requires techniques beyond those covered in this book: the analysis of equation (16.53) requires integration if harvests differ through time, and formal analysis of the optimal choice of harvest rates requires dynamic programming techniques[8] since the choice variable is the entire path of harvests through time. The basic outlines of the management decision are clear, however. Management should decide on a desired harvest rate, or on a desired population size, and choose harvests while monitoring population. If the population starts to fall below the desired level, or the level consistent with the desired harvest, harvests should be cut back; if the population starts to grow too large, harvests should be increased. The economic decision of desired harvest level will be a tradeoff between the benefit and cost of waiting for the population to grow before starting to fish. The benefit is that the steady-state harvest level will be larger; the cost is that there will be a longer period of no profit before the fishing begins. Fishery management in the real world is also complicated by unpredictable shocks to the population size, uncertainty about the natural growth rate of biomass, and uncertainty about and changes in prices and costs; in addition to biomass, managers typically care about other issues also.

16.7.2   Logistic Growth

A more realistic model of the net natural increase in biomass is the logistic growth given by equation (16.55). This function has been found to be a good approximation of the natural growth of various biological populations and also leads to more interesting economic analysis. The logistic function is graphed in Figure 16.9; the shape of

[7]"The Tragedy of the Commons," Garrett Hardin, *Science*, 162(1968):1243–1248. Hardin coined this phrase to describe the problem that an open-access resource will be overused.

[8]A classic reference is *Mathematical Bioeconomics*, Colin Clark, New York: Wiley (1976); the fisheries dynamic programming problem is related to the material in Appendix 15.2.

the function can be confirmed by taking the first and second derivatives of the logistic function with respect to x_t.

With logistic growth and a constant harvest, the differential equation for fish population is

$$\dot{x}_t = r x_t \left(1 - \frac{x_t}{k} \right) - h \tag{16.63}$$

for which the biological equilibrium can be found by setting $h = 0$. The steady-state values for x_t can be found by setting the left-hand side of equation (16.63) equal to zero. If $h = 0$ this implies

$$0 = r x^* \left(1 - \frac{x^*}{k} \right) \Rightarrow x^* = 0 \quad \text{or} \quad x^* = k. \tag{16.64}$$

The stability of these steady states can be seen from the phase diagram of equation (16.63). When $h = 0$ and $\dot{x}_t = F(x_t)$, the phase diagram looks exactly like Figure 16.9, indicating that the steady state $x^* = 0$ is unstable but the steady state $x^* = k$ is stable. This can be confirmed by taking the derivative of $\dot{x}_t$ with respect to x_t:

$$\left. \frac{\partial \dot{x}_t}{\partial x_t} \right|_{h=0} = r \left(1 - \frac{x_t}{k} \right) - \frac{r x_t}{k} = r \left(1 - \frac{2}{k} x_t \right) \tag{16.65}$$

which is positive if $x_t < k/2$ and negative if $x_t > k/2$. Natural net population growth is maximized when $x_t = k/2$, while left to its own the population will stabilize at the upper steady state, $x^* = k$, starting from any positive population size.[9]

With a positive constant harvest, there are two possible steady states:

$$0 = r x^* \left(1 - \frac{x^*}{k} \right) - h \Rightarrow -\frac{r}{k} x^{*2} + r x^* - h \tag{16.66}$$

which implies

$$x^* = \frac{-r \pm \sqrt{r^2 - 4 \left(-r/k \right) \left(-h \right)}}{2(-r/k)} = \frac{k}{2} \pm \sqrt{\frac{k^2}{4} - \frac{kh}{r}} = \frac{k}{2} \pm \sqrt{k \left(\frac{k}{4} - \frac{h}{r} \right)}. \tag{16.67}$$

Clearly the harvest rate must be low enough or there is no real-valued steady state: if $h/r > k/4$, then both solutions for x^* will be imaginary numbers.[10] This turns out to be related to the maximum sustainable harvest, as shown later.

As was true in the no-harvest case, the lower steady-state value in (16.67) is unstable while the upper steady-state value is stable. The positive harvest rate does not affect the slope of the differential equation, which is still given by equation (16.65). So it is still true that for $x_t < k/2$ the derivative will be positive, while for $x_t > k/2$ the derivative will be negative. Since the higher steady state in (16.67) is clearly greater than $k/2$, it is stable; since the lower steady state is clearly less than $k/2$, it is unstable. The phase

---

[9]Note that there is no minimum threshold population size, which makes the logistic function less than totally realistic for many species.

[10]We also want the lowest solution for x^* to be nonnegative, but this is guaranteed since even if $h = 0$ the lowest value of x^* is zero.

FIGURE 16.10

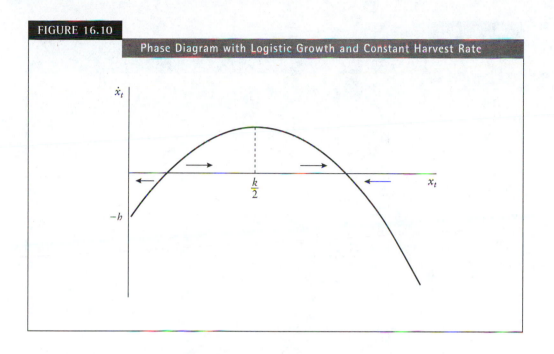

diagram for the positive-harvest case is shown in Figure 16.10, which is identical to the no-harvest case except for a parallel shift down by h.

The maximum sustainable harvest is the largest value of h consistent with a positive steady-state population. From Figure 16.10 it is clear that this implies that the harvest must be such that the graph of the differential equation is tangent to the horizontal axis. That is, the maximum value of $\dot{x}_t$ must equal zero. From equation (16.65) we know that $\dot{x}_t$ is maximized when $x_t = k/2$; we want to choose h such that the corresponding value of $\dot{x}_t$ is zero:

$$0 = \dot{x}_t|_{x_t=k/2} = r\left(\frac{k}{2}\right)\left(1 - \frac{k/2}{k}\right) - h = \frac{rk}{2}\left(1 - \frac{1}{2}\right) - h = \frac{rk}{4} - h \quad \textbf{(16.68)}$$

so the maximum sustainable harvest is

$$h_{\max} = \frac{rk}{4}. \quad \textbf{(16.69)}$$

This is a precarious management decision, though, since with this constant harvest level, there is only one steady-state population level, $x^* = k/2$, which is not globally stable. The phase diagram for the maximum sustainable harvest case is shown in Figure 16.11. If the population starts above the steady-state value, the population will decline smoothly to the steady state. If the population ever falls below the steady state, though, the fish population will decline to zero unless the harvest level is changed. Moreover, a harvest level even barely greater than the maximum sustainable harvest will drive the fish population to zero. There is no steady-state population level in this case, as shown by equation (16.67): both solutions for x^* are imaginary numbers.

If harvest levels are not chosen directly but, instead, by the interaction of population size and effort as shown in equation (16.56), then the differential equation for population size is

$$\dot{x}_t = r x_t\left(1 - \frac{x_t}{k}\right) - a x_t E \quad \textbf{(16.70)}$$

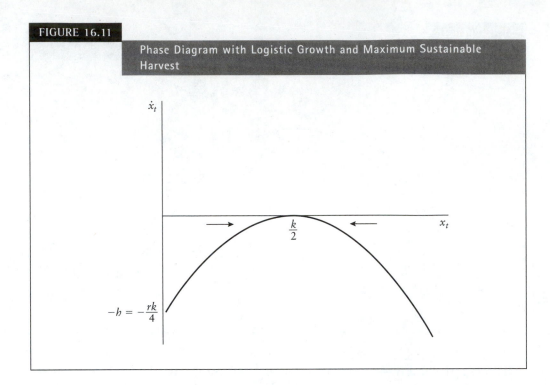

FIGURE 16.11

Phase Diagram with Logistic Growth and Maximum Sustainable Harvest

which implies that in the steady state with positive population size,

$$ax^*E = rx^*\left(1 - \frac{x^*}{k}\right) \Rightarrow E = \frac{r}{a}\left(1 - \frac{x^*}{k}\right) \quad \text{or} \quad x^* = k\left(1 - \frac{aE}{r}\right). \quad \textbf{(16.71)}$$

A population of zero is a steady state also, but it is unstable. Unlike the proportional population growth case, with logistic growth there are many different combinations of effort and population size consistent with a steady state. There is an inverse relationship between effort and the steady-state population size, as shown in (16.71): a higher fish population requires lower effort; higher fishing effort results in a lower steady-state population. In particular,

$$\frac{\partial x^*}{\partial E} = -k\frac{a}{r} < 0. \quad \textbf{(16.72)}$$

The phase diagram for equation (16.70) looks similar to the biological phase diagram shown in Figure 16.9, but instead of being shifted down by a constant amount (as was the case in Figure 16.10), the downward shift is equal to ax_tE and therefore gets larger with population size. The phase diagram for equation (16.70) is shown in Figure 16.12 for an arbitrary positive level of effort, with the graph of the no-harvest (no-effort) differential equation added for comparison. The derivative of $\dot{x}_t$ with respect to x_t is

$$\frac{\partial \dot{x}_t}{\partial x_t} = r\left(1 - \frac{2x_t}{k}\right) - aE \quad \textbf{(16.73)}$$

which, evaluated when effort and population size are in the steady-state relationship shown in (16.71), equals

$$\left.\frac{\partial \dot{x}_t}{\partial x_t}\right|_{\text{steady state}} = r\left(1 - \frac{2x^*}{k}\right) - r\left(1 - \frac{x^*}{k}\right) = -r\frac{x^*}{k} < 0 \quad \textbf{(16.74)}$$

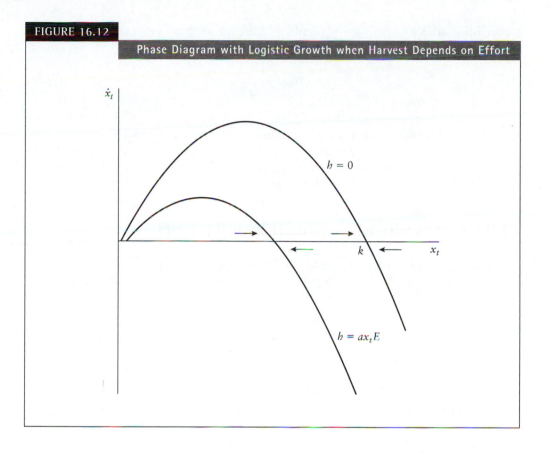

FIGURE 16.12

Phase Diagram with Logistic Growth when Harvest Depends on Effort

as long as population size is positive. Thus if effort is chosen to achieve a positive steady-state population, the steady state will be stable.

In the steady state, profit from the fishery will equal

$$\Pi = Pax^*E - cE = Pa\left(k\left(1 - \frac{aE}{r}\right)\right)E - cE. \qquad \textbf{(16.75)}$$

The optimal level of effort will maximize this profit. The reader can confirm that the first-order condition for this maximization problem results in the optimal effort level of

$$E^* = \frac{r}{2a}\left(1 - \frac{c}{aPk}\right) \qquad \textbf{(16.76)}$$

which implies a fish population size of

$$x^*|_{E^*} = \frac{k}{2} + \frac{c}{2aP} > \frac{k}{2}. \qquad \textbf{(16.77)}$$

We have derived an important economic result: Taking into account the cost of fishing effort, it is not profitable to harvest the maximum sustainable yield.

An open-access fishery will result in increasing effort until the profit from the fishery is driven to zero. Using equation (16.75), the resulting level of effort is

$$E|_{\Pi=0} = \frac{r}{a}\left(1 - \frac{c}{aPk}\right) = 2E^* \qquad \textbf{(16.78)}$$

so the open access to the fishery results in more effort than is optimal. Open access leads to a steady-state fish population of

$$x^*|_{\Pi=0} = \frac{c}{aP} \tag{16.79}$$

which could be either greater or smaller than the population level consistent with the maximum sustainable yield. Since fishing effort is greater than optimal, however, equation (16.72) shows that the open-access population level will be less than the level consistent with optimal fishing effort. Although open access does not drive the population to zero, it does result in a smaller than optimal population.

16.8 IS–LM WITH FED REACTION FUNCTION

In this application we take a linear IS–LM model and add to it a function showing how the Fed adjusts the money supply when the interest rate does not equal a target interest rate. Three equations comprise the model:

$$
\begin{aligned}
Y_t &= C_t + I_t + G_t = a + bY_t + d - er_t + g \\
M_t &= hY_t - kr_t \\
M_t &= M_{t-1} + \gamma(r_{t-1} - \hat{r})
\end{aligned} \tag{16.80}
$$

where the last equation is a Fed reaction function, showing that the Fed will change the money supply when last period's interest rate differs from a target interest rate $\hat{r}$. We use the first two equations in (16.80) to solve for the equilibrium relationship between interest rates and the money supply in each time period:

$$
\begin{pmatrix} 1-b & e \\ h & -k \end{pmatrix} \begin{pmatrix} Y_t \\ r_t \end{pmatrix} = \begin{pmatrix} Y_0 \\ M_t \end{pmatrix} \tag{16.81}
$$

where $Y_0 = a + d + g$ is autonomous aggregate demand. So

$$
r_t = \frac{\begin{vmatrix} 1-b & Y_0 \\ h & M_t \end{vmatrix}}{\begin{vmatrix} 1-b & e \\ h & -k \end{vmatrix}} = \frac{M_t(1-b) - hY_0}{-k(1-b) - eh}. \tag{16.82}
$$

Using equation (16.82) to substitute in for r_{t-1} in the Fed reaction function, we get

$$
\begin{aligned}
M_t &= M_{t-1} + \gamma\left(M_{t-1}\left(\frac{1-b}{-k(1-b) - eh}\right) + \frac{hY_0}{k(1-b) + eh} - \hat{r} \right) \\
&= \left(1 - \gamma\left(\frac{1-b}{k(1-b) + eh}\right)\right) M_{t-1} + \gamma\left(\frac{hY_0}{k(1-b) + eh} - \hat{r}\right).
\end{aligned} \tag{16.83}
$$

With the usual assumptions about the parameters in the IS and LM equations, namely that they are all positive, it is clear from equation (16.83) that there will be a stable steady state of the money supply only if $\gamma > 0$, which makes sense since if the last period's interest rate is higher than the target interest rate, the Fed will want to increase the money supply to drive the interest rate down. More specifically, there will be a stable steady state as long as $-1 < 1 - \gamma((1 - b)/(k(1 - b) + eh)) < 1$ which, assuming that all parameters are positive and $b < 1$, implies that $0 < \gamma < 2(k(1 - b) + eh)/(1 - b)$. The money supply process will converge monotonically to its steady state if $0 < 1 - \gamma((1 - b)/(k(1 - b) + eh)) < 1$, implying that $0 < \gamma < (k(1 - b) + eh)/(1 - b)$. If the Fed reacts too strongly to interest rates, the money supply will change by too much each period, leading to oscillations. If the Fed reacts so strongly that $\gamma > 2(k(1 - b) + eh)/(1 - b)$, it will cause an explosive oscillation and instability.

The steady-state money supply is

$$M^* = \frac{\gamma(hY_0/(k(1 - b) + eh) - \hat{r})}{1 - (1 - \gamma((1 - b)/(k(1 - b) + eh)))} = \frac{-\hat{r}(k(1 - b) + eh)}{1 - b} + \frac{hY_0}{1 - b}. \qquad \textbf{(16.84)}$$

Substituting this into equation (16.82), we get the steady-state interest rate:

$$r^* = \frac{M^*(1 - b) - hY_0}{-k(1 - b) - eh} = \frac{hY_0 - (-\hat{r}(k(1 - b) + eh) + hY_0)}{k(1 - b) + eh} = \hat{r}. \qquad \textbf{(16.85)}$$

The steady-state interest rate is, of course, the target interest rate, since the Fed will change the money supply as long as the interest rate equals anything else. Finally, the steady-state value of output is easily obtained from the equation for the IS curve (the first equation in (16.80)):

$$Y^* = Y_0 + bY^* - er^*$$
$$Y^* = \frac{Y_0 - e\hat{r}}{1 - b}. \qquad \textbf{(16.86)}$$

The speed of convergence to these steady-state values is driven by the magnitude of γ. If γ is close to 0, the Fed does not change the money supply by much even if the interest rate is far from the target value, so the economy takes a long time reaching its steady state. As the Fed reacts more strongly and γ increases, approaching $(k(1 - b) + eh)/(1 - b)$, the economy adjusts to its steady state more quickly. If $\gamma = (k(1 - b) + eh)/(1 - b)$, the coefficient on M_{t-1} in the difference equation (16.83) equals 0 and the money supply always equals its steady-state value. But if the Fed reacts too strongly, it can cause oscillations, slowing convergence or even creating instability.

Problems at the end of the chapter explore how this model is affected by changes in various parameters. Other problems investigate a continuous-time version of the model.

16.9   DYNAMICS OF IS–LM

The simple textbook IS–LM model can be extended easily to incorporate the dynamics of how the equilibrium values of the real interest rate and real output are reached from an arbitrary starting point. The IS curve shows all the values of real gross domestic product (GDP) Y and the real interest rate r for which the goods market is in equilibrium. For a closed economy, this means that aggregate demand, which is the sum of planned consumption, planned investment, and government purchases of goods and services, equals aggregate supply, which is actual (planned plus unplanned) GDP. A simple model of the dynamics of the goods market is that output will rise (fall) whenever there is an excess demand for (supply of) goods. In continuous time, a mathematical formulation of this model is

$$\dot{Y} = \delta(C(Y) + I(r) + G - Y), \quad \delta > 0. \qquad \textbf{(16.87)}$$

When $\dot{Y} = 0$, equation (16.87) is the equation for a simple version of the IS curve, where planned consumption is a (positive) function of real GDP, planned investment depends (negatively) on the real interest rate, and government purchases are exogenous.

Similarly, the LM curve shows all the combinations of real GDP and real interest rates for which the money market is in equilibrium: money demand, which is a function of output and the interest rate, equals money supply, which is exogenous. A simple model of the dynamics of the money market is that the interest rate will rise (fall) when money demand exceeds (is less than) money supply:

$$\dot{r} = \gamma(L(Y, r) - M), \quad \gamma > 0. \tag{16.88}$$

Equations (16.87) and (16.88) form a system of two first-order differential equations; depending on the forms of the various functions, these differential equations may or may not be linear. The steady state of this system is when $\dot{Y} = \dot{r} = 0$. Since $\dot{Y} = 0$ when the goods market is in equilibrium, the graph in (r, Y) − space of this level curve of equation (16.87) is just the IS curve. Similarly, the graph in (r, Y) − space of the level curve $\dot{r} = 0$ of equation (16.88) is the LM curve. So the phase diagram of the system of equations (16.87) and (16.88) is the usual IS–LM diagram with arrows of motion added. This phase diagram is shown in Figure 16.13.

The arrows of motion shown in Figure 16.13 are determined as follows. From equation (16.87), and assuming that the marginal propensity to consume $\partial C / \partial Y$ is less than one, any increase in either r or Y will make $\dot{Y}$ smaller. Thus any (r, Y) combination that lies above and to the right of the IS curve will result in $\dot{Y} < 0$, so in this region of the phase diagram the horizontal arrows of motion point to the left. Similar reasoning shows that the horizontal arrows of motion point to the right in the region below and to the left of the IS curve. From equation (16.88) and the usual assumptions about

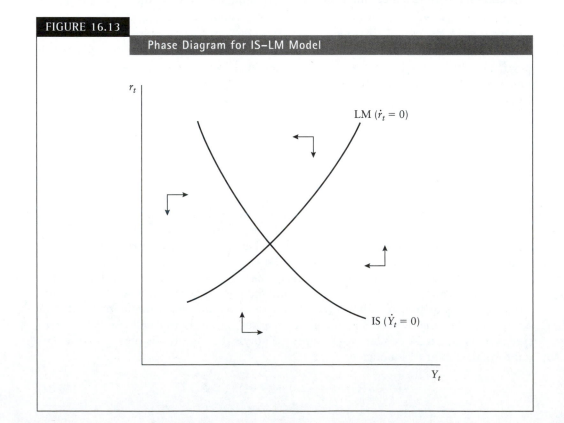

FIGURE 16.13

Phase Diagram for IS–LM Model

LM ($\dot{r}_t = 0$)

IS ($\dot{Y}_t = 0$)

r_t

Y_t

money demand—that it depends positively on output and negatively on the interest rate—any increase in r or decrease in Y will make $\dot{r}$ smaller. So for any (r, Y) combination above and to the left of the LM curve, $\dot{r} < 0$ and the vertical arrows of motion point down. The arrows of motion point up for any point below and to the right of the LM curve.

The arrows of motion in the phase diagram indicate that the steady state, which is where the IS and LM curves intersect, is globally stable. (Compare this to Figure 15.13 in Chapter 15.) Whether transition paths go directly to the steady state or whether they cross from one region to another along the way to the steady state depends on the relative magnitudes of δ, γ, and parameters in the IS and LM functions. An end-of-chapter problem includes simulations showing the effects of various parameters on the transition paths.

16.10 EXPECTATIONS-AUGMENTED PHILLIPS CURVE WITH ADAPTIVE EXPECTATIONS

The Phillips curve, which shows the relationship between the inflation rate and the unemployment rate, is a mainstay of macroeconomic theory. The most frequently used version is the expectations-augmented Phillips curve,

$$\pi_t = \pi_t^e - \alpha(u_t - u_n) \tag{16.89}$$

where π_t is the inflation rate at time t, π_t^e is the inflation rate people expect (based on information available before time t), α is a positive constant, u_t is the unemployment rate at time t, and u_n is the natural rate of unemployment,[11] which is assumed to be constant through time. In this section we will work with a simple model of inflationary expectations: people expect inflation to be the same as it was in the previous period:

$$\pi_t^e = \pi_{t-1}. \tag{16.90}$$

Thus people's inflationary expectations adapt to changes in reality, though with a one-period lag.

The dynamics of inflation appear once the Phillips curve is embedded into a dynamic version of an aggregate demand–aggregate supply model. One way of completing the model[12] is to add to the Phillips curve a dynamic aggregate demand curve based on the quantity theory of money and another equation, called Okun's Law, that relates (through an aggregate production function) changes in the unemployment rate to the growth of real output.

If velocity is constant, the equation of exchange of the quantity theory of money says that the growth rate of the money supply must equal the growth rate of real GDP plus the growth rate of prices (that is, inflation). Rewritten as an aggregate demand equation, this implies that the growth rate of real GDP, g_{Yt}, must equal the growth rate of the money supply, g_M, minus inflation:

$$g_{Yt} = g_M - \pi_t \tag{16.91}$$

where the growth rate of money is assumed to be constant through time.

---

[11]The natural rate of unemployment is the unemployment rate consistent with "full" employment and therefore is the unemployment rate when GDP equals potential GDP. It is also called the NAIRU: the Non-Accelerating Inflation Rate of Unemployment. As equation (16.89) shows, any unemployment rate below the natural rate will result in higher inflation (the aggregate price level will accelerate).

[12]This section is based on *Macroeconomic Theory: A Short Course*, Thomas R. Michl (Sharpe, 2002), Chapter 10.

Okun's Law says that, to a first approximation, the change from one time period to the next in the unemployment rate should be proportional to the growth rate of real GDP:

$$u_t - u_{t-1} = -\beta g_{Yt} \tag{16.92}$$

where β is a positive constant, typically assumed to be between zero and one.

Substituting (16.90) into (16.89) and combining with equations (16.91)–(16.92) yields a three-equation system in the three unknowns π_t, u_t, and g_{Yt}. Since past values of the inflation and unemployment rates are also in the system of equations, it is a system of difference equations. The system can be simplified in order to make it easier to analyze inflation dynamics: first, difference the Phillips curve (16.89) to get

$$\Delta \pi_t = \Delta \pi_{t-1} - \alpha \Delta u_t \tag{16.93}$$

where the symbol Δ stands for a first difference: for example, $\Delta \pi_t = \pi_t - \pi_{t-1}$.[13] Since $\Delta u_t = u_t - u_{t-1}$, substitute in equation (16.92) to get

$$\Delta \pi_t = \Delta \pi_{t-1} + \alpha \beta g_{Yt} \tag{16.94}$$

and then substitute in equation (16.91). The result, after rewriting and rearranging the terms having to do with inflation, is the second-order difference equation

$$\pi_t = \frac{2}{1 + \alpha\beta} \pi_{t-1} - \frac{1}{1 + \alpha\beta} \pi_{t-2} + \frac{\alpha\beta g_M}{1 + \alpha\beta}. \tag{16.95}$$

The steady state of this dynamic process is found by setting inflation in all periods to the steady-state value and then solving:

$$\pi^* = \frac{\alpha\beta g_M/(1 + \alpha\beta)}{1 - (1/(1 + \alpha\beta))(2 - 1)} = \frac{\alpha\beta g_M}{1 + \alpha\beta - 1} = g_M. \tag{16.96}$$

So in the steady state, the inflation rate will equal the growth rate of the money supply or, as Milton Friedman put it so memorably, "inflation is always and everywhere a monetary phenomenon."[14]

The stability of this steady state and the nature of the dynamics of equation (16.95) depend on the coefficients. Using the notation of Section 15.2.6,

$$A_1, A_2 = \frac{2/(1 + \alpha\beta) \pm \sqrt{(2/(1 + \alpha\beta))^2 + 4(-1/(1 + \alpha\beta))}}{2}$$

$$= \frac{1}{1 + \alpha\beta} \pm \sqrt{\left(\frac{1}{1 + \alpha\beta}\right)^2 - \left(\frac{1}{1 + \alpha\beta}\right)} \tag{16.97}$$

$$= \frac{1}{1 + \alpha\beta} \pm \sqrt{\left(\frac{1}{1 + \alpha\beta}\right)\left(\frac{1}{1 + \alpha\beta} - 1\right)} = \frac{1}{1 + \alpha\beta} \pm \sqrt{\frac{-\alpha\beta}{(1 + \alpha\beta)^2}}.$$

---

[13]Equation (16.93) can be derived as follows: from (16.89), $\pi_t = \pi_{t-1} - \alpha(u_t - u_n)$ and $\pi_{t-1} = \pi_{t-2} - \alpha(u_{t-1} - u_n)$. So $\pi_t - \pi_{t-1} = \pi_{t-1} - \pi_{t-2} - \alpha(u_t - u_{t-1})$.

[14]This quote can be found in various places, including Friedman's 1970 Wincott Memorial Lecture, *The Counter-Revolution in Monetary Theory* (Transatlantic Arts, 1970).

FIGURE 16.14

Transition Paths for Inflation

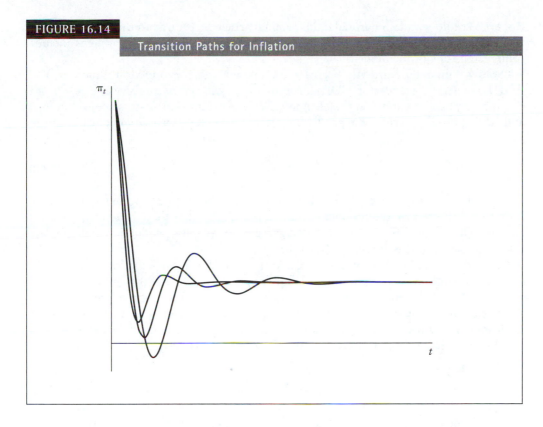

Thus A_1 and A_2 are complex numbers, indicating that the dynamic path of inflation is oscillatory. The condition for the steady state to be stable is that the negative of the coefficient on π_{t-2}, or equivalently half of the coefficient on π_{t-1}, must be less than one. This is true for equation (16.95), so the steady state given by equation (16.96) is stable.

Figure 16.14 shows transition paths for inflation from a high-inflation (8%) initial rate to a steady-state value of 2%; the paths differ because of different assumptions about the slope of the Phillips curve. These paths illustrate some possibilities of what would happen if, in a period of high inflation, the monetary authorities changed permanently to a lower money growth rate. In each case inflation oscillates but the oscillations damp down to the steady-state value. The steeper is the Phillips curve, the smaller are the oscillations and the more quickly the steady state is reached. This is because a steep Phillips curve means that the unemployment rate will not vary much; from Okun's Law, this means that the growth rate of output will be close to zero; from the quantity theory, this means that there will be a close relationship between inflation and the growth rate of the money supply even in the short term. With a flatter Phillips curve, the unemployment rate will vary more, leading to a looser relationship between inflation and money growth.

16.11 SOLOW GROWTH MODEL

Models of economic growth are probably the most frequent application of differential equations in macroeconomics. Although theories of economic growth have a very long history in economics, the foundation for most contemporary models of economic growth is the Solow growth model, named after the American economist

Robert M. Solow and, in particular, his seminal article in 1956.[15] In this section we develop a simple version of the Solow growth model; end-of-chapter problems include some extensions of the model.

Solow's model of growth begins with an aggregate production function $Y = F(L, K)$ where Y stands for GDP, L for the aggregate labor supply, and K for the amount of physical capital in the economy. With constant-returns-to-scale technology and adding time subscripts, the production function can be rewritten in per-capita form as

$$y_t = f(k_t) \tag{16.98}$$

where y stands for per-capita GDP, k for the capital-labor ratio, and the fact that there is no time subscript on the production function indicates that there is no technological change. The usual assumptions for the aggregate per-capita production function are that $f(0) = 0$, $f'(k_t) > 0$, and $f''(k_t) < 0$.

We next assume that the population and the labor force participation rate are constant, so the only way that the capital-labor ratio can change is if the aggregate capital stock changes. This can happen in two ways: physical investment and depreciation. Assuming a closed economy, gross investment equals national saving. If the savings rate is a constant s and the depreciation rate is a constant δ, then net investment per capita, which is the change over time in the capital-labor ratio, is given by the first-order differential equation

$$\dot{k}_t = sf(k_t) - \delta k_t. \tag{16.99}$$

Given the assumptions about the production function, there are two steady states for this dynamic process: one is when $k = 0$ and the other is when $k = k^* > 0$ and $sf(k^*) = \delta k^*$. The steady states are shown in Figures 16.15 and 16.16. Figure 16.15 is often used in macroeconomic textbooks to illustrate the Solow growth model. It graphs gross investment and depreciation (both measured on a per-capita basis) against the capital-labor ratio. The steady states for the capital-labor ratio are the two intersections of these two curves: one at the origin and one at $k^* > 0$. Figure 16.16 is the phase diagram of equation (16.99): it graphs net investment (per capita) against the capital-labor ratio. The steady states are the two points where net investment ($\dot{k}_t$) is zero: at the origin and at $k^* > 0$.

The dynamics of capital formation are clear from Figure 16.16. The steady state at $k = 0$ is unstable while the steady state at $k^* > 0$ is stable. So regardless of initial conditions, the economy will move monotonically[16] toward the upper steady state. Changes in the parameters of the model, for example, changes in the gross saving rate s, the depreciation rate δ, or in parameters of the production function, will change the steady-state level of the capital-labor ratio and the economy will adjust monotonically along a transition path to the new steady state.

Consumption per capita is often used in macroeconomics as a measure of economic welfare. Consideration of the steady state of consumption per capita leads to the "golden rule" of economic growth: the rule that leads to the highest steady-state value of consumption per capita, assuming that the saving rate can be chosen optimally.

---

[15]"A Contribution to the Theory of Economic Growth," Robert M. Solow, *Quarterly Journal of Economics* (February 1956), pp. 65–94.

[16]In continuous-time models, dynamic processes described by first-order differential equations always adjust monotonically.

FIGURE 16.15

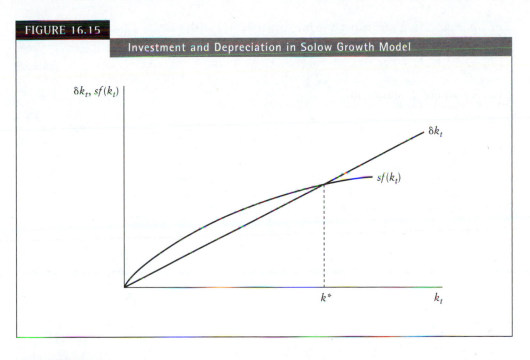

FIGURE 16.16

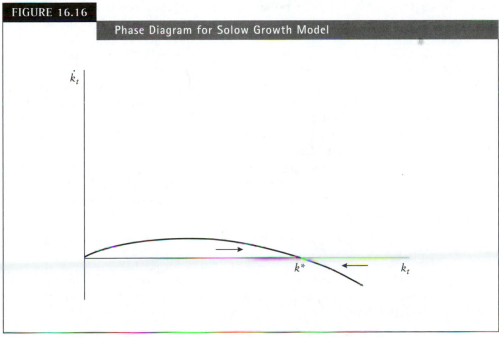

Consumption per capita is equal to the difference between output per capita and savings per capita: $c_t = y_t - sy_t = (1-s)y_t$ so consumption in the steady state is

$$c^* = (1-s)y^* = (1-s)f(k^*) \qquad \textbf{(16.100)}$$

or, since in the steady state $sf(k^*) = \delta k^*$,

$$c^* = f(k^*) - \delta k^*. \qquad \textbf{(16.101)}$$

The optimal saving rate is the one for which $\partial c^*/\partial s = 0$, or

$$f'(k^*)\frac{\partial k^*}{\partial s} - \delta\frac{\partial k^*}{\partial s} = 0. \qquad \text{(16.102)}$$

So the golden-rule level of capital is the one for which

$$f'(k^*) = \delta \qquad \text{(16.103)}$$

as long as $\partial k^*/\partial s \neq 0$, which is true at the positive steady-state value.[17] In Problem 16.25, graphs like Figures 16.15 and 16.16 are used to show the transition from a sub-optimal steady state to the golden-rule steady state.

Problems

16.1 The government is regulating pollution x with an emissions tax of τ per unit of emissions. The goal is to achieve efficiency, which occurs when the marginal damage caused by the pollution, MD, equals the marginal control cost, MCC. But, perhaps because the government does not know the MCC or MD functions, the government takes a trial-and-error approach to setting the tax rate: the tax rate in period t is set equal to the tax rate in the previous period plus an adjustment based on the difference in the previous period between MD and MCC: $\tau_t = \tau_{t-1} + \gamma(MD_t - MCC_t)$. MD is a linear function of the amount of pollution: $MD_t = a + bx_t$; MCC is a linear function of the amount of pollution control, θ: $MCC_t = c + d\theta_t$. The amount of pollution equals $\bar{x} - \theta_t$, where $\bar{x}$ is the amount of pollution that would occur with no control. Each period, firms choose θ_t in order to minimize their costs (pollution control costs plus pollution tax payments, that is, $C_t = CC(\theta_t) + \tau_t x_t = CC(\theta_t) + \tau_t(\bar{x} - \theta_t)$). Assume that all parameters are nonnegative.

(a) Noting that $MCC_t = \partial CC_t/\partial\theta_t$, show that firms will choose θ_t such that $MCC_t = \tau_t$.

(b) Use the information given to write τ_t as a function of τ_{t-1} and the exogenous parameters a, b, c, d, $\bar{x}$, and γ; that is, write a first-order difference equation for τ_t.

(c) Under what conditions is the process described by the difference equation you derived in part (b) stable?

(d) Assuming the conditions you derived in part (c) hold, what are the steady state values of τ?

(e) For what values of γ will the process converge monotonically?

---

[17]This can be proved using the implicit function theorem.

(f)   If the *MD* curve becomes steeper (*b* increases),

   (i)   Is the process more or less likely to be stable?

   (ii)   Is the process more or less likely to oscillate?

   (iii)   How are the steady-state values of τ affected, assuming at least one steady state exists and the cost-minimizing choice of θ is less than $\bar{x}$? [Hint: To sign this comparative static derivative, it may help to sketch the *MD* and *MCC* curves (with pollution x on the horizontal axis) and confirm that the (vertical) intercept of the *MCC* curve must be above the intercept of the *MD* curve.]

16.2   The government regulates pollution by issuing pollution permits. The government's goal is efficiency, which is achieved when the marginal damage done by the pollution, *MD*, equals the marginal control cost *MCC*. Perhaps because they do not know the *MD* or *MCC* functions, the government uses a trial-and-error method for setting permits. So the number of permits issued in any time period, which equals the amount of pollution allowed, equals the number of permits issued in the previous period plus an adjustment based on the difference between *MD* and *MCC*: $x_t = x_{t-1} - \gamma(MD_t - MCC_t)$ where $\gamma > 0$. The marginal damage and marginal control cost functions are $MD_t = a + bx_t^2$ and $MCC_t = c - dx_t^2$, where $a \geq 0, b > 0, c > a$, and $d > 0$.

(a)   Use the information given to write a first-order difference equation for x_t.

(b)   Find the steady-state values of x.

(c)   For what values of γ is there a stable steady state?

(d)   For what values of γ will x_t converge monotonically to a steady state?

(e)   If the *MD* curve shifts up vertically (*a* increases),

   (i)   How are the steady states affected?

   (ii)   Is the process more or less likely to be stable?

   (iii)   Is the process more or less likely to oscillate?

   (iv)   Does the process converge more or less quickly (for a given difference between starting value and steady state)?

(f)   Repeat part (e) for an increase in the slope of the *MD* curve instead of a vertical shift.

16.3   Solve Problem 16.2 with linear *MD* and *MCC* functions and with a nonlinear adjustment function, so $x_t = x_{t-1} + f(MD_t - MCC_t)$, where $f' < 0$ and $f(0) = 0$.

16.4   In a very simple macroeconomic model, output equals consumption, which is a linear function of income, plus government purchases of goods and services: $Y_t = a + bY_t + G_t$. Each period, the government adjusts purchases depending on the difference between output and a target level of output: $\dot{G}_t = \delta(Y_t - \hat{Y})$.

(a)   What restrictions would you impose on the parameters a, b, and δ for the model to make sense economically?

(b)   Under what conditions will there be a stable steady state?

(c)   Assuming the conditions in part (b) hold, find the steady-state values of G and Y.

(d)   How will an increase in the marginal propensity to consume affect the dynamics of this model (likelihood of stability, likelihood of monotonic convergence, values of steady states, and speed of convergence)?

16.5 Solve Problem 16.4 using the discrete-time adjustment function $G_t = G_{t-1} + \gamma(Y_{t-1} - \hat{Y}), \gamma < 0$.

16.6 For a linear supply-and-demand model with Walrasian (price) adjustment $P_t = P_{t-1} + \gamma(Q^D - Q^S)$, derive the conditions under which the process will oscillate explosively. Relate these conditions to the steepness or flatness of demand and supply curves. Also, when the process does converge, show how the slopes of the demand and supply curves affect the speed of convergence.

16.7 Analyze the dynamics (find the steady states, conditions under which the process is stable, and conditions under which the process oscillates) of a linear cobweb model in which supply depends on expected price, $Q_t = c + d\,E[P_t]$ and expectations are formed according to the expectations process $E[P_t] = P_{t-1} + \gamma(P_t - P_{t-1})$, where $0 \le \gamma \le 1$. (When $\gamma = 0$ this leads to the model of Section 16.4; when $\gamma = 1$ this is a perfect-foresight model, in which the expected price level always turns out to equal the actual price level. When $0 < \gamma < 1$, firms correctly anticipate the direction in which prices will be changing but underestimate the amount of the change.) Explain how the value of γ affects the speed of convergence (for a given difference between starting and steady-state values).

16.8 How would the results of the model in Section 16.5 change if the adjustment equation is $n_t = n_{t-1} + f(\Pi^*_{i,t-1})$, where each firm has to pay a fixed cost F to enter the market, so no entry will happen until the typical profit exceeds F? Specifically, if

$$f(\Pi^*_{i,t-1}) = \begin{cases} \gamma \Pi^*_{i,t-1} & \text{for} & \Pi^*_{i,t-1} < 0 \\ 0 & \text{for} & 0 \le \Pi^*_{i,t-1} \le F \\ \gamma \Pi^*_{i,t-1} & \text{for} & \Pi^*_{i,t-1} > F \end{cases}$$

sketch the phase diagram and analyze whether one or more steady states exist, under what conditions the process converges, and under what condition the process oscillates. How do the dynamics change as F gets larger?

16.9 Solve Problem 16.8 in continuous time.

16.10 Using a spreadsheet computer program, simulate the dynamics of the duopoly model of Section 16.6.1. Begin with $a = 10, b = 2$, and $c = 4$, implying $q^* = 1$. Let $\gamma = .6$. Then simulate for 20 time periods the effects of each of the following changes on the steady-state value of q^*, the stability of the steady state, whether the process oscillates or is monotonic, and the speed of convergence:

(a)   An upward shift of the demand curve, to $a = 11$

(b)   A steeper demand curve, $b = 3$

(c)   A decrease in marginal cost, to $c = 3$

(d)   An increase in the speed of adjustment, to $\gamma = 1$

16.11 How will your answers to parts (a), (b), and (c) of Problem 16.10 be affected if γ changes? Specifically, predict the effects, and then simulate the process to confirm your predictions, of changing γ from .6 to

(a) .9

(b) .4

(c) .1

16.12 In the continuous-time model of Section 16.6.2, suppose the market (inverse) demand function is $P_t = a - bQ_t^2 = a - b(q_{1,t} + q_{2,t})^2$.

(a) Find firm 1's reaction function.

(b) Using symmetry, find firm 2's reaction function and then substitute both firms' reaction functions into the adjustment equations (16.47).

(c) Draw the phase diagram with arrows of motion.

(d) Based on your phase diagram, characterize the system as globally stable, globally unstable, or stable along a saddlepath.

(e) Solve for the steady-state value $q^* = q_1^* = q_2^*$.

16.13 For the fisheries model of Section 16.7, assume that the biological growth function is $F(x_t) = ax_t^2 + bx_t$, where $a > 0$.

(a) Find the steady-state values of x.

(b) For each steady-state value, find out whether it is stable.

16.14 Redo the proportional growth analysis of Section 16.7.1 if the harvest function is $h = ax_t E^\beta$, where $0 < \beta < 1$ (so there are positive but decreasing returns to increasing effort).

16.15 For aquaculture operations, managers can affect the biological growth function by, for example, providing nutrients. Let α be a parameter that represents these sorts of management efforts. So let the proportional logistic growth function be $F(x_t, \alpha) = rx_t(1 - (x_t/k))\alpha^\beta$ where $0 < \beta < 1$ (so there are positive but decreasing returns to increasing α). Redo the analysis of Section 16.7.2 using this growth function.

16.16 Redo the logistical growth analysis of Section 16.7.2 if the harvest function is $h = ax_t E^\beta$, where $0 < \beta < 1$ (so there are positive but decreasing returns to increasing effort). When β gets larger, is convergence faster or slower?

16.17 Analyze the linear IS–LM model of Section 16.8 in continuous time, with the partial adjustment equation $\dot{M} = \delta(r_t - \hat{r})$. In particular,

(a) What sign must δ have for the model to make economic sense?

(b) Find the steady-state value of M.

(c) Find the steady-state value of Y.

(d) Is the steady state stable?

(e) Find the definite solution for M_t given a starting value M_0.

(f) How does an increase in e affect the steady state and the dynamics (stability and speed of convergence)?

(g) Explain the economics of your answer to part (f).

16.18 Analyze the linear IS–LM model of Section 16.8 in continuous time, with the partial adjustment equation $\dot{M} = \delta(r_t - \hat{r})$ and a proportional income tax τ, so $C_t = a + b(Y_t(1 - \tau))$. In particular, how does an increase in the tax rate affect the steady-state values of M and Y, stability, and the speed of convergence? Explain the economics of your results.

16.19 Solve Problem 16.18 using the discrete-time adjustment function of Section 16.8.

16.20 Using the model of Section 16.8, describe how the steady-state values of M and Y, the stability of the steady states, the likelihood of oscillations, and the speed of convergence are affected by each of the following. Explain the economics in each case.

(a) An increase in the marginal propensity to consume

(b) An increase in government purchases of goods and services

(c) An increase in k

16.21 For the nonlinear IS–LM system, where $Y_t = C(Y_t) + I(r_t) + G$ and $M_t = L(Y_t, r_t)$, assume the functions have derivatives with the usual signs.

(a) Find the sign of $\partial r_t^* / \partial M_t$, where r_t^* is the equilibrium interest rate in time t, as implicitly defined by the IS–LM system.

(b) If the Fed has the reaction function $\dot{M}_t = \delta(r_t - \hat{r})$, where $\hat{r}$ is a target interest rate and $\delta > 0$, use your answer to part (a) to find the sign of $\partial \dot{M}_t / \partial M_t$ and determine whether the process converges to a stable steady state.

(c) Sketch the phase diagram for the process and use it to illustrate how the steady state and dynamics of the system are affected when the government increases its purchases of goods and services.

16.22 Using a computer spreadsheet program, simulate the dynamic IS–LM model of Section 16.9 when $C_t = Y_t^\alpha$, $I_t = \bar{I}/(1 + r_t)$, and the money-demand curve is linear. Begin with the following values: $\alpha = .9$, $M = 20$, $\bar{I} = 40$, $k = 6$, $h = .2$, $\gamma = .6$, $\delta = .4$, and starting values of $Y_0 = 100$ and $r_0 = .04$. Then see how the steady state and dynamics of the system are affected by each of the following changes:

(a) Change k to 5.5.

(b) Change h to .21.

(c) Change γ to .5.

(d) Change δ to .5.

(e) Change $\bar{I}$ to 39.

(f) Change α to .89.

16.23 How are the results of Section 16.10 affected if inflationary expectations are formed according to the equation $\pi_t^e = \pi_{t-1} + \gamma(\pi_t - \pi_{t-1})$ where $0 \leq \gamma \leq 1$? (When $\gamma = 0$ this leads to the model of Section 16.10; when $\gamma = 1$ this is a perfect-foresight model, in which expected inflation always turns out to equal actual inflation. When $0 < \gamma < 1$, agents correctly anticipate the direction in which inflation will be changing but underestimate the amount of the change.)

16.24 In a simpler version of the model in Section 16.10, inflation evolves through time based on a dynamic aggregate supply function, $\pi_t = \pi_{t-1} + \gamma(y_t - y_P)$ where y_t is the growth rate of real GDP, y_P is the growth rate of potential GDP, and $\gamma > 0$. This dynamic aggregate supply function is matched with a dynamic aggregate demand function based on Okun's Law: $y_t = g_M - \pi_t$. Use these two equations to derive a linear first-order difference equation for inflation and determine the steady-state values of inflation and output growth, whether the difference equation is stable, whether the inflation process is monotonic or if it oscillates, and how an increase in γ affects the speed of convergence.

16.25 Using the model of Section 16.11, draw graphs like Figures 16.15 and 16.16 to show what happens to gross per-capita investment and the capital-labor ratio if an economy starts in a steady state at a capital-labor ratio of k^* that is less than the golden-rule capital-labor ratio, and then the savings rate is increased to the golden-rule savings rate. Describe in words the transition from the original steady state to the golden-rule steady state.

16.26 Analyze a Solow growth model like that in Section 16.11, but in discrete time, so $k_t = k_{t-1} + sy_t - \delta k_t$. Use the per-capita production function $f(k_t) = k_t^\alpha$ where $\alpha < 1$ and show that there will be monotonic convergence to a positive steady state.

16.27 Extend the Solow growth model to allow labor force growth. Let $Y_t = F(K_t, L_t) = K_t^\alpha L_t^{1-\alpha}$ and let the labor supply grow at exponential rate n: $L_t = L_0 e^{nt}$.

    (a) Take the time derivative of L_t and use it to show that the proportional growth rate of L, $\dot{L}_t/L_t$, equals n.

    (b) Show that per-capita output equals k_t^α, just as when there was no labor force growth.

    (c) Take the time derivatives of K_t and L_t and use them to show that $\dot{k}_t = d(K_t/L_t)/dt$ equals $sy_t - (\delta + n)k_t$.

    (d) Use the result from part (c) to derive the differential equation relating $\dot{k}_t$ to k_t.

    (e) Draw the phase diagram and use it to characterize the steady states and stability of this growth model.

    (f) Illustrate using your phase diagram how the steady states and dynamics of the growth model are affected by a decrease in the labor growth rate.

16.28 Extend the model of Problem 16.27 to add growth in the *quality* of the labor force, caused, for example, by education. Let the production function be $Y_t = F(K_t, L_t, Q_t) = K_t^\alpha (Q_t L_t)^{1-\alpha}$ where Q is labor quality, which grows at a rate depending on education E: $Q_t = Q_0 e^{m(E)t}$. Define y and k in terms of *effective*, or *quality-adjusted*, labor: $y_t = Y_t/(Q_t L_t)$ and $k_t = K_t/(Q_t L_t)$.

    (a) Take the time derivative of Q_t and use it to show that the proportional growth rate of Q, $\dot{Q}_t/Q_t$, equals $m(E)$.

    (b) Show that y_t equals k_t^α.

    (c) Take the time derivatives of K_t, L_t, and Q_t and use them to show that $\dot{k}_t = d(K_t/Q_t L_t)/dt$ equals $sy_t - (\delta + m(E) + n)k_t$.

(d) Use the result from part (c) to derive the differential equation relating $\dot{k}_t$ to k_t.

(e) Show that this is a steady-state-growth model: when k is in a steady state, output, Y, per-capita output, (Y/L), and per-capita consumption, $(1-s)Y/L$, all grow at constant proportional rates through time. (For example, show that $\dot{Y}/Y$ is a constant.) Calculate these growth rates. [Hint: Use your answer to part (b) to show that $\dot{y} = 0$ in the steady state; then write out the time derivative $\dot{y} = d(Y_t/Q_t L_t)/dt$.]

(f) Assuming $m' > 0$, how does an increase in E affect the steady-state growth rates from part (e)?

Static Games with Complete Information: Theory

17.1   INTRODUCTION

Most of microeconomic theory examines maximizing behavior by economic agents. In some situations agents can make decisions in isolation without reference to the actions or decisions of other agents. For example, an individual consumer's utility-maximizing consumption choices are based on prices and income. The consumer does not need to monitor or respond to decisions by other consumers in making his/her own choices. The same principle holds for firms in perfectly competitive or monopoly markets. In other cases, however, there will be interdependence: Each agent's decision-making will directly depend on other agents' actions.

Cases where agents' choices are interdependent are very common in economics. Consider the following examples:

1. There are two firms selling differentiated products in a duopoly market. Each firm must make decisions regarding price, product quality, and advertising.

2. Two countries must choose trade policies, such as import tariffs and export subsidies, with welfare effects depending on both countries' choices.

3. A firm must choose compensation policies for its managers that specify how compensation will depend on the firm's performance. The managers' decisions will then reflect the incentives provided by the compensation policy.

The common thread to these cases is interdependence: The best choice by one agent depends on the choices made by the other agent(s).

Game theory is the branch of applied mathematics that models situations with **strategic interactions**—interactions where the payoffs for one agent depend on the strategies chosen by other agents. The mathematics applies to any model of strategic interaction and encompasses both games (e.g., chess or tic-tac-toe) and economic applications.

There are several ways of classifying possible types of games. One distinction is between **static games** and **dynamic games**. Static games are games in which all decisions are made simultaneously, usually within a single time period. Static games are one-shot in the sense that players each choose a single action and then the game ends. Dynamic games are games where decisions are made sequentially. Dynamic games often involve multiple time periods and/or more than one opportunity for players to move.

As an example of the distinction between static and dynamic games consider the Cournot model of oligopoly. The basic Cournot model, as it has been presented in earlier chapters, is a static game. Each firm makes a single output decision and all firms' decisions are made simultaneously. This is not the only way to model an oligopoly. Suppose that some firms make (and announce) their decisions before other firms or suppose that firms repeat the static Cournot game over multiple time periods. These scenarios would be dynamic games.

Games are also classified by the types of information available to the players. All of the games we will consider in this text will be games of **complete information**: games in which all elements of the structure of the game are known to all players. In contrast, games of **incomplete information** are games in which some players have private information that is not known by the other players.[1] A related, but not identical distinction, is made between games of **perfect information** and games of **imperfect information**. In games of perfect information each player knows all moves made prior to the point where the player is called upon to make a decision. In games of imperfect information at least some players have not learned some of the previous moves made in the game.

In addition to the different types of games there are also different ways of representing and solving games. One method of representing games, the **normal form**, focuses on strategies and payoffs.[2] A second method, the **extensive form**, focuses on the sequence of moves or decisions. The two forms are not mutually exclusive; instead, they are different ways of viewing a game, and the choice of form depends on which is easier or more intuitive for a particular application. The normal form (covered in this chapter) is usually used for static games in which players' choices are made simultaneously, whereas the extensive form (covered in Chapter 19) is usually employed for dynamic games in which players' moves are sequential choices.

Finally, the solution concept for finding an equilibrium to a game depends on the type of game. Static games are solved by finding a **Nash equilibrium.** The solution for dynamic games is based on **subgame-perfect Nash equilibrium,** which is a refinement of the static game solution concept. These terms will be defined as we develop the theory of static games, in this chapter, and dynamic games, in Chapter 19.

17.2 GAMES IN NORMAL FORM

Every game has "rules." The rules must specify

- Who the players are
- The nature of possible choices or moves
- The order of moves
- The payoffs that players receive

In static games these elements are represented mathematically by sets and functions. The **normal form** representation of a game consists of a player set, a strategy set for

---

[1]For example, in a game that models competition between two firms, complete information would imply that each firm knew the other's production cost. An incomplete information game would occur if each firm knew its own production cost but not its rival's cost. For an introduction to games of incomplete information, see R. Gibbons, *Game Theory for Applied Economists,* Princeton University Press, 1992. For a more advanced exposition, see D. Fudenberg and J. Tirole, *Game Theory,* MIT Press, 1991.

[2]When strategies are discrete the normal form of the game can be written as a matrix and is sometimes referred to as the **matrix form.**

each player, and a payoff function for each player. Formally, the player set, N, which enumerates the players, is

$$N = \{1, 2, 3, \ldots n\} \quad \text{where } n \text{ is the total number of players.} \qquad \textbf{(17.1)}$$

Each of the n players chooses from a set of strategy options. A strategy specifies a player's actions for all possible contingencies that might arise in the game. In some cases the strategy sets might be quite small and involve just two possible choices, such as charge a high price or charge a low price. In other cases the strategy sets will be quite large, such as in chess where a strategy could specify a player's move for all possible configurations of the board, or even infinite if a strategy choice is a continuous variable. We will let s_{ij} denote the jth possible strategy for player i. The full set of strategies, S_i, for player i is

$$S_i = \{s_{i1}, s_{i2}, s_{i3}, \ldots, s_{it_i}\} \qquad \textbf{(17.2)}$$

where t_i is the total number of strategies available to player i. In some cases it may also be useful to denote the set of all players' strategies. This set of sets, S, is

$$S = \{S_1, S_2, S_3, \ldots, S_n\}. \qquad \textbf{(17.3)}$$

Finally we have the outcomes as specified by the players' payoff functions. A player's payoff from a game will normally depend on the strategy choices of all the players. We can write the payoff function, Π_i, for player i as

$$\Pi_i = \Pi_i(s_1, s_2, s_3, \ldots, s_n), \qquad \textbf{(17.4)}$$

where s_i represents some specific strategy choice by player i (for simplicity we have dropped the second subscript).

17.3 EXAMPLES OF NORMAL FORM GAMES

To illustrate the setup of a game in normal form let's examine a simple example of interaction between firms in a duopoly market. The example will consider firms' advertising choices.

For our model we will make the following assumptions:

1. There are two firms in the market selling a product at some fixed (exogenous) price.
2. Advertising does not affect the overall level of market demand: the total quantity sold in the market is the same regardless of the level of advertising.
3. Firms choose between two advertising levels. We will call these "high" and "low."
4. Each firm's share of the market depends on the relative advertising levels chosen by the firms.

Now let's translate these assumptions into game theory notation. Each of the two players (firms) has a strategy set with two strategy choices: high advertising, A_H, or low advertising, A_L. The strategy set for a firm is

$$S_i = \{A_L, A_H\}, \quad i = 1, 2. \qquad \textbf{(17.5)}$$

To write the payoff functions we need to introduce some additional variables. Let Π_0 be the level of profits for the industry gross of advertising costs and let m_{jk} be a firm's market share when that firm chooses strategy j ($j = L$ or $j = H$) and the other firm chooses strategy k, ($k = L$ or $k = H$). For a particular combination of high- or low-advertising choices market shares sum to one, i.e., $m_{jk} + m_{kj} = 1$.

There are four possible advertising combinations. The payoff function specifies outcomes for all four combinations. Thus for firm 1 we can write

$$
\begin{aligned}
\Pi_1(A_H, A_H) &= m_{HH}\Pi_0 - A_H \\
\Pi_1(A_H, A_L) &= m_{HL}\Pi_0 - A_H \\
\Pi_1(A_L, A_H) &= m_{LH}\Pi_0 - A_L \\
\Pi_1(A_L, A_L) &= m_{LL}\Pi_0 - A_L.
\end{aligned}
\tag{17.6}
$$

The form for each equation is that the firm's profit is its share of the market times market profit minus the amount spent on advertising. A similar set of equations describes the payoff function for firm 2.

Fortunately, there is a simple way to portray strategy choices and payoffs. The payoff combinations can be written in a matrix where rows and columns correspond to strategy choices. First, let's simplify by assigning numerical values to the variables. Let

$$
\begin{aligned}
\Pi_0 &= 1{,}000 \\
A_H &= 400 \\
A_L &= 200 \\
m_{HH} &= 1/2 \\
m_{LL} &= 1/2 \\
m_{HL} &= 4/5 \\
m_{LH} &= 1/5.
\end{aligned}
\tag{17.7}
$$

Thus, when both firms advertise at a high level they split the market 50-50: each firm has a gross profit of 500 and a net profit, after advertising, of 100.

We can calculate the payoffs for each pair of players' strategies. The matrix form of a game in equation (17.8) shows payoffs for each pair of strategies. Each row corresponds to a strategy choice by player 1 and each column corresponds to a strategy choice by player 2. The numbers in the matrix show the two players' payoffs (player 1's payoff is listed before the comma and player 2's payoff is listed after) for the strategy combination. For example, if player 1 chooses L (low advertising) and player 2 chooses H (high advertising) then we are in the upper-right corner where player 1's payoff is 0 and player 2's payoff is 400:

$$
\begin{array}{cc}
 & \text{Firm 2} \\
 & \begin{array}{cc} L & \quad H \end{array} \\
\text{Firm 1} \quad \begin{array}{c} L \\ H \end{array} & \begin{pmatrix} 300,300 & 0,400 \\ 400,0 & 100,100 \end{pmatrix}.
\end{array}
\tag{17.8}
$$

In the game above consider firm 1's decision-making. If firm 2 chooses a low level of advertising then firm 1 needs to consider its best choice. Firm 1 will earn 300 by choosing L and 400 by choosing H. Since $400 > 300$, H is the best choice for firm 1 if firm 2 chooses L. If instead firm 2 chooses H, then a profit comparison again shows that firm 1's best choice is H since it earns 100 by choosing H and 0 if it chooses L. For firm 1 the low-advertising strategy is **strictly dominated**: firm 1 earns a greater payoff from H than from L no matter which strategy is chosen by firm 2. Similar reasoning shows that H is a **dominant strategy** for firm 2.

In the next section we will show how eliminating dominated strategies can, in some cases, solve for the equilibrium to a game. For this game, since neither firm would ever choose L, the equilibrium is for both firms to choose H and earn profits of 100. Notice that the equilibrium profit levels are inferior to those earned when both firms choose low advertising. Yet, even though both firms would do better with (L, L), the low-advertising strategy is not individually rational for either firm.

The general form of this game is known as the **prisoners' dilemma** (the original exposition was for two arrested partners in crime being questioned separately and choosing between silence and confession with payoffs measured as jail sentences). The prisoners' dilemma form is very common in economics, since it arises whenever

1. Agents choose between "cooperation" (in this case, low-advertising levels) and "defection" (in this case, high-advertising levels).

2. Cooperation is jointly optimal, in the sense of giving higher combined payoffs, but defection is individually rational given the incentives inherent in the payoff structure.

3. Agents (following their individually rational incentives) end up in an equilibrium where they earn lower payoffs than would have been earned if they had been able to enforce cooperation.

Our advertising model illustrates just one of many examples of the prisoners' dilemma in competition between firms. The dilemma also arises in other contexts, such as the theory of public goods where cooperating is contributing to the public good and defecting is free-riding on the contributions made by others.

There is another way to characterize the equilibrium in the example above. The combination (H, H) is an equilibrium in the sense that neither firm has any incentive to change strategy given the strategy chosen by the other firm. Put differently, each firm has chosen its best response to the other firm's strategy. This concept of equilibrium is called the **Nash equilibrium** of the game. The Nash equilibrium concept will be described in more detail in Section 17.5.

Not all games have unique, straightforward solutions. As we progress through our analysis of game theory we will develop the tools to solve more complicated games.

17.4 SOLUTION BY ITERATED ELIMINATION OF STRICTLY DOMINATED STRATEGIES

In some games, such as the prisoners' dilemma, it is possible to find an equilibrium simply by eliminating all dominated strategies. If successive, or iterated, elimination of dominated strategies results in a single pair of remaining strategies, then that remaining pair is the equilibrium of the game. Formally, we have the following definition:

DEFINITION

17.1

Strictly Dominated Strategies

*In an n-player game, a strategy, s_i', for player i is **strictly dominated** if there exists some strategy s_i'' such that:*

$$\Pi_i(s_1, s_2, \ldots, s_i'', \ldots s_n) > \Pi_i(s_1, s_2, \ldots, s_i', \ldots s_n)$$

for all possible combinations of the other players' strategies.[3]

---

[3]It is also possible to define weakly dominated strategies with a weak inequality sign.

Our definition simply states that a strategy is dominated if there is some other strategy that always yields a higher payoff. Note that the definition only requires that a strategy be dominated by one other possible strategy.

Since a rational player will never choose a dominated strategy, it can be eliminated from consideration.[4] More precisely, if in an original game G there is a dominated strategy, then we can form a new game G' that is identical to G except that the dominated strategy s' has been eliminated from player i's strategy set. Since the dominated strategy would not be played in any case, any equilibrium in the simplified game G' is also an equilibrium to the original game G. If we can continue this process until we find a game G^* in which each player has a single remaining strategy, then those strategies are an equilibrium to both G^* and to the original game G.

Consider a numerical example where each player has three strategies. The original game, G, is

$$
\begin{array}{c}
\text{Player 2} \\
\begin{array}{ccc}
L & C & R
\end{array} \\
\text{Player 1} \quad
\begin{array}{c}
T \\ M \\ B
\end{array}
\begin{pmatrix}
3,3 & 2,6 & 3,1 \\
2,4 & 1,4 & 0,4 \\
1,5 & 0,2 & 6,0
\end{pmatrix}.
\end{array}
\tag{17.9}
$$

For player 1 the middle strategy M is strictly dominated by the top strategy T: $3 > 2$, $2 > 1$, and $3 > 0$. Note that M is not dominated by B, nor does it need to be. A strategy can be eliminated as long as it is dominated by *one* other strategy.

Eliminating M gives a new game G':

$$
\begin{array}{c}
\text{Player 2} \\
\begin{array}{ccc}
L & C & R
\end{array} \\
\text{Player 1} \quad
\begin{array}{c}
T \\ B
\end{array}
\begin{pmatrix}
3,3 & 2,6 & 3,1 \\
1,5 & 0,2 & 6,0
\end{pmatrix}.
\end{array}
\tag{17.10}
$$

Neither of player 1's remaining strategies is dominated, but for player 2 the strategy R is dominated by either L or C. Eliminating R yields the game G'':

$$
\begin{array}{c}
\text{Player 2} \\
\begin{array}{cc}
L & C
\end{array} \\
\text{Player 1} \quad
\begin{array}{c}
T \\ B
\end{array}
\begin{pmatrix}
3,3 & 2,6 \\
1,5 & 0,2
\end{pmatrix}.
\end{array}
\tag{17.11}
$$

Continuing this process we eliminate B and then L to get the game G^*:

$$
\begin{array}{c}
\text{Player 2} \\
C \\
\text{Player 1} \quad T \quad (2,6).
\end{array}
\tag{17.12}
$$

The equilibrium (and only possible play) for G^* is for player 1 to choose T and for player 2 to choose C. This is also the equilibrium of the original game G.

---

[4]Solving a game through iterated elimination of dominated strategies requires an assumption of common rationality. This means not only that all players are rational, but that all players know that the other players are rational (and will not play dominated strategies) and that all players know that all other players know that all players are rational and so on, *ad infinitum*.

For most games in economics (the prisoners' dilemma is an exception) iterated elimination of dominated strategies fails to reach the final stage, G^*, where each player has a single remaining strategy, and the method therefore fails to specify an equilibrium. For these games we need a more powerful solution concept in order to find an equilibrium.

17.5 NASH EQUILIBRIUM

The concept of Nash equilibrium is simple and intuitive. A strategy combination is a Nash equilibrium if no player has an incentive, *given the strategy choices of the other players,* to switch strategies. Formally, we have the following definition:

DEFINITION 17.2

Nash Equilibrium

*Let a vector $(s_1^*, s_2^*, s_3^*, \ldots, s_n^*)$ represent a particular set of strategy choices for the n players. These strategies form a **Nash equilibrium** if for every player i and for every strategy s_i' available to player i:*

$$\Pi_i(s_1^*, s_2^*, \ldots, s_i^*, \ldots, s_n^*) \geq \Pi_i(s_1^*, s_2^*, \ldots, s_i', \ldots, s_n^*).$$

In a Nash equilibrium no player has an incentive to switch strategies: s_i^* is as good as or better than any other strategy that is available to player i. Note, in particular, two features of this definition. The first is the weak inequality sign. Nash equilibrium only requires that every player not have a positive incentive to change strategies. There may be other strategies that are just as good as s_i^*. Second, there may be more than one vector of players' strategies that satisfies the requirement for Nash equilibrium, i.e., there may be multiple solutions to a game.

The connection between solving by iterated elimination of strictly dominated strategies and the Nash equilibrium is straightforward. The following two theorems summarize the relationship:

THEOREM 17.1

If iterated elimination of strictly dominated strategies yields an equilibrium, then that equilibrium is also the unique Nash equilibrium to the game.

THEOREM 17.2

Any Nash equilibrium survives iterated elimination of strictly dominated strategies.

The theorems can be proved by contradiction. (The proofs are straightforward and are left as an exercise for the interested reader.) The key to understanding both theorems is that, by definition, no strictly dominated strategy can ever be part of a Nash equilibrium and no strategy that is played in a Nash equilibrium can be strictly dominated.

There is an alternative way of presenting the concept of Nash equilibrium. This alternative, in terms of best-response functions, is especially useful when strategy sets are continuous. Consider player 1 in a two-player game.[5] For each possible strategy chosen by player 2 we can calculate player 1's best response. The best response to a given strategy s_2 is found by choosing an s_1 that solves the problem

$$\text{Maximize } \Pi_1(s_1, s_2). \tag{17.13}$$

With continuous strategy spaces and differentiable payoff functions this problem would be solved with calculus. The first-order condition for a maximum would implicitly define player 1's best strategy as a function of s_2. Let us denote player 1's best response by the function $R_1(s_2)$. Similarly, player 2 would have a best-response function $R_2(s_1)$. For a Nash equilibrium both players' strategies must be best responses to each other, i.e., the Nash equilibrium strategy pair, (s_1^*, s_2^*), is the simultaneous solution

$$s_1^* = R_1(s_2^*) \tag{17.14}$$

and

$$s_2^* = R_2(s_1^*). \tag{17.15}$$

This may look familiar. The Cournot model, first presented in Chapter 2, was solved by this method. In the next chapter we will represent the Cournot model as a game theory model.

17.6   EXAMPLES OF NASH EQUILIBRIA

In this section we will solve two numerical examples of static games: one with discrete strategy choices and one with continuous strategy choices. Our first example is a two-player game in which each player has three possible strategies. The matrix form of the game is

$$
\begin{array}{c}
 & & \text{Player 2} \\
 & & \begin{array}{ccc} L & C & R \end{array} \\
\text{Player 1} & \begin{array}{c} T \\ M \\ B \end{array} & \begin{pmatrix} 3,3 & 2,1 & 3,1 \\ 2,4 & 2,4 & 0,4 \\ 1,5 & 0,2 & 6,5 \end{pmatrix}.
\end{array} \tag{17.16}
$$

Notice that neither player has any strictly dominated strategies. To solve the game we use the method of finding best responses and looking for Nash equilibrium strategy pairs in which the strategies are best responses to each other.

To find player 1's best responses we examine each column (a player 2 strategy choice) and find the best row (a player 1 strategy choice) response. In the following matrix the best responses for player 1 are underlined:

$$
\begin{array}{c}
 & & \text{Player 2} \\
 & & \begin{array}{ccc} L & C & R \end{array} \\
\text{Player 1} & \begin{array}{c} T \\ M \\ B \end{array} & \begin{pmatrix} \underline{3},3 & \underline{2},1 & 3,1 \\ 2,4 & \underline{2},4 & 0,4 \\ 1,5 & 0,2 & \underline{6},5 \end{pmatrix}.
\end{array} \tag{17.17}
$$

---

[5]The method presented below is easily extended to multi-player games.

Note that for player 2's strategy C player 1 has two different (identical payoff) best responses.

In a similar vein we find player 2's best responses by examining each row (player 1 strategy choice) and finding the best column (player 2 strategy choice) response. In the following matrix entries in which both responses are underlined are Nash equilibria:

$$
\begin{array}{cc}
& \text{Player 2} \\
& \begin{array}{ccc} L & C & R \end{array} \\
\text{Player 1}\quad
\begin{array}{c} T \\ M \\ B \end{array}
&
\begin{pmatrix}
\underline{3},\underline{3} & 2,1 & 3,1 \\
2,\underline{4} & \underline{2},\underline{4} & 0,\underline{4} \\
1,\underline{5} & 0,2 & \underline{6},\underline{5}
\end{pmatrix}.
\end{array}
\qquad (17.18)
$$

There are three Nash equilibria: (T,L), (M,C), and (B,R). In any game written in a matrix form we can find all pure-strategy Nash equilibria (if any exist) by mechanically underlining best responses.[6]

We now examine a two-player example with continuous strategies. Suppose that the strategy set for each player is

$$
S_i = \{s_i : s_i \geq 0\}. \qquad (17.19)
$$

In other words, each player must choose a non-negative level of the strategy variable. This is, of course, common in economics: quantity, price, consumption choices, etc., cannot normally take on negative values. Next, let's suppose that the payoff functions for players 1 and 2 are

$$
\Pi_1 = 10s_1 - s_1^2 - s_2 s_1 - 3s_1
$$

and

$$
\Pi_2 = 10s_2 - s_2^2 - s_1 s_2 - 2s_2. \qquad (17.20)
$$

Since the game involves continuous strategy sets it is not possible to construct a discrete payoff matrix. Instead we can solve for the best-response functions and use these to find Nash equilibria. The best response for each player can be found by taking the respective first-order conditions. These are

$$
\frac{\partial \Pi_1}{\partial s_1} = 10 - 2s_1 - s_2 - 3 = 0
$$

and

$$
\frac{\partial \Pi_2}{\partial s_2} = 10 - 2s_2 - s_1 - 2 = 0. \qquad (17.21)
$$

Rewriting the first-order conditions gives the best-response functions for the respective players as

$$
s_1 = R_1(s_2) = \frac{1}{2}(7 - s_2)
$$

and

$$
s_2 = R_2(s_1) = \frac{1}{2}(8 - s_1). \qquad (17.22)
$$

---

[6]The terms pure strategy and mixed strategy are defined in the next section.

FIGURE 17.1

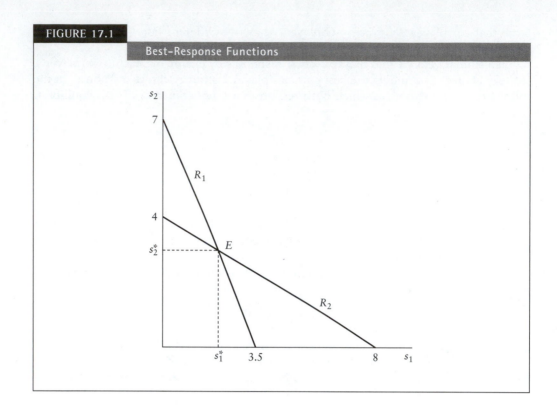

Best–Response Functions

These best-response functions are graphed in Figure 17.1. The Nash equilibrium point, E, is where the best-response functions intersect.

To solve for the Nash equilibrium we find the values (s_1^*, s_2^*) that give the simultaneous solution to the two first-order (or response function) equations. The reader should check that these solutions are

$$s_1^* = 2 \quad \text{and} \quad s_2^* = 3. \tag{17.23}$$

The reader can also verify that, in equilibrium, player 1 receives a payoff of 4 and player 2 receives a payoff of 9.

17.7 AN INTRODUCTION TO MIXED STRATEGIES

Not all games have **pure-strategy** Nash equilibria in which each player chooses a single strategy with probability of one. There are important economic applications that only have solutions in **mixed strategies** in which a player randomizes by choosing the probabilities for playing the possible pure strategies. In this section we examine mixed-strategy equilibria in the simplest possible model: two-player, discrete-strategy games.

Consider a two-player game in which each player has m possible pure strategies. Although the number of strategies is the same for each player the strategies themselves may well be different.[7] Let s_{ij} be the jth pure strategy for player i. A mixed strategy specifies the probability with which player i will play each possible pure strategy. Let p_j be the probability that player 1 chooses pure strategy j and q_k be the probability that player 2 chooses pure strategy k. Using **p** and **q** to denote the vectors of probabilities

---

[7]The extension to different numbers of strategies for the two players adds an extra subscript but is otherwise straightforward.

we can write the mixed-strategy sets for players 1 and 2 as

$$S_1 = \left\{ \mathbf{p} : 0 \le p_j \le 1, \sum_{j=1}^{m} p_j = 1 \right\}$$

$$S_2 = \left\{ \mathbf{q} : 0 \le q_k \le 1, \sum_{k=1}^{m} q_k = 1 \right\}.$$

(17.24)

Pure strategies are a subset of the set of mixed strategies: the pure strategy s_{1j} is equivalent to player 1 choosing the mixed strategy $p_j = 1$ and $p_i = 0$ for $i \ne j$.

To characterize the mixed-strategy Nash equilibrium we must calculate the expected value of the players' payoffs. Let Π_{ijk} be player i's payoff when player 1 chooses pure strategy j and player 2 chooses pure strategy k. The probability of this payoff is the probability, p_j, that player 1 plays strategy j times the probability, q_k, that player 2 chooses strategy k. The expected payoff for player i is the sum, over all possible outcomes, of the probability of an outcome times the payoff from that outcome.[8] Since the probability of a given outcome is the probability that player 1 chooses strategy j times the probability that player 2 chooses strategy k, we can write the expected payoff for player 1 as

$$\begin{aligned} E(\Pi_1) = \ & p_1 q_1 \Pi_{111} + p_1 q_2 \Pi_{112} + \cdots p_1 q_m \Pi_{11m} \\ & + p_2 q_1 \Pi_{121} + p_2 q_2 \Pi_{122} + \cdots p_2 q_m \Pi_{12m} \\ & + \cdots \\ & + p_m q_1 \Pi_{1m1} + p_m q_2 \Pi_{1m2} + \cdots p_m q_m \Pi_{1mm}. \end{aligned}$$

(17.25)

This expression can be simplified by using summation notation:

$$E(\Pi_1) = \sum_{j=1}^{m} \sum_{k=1}^{m} p_j q_k \Pi_{1jk}.$$

(17.26)

The double-summation sums over m pure strategies for each player for a total of m^2 possible outcomes.

The definition of a mixed-strategy Nash equilibrium is similar to that given in definition 17.2 for a pure-strategy Nash equilibrium. We have the following:

DEFINITION 17.3

Mixed–Strategy Nash Equilibrium

The probability vectors $\mathbf{p}^*$ and $\mathbf{q}^*$ are a Nash equilibrium if for all other possible mixed strategies $\mathbf{p}'$ and $\mathbf{q}'$ we have

$$\sum_{j=1}^{m} \sum_{k=1}^{m} p_j^* q_k^* \Pi_{1jk} \ge \sum_{j=1}^{m} \sum_{k=1}^{m} p_j' q_k^* \Pi_{1jk}$$

and

$$\sum_{j=1}^{m} \sum_{k=1}^{m} p_j^* q_k^* \Pi_{2jk} \ge \sum_{j=1}^{m} \sum_{k=1}^{m} p_j^* q_k' \Pi_{2jk}.$$

---

[8]The analysis that follows assumes that players are risk-neutral, which means that their goals can be represented as maximizing expected payoffs (as opposed to expected utility).

In other words, the vectors $\mathbf{p}^*$ and $\mathbf{q}^*$ are a Nash equilibrium if no other mixed strategies would yield higher payoffs for either player.

Now that we have defined a mixed-strategy equilibrium we present, without proof, an important theorem:

THEOREM **17.3**

> *Every n-player game with finite pure-strategy sets has at least one pure or mixed-strategy Nash equilibrium.*[9]

The theorem ensures that every game has a solution. Unfortunately, however, the theorem does not tell us how to find that solution. While we can easily find pure-strategy equilibria (if any exist), there is no straightforward method for calculating a game's mixed-strategy equilibria.[10] In the simplest case of two-player, two-strategy games, however, there is a method for finding mixed-strategy solutions. The method of solving this game illustrates the characteristics of mixed strategies in more complex games.

Suppose we have a two-player game where the payoff matrix is

$$
\begin{array}{c}
& \text{Player 2} \\
& \begin{array}{cc} L & \quad R \end{array} \\
\text{Player 1} \quad \begin{array}{c} T \\ B \end{array} & \left(\begin{array}{cc} A,a & B,b \\ C,c & D,d \end{array} \right).
\end{array}
\tag{17.27}
$$

In this game each player essentially chooses a single probability: if strategy one is played with probability p, then strategy two must be played with probability $1 - p$. Using this we can write player 1's expected payoff as

$$
E(\Pi_1) = pqA + p(1-q)B + (1-p)qC + (1-p)(1-q)D.
\tag{17.28}
$$

Note that, for given q, player 1's payoff is a linear function of p.

Now consider the choice of p. Taking the derivative of player 1's payoff function gives

$$
\frac{\partial E(\Pi_1)}{\partial p} = qA + (1-q)B - qC - (1-q)D.
\tag{17.29}
$$

Unlike the usual first-order condition for a maximum this equation does not contain the choice variable p. Since the derivative depends on q, but not on p, player 1 cannot choose the value of the derivative, which may be positive or negative or zero depending on q and the payoffs. If the derivative is positive, then increases in p uniformly increase

---

[9]Games with continuous strategy sets have an infinite number of pure strategies. In this case, the theorem would state that if the strategy sets are closed and bounded and the payoff functions are continuous, then there exists at least one Nash equilibrium (possibly in mixed strategies). The terms closed and bounded require, respectively, that the strategy set contain its endpoints and that the endpoints be finite.

[10]The complicating factor here is that some strategies may be played with zero probability. Finding which strategies are played with positive probabilities and which with zero probabilities involves a linear programming problem for discrete strategy choices and (even worse) a trial-and-error search for continuous strategy choices.

player 1's payoff. Player 1 should therefore choose the highest-possible level of p. This means that player 1's best response is $p = 1$, which is equivalent to playing pure-strategy one. If the derivative is negative, then player 1's best response is $p = 0$, which means choosing the second pure strategy.

When the derivative equals zero, player 1's payoff is the same for all possible levels of p. This implies that player 1 is indifferent between all possible strategies: any mixed-strategy choice (including the two pure strategies of $p = 0$ and $p = 1$) yields the same payoff. Thus, only when the derivative equals zero can it be an equilibrium for player 1 to play a mixed strategy.

To find a mixed-strategy solution for the game we repeat the process above for player 2. Without giving the full derivation a second time we simply write the derivative of player 2's payoff function with respect to q:

$$\frac{\partial E(\Pi_2)}{\partial q} = pa - pb + (1-p)c - (1-p)d. \qquad \textbf{(17.30)}$$

As above, the derivative of the payoff is independent of the player's probability choice. Player 2 will choose a pure strategy ($q = 1$ or $q = 0$) if the derivative is non-zero and will be indifferent between all mixed strategies when the derivative is zero.

We can now give the solution to this game. A mixed-strategy Nash equilibrium will only exist when

$$\frac{\partial E(\Pi_1)}{\partial p} = 0 \quad \text{and} \quad \frac{\partial E(\Pi_2)}{\partial q} = 0. \qquad \textbf{(17.31)}$$

Solving these equations for the mixed-strategy Nash equilibrium (p^*, q^*) yields

$$p^* = \frac{d - c}{a - b - c + d}$$
$$q^* = \frac{D - B}{A - B - C + D}. \qquad \textbf{(17.32)}$$

Of course a mixed-strategy equilibrium only exists if both solutions are between zero and one.

To illustrate the solution method consider the following game:

$$
\begin{array}{c}
\text{Player 2} \\
\begin{array}{cc}
L & R
\end{array} \\
\text{Player 1} \quad
\begin{array}{c}
T \\
B
\end{array}
\begin{pmatrix}
3,1 & 2,4 \\
2,2 & 3,1
\end{pmatrix}.
\end{array}
\qquad \textbf{(17.33)}
$$

There are no pure-strategy Nash equilibria to this game. Theorem 17.3, however, assures us that the game does have a mixed-strategy equilibrium. To find this solution we start by writing the expected payoffs for the players as

$$E(\Pi_1) = 3pq + 2p(1-q) + 2(1-p)q + 3(1-p)(1-q)$$
$$= 2pq - p - q + 3$$

and

$$\textbf{(17.34)}$$

$$E(\Pi_2) = pq + 4p(1-q) + 2(1-p)q + (1-p)(1-q)$$
$$= -4pq + 3p + q + 1.$$

The derivative of the payoff functions with respect to the player's respective probability choices are

$$\frac{\partial E(\Pi_1)}{\partial p} = 2q - 1$$

and   **(17.35)**

$$\frac{\partial E(\Pi_2)}{\partial q} = -4p + 1.$$

Setting the two equations equal to zero and solving yields the mixed-strategy Nash equilibrium as

$$p^* = \frac{1}{4} \quad \text{and} \quad q^* = \frac{1}{2}.$$ **(17.36)**

Finally, we can find the players' expected payoffs by substituting these equilibrium probabilities into the equations in 17.34. This gives

$$E(\Pi_1) = 2p^*q^* - p^* - q^* + 3 = 2.5$$

and   **(17.37)**

$$E(\Pi_2) = -4p^*q^* + 3p^* + q^* + 1 = 1.75.$$

To end our discussion of this example we note a somewhat problematic feature of a mixed-strategy equilibrium: neither player has any positive incentive to play the equilibrium strategy. By this we mean that if player 2 chooses q^*, then any probability p (including p^*) yields the same expected payoff for player 1. Similarly, given p^* player 2 is indifferent between all possible levels of q. The reason that player 1 (player 2) must choose p^* (q^*) in equilibrium is to ensure that the other player also chooses the equilibrium mixed strategy.

In more complicated games mixed-strategy equilibria are sometimes found in much the same manner as in the two-player, two-strategy model. What complicates the attempt to find equilibria in more general games is that not all strategies need to be played with positive probabilities. Finding which strategies are played with zero probabilities and which are played with positive probabilities is often a difficult analytic exercise.

SUMMARY

Static game theory models strategic interactions between players. Such interactions occur frequently in economic contexts. Iterated elimination of dominated strategies is one solution method, but the class of games for which this method yields a solution is fairly small. A more powerful solution method for static games, the Nash equilibrium, looks for strategy combinations in which no player has an incentive to deviate from the equilibrium. This method can generate solutions to games with both discrete strategy sets (using payoff matrices) and continuous strategy sets (using best-response functions). In some cases, however, games will not have pure-strategy equilibria. In these games we must search for mixed strategies in which players randomize over their pure-strategy choices.

Problems

17.1 Use iterated elimination of strictly dominated strategies to find the equilibrium for the following games:

(a)   Player 1

Player 2

$$\begin{array}{c} & L & R \\ T & \begin{pmatrix} 4,2 & 2,6 \\ 1,3 & 1,0 \end{pmatrix} \\ B & \end{array}$$

(b)   Player 1

Player 2

$$\begin{array}{c} & L & C & R \\ T & \begin{pmatrix} 3,7 & 2,8 & 3,4 \\ 4,3 & 3,9 & 4,2 \\ 5,4 & 6,8 & 3,1 \end{pmatrix} \\ M & \\ B & \end{array}$$

17.2 Find all pure-strategy Nash equilibria in the following games:

(a)   Player 1

Player 2

$$\begin{array}{c} & L & R \\ T & \begin{pmatrix} 4,4 & 2,3 \\ 1,3 & 13,9 \end{pmatrix} \\ B & \end{array}$$

(b)   Player 1

Player 2

$$\begin{array}{c} & L & C & R \\ T & \begin{pmatrix} 3,7 & 2,8 & 3,4 \\ 4,3 & 3,9 & 4,2 \\ 5,4 & 6,8 & 3,1 \end{pmatrix} \\ M & \\ B & \end{array}$$

(c)   Player 1

Player 2

$$\begin{array}{c} & L & C & R \\ T & \begin{pmatrix} 8,8 & 9,8 & 3,4 \\ 4,9 & 3,7 & 1,2 \\ 5,4 & 3,8 & 6,8 \end{pmatrix} \\ M & \\ B & \end{array}$$

(d)   Player 1

Player 2

$$\begin{array}{c} & L & C & R \\ T & \begin{pmatrix} 3,7 & 2,8 & 2,4 \\ 4,3 & 1,4 & 4,9 \\ 5,4 & 6,3 & 3,4 \end{pmatrix} \\ M & \\ B & \end{array}$$

17.3 Consider a game where players are given a chance to split a prize worth X dollars. Player i's strategy choice is to claim a share, s_i, of the prize where $0 \leq s_i \leq 1$.

The payoff for each player depends on the sum of the claims. If the sum of the claims is less than or equal to one, then each player receives a payoff equal to her claim times the total prize, X. If the sum of the claims exceeds one, then each player receives a payoff of zero. Find the pure-strategy Nash equilibria for this game when

(a) there are two players

(b) there are n players

17.4 Consider a two-player game where $s_i \geq 0$ is the strategy for player i and the payoff function for player i is $\Pi_i = a_i s_i - s_i s_j - s_i^2$.

(a) Find and graph the best-response functions for the two players.

(b) Solve for the Nash equilibrium strategies. For what values of a_1 and a_2 are the strategy solutions positive? Show and explain what happens when the simultaneous mathematical solution of the best-response functions yields a negative value for one of the strategies.

(c) Find the comparative static effects of a change in a_1 on the players' equilibrium strategies and payoffs.

17.5 Consider a two-player game where $s_i \geq 0$ is the strategy for player i and the payoff function for player i is $\Pi_i = s_i(s_i + s_j)^{-\alpha} - cs_i$, where α is a positive parameter. Find the Nash equilibrium for this game.

17.6 Consider a three-player game where $s_i \geq 0$ is the strategy for player i and the payoff function for player i is $\Pi_i = as_i - s_i s_j - s_i s_k - s_i^2 - c_i s_i$.

(a) Find the best-response functions for the three players.

(b) Solve for the Nash equilibrium strategies. For what values of c_1, c_2, and c_3 are the strategy solutions positive?

(c) Find the comparative static effects of a change in c_1 on the players' equilibrium strategies and payoffs.

17.7 For the following games find all pure- and mixed-strategy Nash equilibria:

Player 2

$$
\begin{array}{cc}
& L \qquad R \\
\textbf{(a)} \quad \text{Player 1} \quad \begin{array}{c} T \\ B \end{array} & \left(\begin{array}{cc} 5,5 & 0,0 \\ 1,1 & 15,15 \end{array} \right)
\end{array}
$$

Player 2

$$
\begin{array}{cc}
& L \qquad\qquad R \\
\textbf{(b)} \quad \text{Player 1} \quad \begin{array}{c} T \\ B \end{array} & \left(\begin{array}{cc} 5+x,5-x & 0,0 \\ 0,0 & 5,5 \end{array} \right) \quad \text{where } x > 0.
\end{array}
$$

Player 2

$$
\begin{array}{cc}
& L \qquad R \\
\textbf{(c)} \quad \text{Player 1} \quad \begin{array}{c} T \\ B \end{array} & \left(\begin{array}{cc} -5,-5 & 10,0 \\ 0,10 & 0,0 \end{array} \right)
\end{array}
$$

Static Games with Complete Information: Applications

18.1 INTRODUCTION

We can model applications in a variety of fields of economics using the theory of static games. The field of industrial organization, which deals with market structure and firm behavior, has been quite influenced by game theory since the 1980s. Several applications from industrial organization are included in this chapter. Three applications from other branches of economics are also included.

We begin the chapter with a game theory model of two firms that are both considering an investment in a market. The problem that the firms face is that there is only room in the market for one firm to make a profit; if both invest, then both lose money. It turns out that this model yields a rather stark outcome: a mixed-strategy equilibrium in which both firms earn a zero expected profit.

The next three sections focus on quantity and price competition in oligopoly. We first review the Cournot analysis of oligopoly quantity choices by presenting it as a game theoretic model. We then consider two models in which firms choose pricing strategies: first with identical products and then with differentiated products. We will see that there are important differences between oligopoly models where firms choose quantity strategies and oligopoly models where firms choose pricing strategies.

There are also important applications of game theory in other branches of economics. The chapter's next two applications are economic opposites. The first, rent-seeking behavior, considers competition for a share of a fixed economic prize. Players waste resources in a competition in which private expenditures simply serve to shift shares of the prize from one player to another. The second example, the provision of public goods, examines the case where each player's private expenditure creates benefits for others. Since individual players fail to take into account the external benefits generated by their private actions, there is under-spending on the public good.

Our final example is from the economics of the family. We develop a model that considers household time allocation. The model explains how players may be worse off if they become more efficient in performing a task.

18.2 A NATURAL MONOPOLY INVESTMENT GAME

Markets come in many sizes. At one extreme there are perfectly competitive markets, markets in which it is possible for many firms to coexist and earn normal profits. At the other extreme are **natural monopolies**, markets in which the average cost curve slopes downward. Our model captures the essence of natural monopoly in assuming that the market is large enough for one firm to earn a positive economic profit, but the market is not large enough for two firms to earn positive profits.

The size of a market depends on both the level of demand and the level of firms' costs. Figure 18.1 shows a natural monopoly. The demand curve D lies above the average cost curve AC for an individual firm. If, however, there were two firms in the market, so that each firm could capture only half of overall market demand (i.e., the curve $1/2D$), then there is no price at which both firms could cover their average cost.

Suppose that two firms are simultaneously choosing whether to make investments (such as developing a new product, building a factory, or opening a retail outlet) that allow them to enter a market. Each firm has two strategy choices, E or S. E denotes a decision to undertake an investment that allows the firm to enter the market and S denotes a decision to forego the investment and stay out of the market.

We will assume that the size of the market (relative to the initial investment) is such that if one firm enters the market it will earn a profit, $\Pi > 0$, but if both firms enter they each incur a loss, $(-L) < 0$. The game matrix is

$$\text{Firm 1} \quad \begin{array}{c} \\ E \\ S \end{array} \overset{\displaystyle \text{Firm 2}}{\overset{\displaystyle \begin{array}{cc} E & \quad S \end{array}}{\begin{pmatrix} -L,-L & \Pi,0 \\ 0,\Pi & 0,0 \end{pmatrix}}}. \qquad \textbf{(18.1)}$$

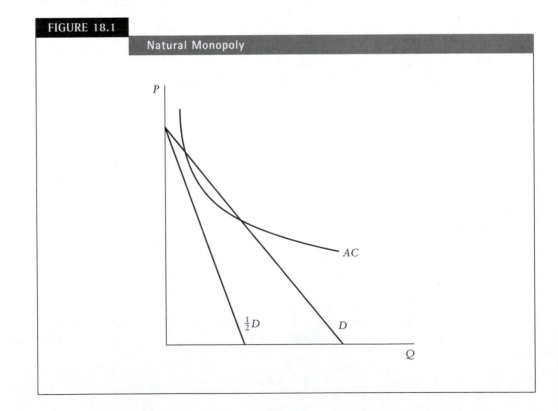

FIGURE 18.1

Natural Monopoly

When we try to solve this game we first discover that neither firm has a dominant (or a dominated) strategy. We next discover that two pairs of strategies are pure-strategy Nash equilibria. Both (E, S) and (S, E) are Nash equilibrium strategy pairs for the two firms: for each combination neither firm has any incentive to switch its strategy. Each firm would, of course, prefer the outcome where it entered and its rival stayed out, but the Nash equilibrium concept does not allow us to specify which of the (mutually exclusive) equilibria would actually occur. There is, however, a third equilibrium to the game. The third equilibrium is a mixed-strategy equilibrium in which firms play each of the pure strategies with some positive probability.

We start our analysis by writing the expected profits for the two firms. Let p and q be the respective probabilities of entry for firms 1 and 2. Expected profits are

$$E(\Pi_1) = -pqL + p(1-q)\Pi + (1-p) \cdot 0 = -pqL + p(1-q)\Pi$$

and   **(18.2)**

$$E(\Pi_2) = -qpL + q(1-p)\Pi + (1-q) \cdot 0 = -qpL + q(1-p)\Pi.$$

Note that $(1-p)$ for firm 1, or $(1-q)$ for firm 2, is the probability that a firm stays out of the market and earns zero payoff regardless of the other firm's decision. We next take the derivatives of the payoff functions with respect to the players' probability choices. These are

$$\frac{\partial E(\Pi_1)}{\partial p} = -qL + (1-q)\Pi$$

and   **(18.3)**

$$\frac{\partial E(\Pi_2)}{\partial q} = -pL + (1-p)\Pi.$$

Note that the two pure-strategy Nash equilibria can be represented as mixed-strategy equilibria. The choices $p^* = 1$ and $q^* = 0$ are best responses to each other, as are the choices $p^* = 0$ and $q^* = 1$. There is also a third equilibrium in which both derivatives equal zero. Setting the derivatives equal to zero and solving gives the (symmetric) mixed-strategy equilibrium of

$$p^* = q^* = \frac{\Pi}{\Pi + L}.$$ **(18.4)**

Both p^* and q^* lie strictly between zero and one, and it is fairly easy to verify that the equilibrium entry probabilities are increasing in Π, the potential profit, and decreasing in L, the potential loss.

An interesting characteristic of this game is that both firms, given their equilibrium probability choices, earn zero expected profits. To check this let's calculate firm 1's expected payoff:

$$E(\Pi_1) = -pqL + p(1-q)\Pi$$

or   **(18.5)**

$$E(\Pi_1) = -\left(\frac{\Pi}{\Pi + L}\right)\left(\frac{\Pi}{\Pi + L}\right)L + \left(\frac{\Pi}{\Pi + L}\right)\left(\frac{L}{\Pi + L}\right)\Pi = 0.$$

Intuitively, what is happening here is that if the market becomes more attractive (a higher profit or lower loss), the firms increase their probabilities of entering. The increased probability of entry exactly offsets the increased attractiveness of the market so that expected profits remain at zero.

This zero profit result follows quite directly from the theoretical discussion of mixed strategies in Chapter 15. Recall that an essential feature of a mixed-strategy equilibrium is that player 2's (1's) mixed strategy leaves player 1 (2) indifferent between possible strategies. In equilibrium, all pure strategies that are played with positive probability must yield equal payoffs. In our example the pure strategy of staying out always yields zero profits. Therefore in the mixed-strategy equilibrium the other pure strategy, enter, will also yield zero profits as will all possible mixed combinations of the two pure strategies. No matter how high the potential profit and no matter how small the potential loss, the expected profit for the two firms will always be zero in the mixed-strategy equilibrium.

18.3   THE COURNOT MODEL REVISITED

In Section 2.5 we solved a model of an n-firm Cournot oligopoly. Here we rewrite the model as a formal game theory model. In the Cournot model a firm's strategy is its choice of output. The strategy set for firm i can be written as

$$S_i = \{q_i : q_i \geq 0\}, \tag{18.6}$$

i.e., the firm's strategy space is the set of all nonnegative outputs. The payoff functions—based on linear demand ($P = a - bQ$), constant marginal costs ($MC = c$), and a per-unit tax ($\$t$/unit)—were derived in equation (2.66). For firm i this payoff function is

$$\Pi_i = P q_i - (c + t) q_i$$
$$\Pi_i = (a - bQ) q_i - (c + t) q_i$$
$$\Pi_i = \left(a - b \left(\sum_{j=1}^{n} q_j \right) \right) q_i - (c + t) q_i. \tag{18.7}$$

For any set of strategy choices by the other firms, firm i will have a best-response function. To find the best-response functions we maximize profits holding other firms' strategies constant. The result, derived in equation (2.68), is

$$q_i = R_i(q_1, q_2 \ldots q_i, q_{i+1}, \ldots q_n) = \frac{(a - c - t)}{2b} - \frac{1}{2} \left(\sum_{\substack{j=1 \\ j \neq i}}^{n} q_j \right), \tag{18.8}$$

where the summation runs from 1 through n, but excludes firm i.

The Nash equilibrium is the simultaneous solution to the n best-response equations. From equation (2.70) each firm plays a Nash equilibrium strategy of

$$q_i^* = \frac{(a - c - t)}{(n + 1)b}. \tag{18.9}$$

Substituting the Nash equilibrium strategies into equation (18.7), we can calculate each firm's equilibrium payoff. The steps of the derivation and the simplified payoff functions are

$$\Pi_i^* = \left(a - b\left(\sum_{j=1}^{n} q_j^*\right)\right) q_i^* - (c + t)\, q_i^* = (a - bnq_i^*)q_i^* - (c + t)q_i^*$$

$$\Pi_i^* = \left(a - bn\frac{(a - c - t)}{(n + 1)b}\right)\frac{(a - c - t)}{(n + 1)b} - (c + t)\frac{(a - c - t)}{(n + 1)b}$$

$$\Pi_i^* = \frac{(a - c - t)^2}{(n + 1)b} - \frac{n(a - c - t)^2}{(n + 1)^2 b} = \frac{(n + 1)(a - c - t)^2}{(n + 1)^2 b} - \frac{n(a - c - t)^2}{(n + 1)^2 b}$$

$$\Pi_i^* = \frac{(a - c - t)^2}{(n + 1)^2 b}.$$

$$(18.10)$$

The solution to this model was originally derived under the Cournot assumption that each firm treated the other firms' outputs as fixed or exogenous. This assumption is consistent with the Nash equilibrium concept, which requires that a firm's equilibrium strategy choice be a best response to the other firms' equilibrium strategy choices. Because of the equivalence of the concepts the solution to this model is referred to as the Cournot-Nash or Nash-Cournot equilibrium.

The Cournot oligopoly model is often used to analyze competition among firms in an oligopoly. We will see further applications in Chapter 20 when we embed the static Cournot model solutions in the richer dynamic contexts of trade policy choices and repeated interactions between firms.

18.4 THE BERTRAND DUOPOLY MODEL

Cournot's model of oligopoly assumed that firms choose output as their strategy variable. The market price was then based on the sum of the output choices. Forty-five years later, Joseph Bertrand published a review of Cournot's book.[1] In his review Bertrand argued that Cournot's emphasis on firms' output choices was misplaced. Instead, Bertrand argued that firms in imperfectly competitive markets choose prices and then respond to the market demand by producing whatever quantity consumers wish to purchase. Given that the market demand curve fixes a definite relationship between quantity demanded and price, the difference in approaches may seem semantic. Nevertheless, as we shall see next, the change in the strategy variable drastically changes the nature of the market equilibrium.

18.4.1 Identical Products

The original papers by both Cournot and Bertrand assumed that firms' products were identical or homogeneous. This section maintains that assumption; the next section considers a market with differentiated products. In the Cournot model the identical-products assumption ensures that there is a single market price, determined by total output, at which all firms sell their product. The effect of this assumption in the Bertrand model, however, is somewhat different. Suppose that two firms in a duopoly charge different prices. Then, because the products are identical, all consumers will buy from the

---

[1]For translations of the section on oligopoly of Cournot's 1838 book and of Bertrand's 1883 review (both originals were in French), see *Cournot Oligopoly: Characterizations and Applications*, Andrew Daughety (ed.), Cambridge University Press, New York, 1988.

low-price firm, and the high-price firm will have zero sales. This is the key to deriving the Nash equilibrium of the model.

In the Bertrand model each firm chooses a pricing strategy. The strategy space for firm i is

$$S_i = \{p_i : p_i \geq 0\}. \tag{18.11}$$

To write the payoff functions we need to introduce notation for the market demand function. First, define the variable p as

$$p = \min(p_1, p_2). \tag{18.12}$$

This is read as p equals the minimum of p_1 and p_2. Since consumers buy from the firm charging the lowest price, the total quantity demanded is given by the demand function:

$$Q = Q(p). \tag{18.13}$$

For an individual firm there are three possibilities for quantity demanded. Firm i's demand is

$$
\begin{aligned}
q_i &= Q(p) && \text{if } p_i < p_j \\
q_i &= \frac{Q(p)}{2} && \text{if } p_i = p_j \\
q_i &= 0 && \text{if } p_i > p_j.
\end{aligned}
\tag{18.14}
$$

If prices are different, then the low-price firm serves the entire market while the high-price firm sells nothing. If prices are equal, then consumers are indifferent, and the two firms are assumed to split the market equally. Assuming that both firms operate with the same constant marginal cost of c, the profit function for firm i can be written as

$$
\begin{aligned}
\Pi_i &= p_i Q(p_i) - c Q(p_i) && \text{if } p_i < p_j \\
\Pi_i &= \frac{1}{2}(p_i Q(p_i) - c Q(p_i)) && \text{if } p_i = p_j \\
\Pi_i &= 0 && \text{if } p_i > p_j.
\end{aligned}
\tag{18.15}
$$

Since the profit function is discontinuous (and its derivatives are undefined) at $p_i = p_j$, we cannot use calculus to solve for best-response functions or the Nash equilibrium. Instead we must search for Nash equilibria in the space of all possible outcomes: potential equilibria initially include all positive combinations of the two firms' prices. Recall that Nash equilibria require that neither firm have an incentive to change price. Thus, the search for an equilibrium can be narrowed somewhat by realizing that neither outcomes where both firms charge more than the monopoly price, P_m, nor outcomes where both charge less than marginal cost, c, can be equilibria.[2] The remaining possibilities for Nash equilibria are shown as the shaded area (including the boundary) in Figure 18.2.

We solve the model by ruling out two other types of equilibria. First, outcomes with positive profits are not equilibria. If prices are equal, then the firms are splitting the market. Either firm could capture the entire market and earn twice as much profit

---

[2]If both charged a price higher than the monopoly price, then one firm could lower its price to the monopoly level and earn higher profits. If both prices are below-cost, then at least one firm has negative profits and could earn zero profits by raising price to cost.

FIGURE 18.2

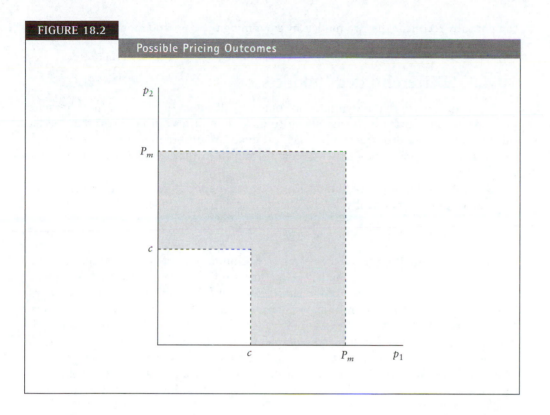

with a slight decrease in price. If prices are unequal, then the higher-price firm is selling nothing and could improve from a zero profit to a positive profit by setting price slightly below the lower-price firm. In either case, no solution with positive profits satisfies the no-incentive-to-change-strategy condition for a Nash equilibrium. Similarly, no negative-profit outcome could be a Nash equilibrium: a negative-profit firm could earn zero profits by increasing price to marginal cost or higher.

Thus, equilibrium requires that both firms earn zero profits. This implies that the lowest-price firm must charge a price equal to marginal cost. Now if the higher-price firm charged a price greater than marginal cost, then there would be a range of prices greater than marginal cost and less than this higher price at which the first firm could earn positive profits; thus, this is not a Nash equilibrium. The only remaining candidate for a Nash equilibrium is

$$p_1^* = p_2^* = c \quad \text{and} \quad \Pi_1^* = \Pi_2^* = 0. \tag{18.16}$$

To verify that this is an equilibrium, note that if a firm lowered price it would earn negative profits, and if it raised price profits would remain at zero. Since Nash equilibrium only requires that no other strategy give a payoff higher than the equilibrium, equation (18.16) is the solution to the model.

The Bertrand-Nash equilibrium prediction is quite different than the Cournot-Nash solution. The Bertrand prediction is that, even with only two firms, market outcomes will be competitive, while the Cournot result is that markets approach perfect competition only as the number of firms becomes large. Public policy decisions involving oligopoly, e.g., trade or antitrust policy, depend on the way that market structures, or number of firms, determine the nature of market outcomes. Economists' advice on these public policy issues therefore hinges on which model is used. A major change in results from a seemingly small change in assumptions is disconcerting. Fortunately, we can partially reconcile the two models by examining the Bertrand model when products are differentiated.

18.4.2 Differentiated Products

Many products are differentiated in some way. Product quality, product characteristics, packaging, and advertising all act to make one firm's output a less than perfect substitute for other firms' outputs. When products are differentiated it is no longer the case that all consumers buy from the low-price firm or that a higher-price firm makes no sales.

We will develop a duopoly model in which demand for firm i's output is a linear function of prices. Specifically, let the quantity demanded for firm i be

$$q_i = a - p_i + bp_j. \tag{18.17}$$

In this demand equation there is a negative relation between a firm's own price and its quantity demanded. The relation between a firm's quantity demanded and its competitor's price depends on the sign of the parameter b. If b is positive, then increases in the competitor's price raise a firm's quantity demanded, and the two products are substitutes. If b is negative, then increases in the competitor's price lower a firm's quantity demanded, and the two products are complements. Our analysis will assume that the products are substitutes, which is the usual assumption in defining a market where firms compete. We will also assume that $0 < b < 1$, since if b were greater than one, then when both firms raised prices by equal amounts the total quantity demanded would actually rise.

As in the previous models we assume that marginal costs are constant (and the same for the two firms) and that fixed costs are zero. Profits for firm i are

$$\Pi_i = p_i q_i - c q_i = (p_i - c) q_i = (p_i - c)(a - p_i + bp_j). \tag{18.18}$$

The first-order condition (using the product rule) for profit maximization is

$$\frac{\partial \Pi_i}{\partial p_i} = (a - p_i + bp_j) - (p_i - c) = 0. \tag{18.19}$$

We can solve the two firms' first-order conditions simultaneously for equilibrium prices. Before doing this we graph the reaction functions, or best-response functions. From the first-order condition the best-response function for firm i is

$$p_i = \left(\frac{a+c}{2}\right) + \left(\frac{b}{2}\right) p_j. \tag{18.20}$$

The two reaction functions are graphed in Figure 18.3. Note that (assuming $a > c$) both prices always exceed marginal cost. Note also that, unlike the Cournot model, reaction functions are upward-sloping: a firm reacts to a rival's increase in price by raising its own price. The general pricing strategy for a firm is to charge more (ceding market share, but maintaining profitability) when its rival charges a low price, but to undercut when its rival charges a high price.

Finally note that (because we have assumed identical production costs) the two reaction functions intersect at an equilibrium point on the 45-degree line. To solve for the symmetric equilibrium we write the two firms' first-order conditions in matrix form as

$$\begin{pmatrix} 2 & -b \\ -b & 2 \end{pmatrix} \begin{pmatrix} p_1 \\ p_2 \end{pmatrix} = \begin{pmatrix} a+c \\ a+c \end{pmatrix}. \tag{18.21}$$

FIGURE 18.3

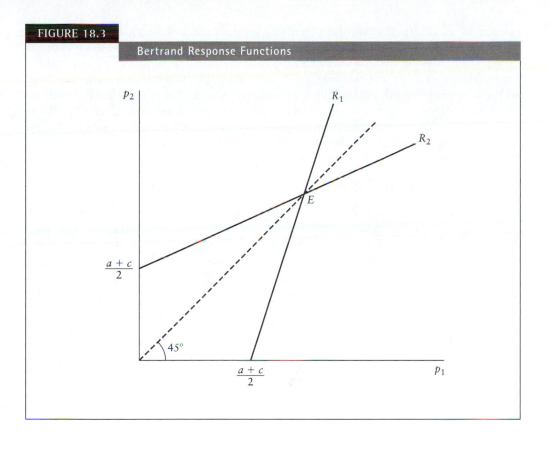

Using Cramer's rule gives the firms' equilibrium prices as

$$p_1^* = p_2^* = \frac{(2+b)(a+c)}{4-b^2} = \frac{a+c}{2-b}.$$ (18.22)

What is important about this result is how it contrasts with the identical-product Bertrand model. Instead of prices equal to marginal cost and zero profits, we get a result that is similar to the Cournot model: duopolists charge prices in excess of costs and earn positive profits. Nevertheless, even though the Cournot and differentiated-product Bertrand models are similar in their predictions about market outcomes, the two models often yield differing conclusions in dynamic-game models of public policy issues. For example, in a model of firm merger decisions the Cournot model predicts that mergers are often unprofitable and need not be regulated, while the Bertrand model predicts that mergers are always profitable and will raise the prices paid by consumers so that antitrust regulation is needed. The difference in results in dynamic models stems from whether reaction functions slope down (Cournot model) or up (Bertrand model).

18.5 RENT–SEEKING BEHAVIOR

In this section we consider the problem of the competitive division of a fixed economic pie. In many situations economic agents vie for some level of rents or economic gains. Agents are willing to expend resources in pursuing these gains. Examples of rent-seeking behavior include lobbying for tax breaks or trade protection, localities offering tax breaks in an attempt to influence firms' location choices, service and advertising competition by firms, and tournaments where the chance of winning the prize depends on effort or expenditure. Although total rents will often depend on total rent-seeking expenditures, we limit this example to the more tractable case in which rents are fixed and rent-seeking behavior is socially unproductive. Our model is essentially a continuous version of the discrete-choice prisoners' dilemma presented in Section 17.3: Rent-seeking expenditures are individually rational, but collectively payoff-reducing. We first solve a two-player model and then extend this to the n-player case.

18.5.1   The Two–Player Model

Consider two players who split an economic rent of R dollars. Each player spends money in vying for the rents. Let x_1 and x_2 represent the two players' respective expenditures. We will assume that each player's share of the rents is equal to the player's share of total rent-seeking expenditure. The formula for the share for player i is

$$s_i = \frac{x_i}{x_i + x_j}. \tag{18.23}$$

Player i's payoff function is

$$\Pi_i = s_i R - x_i = \left(\frac{x_i}{x_i + x_j} \right) R - x_i. \tag{18.24}$$

Using the quotient rule, the first-order condition is

$$\frac{\partial \Pi_i}{\partial x_i} = \left(\frac{(x_i + x_j) - x_i}{(x_i + x_j)^2} \right) R - 1 = 0$$

or $\tag{18.25}$

$$(x_i + x_j)^2 = x_j R.$$

Taking positive square roots of both sides and solving for player i's best-response function we get

$$x_i = \sqrt{x_j R} - x_j. \tag{18.26}$$

Since the players are symmetric, spending levels will be equal in equilibrium. Dropping the subscript, the equilibrium expenditure, x^*, is found, in several steps, as

$$x = \sqrt{xR} - x$$
$$2x = \sqrt{xR}$$
$$4x^2 = xR$$

or $\tag{18.27}$

$$x^* = \frac{R}{4}.$$

Figure 18.4 shows the players' best-response functions and the equilibrium. Note that the response functions, unlike those in earlier examples, are nonlinear. Although many models in economics employ simplifying assumptions that generate linear response functions, more complex models may well have nonlinear responses.

FIGURE 18.4

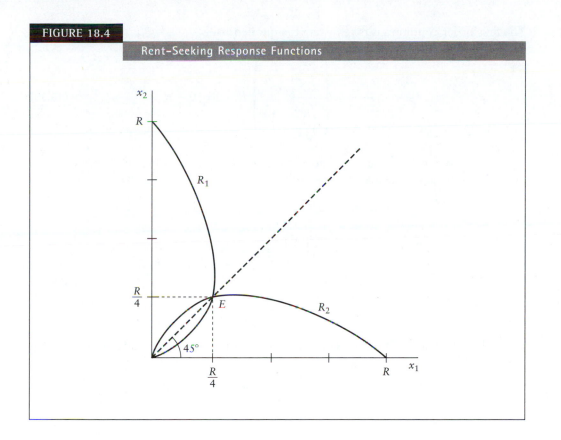

To interpret the response functions, consider player two. The upward-sloping section of the best-response function R_2 is where $x_2 > x_1$ and player two gets more than half of the rents. The inequality and the shares are reversed on the downward-sloping section of the best-response function. In equilibrium the combined payoffs total half of the available rents; the other half is dissipated in unproductive rent-seeking expenditures.

18.5.2   The n-Player Model

In the n-player model a player's share of the rents is given by

$$s_i = \frac{x_i}{X}$$

where

$$X = \sum_{j=1}^{n} x_j.$$

(18.28)

The payoff function for player i is

$$\Pi_i = \left(\frac{x_i}{\sum_{j=1}^{n} x_j} \right) R - x_i.$$

(18.29)

The first-order condition is taken using the quotient rule

$$\frac{\partial \Pi_i}{\partial x_i} = \left(\frac{\displaystyle\sum_{j=1}^{n} x_j - x_i}{\left(\displaystyle\sum_{j=1}^{n} x_j\right)^2} \right) R - 1 = 0. \tag{18.30}$$

To solve the model we use symmetry. Letting x (without a subscript) represent each player's equilibrium expenditure, we derive x^* as

$$\frac{nx - x}{(nx)^2} R = 1$$

$$\frac{(n-1)}{n^2 x} R = 1$$

or   (18.31)

$$x^* = \frac{(n-1)}{n^2} R.$$

From this we can find total rent-seeking expenditures and each player's equilibrium payoffs. These are

$$X^* = nx^* = \frac{(n-1)}{n} R$$

and   (18.32)

$$\Pi^* = \left(\frac{x^*}{X^*} \right) R - x^* = \left(\frac{1}{n} \right) R - \frac{(n-1)}{n^2} R = \frac{R}{n^2}.$$

Note that setting $n = 2$ yields (as it should) the two-player solutions derived earlier.

The main comparative static results for this model relate to the effect of the number of players on the equilibrium outcomes. Taking the comparative static derivatives gives

$$\frac{\partial x^*}{\partial n} = \frac{n^2 - 2n(n-1)}{n^4} R = \frac{(2-n)}{n^3} R < 0 \quad \text{if } n > 2$$

$$\frac{\partial X^*}{\partial n} = \frac{n - (n-1)}{n^2} R = \frac{1}{n^2} R > 0$$

and   (18.33)

$$\frac{\partial \Pi^*}{\partial n} = \frac{-2R}{n^3} < 0.$$

Thus, increases in the number of players decrease spending per player, but total unproductive rent-seeking expenditures increase and payoffs decrease. Competition is, therefore, socially wasteful.

18.6   PUBLIC GOODS

In the previous section we examined a model with a fixed economic benefit divided up between the players. In this section we assume that the benefit depends on the player's expenditures and that the benefit is public. Pure public goods are goods, such as lighthouses or public television broadcasts, where (1) one player's consumption of the good does not diminish other players' consumptions and (2) it is not possible to exclude non-contributors from consuming the good. We will use the model to answer two questions. First, are individually chosen Nash equilibrium contributions to the public good socially optimal? Second, how do changes in the number of players affect the (Nash equilibrium and socially optimal) levels of the public good?

We will let x_i denote the expenditure level, or contribution, of player i. The total level, or benefit, of the public good, B, is given by some function of the total spending. Let us denote this as

$$B = B(X)$$

where

$$X = \sum_{j=1}^{n} x_j.$$

(18.34)

We will assume that the benefit function is concave, i.e., there are diminishing marginal returns to expenditures on the public good, so that

$$\frac{\partial B}{\partial X} = B'(X) > 0$$

and

(18.35)

$$\frac{\partial^2 B}{\partial X^2} = B''(X) < 0.$$

An individual player's payoff is the benefit of the public good minus the amount spent by the player in providing the public good, or

$$\Pi_i = B(X) - x_i.$$

(18.36)

The first- and second-order conditions (using the chain rule with $\partial X/\partial x_i = 1$) for a maximum are

$$\frac{\partial \Pi_i}{\partial x_i} = B'(X) - 1 = 0$$

and

(18.37)

$$\frac{\partial^2 \Pi_i}{\partial x_i^2} = B''(X) < 0.$$

The players' (identical) first-order conditions implicitly define X^*, the level of public good provision by individually rational agents. The implicit solution requires that $B'(X^*) = 1$. Since n does not appear in this equation, the total expenditure and the equilibrium level of the public good are independent of the number of players.

We now compare the equilibrium of the game to the socially optimal outcome. The optimal solution would maximize the total net benefits generated by the public good. This is the sum of the individual payoffs, or

$$\Pi = \sum_{j=1}^{n} \Pi_j = \sum_{j=1}^{n} (B(X) - x_j) = nB(X) - X.$$

(18.38)

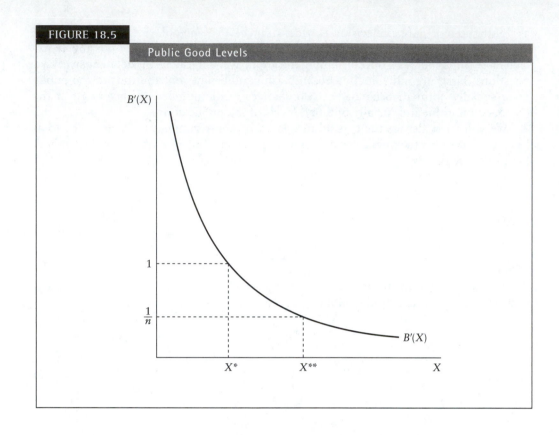

FIGURE 18.5

Public Good Levels

The first- and second-order conditions (with respect to X) for a social optimum are

$$\frac{\partial \Pi}{\partial X} = nB'(X) - 1 = 0$$

and   **(18.39)**

$$\frac{\partial^2 \Pi}{\partial X^2} = nB''(X) < 0.$$

This first-order condition implicitly defines the optimal level of spending, X^{**}, as the solution to $B'(X^{**}) = (1/n)$. Contrast this with the Nash equilibrium where $B'(X^*) = 1$. Figure 18.5 shows the graph of the marginal benefit function $B'(X)$. Since $B(X)$ is concave, the marginal benefit function, $B'(X)$, is downward-sloping. It is clear that the private Nash equilibrium level of spending, X^*, falls short of the optimal level X^{**}. The reason for this is that each individual's contribution to the public good creates external benefits for the other players, yet the individual maximizing choices made in the Nash equilibrium ignore these external benefits.

Note that the degree of inefficiency, the gap between the two solutions, rises as n rises. Economically, as n increases there are more consumers who could benefit from the public good so the optimal provision level rises. Yet, larger numbers of consumers increase an individual consumer's ability to free-ride on the contributions of others. The result is a growing gap between the optimal provision and the actual level.

18.7 HOUSEHOLD TIME ALLOCATION

Our final application models a household as a two-player non-cooperative game. We will label the players H and W. As you will see below, the subscripts might be interpreted as a husband and wife from a 1950s sitcom (think Ward and June Cleaver in the TV show, *Leave It to Beaver*). The model is broad enough, however, to encompass a range of situations (including the case of two college roommates) where two players have a choice of allocating time either to leisure or to production of a household public good. The focus of the model is on how an improvement in one player's efficiency in providing the household public good can affect time allocation and payoffs.

We start with some simplifying assumptions. First, each player has an exogenous amount of time available. We will label total time available for players H and W as T_H and T_W, respectively. Players may allocate their total available time between leisure (ℓ_H and ℓ_W) or household production time (z_H and z_W). Thus, the two time constraints are

$$\ell_H + z_H = T_H$$

and **(18.40)**

$$\ell_W + z_W = T_W.$$

Household production is a public good that depends on time allocated and each player's efficiency in the activity. (You might want to think of household production as including activities such as vacuuming or washing dishes.) We assume that the level of production of this public good is additive in time allocated:

$$Z = z_W + hz_H,$$ **(18.41)**

where h is a parameter that indicates player H's relative efficiency in household activities. We will assume that player H is less efficient in producing the public good than is player W, i.e., $h < 1$.

Finally, we keep the model simple by assuming that each player's payoff is given by a multiplicative utility function that depends only on household production and on leisure time. The utility functions are

$$U^H = Z\ell_H$$

and **(18.42)**

$$U^W = Z\ell_W.$$

Substituting in for Z and using the time constraint that $\ell_i = T_i - z_i$ allows us to write the utility functions as

$$U^H = (z_W + hz_H)(T_H - z_H)$$

and **(18.43)**

$$U^W = (z_W + hz_H)(T_W - z_W).$$

To derive the best-response functions, we take the first-order conditions for each player. These are

$$\frac{\partial U^H}{\partial z_H} = h(T_H - z_H) - (z_W + hz_H) = 0$$

and   **(18.44)**

$$\frac{\partial U^W}{\partial z_W} = (T_W - z_W) - (z_W + hz_H) = 0.$$

Rearranging gives the best-response functions for players H and W as

$$z_H = \frac{T_H}{2} - \frac{z_W}{2h}$$

and   **(18.45)**

$$z_W = \frac{T_W}{2} - \frac{hz_H}{2}.$$

Both response functions are downward-sloping. Higher levels of household time for one player lead to reductions in household production time by the other player. Note also that what matters for player W is player H's effective time, i.e., hz_H.

   The Nash equilibrium is the simultaneous solution to the two best-response functions. Solving gives

$$z_H^* = \frac{2hT_H - T_W}{3h}$$

and   **(18.46)**

$$z_W^* = \frac{2T_W - hT_H}{3}.$$

Note that lower values of the parameter h lead to lower values of z_H^* and higher values of z_W^*. In a sense, as we will explore further in a moment, there are advantages to being relatively inefficient since the less-efficient player gets more leisure time in equilibrium. Indeed, corner solutions are possible where player H contributes nothing and spends all time in leisure (e.g., in a 1950s sitcom where the husband has limited total, afterwork, time in the household and is not very efficient at household production so that z_H would equal zero). For the comparative statics below, however, we will assume interior solutions.

   We now turn to the question of how changes in the parameter h affect the players' equilibrium payoffs. Substituting the equilibrium values of z_H^* and z_W^* into the utility functions (equation (18.43)) and simplifying gives

$$U^H = \frac{(hT_H + T_W)^2}{9h}$$

and   **(18.47)**

$$U^W = \frac{(hT_H + T_W)^2}{9}.$$

It is apparent from the simple functional form that player W's payoff is increasing in h. For player H we use the quotient rule to find the derivative:

$$\frac{\partial U^H}{\partial h} = \frac{18hT_H(hT_H + T_W) - 9(hT_H + T_W)^2}{81h^2}$$

or

(18.48)

$$\frac{\partial U^H}{\partial h} = \left(\frac{hT_H + T_W}{9h^2}\right)(hT_H - T_W)$$

Since the first term is positive, the sign of the derivative is the same as the sign of $(hT_H - T_W)$. Since $h < 1$ by assumption the derivative will be negative so long as the (exogenous) level of T_H is not too high relative to T_W. In particular if $T_H \leq T_W$, then player H is definitely worse off when his efficiency in household production increases. The intuition here is that player H's efficiency affects his payoff both directly and indirectly. The direct effect is positive: more work gets done for a given amount of time. The indirect effect, however, is negative: player H's increased efficiency means that player W reduces her household production time, and this tends to reduce player H's utility. For the appropriate parameter values, the indirect effect dominates and player H is worse off.

Problems

18.1 Consider an alternative version of the entry model of Section 18.2. Suppose that the market is a natural monopoly, but that firm 1 has a lower cost for entering the market. Let the positive payoffs to firm 1 and 2 be given by $\Pi_1 > \Pi_2 > 0$ and the losses by $L_2 > L_1 > 0$. Find all Nash equilibria to this game. Does the symmetric model result of zero expected profit still hold?

18.2 Consider a Cournot duopoly where demand is $P = a - Q$, $Q = q_1 + q_2$, but the firms have asymmetric marginal costs: c_1 for firm 1 and c_2 for firm 2.

    (a) Find the Nash equilibrium when $2c_i < a + c_j$ for each firm.

    (b) Find the Nash equilibrium when $c_1 < c_2 < a$ but $2c_2 > a + c_1$. Show the graph of the reaction functions for this case.

18.3 Consider a two-country, X and Y, model with one firm located in each country. Suppose that demand in each country is $P = a - Q$, where Q is the total quantity being sold in the relevant country. Suppose that firm X, the firm in country X, sells in both markets (its home market X and the foreign market Y) while firm Y, the firm in country Y, sells only in its own home market. Also suppose that firm Y operates with total costs of $TC_y = c_y q_y^3$ and that firm X operates with total costs of $TC_x = c_x q_x^3$, where $q_x = q_{xx} + q_{xy}$ is the sum of firm X's sales in its home market plus firm X's exports to country Y.

    (a) Write the expression for each firm's profit. Find the first- and second-order condition(s) for each firm's profit-maximization problem.

    (b) Find the changes in the firms' output choices when c_y increases.

18.4 Consider a Cournot duopoly with the constant-elasticity demand function $P = Q^{-\alpha}$, where $Q = q_1 + q_2$ and α is a positive parameter. Assume that both firms operate with a constant marginal cost of $\$c$/unit. Find the Nash equilibrium.

18.5 Consider an n-firm Cournot oligopoly with the constant-elasticity demand function $P = Q^{-\alpha}$, where $Q = \sum_{i=1}^{n} q_i$ and α is a positive parameter. Assume that all firms operate with a constant marginal cost of $\$c$/unit. Find the Nash equilibrium. Find the effect of an increase in n on the equilibrium market price.

18.6 Consider a model with two firms. The two firms both sell output in a first market, but firm 1 also sells output in a second market. Let the production costs for firm 1 be $TC_1 = 0.5(q_{11} + q_{12})^2$ where q_{1j} is the quantity sold by firm 1 in market j. Let production costs for firm 2 be $TC_2 = 0.5(q_2)^2$. Suppose that demand in the first market is $P = 30 - q_{11} - q_2$ and that in the second market firm 1 can sell unlimited quantities at a price of $\$10$.

  (a) Write the expression for each firm's profit. Find the first-order conditions.

  (b) Solve for the Nash equilibrium quantity choices. Find the equilibrium payoff for each firm. Which firm earns higher profits?

18.7 Consider a Bertrand duopoly where firm 1 operates with a marginal cost of c_1 and firm 2 operates with a marginal cost of $c_2 > c_1$. Assume that the products are identical.

  (a) Assume that prices are continuous variables. Does the model, strictly speaking, have a Nash equilibrium? Explain your reasoning. Show that there is "almost" a Nash equilibrium in which firm 1 sets $p_1 = c_2 - \varepsilon$, where ε is an arbitrarily small number, and firm 2 chooses $p_2 = c_2$.

  (b) Now assume that prices are discrete variables measured in pennies. Find all Nash equilibria to the model. Would the set of Nash equilibria be diminished if we eliminated weakly dominated strategies? Explain.

18.8 Consider a Bertrand duopoly model with differentiated products. Let the basic form of the demand equation for firm i be $q_i = a_i - p_i + b_i p_j$. Let total costs for firm i be $TC_i = c_i q_i$.

  (a) Let $a_1 = a_2 = a$ and $b_1 = b_2 = b$, but assume that $c_1 > c_2$. Solve for the Nash equilibrium prices and quantities. Which firm charges the higher price? Which sells a larger quantity? Using the graph of the best-response functions, show the effect of an increase in c_1.

  (b) Let $c_1 = c_2 = c$ and $b_1 = b_2 = b$, but assume that $a_1 > a_2$. Solve for the Nash equilibrium prices and quantities. Which firm charges the higher price? Which sells a larger quantity? Using the graph of the best-response functions, show the effect of an increase in a_1.

  (c) Let $a_1 = a_2 = a$ and $c_1 = c_2 = c$, but assume that $b_1 > b_2$. Solve for the Nash equilibrium prices and quantities. Which firm charges the higher price? Which sells a larger quantity? Using the graph of the best-response functions, show the effect of an increase in b_1.

18.9 Consider two firms that play the following game. Each firm must decide whether to be in a line of business. If a firm decides to sell the product, it must spend $\$F$ printing and distributing a catalog. The marginal cost of actually producing and shipping the product is $\$c$/unit. Since the price must be printed in the catalog, the decisions on whether to be in the business and what price to charge are made at the same time. Also assume that the two firms' decisions are made simultaneously and independently, so that neither firm knows the other's decision when making its own choice. Finally, assume that the two firms' products are identical. Is there a pure-strategy Nash equilibrium for this game? Explain.

18.10 Consider an n-player rent-seeking model in which total rents depend on the level of rent-seeking expenditure. Specifically, assume that total rents are $R = R_0 X^{0.5}$ where X is the sum of the players' rent-seeking expenditures. Assume, as in Section 18.5, that each player's share of the total rents equals the player's share of total rent-seeking expenditures. Find the Nash equilibrium. How do the equilibrium solutions for x_i, X, and Π_i depend on the number of players?

18.11 Consider the problem of a common resource. Suppose that each agent i chooses a private activity level x_i. The private activity draws on some publicly used resource and production efficiency declines as the total usage of the common resource increases. Let $B(X)x_i$ be the benefit of the activity to player i, where X is the sum of the n agents' activities. Assume $B'(X) < 0$ and $B''(X) < 0$. Let the cost of the activity to a player be cx_i, so that the net payoff is $\Pi_i = B(X)x_i - cx_i$. Compare the Nash equilibrium to the social optimum. [Hint: Use the concavity of the function $B(X)$.]

18.12 Consider the public good application in Section 18.6. Suppose that the benefit function takes the form $B(X) = bX^\alpha$ where b and α are positive parameters. Solve for the Nash equilibrium and the social optimum.

18.13 Consider a beach on a hot summer day. Let the length of the beach be normalized to one and assume that consumers are uniformly distributed along the length of the beach. Suppose that

(i)   There are two ice cream vendors

(ii)   Both vendors operate with a marginal cost of c

(iii)   The price at which the vendors sell ice cream is exogenously fixed

(iv)   Consumers at the beach buy from the nearest vendor

(v)   Each vendor's strategy is a location choice

(a)   Find the Nash equilibrium locations.

(b)   Now suppose that there are three vendors. Prove that there is no pure-strategy Nash equilibrium.

18.14 (Difficult) Consider a model with the following assumptions:

(i)   There are two countries X and Y

(ii)   There is one firm in each country; call these firm X and firm Y

(iii)   Both firms operate with zero production costs, produce identical products, and operate under the Cournot assumption with respect to outputs in the market where the firms compete

(iv)   All consumers (in both countries) are identical and each individual consumer demands a quantity of $q = (1 - p)$

(v)   The number of consumers in country X equals a parameter x and the number of consumers in country Y equals a parameter y. Let $n = x + y$ represent the total number of consumers.

(vi) Country X does not allow any imports at all.

(vii) Country Y allows imports but bans dumping: assume that this means that country Y requires foreign firms to charge the same price in Y as they charge in their own country.

(a) Let q_{xh} represent the quantity that firm X sells to its home consumers, let q_x represent the quantity that firm X sells to consumers in country Y, and let q_y represent the quantity sold by firm Y. Write the profit function for each firm.

(b) Firm X is constrained by the fact that it cannot price-discriminate between the two groups of consumers. Find the first-order conditions for firm X's (constrained) profit-maximization problem. Find the first-order condition for firm Y's profit-maximization problem.

(c) Solve for the firms' quantities and the market price. [Hint: Check your work by explaining why the price solution should be an equation such that $x = n$ generates the monopoly price and $x = 0$ generates the standard Cournot duopoly price.]

(d) Suppose that x increases and y decreases by an identical amount (i.e., n is held constant). What effect does this have on the market price?

18.15 Use the model of household time allocation from the text, but assume that this is a "loving" marriage where time is allocated to maximize the sum of the players' utilities. Find the time allocations. How do they compare to the solutions in the non-cooperative game?

Dynamic Games with Complete Information: Theory

19.1 INTRODUCTION

The two previous chapters explored static games. We now turn to a consideration of dynamic games. Dynamic games are used whenever decisions occur sequentially rather than simultaneously. Examples include bargaining relationships with a series of offers and counteroffers, repeated interactions over time between firms in a market, decisions by existing firms that affect whether new firms subsequently enter a market, and government policy decisions to which firms subsequently react. Because these contexts all depend on the order and timing of decisions, we cannot simply model them as static games.

Dynamic games require different forms for representing games and different solution concepts. In static games we use the normal form representation, which can be represented by either a payoff matrix (for discrete strategy spaces) or best-response functions (for continuous strategy spaces). This method works well when players' equilibrium strategies are equivalent to choosing a single action. In a dynamic game, however, a strategy is not simply an action, but instead a full specification of a player's moves for all possible situations that may occur in the play of the game. Although dynamic games can be represented by the normal form, that form often obscures important issues related to the sequence of moves. Instead, dynamic games are usually represented in *extensive* form. The extensive form uses a diagram called a *game tree* to emphasize the sequence of moves in a game.

The differences between static and dynamic games extend beyond the method of presentation. The solution concepts used to define equilibrium are also different. In static games we found solutions using the concept of Nash equilibrium. We will see that this concept turns out to be too "weak" for dynamic games: Nash equilibria in dynamic games can include outcomes that are unreasonable in the sense that players are allowed to pick dominated actions (but not dominated strategies). A stronger version of Nash equilibrium, called *subgame-perfect Nash equilibrium*, is needed in dynamic games.

As in the previous two chapters we confine our discussion to games of complete information: games where all relevant information about payoffs is public rather than private—what one player knows about the structure of the payoff functions, all players know.

19.2 GAMES IN EXTENSIVE FORM

The sequential nature of dynamic games is best represented by the extensive form. We will first give a definition of the extensive form of a game, and then, in the following subsections, flesh out the terms and meaning of this definition. The players in our definition will be numbered $0 \ldots n$. Player 0 is "nature" while the other n players are actual decision makers. Nature is introduced as a player as a convenience in modeling random events.

DEFINITION **19.1**

The Extensive Form

The extensive form of a game specifies:

(a) *The set of players,*

(b) *The order of moves,*

(c) *The possible actions that a player can take at each move, and for moves by nature a probability distribution over possible actions,*

(d) *The information a player has at each move, and*

(e) *The payoffs, to players $1 \ldots n$, for every possible combination of moves.*

In our discussion of static games we employed two methods: a game matrix for discrete-choice games and best-response functions for continuous-choice games. A similar distinction applies to extensive form games. Discrete games can be modeled using game trees while continuous choice games will involve best-response functions to other players' earlier choices.

19.2.1 An Introduction to Game Trees

Let us start with the simplest possible situation: a one-player game with two choices. The game tree of this game is represented by a collection of nodes and branches (branches are also called arcs or line segments). Nodes represent either decision points or outcomes, while branches represent available choices. In Figure 19.1, player 1 has

FIGURE 19.1

A One-Player Game Tree

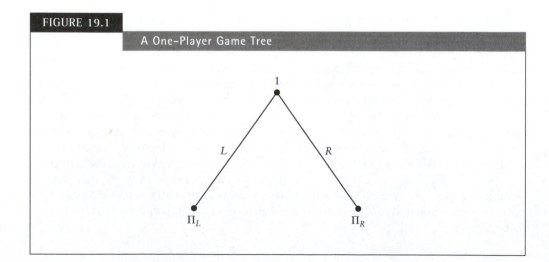

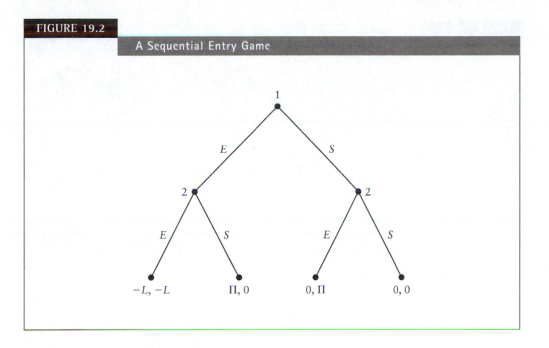

FIGURE 19.2

A Sequential Entry Game

two possible choices, left or right. The initial node is labeled 1 to indicate that this is a decision point for player 1 and the two arcs are labeled L and R for player 1's two options. The bottom two nodes are called terminal nodes and the outcomes, in terms of payoffs, are labeled as Π_L and Π_R. The equilibrium for this very simple game is for the player to choose whichever action yields the higher payoff.

Two-player games can be diagrammed in a similar manner. Let us consider an investment decision game. Each of two firms chooses whether to make an investment that allows it to enter a market. A firm's alternative choice is to stay out of the market and earn zero profits. In Section 18.2 we modeled this as a static game with simultaneous choices. Now let us examine what happens when the choices are sequential. Suppose that firm 1 makes its decision (and acts on this decision) first, then firm 2 learns what firm 1 has done and makes its own investment decision. The game tree is shown in Figure 19.2. The decision branches of the tree are labeled E, for enter, and S, for stay out. The outcomes at the bottom of the game tree list each firm's profit, $\Pi > 0$, loss, $-L < 0$, or zero profit payoff for the different combinations of strategies. By convention, the first payoff number is firm 1's payoff and the second is firm 2's payoff.

How do we solve this game? To find an equilibrium for this game in extensive form we use a technique called **backwards induction**. Backwards induction involves examining the last decision points in a game, eliminating actions that would not be played, erasing these eliminated actions, redrawing the game tree, and then repeating this process. In our example the last decision points are those of firm 2. On the left branch of the game tree (where firm 1 has chosen to enter) firm 2's best choice (since $0 > -L$) is to stay out of the market. On the right branch firm 2's optimal choice is to enter since $\Pi > 0$. Figure 19.3 shows the truncated game tree without firm 2's dominated actions.

We now can solve the game. We assume that both players are rational and that this rationality is common knowledge. (Common rationality is crucial for solving games and is always implicitly assumed.) Firm 1 (looking ahead and anticipating firm 2's rational choices) earns a positive profit if it enters and zero profit if it stays out of the

FIGURE 19.3

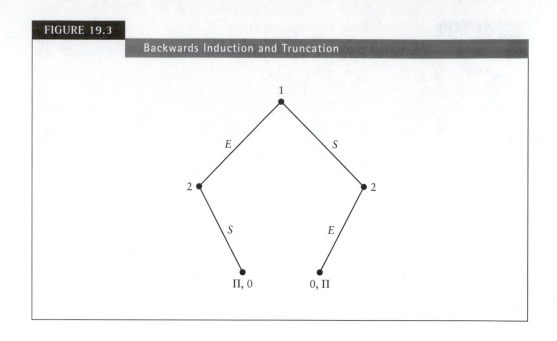

Backwards Induction and Truncation

market. Thus, the equilibrium is that firm 1 enters and earns a profit of Π, while firm 2 stays out and earns zero profit.

19.2.2 Information Sets

In the previous example firm 2 knew, prior to making its decision, firm 1's choice. What if this decision were not known in advance?[1] To draw the game tree for this game we introduce the concept of an information set.

DEFINITION # 19.2

Information Sets

*An information set is a collection of decision nodes for a player where (at the move corresponding to these nodes) at each node in the set the player has the same decision branches and the player does not know which node has or has not actually been reached in the play of the game. Games in which all information sets are singletons, or single nodes, are called games of **perfect information**; while games where some information sets are not singletons are called games of **imperfect information**.*

Figure 19.4 shows the simultaneous move version of the entry game. This is a static game of imperfect information represented in extensive form. The dotted line indicates an information set for firm 2, i.e., the dotted line indicates that firm 2 knows that it is

---

[1]In game theory no distinction is generally made between decisions that are not known in advance and decisions that are made simultaneously. Thus, a game in which firm 1 chooses first chronologically but firm 2 does not know this decision is treated the same as the version of the entry game in which both firms make their entry decisions simultaneously.

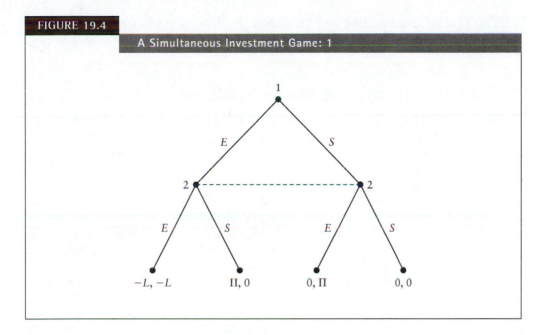

FIGURE 19.4

A Simultaneous Investment Game: 1

1

E   S

2 - - - - - - - - - - - - - 2

E   S   E   S

$-L, -L$ $\Pi, 0$ $0, \Pi$ $0, 0$

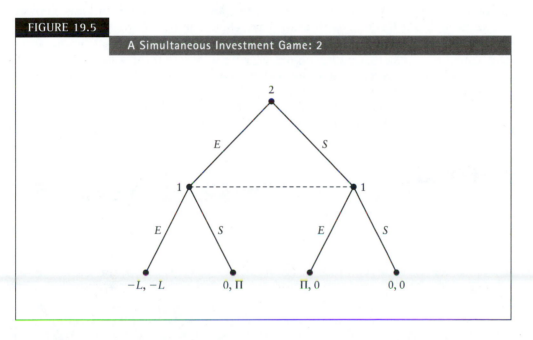

FIGURE 19.5

A Simultaneous Investment Game: 2

2

E   S

1 - - - - - - - - - - - - - 1

E   S   E   S

$-L, -L$ $0, \Pi$ $\Pi, 0$ $0, 0$

at a node where there is a choice of enter or stay out, but firm 2 does not know whether firm 1 has chosen enter or stay out.[2]

One point that should be made here is that static simultaneous move games can be represented in different, but equivalent, game trees. Consider Figure 19.5. Here firm 2 is represented as moving first, and it is firm 1 that, without knowing firm 2's move, faces a two-node information set.[3]

---

[2]Some texts use a "balloon" (instead of a dotted line) around nodes to indicate an information set.

[3]Note that the terminal node payoffs still follow the convention that player 1's payoff is listed first and player 2's payoff second.

The games in Figures 19.4 and 19.5 are equivalent. In the extensive form representation of simultaneous move games the order of moves is a matter of modeling convenience.

19.2.3 Uncertainty and Moves by Nature

We now introduce uncertainty into our model. Suppose that in our example there is uncertainty about the potential consumer demand. For simplicity let there be two market sizes: large and medium. We will assume that in a large market both firms could enter and earn positive profits; while in a medium-sized market there is only room for one firm to earn profits. Let q be the probability that the market is large and $1 - q$ the probability that the market is of medium size. Figures 19.6–19.9 show four possible versions of the game: the first two are sequential (firm 1 moves first and announces its decision) and the latter are simultaneous. In all four diagrams we simplify notation by using a plus sign for positive profits and a negative sign for negative profits, rather than trying to specify actual profit levels.

Let us now interpret the four different versions of the game. Figure 19.6 shows a game in which all information sets are singletons (single nodes). Thus, in this game: (1) nature randomly picks a market size, (2) the firms learn the size of the market, (3) firm 1 chooses to enter or stay out, and (4) firm 2 learns this decision and then chooses to enter or stay out. Figure 19.7 again shows a game in which firms' choices are sequential, but this time both firms make choices in ignorance of the actual size of the market. We show this by having both nodes for firm 1 in a single information set; while firm 2 has two information sets. Firm 2 has an information set where it learns firm 1 has entered, but does not know the size of the market, and a second information set where it learns firm 1 has stayed out, but again does not know the size of the market. The second set of diagrams shows the simultaneous move games in which nature's move is either known (Figure 19.8) or unknown (Figure 19.9).

FIGURE 19.6

Random (Known) Demand

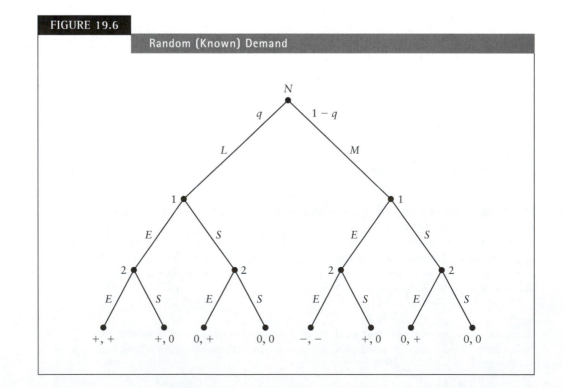

FIGURE 19.7

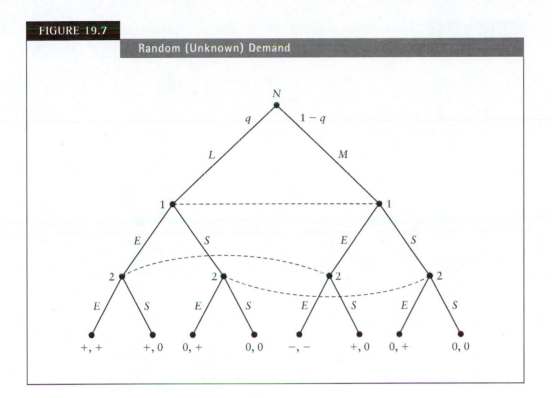

FIGURE 19.8

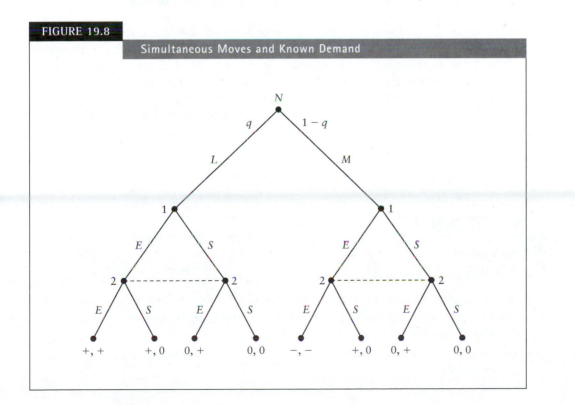

FIGURE 19.9

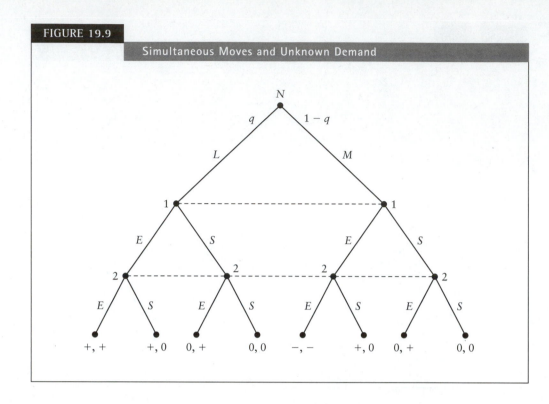

Simultaneous Moves and Unknown Demand

19.3 EQUILIBRIUM IN EXTENSIVE FORM GAMES

Static games are solved using the concept of Nash equilibrium. Dynamic games also have Nash equilibria, but, as we will see next, the Nash equilibrium concept is too weak. By this we mean that some Nash equilibria in extensive form games are unreasonable. To sort out reasonable versus unreasonable Nash equilibria we need to introduce the concept of a *subgame* and also to define a strategy in an extensive form game. Once we have these tools in place we will be able to characterize equilibria in dynamic games.

19.3.1 Subgames

A subgame is a game within a game. More formally,

DEFINITION   **19.3**

Subgames

A subgame is a portion of a game that

(a) *Starts at a node (other than a terminal node) that is a singleton information set,*

(b) *Contains all nodes and branches of the original game that follow the subgame's starting node, and*

(c) *Does not cut any (of the dotted lines that indicate) information sets of the original game.*

We can apply this concept of a subgame to the examples in the previous section. By our definition all games have at least one subgame—the game itself. Let us count the

FIGURE 19.10

(Im)perfect Information for Firm 1 (2)

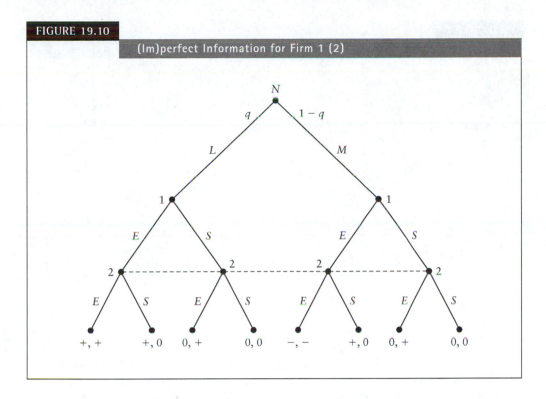

number of subgames (other than the game itself) in our examples. In Figure 19.6 there are six subgames: one for each of firm 1's single decision nodes and one for each of firm 2's single decision nodes. Figure 19.8 has two subgames: one for each of firm 1's decision nodes. Since Figures 19.7 and 19.9 contain no singleton decision nodes, neither of these games has any subgames.

To illustrate the impact of requirement (c) in the definition of a subgame, consider one final version of the entry game. Suppose firm 1 learns the size of the market, but that firm 2 learns neither the size of the market nor firm 1's decision. This game is shown in Figure 19.10. Firm 1 does have singleton information sets, but there are no subgames (other than the game itself) because any attempt to form a subgame would have to cut the information set for firm 2.

19.3.2 Strategies

In a simultaneous move game a strategy is simply an action. For example, in the Cournot duopoly game a firm's strategy is its choice of an output level. In dynamic games, however, players do not simply act: they also react. Furthermore, they must have plans for all contingencies, plans on how they will react for all situations that could conceivably arise in the game. Based on this we can define a strategy for an extensive form game.

DEFINITION 19.4

A Strategy in an Extensive Form Game

A strategy in an extensive form game is a comprehensive plan that specifies the collection of actions a player will take at all possible decision nodes in the game tree. Such a plan may be dependent on the history of play and may include mixed strategies.

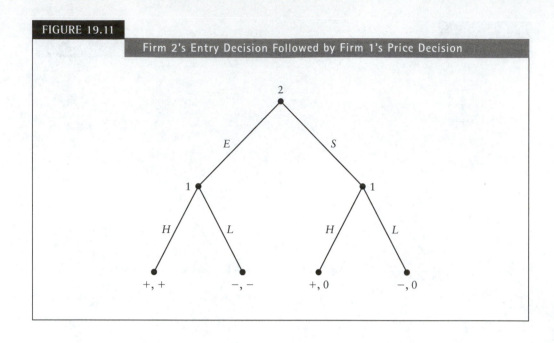

FIGURE 19.11

Firm 2's Entry Decision Followed by Firm 1's Price Decision

To illustrate the concept of a strategy in an extensive form game let us consider a different kind of entry game. Suppose that firm 1 is currently in a market and that firm 2 is choosing whether to enter, E, or stay out, S. Let us assume that, following the entry decision by firm 2, firm 1 can choose between two pricing strategies: a high price, H, or a low price L.[4] The firms then earn profits, which will be positive at a high price and negative at a low price. (Firm 2, of course, earns zero if it does not enter.) The game tree is shown in Figure 19.11. Since numerical payoffs don't matter in our analysis below, the terminal nodes give only the signs of firm 1's and firm 2's respective profits.

Firm 2 has one decision node and two possible actions. Its two strategy choices are E and S. Firm 1 also has two possible actions, but, because firm 1 has two decision nodes, it has four (two nodes times two actions) possible strategies. A strategy specifies an action for every possible contingency. In this case a strategy for firm 1 is a pair of contingent actions: the first element specifies an action if firm 2 enters and the second element an action if firm 2 stays out. The set of possible strategies for firm 1 is given by

$$S_1 = \{(H, H), (H, L), (L, H), (L, L)\}, \tag{19.1}$$

where the first element in each pair indicates firm 1's action if firm 2 has chosen E and the second element indicates firm 1's action if firm 2 has chosen S. Thus, (L, H) indicates that firm 1's strategy is to choose a low price if firm 2 enters but a high price if firm 2 stays out of the market. *To avoid confusion, note that a strategy specifies actions for all possible contingencies but that in the actual play of the game firm 1 only takes one action: the other half of the strategy pair is what firm 1 would have hypothetically done if firm 2 had made a different decision.*

---

[4]We don't explicitly model firm 2's pricing level (assuming firm 2 enters). We can skip this step because only the signs of the firms' profits matter in our analysis and these signs are determined by firm 1's choice of a high or low price.

19.3.3   Subgame–Perfect Nash Equilibria

The definition of a Nash equilibrium for a dynamic game is the same as that for a static game: A Nash equilibrium is a set of strategies such that no player has an incentive to switch to a different strategy. To find the Nash equilibria let us show this game in normal form. Letting a plus sign indicate positive payoffs and a minus sign negative payoffs, the matrix that gives the players' payoffs as a function of strategy choices is

$$
\text{Firm 1}
\begin{array}{c}
\\
\\
H,H \\
H,L \\
L,H \\
L,L
\end{array}
\begin{array}{c}
\text{Firm 2} \\
\begin{array}{cc}
E & S
\end{array} \\
\left(
\begin{array}{cc}
\underline{+},\underline{+} & \underline{+},0 \\
\underline{+},\underline{+} & -,0 \\
-,- & \underline{+},\underline{0} \\
-,- & -,\underline{0}
\end{array}
\right)
\end{array}
\qquad (19.2)
$$

where each player's best responses to the other player's strategy choices are underlined. Note again that firm 1 takes a single action, but firm 1's strategy specifies both what firm 1 does for firm 2's actual decision and what firm 1 would have done if firm 2 had (hypothetically) made the opposite decision. For example, in the bottom left-hand cell, firm 1's strategy is to choose a low price if firm 2 enters and a low price if firm 2 stays out, firm 2's strategy is to enter, and both firms earn negative payoffs.

The entries where both players' strategies are underlined indicate Nash equilibria. The three Nash equilibria, *E*1, *E*2, and *E*3, are

	Firm 1	Firm 2
*E*1:	(*H, H*)	*E*
*E*2:	(*H, L*)	*E*
*E*3:	(*L, H*)	*S*

Although there are three Nash equilibria in this game, two of the equilibria, *E*2 and *E*3, are problematic. To see this let us examine Figure 19.11 and the two subgames that begin at firm 1's singleton decision nodes. At the left-hand node, where firm 2 has entered, it is clear that *H* is the best subgame action for firm 1. The same is true for the subgame where firm 2 has chosen to stay out of the market. *H* is a dominant action for each of the subgames (it is never profitable for firm 1 to charge a low price), yet two of the Nash equilibria posit an action of *L* for at least one subgame.

The problem with *E*2 and *E*3 is that these Nash equilibria involve potentially playing the dominated low-price actions. *E*2 and *E*3 survive as Nash equilibria because the dominated actions lie "off the equilibrium path," i.e., in the equilibria play proceeds down the other path so that the dominated actions are never actually reached or played. To put it differently, *E*2 and *E*3 posit actions that are not credible: if, for whatever reason, firm 2 deviated from its equilibrium strategy, firm 1 would not rationally choose to carry through its supposed strategy choice. For example, *E*3 involves a threat, that firm 1 will charge a low price if firm 2 enters; nevertheless it would not be in firm 1's best interest to carry out that threat if firm 2 called firm 1's bluff by actually entering the market.

We now refine the Nash equilibrium concept in order to apply it to extensive form games. The refinement is to require that all strategies be credible in the sense that actions should be optimal in all potential subgames, both on and *off* the equilibrium path.

For threats or promises to be believed it must actually be in a player's best interest to carry through with the threat or promise. Formally, we define a subgame-perfect Nash equilibrium as:

DEFINITION **19.5**

Subgame–Perfect Nash Equilibrium

A set of strategies in an extensive form game is a subgame-perfect Nash equilibrium if and only if the actions posited by these strategies constitute Nash equilibria in all subgames of the game.

Just as the Nash equilibrium concept was used for static games, subgame-perfect Nash equilibrium is used for dynamic games. The alert reader might wonder about the relationship between the backwards induction method described above and subgame-perfect Nash equilibria. They are in fact the same in any game of perfect information: any subgame-perfect Nash equilibrium survives the backwards induction argument, and any equilibrium found by backwards induction is also a subgame-perfect Nash equilibrium. Subgame perfection, however, is more general: it can be used even in games of imperfect information where the backwards induction argument is unavailable.

Before proceeding to the next section we conclude with two caveats. First, as with static games, dynamic games may have mixed-strategy equilibria. Second, while backwards induction and subgame perfection allow us to solve any game of perfect information in an essentially mechanical fashion, there is no straightforward method for solving games of imperfect information. When information sets contain multiple nodes, the question of what players believe about the probability of being at a given node comes to the fore. Such games are solvable but are beyond the scope of this text.[5]

19.3.4 Subgame–Perfect Nash Equilibrium: An Example

So far we have been examining games with discrete action and strategy spaces. We now turn to a simple example of a two-player game with continuous choice variables. The structure of our game is simple: (1) player 1 chooses an action x_1, (2) player 2 observes this choice and then chooses an action x_2, and (3) the players receive payoffs defined by the functions $\Pi_1(x_1, x_2)$ and $\Pi_2(x_1, x_2)$.

To solve this game we start with the subgame in which player 2 chooses x_2. The first-order condition for player 2 is

$$\frac{\partial \Pi_2(x_1, x_2)}{\partial x_2} = 0. \tag{19.3}$$

This equation implicitly defines player 2's optimal strategy. The optimal strategy is a rule that specifies x_2 as a function of x_1. Let us write player 2's best response to x_1 as

$$x_2^* = R_2(x_1). \tag{19.4}$$

---

[5]For a more complete discussion see R. Gibbons, *Game Theory for Applied Economists,* Princeton University Press, 1992. For a more advanced exposition see D. Fudenberg and J. Tirole, *Game Theory*, MIT Press, 1991.

Now that we have player 2's strategy in the subgame we turn to player 1's decision. Assuming common rationality, so that player 1 anticipates a rational decision by player 2, we can write player 1's payoff as a function of x_1 and of x_2^*:

$$\Pi_1 = \Pi_1(x_1, R_2(x_1)). \tag{19.5}$$

Player 1's best strategy choice is found by taking the total derivative of this payoff function with respect to x_1. The first-order condition is

$$\frac{d\Pi_1(x_1, R_2(x_1))}{dx_1} = \frac{\partial \Pi_1}{\partial x_1} + \frac{\partial \Pi_1}{\partial x_2}\frac{\partial R_2}{\partial x_1} = 0. \tag{19.6}$$

Note that this equation contains both direct and indirect effects. The direct effect is just $\partial \Pi_1 / \partial x_1$. The indirect effect, $(\partial \Pi_1 / \partial x_2)(\partial R_2 / \partial x_1)$, is a strategic effect that occurs because changes in x_1 change the choice of x_2 which in turn changes Π_1. This strategic effect is the essential difference between the static game equilibria in which players' choices are simultaneous and dynamic game equilibria with sequential choices.

To illustrate the difference between static and dynamic equilibria let's return to the basic Cournot duopoly model from Chapter 2, which was represented as a game theory model in Chapter 18. With linear demand, $P = a - Q$ (we set the demand slope equal to 1 for simplicity) and constant marginal cost, c, the firms have payoff functions of

$$\Pi_1 = aq_1 - q_1^2 - q_2q_1 - cq_1$$

and

$$\tag{19.7}$$

$$\Pi_2 = aq_2 - q_2^2 - q_1q_2 - cx_2.$$

The best-response functions in the static game were

$$q_1 = R_1(q_2) = \frac{1}{2}(a - c - q_2)$$

and

$$\tag{19.8}$$

$$q_2 = R_2(q_1) = \frac{1}{2}(a - c - q_1).$$

The static Nash equilibrium was found by solving the best-response functions simultaneously. The equilibrium actions and payoffs were

$$q_1^* = \frac{a - c}{3}, \quad \Pi_1^* = \frac{(a - c)^2}{9}$$

and

$$\tag{19.9}$$

$$q_2^* = \frac{a - c}{3}, \quad \Pi_2^* = \frac{(a - c)^2}{9}.$$

Now let us consider the dynamic game where firm 1 moves first, a leadership model known as the Stackelberg model. Equation 19.8 gives firm 2's strategy rule, i.e., q_2 as a function of q_1. There is a subtle difference, however, between the static Cournot game and the dynamic Stackelberg game. In the static game a Nash equilibrium strategy choice for firm 2 is a single output level expressed as a function of parameters. In the dynamic leadership (sub)game the Nash equilibrium strategy for firm 2 is a rule that specifies how firm 2 will respond to any given choice by firm 1. In the (full) game equilibrium firm 2 will, of course, choose a single output, but its strategy for playing the game is still *an output rule rather than an output level*.

Since in a leadership model firm 2 will react to firm 1's output choice and firm 1 will look ahead to anticipate this reaction, we cannot simply take a first-order condition for firm 1 that treats firm 2's output as exogenous. Instead, we must find firm 2's response to any output choice by firm 1. This is simply firm 2's reaction function. To find an equilibrium we substitute this solution into firm 1's payoff function, which gives

$$\Pi_1(q_1, R_2(q_1)) = aq_1 - q_1^2 - \frac{1}{2}(a - c - q_1)\, q_1 - cq_1. \qquad \textbf{(19.10)}$$

The first-order condition for firm 1 is

$$\frac{d\Pi_1(q_1, R_2(q_1))}{dq_1} = \frac{a - c}{2} - q_1 = 0. \qquad \textbf{(19.11)}$$

Solving this equation, substituting into player 2's response function, and then substituting these solutions into the payoff functions gives the solution to the dynamic game:

$$q_1^* = \frac{a - c}{2}, \quad \Pi_1^* = \frac{(a - c)^2}{8}$$

and

$$\qquad \textbf{(19.12)}$$

$$q_2^* = R_2(q_1) = \frac{a - c}{4}, \quad \Pi_2^* = \frac{(a - c)^2}{16}.$$

Comparing these dynamic solutions with the static game solutions shows that there is a first-move advantage for firm 1. Compared to the static Cournot solution, firm 1 produces a higher output and earns a higher payoff. Firm 2 produces less and receives a lower payoff in the sequential game. (Note that market output is higher and market price is lower compared to the static Cournot equilibrium.)

A leader in a dynamic game must receive a payoff at least as high as the Nash equilibrium payoff in the static game. This follows from the fact that the leader always has the option of choosing the static Nash equilibrium strategy. Nevertheless, not all games have first-move advantages. The follower may do better or worse than in the static game. Indeed, it is sometimes the case that the follower earns a higher payoff than the leader. Whether a follower does better or worse depends on the partial and cross-partial derivatives of the payoff functions.[6] The main point, however, is that the structure of a game matters: In most games there will be different choices and payoffs when moves are sequential instead of being simultaneous.

We conclude our theoretical discussion of dynamic games by examining two special classes of dynamic games: two-stage games and repeated games.

19.4 TWO-STAGE GAMES

Two-stage games are games in which one game follows another. In most economic applications the first game sets the stage, or defines the environment, in which the second game is played. For example, the first stage may be one in which firms choose product

[6]For references that provide a fuller explanation of these issues see J. Bulow, J. Geanakoplos, and P. Klemperer, "Multimarket Oligopoly: Strategic Substitutes and Complements," *Journal of Political Economy*, 93 (1985), pp. 488–511, and D. Fudenberg and J. Tirole, "The Fat Cat Effect, the Puppy Dog Ploy, and the Lean and Hungry Look," *American Economic Review: Papers and Proceedings*, 74 (1984), pp. 361–368.

quality (or production capacities) and the second stage one in which firms choose prices. The first-stage choices will determine the profit consequences of second-stage pricing decisions. Two-stage games are often used to model the effects of government policies, such as trade policies or regulatory policies. In the first stage, government(s) choose policies, e.g., tariff levels or pollution taxes. This sets the environment for firms' second-stage choices.

The analysis in this section will focus on games that are simultaneous within stages, but sequential across stages. We will also assume continuous (as opposed to discrete) action spaces. In stage one players 1 and 2 simultaneously choose actions x_1 and x_2. Players 3 and 4 observe these choices and then, in stage two, simultaneously choose their actions x_3 and x_4.[7]

Subgame perfection requires that the equilibrium to the game as a whole includes a Nash equilibrium to the second-stage subgame. This second-stage subgame takes x_1 and x_2 as given. Thus players 3 and 4 choose their actions, x_3 and x_4, to maximize their respective payoff functions, which are

$$\Pi_3 = \Pi_3(x_1, x_2, x_3, x_4)$$

and   **(19.13)**

$$\Pi_4 = \Pi_4(x_1, x_2, x_3, x_4).$$

The first-order conditions for these players are

$$\frac{\partial \Pi_3}{\partial x_3} = 0$$

and   **(19.14)**

$$\frac{\partial \Pi_4}{\partial x_4} = 0.$$

Solving these first-order conditions simultaneously gives the second-stage equilibrium as functions that are contingent on the first-stage actions. These solutions can be written as

$$x_3^* = R_3(x_1, x_2)$$

and   **(19.15)**

$$x_4^* = R_4(x_1, x_2).$$

Note that these solutions have asterisks. Since the first-stage variables are endogenous to the model this may seem to contradict our convention that asterisked solutions must have only exogenous variables on the right-hand side. We resolve this (seeming) inconsistency by noting that equation 19.15 gives the second-stage players' *equilibrium strategies*. The *equilibrium actions* for the second stage will be found once we find the equilibrium first-stage choices.

---

[7]We can easily drop a player (if there is only one player in the first or second stage). The theory will also cover cases where the first-stage players are the same as the second-stage players. Finally, extensions to more than two players per-stage are straightforward.

We now turn to the first stage. Players 1 and 2 anticipate that the second-stage equilibrium depends on the first-stage choices. Thus, we write the payoffs for these players as

$$\Pi_1 = \Pi_1(x_1, x_2, R_3(x_1, x_2), R_4(x_1, x_2))$$

and   (19.16)

$$\Pi_2 = \Pi_2(x_1, x_2, R_3(x_1, x_2), R_4(x_1, x_2)).$$

To take the first-order conditions we must realize that, for each of these players, the player's action enters the payoff function three times: once directly and twice indirectly. The first-order conditions are

$$\frac{d\Pi_1}{dx_1} = \frac{\partial \Pi_1}{\partial x_1} + \frac{\partial \Pi_1}{\partial x_3^*}\frac{\partial R_3}{\partial x_1} + \frac{\partial \Pi_1}{\partial x_4^*}\frac{\partial R_4}{\partial x_1} = 0$$

and   (19.17)

$$\frac{d\Pi_2}{dx_2} = \frac{\partial \Pi_2}{\partial x_2} + \frac{\partial \Pi_2}{\partial x_3^*}\frac{\partial R_3}{\partial x_2} + \frac{\partial \Pi_2}{\partial x_4^*}\frac{\partial R_4}{\partial x_2} = 0.$$

The first-stage equilibrium is the pair of strategies (x_1^*, x_2^*) that simultaneously solve these two first-order conditions. Note the difference here between the simple static game between players 1 and 2 versus the dynamic game in which players 1 and 2 make choices in the first stage. The dynamic game includes the indirect, or strategic, effects of the first-stage players' choices on the second-stage decisions.

The overall solution for the game is the subgame-perfect Nash equilibrium set of strategies that satisfies the four players' first-order conditions. We can write this equilibrium as

$$E = \{x_1^*, x_2^*, x_3^* = R_3(x_1^*, x_2^*), x_4^* = R_4(x_1^*, x_2^*)\},$$ (19.18)

where the notation makes explicit the idea that the second-stage equilibrium depends directly on the first stage.

19.5 REPEATED GAMES

Another important type of dynamic game is a repeated game. Consider, for example, the prisoners' dilemma or the Cournot oligopoly model. These models treat the players as interacting once in a static game. Yet few actual markets last for only a single time period. Instead, firms interact in repeated time periods. In other words, the firms engage in a dynamic game that consists of repeated plays of the static game. Does this repetition matter or is the equilibrium in the repeated game simply a repetition of the static equilibrium? This section develops the tools to answer this question.

We start with the question of why repetition might make a difference. The answer is that repetition opens up the possibilities of threats and promises. Consider the prisoners' dilemma game. In the static prisoners' dilemma each player chooses between only two strategies: cooperate or defect. In the repeated prisoners' dilemma strategy spaces become more complex. A player is able to condition actions (cooperate or defect) on the past actions of other players. A player might promise to cooperate if other players have cooperated in past plays of the game or a player might threaten to defect if other players have defected in previous plays. The issue of subgame perfection, the credibility of threats and promises, turns out to be crucial in defining the equilibria of repeated games.

To formalize these notions, let G be a static game.[8] Let $G(T)$ be the dynamic game where the static game G is repeated T times. To make the analysis concrete, assume that G is a two-person prisoners' dilemma with the following payoffs:[9]

$$
\begin{array}{cc}
 & \text{Player 2} \\
 & \begin{array}{cc} C & \quad D \end{array}
\end{array}
$$

$$
\text{Player 1} \quad
\begin{array}{c} C \\ D \end{array}
\left(
\begin{array}{cc}
R,R & L,W \\
W,L & P,P
\end{array}
\right). \tag{19.19}
$$

where C is cooperate and D is defect. The payoffs, R for reward, L for lose, W for win, and P for punishment are assumed to satisfy the following inequalities:

$$
W > R > P > L \tag{19.20}
$$

and

$$
R > \frac{W + L}{2}.
$$

The first set of inequalities defines the prisoners' dilemma game, and the last inequality ensures that the payoffs from cooperation exceed the payoffs that the players could earn by alternating cooperation and defection.

We know, from Chapter 17, that the unique Nash equilibrium to this static game is for both players to defect. Is the equilibrium different when the game is repeated? In particular, are promises of mutual cooperation credible?

Suppose, initially, that the game is played for a specified finite number of repetitions T. Consider the last time period. In this time period promises or threats concerning future behavior are irrelevant. The final period subgame is simply the static prisoners' dilemma. Thus, the only equilibrium for this final subgame is for both players to defect. Now consider time period $T - 1$. Promises to cooperate in time T are not credible since both players will defect at time T. Thus, the only equilibrium behavior at time $T - 1$ is also for both players to defect.

This line of reasoning by backwards induction can be extended all the way back to the initial time period in which both players will defect. We conclude that the only subgame-perfect Nash equilibrium for a dynamic game in which the prisoners' dilemma is repeated T times is for both players to defect in every period. More generally, we offer the following theorem:

THEOREM | **19.1**

Subgame–Perfect Nash Equilibrium in a Finitely Repeated Game:

Let $G(T)$ represent a game G repeated T times. If G has a unique Nash equilibrium, then the unique subgame-perfect Nash equilibrium for $G(T)$ is for each player to play the static equilibrium strategy of G in all T time periods.

---

[8]Although we assume G is static, the analysis that follows would apply to cases where G is a dynamic game.
[9]The assumption of a prisoners' dilemma game is not very restrictive. All of the conclusions derived in the text extend to any game G which has a unique Nash equilibrium.

The proof, by backwards induction, is essentially the same as the proof given above for the prisoners' dilemma game.

We now turn to infinitely repeated games. To analyze such a game we need to introduce the idea of discounting.[10] Monetary payoffs in future time periods have a present equivalent. The term **present value** is used to denote this equivalence. The present value of a future payoff at time t is the amount of money a player would have to bank now so as to generate that payoff at time t. Let r represent the interest rate. A bank balance of $\$X$ will grow into $\$(1 + r)X$ after one period, $\$(1 + r)^2 X$ after two periods and $\$(1 + r)^t X$ after t periods. Reversing this we can say that the present value of $\$X$ received t years in the future is $\$X/(1 + r)^t$, where this last term is the amount of money needed now to yield $\$X$ at some future time t. Below, we will let V denote a present value. For example the present value of receiving a $\$X$ in each time period from $t = 0 \ldots T$ is

$$V = \sum_{t=0}^{T} \frac{X}{(1 + r)^t}.$$ (19.21)

In an infinitely repeated game a player's payoff is defined as the sum of the present values of the payoffs received in each time period. Let Π_t be a player's payoff in time period t. Then the player's total payoff, in present value terms, is

$$V = \sum_{t=0}^{\infty} \delta^t \Pi_t.$$ (19.22)

where δ, called the discount factor, is defined as

$$\delta = \frac{1}{1 + r} < 1.$$ (19.23)

In our analysis below we will be using infinite sums. It may be helpful at this point to recall the rules for the infinite sum of a variable (less than one) raised to a sequence of powers. These rules are

$$\sum_{t=0}^{\infty} \delta^t = \frac{1}{1 - \delta}$$

$$\sum_{t=1}^{\infty} \delta^t = \delta \sum_{t=0}^{\infty} \delta^t = \frac{\delta}{1 - \delta}$$ (19.24)

$$\sum_{t=\tau}^{\infty} \delta^t = \delta^\tau \sum_{t=0}^{\infty} \delta^t = \frac{\delta^\tau}{1 - \delta}.$$

To find equilibria in an infinitely repeated prisoners' dilemma game we first note that the backwards induction argument that ruled out cooperation in the finitely repeated game does not apply here. The reason is quite simple: there is no final time period in an infinitely repeated game.

Since arguments from backwards induction are not applicable we need to use a different method to find equilibria. This is, however, a rather complicated process. In a dynamic game a strategy for a player specifies what the player will do at every time t. Actions at time t can be conditioned on the entire history of the game up to that point.

---

[10]We could also introduce discounting for finitely repeated games, but discounting would not alter any of the conclusions given above.

For example, a strategy might be to cooperate at time t if the other player has cooperated in every previous even-numbered time period or if the other player has cooperated in half of the odd-numbered time periods and to defect otherwise. Such a strategy may not be very sensible, but if we search for all equilibria then we must examine every possible strategy—and there are infinitely many strategies. Even worse, there will often be an infinite number of equilibria in an infinitely repeated game!

Game theorists generally skirt this conundrum by focusing on an outcome of particular interest and examining whether there is some set of equilibrium strategies that will yield this outcome. It is often the case that the most interesting outcome is the Pareto optimal outcome, an outcome where it is not possible to increase any player's payoff without decreasing at least one other player's payoff. In the prisoners' dilemma this means an outcome in which both players cooperate in every period. There may be many strategies that are equilibrium strategies and yield this result, but we will focus on one particular type of strategy, a **trigger strategy,** that is especially simple to analyze.

DEFINITION | ## 19.6

Trigger Strategies

A trigger strategy in an infinitely repeated game takes the following form:

(a) *Cooperate at time $t = 0$.*

(b) *At time $t > 0$ cooperate if* all *players have chosen cooperation in* all *previous time periods, otherwise defect.*

This trigger strategy is sometimes called a "grim" strategy since cooperation is totally abandoned forever after even a single instance of defection. In games other than the prisoners' dilemma a trigger strategy "cooperate" will be some strategy that yields payoffs higher than the static Nash equilibrium and "defect" will mean reverting to the static Nash equilibrium strategy.

To check whether a trigger strategy is a Nash equilibrium we examine whether either player has an incentive to deviate from this strategy. Consider player 1. If both players play trigger strategies, then both cooperate in all time periods, and from equation 19.19 both earn a per-period payoff of R. The present value of player 1's payoff is

$$V_1^C = \sum_{t=0}^{\infty} \delta^t R = \frac{R}{1-\delta} \qquad (19.25)$$

where the C superscript indicates cooperation and the last equality uses the formula for an infinite series.

Now consider the decision to defect in time zero.[11] If player 1 defects, then player 1 earns W in time zero and P (since the players will revert to defection) in all subsequent time periods. The payoff for defection is therefore

$$V_1^D = W + \sum_{t=1}^{\infty} \delta^t P = W + \frac{\delta P}{1-\delta}. \qquad (19.26)$$

---

[11]It is a feature of infinitely repeated (identical) static games that the dynamic game at any positive time $t > 0$ looks just like the dynamic game at time zero, i.e., there are an infinite number of time periods remaining. Thus, the decision to defect at time zero is based on the same comparisons as the decision to defect at a later time, and if a player chooses not to defect at time zero the player will also choose not to defect at all future times.

For the trigger strategy to be a Nash equilibrium it must yield a higher payoff than defection. This requires that

$$V_1^C \geq V_1^D$$

or   **(19.27)**

$$\frac{R}{1-\delta} \geq W + \frac{\delta P}{1-\delta}.$$

Since $R/(1-\delta) = R + \delta R/(1-\delta)$, the last inequality can also be written as

$$W - R \leq \frac{\delta(R-P)}{1-\delta}.$$ **(19.28)**

Equation 19.28 has a straightforward economic interpretation: the left-hand side is the immediate gain (inherent in the prisoners' dilemma) from defecting while other players are choosing to cooperate, and the right-hand side is the present value of the future loss that occurs when the game reverts to mutual defection in all subsequent time periods. Thus, the trigger strategy constitutes an equilibrium if the immediate gain from defection is smaller than the future losses.

Whether the trigger strategy satisfies the condition for a Nash equilibrium depends on the discount factor. Solving equation 19.28 for δ yields, in several steps,

$$\delta(R-P) \geq (1-\delta)(W-R)$$

$$\delta(R-P) + \delta(W-R) \geq (W-R)$$ **(19.29)**

$$\delta \geq \frac{(W-R)}{(W-P)}.$$

Note that the right-hand term in the last line is, by the definition of the prisoners' dilemma, less than one. Let δ^* be the value of δ that solves the last line of equation 19.29 as an equality. Then the trigger strategy is a Nash equilibrium if and only if the actual discount rate, δ, satisfies

$$\delta \geq \delta^* = \frac{(W-R)}{(W-P)}.$$ **(19.30)**

The economic interpretation of this result hinges on the interpretation of δ. When δ is small the future is heavily discounted, i.e., the present value of future payoffs is small. In the trigger strategy model the cost of defection is in terms of forgoing the benefits of future cooperation. Thus, for small values of δ, where the future costs don't count as heavily, players defect to gain an immediate advantage. For larger values of δ the future loss outweighs the present gain and the trigger strategies constitute a Nash equilibrium.

The comparative statics of the critical value of the discount rate are straightforward. The partial derivatives with respect to the static payoff values are

$$\frac{\partial \delta^*}{\partial R} = \frac{-1}{(W-P)} < 0$$

$$\frac{\partial \delta^*}{\partial P} = \frac{(W-R)}{(W-P)^2} > 0$$ **(19.31)**

$$\frac{\partial \delta^*}{\partial W} = \frac{(W-P)-(W-R)}{(W-P)^2} = \frac{(R-P)}{(W-P)^2} > 0.$$

Higher values of δ* *shrink* the range of discount factors for which trigger strategies are a Nash equilibrium; lower values *expand* the possibilities for cooperation. Thus, parameter changes that lower (raise) δ* make it more likely that a cooperative trigger strategy will pass (fail) the test for Nash equilibrium. The signs of the derivatives should, therefore, coincide with your economic intuition: higher values for cooperation, R, raise the likelihood of a trigger strategy equilibrium; higher gains for defecting, W, lower this likelihood, and higher punishment payoffs, P, also lower the likelihood of a trigger strategy equilibrium.

We conclude with two observations about the basic model. First, most games are not truly infinite, yet still have no definite final period. For example, two firms competing in a market do not really expect to compete to the end of time, but neither do they know exactly when changes in technology or consumer preferences will render the product obsolete. A simple way to handle this situation is to assume that, at each time t, there is a probability q that the game will be played at time $t+1$ and a probability $(1-q)$ that time t will be the last time period. Players now face an expected present value since the end of the game is uncertain. We can model this by redefining δ as

$$\delta' = \frac{q}{(1+r)}. \qquad (19.32)$$

All of the results derived above now hold using δ' in place of δ. The effect of uncertainty as to when the game ends is equivalent to a lower discount rate and a lower likelihood that a trigger strategy will satisfy the requirement for a Nash equilibrium.

Finally, we return to the issue of multiple equilibria. Solving for a trigger strategy equilibrium does not in any way preclude the existence of other equilibria. Indeed, there may be many equilibria as indicated by the following theorem.

THEOREM 19.2

Equilibria in Infinitely Repeated Games

Let G(∞) represent an infinitely repeated game G with discounting, then:

(a) *G(∞) always has at least one equilibrium in which each player plays the static equilibrium strategy of G in each time period.*

(b) *As δ approaches 1 any outcome in which each player earns (on average) a per-period payoff greater than the static Nash equilibrium payoff can be supported as an equilibrium of G(∞).*

Note that the theorem as stated applies not only to the prisoners' dilemma, but to all repeated games.

The first part of the theorem asserts that the repetition of the static Nash strategies is an equilibrium to G(∞). This must be true since, by the definition of Nash equilibrium, if all other players play this strategy, then no player can improve on this by altering actions in any time period. The second part describes equilibria as δ approaches 1. In this case the future is essentially undiscounted and any immediate one-period gain from defecting is offset by an infinite sum of undiscounted future losses so that defection cannot be optimal. Thus, as long as there is some future cost to defecting, which is true whenever cooperation yields payoffs higher than the static Nash payoffs, defecting cannot be optimal. The result that any type of cooperation can be sustained as an equilibrium when δ approaches 1 is often referred to as the Folk theorem.[12]

The Folk theorem is actually quite problematic. It essentially tells us that, for a sufficiently high discount rate, any outcome is possible in an infinitely repeated game. This is an embarrassment of riches. If any equilibrium is possible, then it is difficult to predict which will occur or how changes in the economic environment will alter equilibrium outcomes. One way around this problem is to focus on particular equilibria that seem natural in the context of a given model (for example, collusive pricing in an oligopoly model). This, of course, still skirts the question of other possible equilibria.

SUMMARY

In this chapter we have extended our mathematical tools to cover a variety of games in which decisions are made sequentially. We considered the extensive form, including game trees, as a natural method for presenting the sequencing and information issues that arise in dynamic games. Within the extensive form there may be games within games, or subgames. This concept of subgames helped us to revise the equilibrium solution concept for dynamic games. Subgame perfection was needed because the Nash equilibrium concept used in static games sometimes allowed strategies that posited dominated actions in (off the equilibrium path) subgames and, as a result, the static Nash equilibrium concept could yield unreasonable equilibria in dynamic games.

Within the broad framework of dynamic games we examined two classes of games that are widely used in economic analysis. Two-stage games apply when there are sequential subgames, but the moves within the subgames are simultaneous. The solution method for these two-stage games was to solve for the static Nash equilibrium of the second subgame as a function of first-stage choices and then to solve for the first-stage choices.

For repeated games we discovered that finite repetition made little difference: the dynamic equilibrium was simply a repeated static equilibrium. In infinitely repeated games (or games where the number of repetitions was probabilistic) repetition did alter the range of possible outcomes. Although repetition of the static equilibrium is an equilibrium in an infinitely repeated game, other equilibria are also possible. In particular, we found that trigger strategies—which promise cooperation if other players cooperate, but threaten punishment if other players defect—were possible equilibria in these games. The key to a trigger strategy equilibrium was that the future counts. For higher discount rates, which place a higher present value on future payoffs, there may be many strategies that yield payoffs better than what the players could earn by static Nash behavior.

In the next chapter we will use game trees, two-stage games, and infinitely repeated games as frameworks for analyzing economic interactions that take place sequentially.

Problems

19.1 For the following games in extensive form identify the information sets and the subgames. Which games are games of perfect information?

---

[12]The theorem was not discovered by an economist named Folk! Instead, the theorem seems to have been widely known, but unpublished—it was part of the folklore of game theorists.

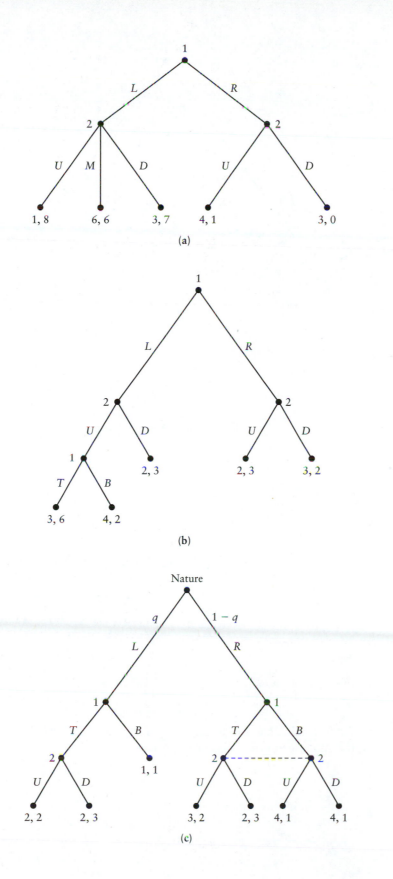

(a)

(b)

(c)

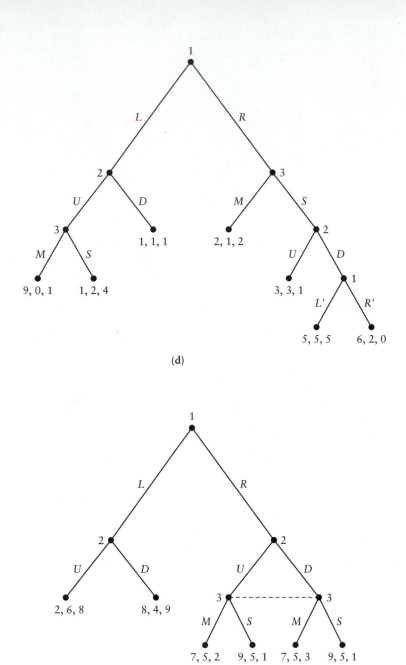

(d)

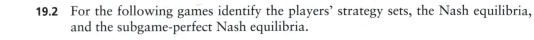

(e)

19.2 For the following games identify the players' strategy sets, the Nash equilibria, and the subgame-perfect Nash equilibria.

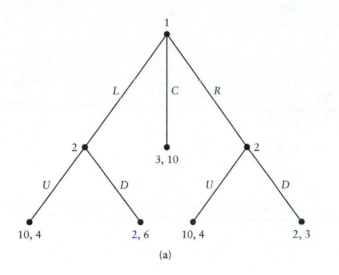

(a)

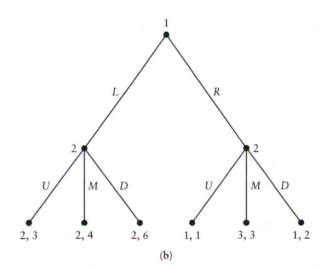

(b)

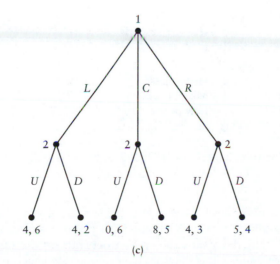

(c)

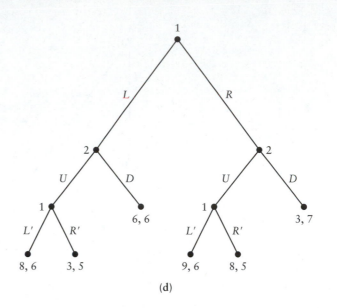

(d)

19.3 Find the subgame-perfect Nash Equilibria for the games in problem 19.1.

19.4 Consider the following one-period duopoly game: Demand for a product is high with a probability 2/3 and low with a probability 1/3; each firm may choose between a high price and a low price.

(a) Show the game tree for the case where the level of demand is known, and firm 2 knows firm 1's price before firm 2 makes its pricing decision.

(b) Same as (a) except that the demand level is not known in advance.

(c) Same as (b) except that firm 2 does not know firm 1's pricing decision.

(d) Show the game tree in which Nature moves first: game (b) is played with probability q and game (c) is played with probability $1 - q$.

(e) Suppose you are given the following information concerning industry profits:

	Industry Profits	
	High D	Low D
High prices	240	120
Mixed prices	180	90
Low prices	120	60

Market shares are (1/2, 1/2) if prices are equal and (5/6, 1/6) if prices are equal (the lower-price firm gets the 5/6).

Find the equilibria for games described above.

19.5 Consider a two-player game where players are given a chance to split a prize worth X dollars. Player i's choice is to claim a share, s_i, of the prize where $0 \le s_i \le 1$. The payoff for each player depends on the sum of the claims. If the sum of the claims is less than or equal to one, then each player receives a payoff equal to her claim times the total prize, X. If the sum of the claims exceeds one, then each player receives a payoff of zero. Suppose the players move in sequence. Find the Nash equilibria and the subgame-perfect Nash equilibrium for this game.

19.6 Consider a two-player game where $x_i \geq 0$ is the strategy for player i and the payoff function for player i is $\Pi_i = a_i x_i - x_i x_j - x_i^2$.

(a) Solve for the subgame-perfect Nash equilibrium strategies. For what values of the parameters a_1 and a_2 are these solutions positive? Show and explain what happens when the simultaneous mathematical solution of the best-response functions yields a negative value for one of the strategies.

(b) Find the comparative static effects of a change in a_1 on the players' equilibrium strategies and payoffs.

19.7 Consider a three-player game where x_i is the choice variable for player i and the payoff function for player i is $\Pi_i = 24x_i - x_i x_j - x_i x_k - x_i^2 - c_i x_i$. Suppose the players move in order.

(a) Solve for the subgame-perfect Nash equilibrium.

(b) Find the comparative static effects of a change in c_1 on the players' equilibrium strategies and actions.

19.8 Consider a two-stage game. In stage one player 1 chooses x_1. In stage two players 2 and 3 simultaneously choose x_2 and x_3. Find the subgame-perfect Nash equilibrium when the players' payoff functions are

$$\Pi_1 = 0.5(x_2 + x_3)^2 + x_1 x_2 + x_1 x_3$$
$$\Pi_2 = (12 - x_2 - x_3) x_2 - x_1 x_2$$
$$\Pi_3 = (12 - x_2 - x_3) x_3 - x_1 x_3.$$

19.9 Consider a two-stage game. In stage one players 1 and 2 simultaneously choose x_1 and x_2. In stage two players 3 and 4 simultaneously choose x_3 and x_4. Find the subgame-perfect Nash equilibrium when the players' payoff functions are

$$\Pi_1 = 3x_1(x_3 + x_4)$$
$$\Pi_2 = x_2(x_3 + x_4)$$
$$\Pi_3 = (6 - x_3 - x_4) x_3 - x_1 x_3 - x_2 x_3$$
$$\Pi_4 = (6 - x_3 - x_4) x_4 - x_1 x_4 - x_2 x_4.$$

19.10 Consider a two-player two-stage game. In stage one each player i chooses to spend an amount a_i where a_i is drawn from the set $S = \{0,3\}$, i.e., each player chooses one of two possible levels of a_i. Stage two is a static game in which each player i chooses an action from the set $T = \{H, L\}$. The payoff matrix for the second stage game is

Player 2

		L	H
Player 1	L	$12,12$	$0, 13 - a_2$
	H	$13 - a_1, 0$	$1,1$

Assume that there is no discounting so that a player's total payoff is the second-stage payoff minus the amount spent in the first stage. Find the subgame-perfect Nash equilibrium.

19.11 Consider a finitely repeated two-player game. Suppose that the static game is described by the payoff matrix

$$
\begin{array}{c} \\ \\ \text{Player 1} \quad \begin{array}{c} E \\ S \end{array} \end{array}
\begin{array}{c} \text{Player 2} \\ \begin{array}{cc} E & S \end{array} \\ \left(\begin{array}{cc} -L,-L & \Pi,0 \\ 0,\Pi & 0,0 \end{array} \right). \end{array}
$$

(a) Describe the static pure-strategy equilibria for this game.

(b) Suppose the game is played twice. Describe the pure strategies available to player 1 (this may be more complex than it appears). Describe the subgame-perfect Nash equilibrium strategies that yield positive payoffs for both players.

19.12 Consider the repeated prisoners' dilemma game in Section 19.5. Suppose that the payoff matrix and the structure of the model are the same as in the text except that defection is detected with a time lag. Specifically, suppose that a player can defect twice, in time periods 0 and 1, before the punishment of reversion to the static Nash equilibrium begins in time period 3. Show that this reduces the range of discount factors for which a trigger strategy is a Nash equilibrium.

19.13 Consider an infinitely repeated two-player game with discounting. Let the one period payoffs for the players be

$$
\Pi_1 = 120s_1 - s_1 s_2 - 2s_1^2
$$
$$
\Pi_2 = 120s_2 - s_1 s_2 - 2s_2^2.
$$

(a) Find the Nash equilibrium actions, s_1^N and s_2^N, for the static one-period game.

(b) Find the cooperative actions, s_1^C and s_2^C, that maximize the sum of the two players' payoffs for the static game.

(c) Define a trigger strategy based on the cooperative actions. Find the present value of the trigger strategy.

(d) Let s_1^D be the action that player 1 takes if she defects from the trigger strategy. Find s_1^D [Hint: The best defection action is the one that maximizes player 1's payoff given that player 2 has chosen s_2^C.] Find the present value for player 1 when player 1 defects.

(e) For what values of the discount rate will the trigger strategy be an equilibrium? [Hint: You will want a calculator for this problem.]

Dynamic Games with Complete Information: Applications

20.1   INTRODUCTION

Any situation in which one agent makes a decision and another agent reacts to that decision can be modeled as a dynamic game. Not surprisingly then, there are a wide variety of applications in economics. In this chapter we touch on just a few of many possible applications.

Our first model considers two-player bargaining. This type of model might apply to negotiations between a firm and a union, negotiations between two partners in a business, or negotiations between two parties to a lawsuit. We start the model with a very simple situation: one player makes a take-it-or-leave-it offer. We then extend the model to the case where both players are given an opportunity to make an offer, and, finally, to the case where players alternate offers indefinitely until an agreement is reached.

The second application in this chapter is drawn from international trade theory. Trade policy can often be modeled as a two-stage game. In the first stage governments choose trade polices, while in the second stage firms react to the trade policy. The first-stage analysis can cover several types of trade policies, e.g., tariffs, quotas, antidumping rules, and subsidies. The second stage might encompass many aspects of firm decision-making, including output, price, research spending, and plant location. In our particular example we consider a game in which governments' first-stage export subsidy policies affect firms' second-stage output choices. The model shows how trade policy games can result in a type of prisoners' dilemma—each country has an individual incentive to subsidize exports, but when both governments subsidize, both countries end up with lower economic welfare.

The third application is another two-stage game. We examine a first stage in which two firms each choose production technology followed by a second stage Cournot duopoly. The model predicts a prisoners' dilemma: each firm has an incentive to adopt a lower-marginal cost, higher-fixed cost technology; however, when both firms respond to this incentive the result is a fall in market price and lower profits.

The final applications extend static oligopoly theory, using both the Bertrand and Cournot models, to the more realistic case where oligopolists compete repeatedly in the same market. We examine the conditions under which trigger strategies can allow firms to achieve higher profits than in the static Nash equilibrium. As in Chapter 19, the discount rate plays a critical role in determining whether repeating the static game alters equilibrium outcomes.

Our coverage in this chapter is far from complete. There are many other applications of dynamic game analysis in economics including executive compensation, government regulatory policy, and decisions by firms to enter new markets. Still, the applications presented in this chapter provide a strong foundation for understanding the economics of dynamic games of complete information.

20.2   SEQUENTIAL BARGAINING MODELS

This section examines bargaining models in which players alternate offers over time. As an economic example, suppose that two firms are considering a joint venture. Each firm might bring its own technology or special expertise to the joint venture, and each firm's contribution might be crucial for the project's success. Let us assume that the joint venture will generate a certain profit and that this profit is known to the firms. The issue for the firms is how to divide the profits. If the firms can reach agreement on a division of the joint venture's profits, then the project can go ahead. If, however, the firms fail to agree then the project (and its profits) will be delayed or canceled. The following model considers the sequence of offers and counteroffers in this type of bargaining situation.

20.2.1   A Single-Offer Bargaining Model

Consider the first period of a bargaining model. There are two players, x and y, bargaining over how to split a fixed prize. For simplicity we set this prize equal to one dollar so that bargaining positions can be expressed as a share of the dollar. In time period zero player x asks for a share s_0 and offers player y a share of $1 - s_0$. In our notation s will denote player x's share and the subscript will denote the time at which an offer is made. Once the offer is made, player y can accept this offer or reject it. If the offer is accepted, then the payoffs are s_0 and $1 - s_0$, respectively. If the offer is rejected, then time period zero ends without an agreement. The game then moves to time period one. The move to a later time period means that later payoffs must be discounted. Thus, the time value of money penalizes any failure to reach agreement.

Before we consider bargaining in later periods, let us first illustrate a game in which the game ends at the start of time period one. Assume that if no agreement is reached in time zero the players receive some exogenous payoffs of Π_x and Π_y such that $\Pi_x + \Pi_y \leq 1$. Since the players receive these payoffs in the later time period their payoffs must be discounted. The present values (as evaluated at the start of the game) of the exogenous split are $\delta\Pi_x$ and $\delta\Pi_y$ (for players x and y, respectively), where δ is the discount factor.

This game is illustrated in Figure 20.1. Acceptance is denoted by A and rejection by R. To solve the game we use backwards induction. At the last decision point player y can earn $1 - s_0$ by accepting the offer or $\delta\Pi_y$ by rejecting the offer. Following the usual convention, that in cases of indifference an offer is accepted, player y will accept if

$$1 - s_0 \geq \delta\Pi_y$$

or   **(20.1)**

$$s_0 \leq 1 - \delta\Pi_y.$$

Figure 20.2 is the truncated game diagram based on player y's choice rule. If player x asks for a share $s_0 > 1 - \delta\Pi_y$ then the offer is rejected. Any demand $s_0 \leq 1 - \delta\Pi_y$ is accepted. Therefore, if player x chooses to make an offer that will be accepted, then the

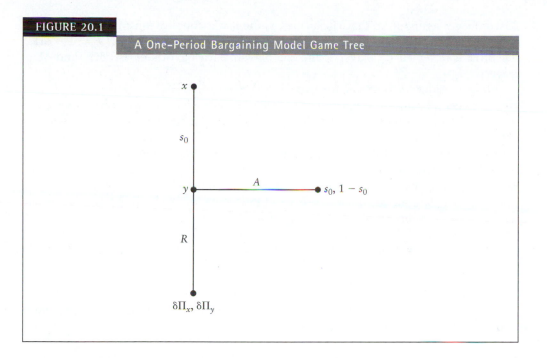

FIGURE 20.1

A One-Period Bargaining Model Game Tree

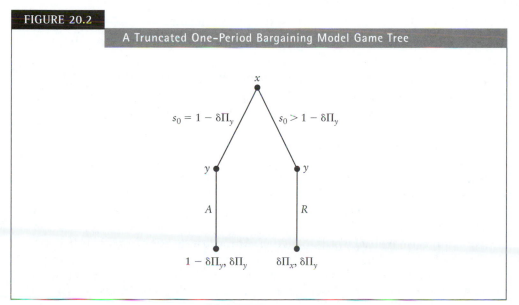

FIGURE 20.2

A Truncated One-Period Bargaining Model Game Tree

optimal (from the point of view of player x) demand is $s_0 = 1 - \delta\Pi_y$. These choices and the associated payoffs are shown in the diagram. Player x will choose to make an acceptable offer if

$$s_0 \geq \delta\Pi_x$$

or

$$1 - \delta\Pi_y \geq \delta\Pi_x \qquad (20.2)$$

or

$$\delta(\Pi_x + \Pi_y) \leq 1.$$

Based on our assumption that $\Pi_x + \Pi_y \leq 1$, the last inequality must hold strictly. The equilibrium for the game therefore is that player x makes an offer $s_0 = 1 - \delta\Pi_y$, the offer is accepted by player y and the players earn payoffs of $1 - \delta\Pi_y$ and $\delta\Pi_y$, respectively.

Now consider the outcome for two special cases. If $\Pi_x = \Pi_y = 0$, then $s_0 = 1$, i.e., player x gets the entire prize. This occurs because player y is in a weak bargaining position: after a rejection by player y the prize disappears and there is a zero payoff for player y. Next consider the other extreme where $\Pi_x + \Pi_y = 1$, i.e., the prize does not shrink. Here player x still gains a first-move advantage: Player y is put in a position of accepting the offer from x or waiting, and waiting has a cost. Player y earns a payoff exactly equal to the present value, $\delta\Pi_y$, of the exogenous payoff, but player x earns a payoff higher than the present value, $\delta\Pi_x$, of the exogenous split. To see this last result we write the equilibrium payoff for player x as

$$s_0 = 1 - \delta\Pi_y = 1 - \delta(1 - \Pi_x) = (1 - \delta) + \delta\Pi_x > \delta\Pi_x. \tag{20.3}$$

20.2.2 A Two-Offer Bargaining Model

We now extend the model to the game in which both players have an opportunity to make offers. The sequence of moves is

1. Player x makes an offer that specifies player x's share as s_0.

2. Player y accepts (and the game ends) or rejects this offer.

3. If player y rejects s_0, then player y is given an opportunity to offer a share, s_1, to player x.

4. Player x accepts or rejects player y's counteroffer.

5. If player x rejects the counteroffer, then the players receive an exogenous split of the prize with a share s going to player x and $1 - s$ going to player y.

Step 5 simplifies the model by assuming that the amount of prize does not diminish over time. Although the amount of the prize does not diminish, the present value of a settlement does decline if agreement is delayed. In the game tree, shown in Figure 20.3, s_t represents player x's share of the settlement, and the subscript refers to the time, t, at which an offer is made. The payoff values are discounted according to whether they occur in time 0, 1, or 2.

We again solve by backwards induction. At the last decision point player x will accept any offer s_1 such that $s_1 \geq \delta s$. Player y has no incentive to offer more than is necessary and will therefore offer $s_1 = \delta s$ (students should check their comprehension of the model by verifying that player y does better by making an acceptable offer rather than an unacceptable offer).

Figure 20.4 shows the game tree after we have solved for the last two decisions. The next step is to determine which initial offers s_0 will be accepted. Player y will accept an offer of $1 - s_0$ if

$$1 - s_0 \geq \delta(1 - \delta s)$$

or

$$s_0 \leq 1 - \delta + \delta^2 s. \tag{20.4}$$

The maximum value of s_0 exceeds player x's payoff, $\delta^2 s$, from an unacceptable initial offer. The equilibrium to the game is therefore: Player x offers $s_0^* = 1 - \delta + \delta^2 s$, player y accepts, and the players earn payoffs of $1 - \delta + \delta^2 s$ and $\delta - \delta^2 s$.

FIGURE 20.3

A Two–Period Bargaining Model Game Tree

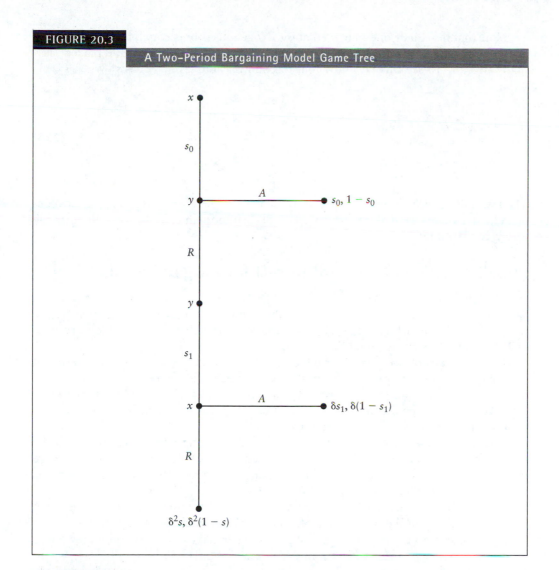

FIGURE 20.4

A Truncated Two–Period Bargaining Model Game Tree

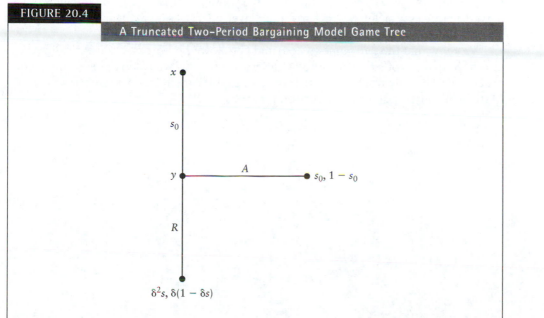

Note that if s equals one-half, so that the exogenous split is equal, then the outcome of the bargaining equilibrium is that player x earns more than one-half and player y less than one-half. To see this we show that discounting implies $s_0^* \geq 1 - s_0^*$ when $s = \frac{1}{2}$:

$$s_0^* \geq 1 - s_0^*$$
$$1 - \delta + \delta^2 s \geq \delta - \delta^2 s$$
$$1 - \delta + \frac{1}{2}\delta^2 \geq \delta - \frac{1}{2}\delta^2 \qquad \textbf{(20.5)}$$
$$1 - 2\delta + \delta^2 \geq 0.$$

The last line holds as a strict inequality when $\delta < 1$. Therefore, even when the exogenous final division is equal, the first player will earn more in equilibrium as long as the future is discounted.

20.2.3 A (Potentially) Infinite-Offer Bargaining Model

The final version of this model is one in which players alternate offers until an agreement is reached. There is no limit on the number of offers, so the game is potentially infinite. Nonetheless, the equilibrium for the game is that the first offer is accepted and the game ends in the initial time period.

Figure 20.5 shows the two-period game of Figure 20.3 except that the terminal node payoffs have been removed from the diagram. Call this game segment G. The infinitely repeated bargaining game is a sequence $G(\infty)$ in which G is infinitely repeated. Since $G(\infty)$ has no final endpoint we cannot use backwards induction. Instead we use an alternative method to find an equilibrium.

Consider two ways of viewing the repeated bargaining game. The first is as the game $G(\infty)$. The second is as the game G followed by $G(\infty)$. These two perspectives are identical. It is a minor paradox of infinity that an infinite repetition of G is the same as G followed by an infinite repetition of G.

Now suppose, for a moment, that we know the equilibrium for the first version. Let s and $1 - s$ represent the players' payoffs in this equilibrium. We could then write the second version as in Figure 20.3, i.e., the players each make one offer and if they fail to agree, then they play a game, $G(\infty)$, in which their payoffs will be s and $1 - s$. We already found (in the last section), however, the equilibrium for the game in Figure 20.3: player x receives a payoff $1 - \delta + \delta^2 s$ and player y receives a payoff of $\delta - \delta^2 s$. Let $f(s)$ denote this payoff for player x, i.e., $f(s) = 1 - \delta + \delta^2 s$.

We know that the two games, $G(\infty)$ and G followed by $G(\infty)$, are identical; yet, we also have two payoffs that seem to be different. How do we resolve this seeming contradiction? The only resolution is that the two payoffs are in fact the same, that $s = f(s)$. Mathematically, we solve this equation for s^*:

$$s = f(s)$$
$$s = 1 - \delta + \delta^2 s$$
$$(1 - \delta^2)s = 1 - \delta$$
$$s = \frac{1 - \delta}{1 - \delta^2} = \frac{1 - \delta}{(1 - \delta)(1 + \delta)} \qquad \textbf{(20.6)}$$

or

$$s^* = \frac{1}{(1 + \delta)}.$$

FIGURE 20.5

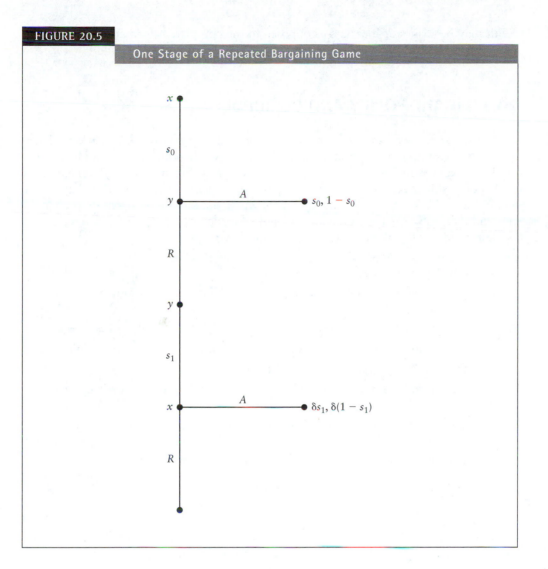

The equilibrium to the repeated game is that player x initially asks for s^*, player y immediately accepts the offer, and the players receive payoffs of $s^* = 1/(1 + \delta)$ and $1 - s^* = \delta/(1 + \delta)$. Note that even with a potentially infinite sequence of offers there is still an advantage to making the first offer. The relative payoffs depend on the discount factor. Since the discount factor is less than or equal to one, player x receives a higher payoff than player y. Note however, the interest rate and discount factor depend on the length of a time period: an annual interest rate of 12% is roughly equivalent to a monthly interest rate of 1%. As the length of time periods between offers becomes arbitrarily smaller, interest rates approach zero; the discount factor approaches one; and the two players' payoffs approach equality.

The bargaining model we have just solved is premised on complete information: players know each other's payoff functions. In many real-world situations there may be asymmetric or incomplete information. For example, in union-firm bargaining a firm may have better information about future demand and profits than does a union. Yet, the firm may be unable to credibly share this information with the union—any firm has an incentive to understate expected profitability so as to reach a better bargain. The result may be strikes or lockouts as opposed to the complete information model result that agreements are reached instantaneously.

20.3 TRADE POLICY AND OLIGOPOLY

In this section we examine a model of trade policy in oligopolistic markets. There are many possible configurations of trade flows and trade policies. Here we focus on a model of export subsidies.[1] We develop a two-stage game. In stage one governments set trade policies. In stage two firms make export decisions. As in the previous chapter, the game is solved backwards. We first model the firms' decisions contingent on trade policies and then solve the first stage for the optimal trade policies.

Our model will make the following assumptions:

1. There are three countries, x, y, and z.
2. Countries x and y each contain one firm (firm x and firm y) producing a homogeneous good.
3. Each firm operates with a total cost function of $TC_i = c_i q_i$, where $i = x$ or y.
4. Countries x and y do not consume the good; all output is exported to country z.
5. Demand in country z is $P = a - Q$, where $Q = q_x + q_y$.
6. Countries x and y grant per-unit export subsidies of s_x and s_y to their respective firms.
7. Country z practices free trade, i.e., it does not tax or subsidize imports.

Under these assumptions we can write the profits for firm x as

$$\Pi_x = Pq_x - c_x q_x + s_x q_x$$

or

$$\Pi_x = (a - q_x - q_y)q_x - c_x q_x + s_x q_x. \tag{20.7}$$

A similar derivation gives the profit function for firm y as

$$\Pi_y = (a - q_x - q_y)q_y - c_y q_y + s_y q_y. \tag{20.8}$$

The respective first-order conditions for profit maximization are

$$\frac{\partial \Pi_x}{\partial q_x} = a - 2q_x - q_y - c_x + s_x = 0$$

and

$$\frac{\partial \Pi_y}{\partial q_y} = a - q_x - 2q_y - c_y + s_y = 0. \tag{20.9}$$

---

[1]See J. Brander and B. Spencer, "Export Subsidies and International Market Share Rivalry," *Journal of International Economics*, 18 (1985), pp. 83–100, for the original derivation of this model.

Solving these equations simultaneously (e.g., writing the equations in matrix form and applying Cramer's rule) yields solutions for the firms' outputs. Substitution then gives solutions for total output and price. The reader can verify that these solutions are

$$q_x^* = \frac{a - 2c_x + 2s_x + c_y - s_y}{3}$$

$$q_y^* = \frac{a - 2c_y + 2s_y + c_x - s_x}{3}$$

$$Q^* = q_x^* + q_y^* = \frac{2a - c_x - c_y + s_x + s_y}{3}$$

$$P^* = a - Q^* = \frac{a + c_x + c_y - s_x - s_y}{3}.$$

(20.10)

These solutions are the equilibrium second-stage strategy rules for the firms as functions of the trade policies and parameters. We next turn to the first-stage choices of the trade policy variables. Countries x and y each choose trade policy in order to maximize their respective levels of economic welfare. The private gain for each country from exporting the product is in terms of its firm's profits. The net welfare gain is this profit level minus the total cost of the export subsidy.

The economic welfare function for country x is

$$W_x = \Pi_x - s_x q_x = (P - c_x + s_x)q_x - s_x q_x = (P - c_x)q_x. \qquad \textbf{(20.11)}$$

Note that welfare for each country is the profit margin (price minus cost) times the volume of exports. Substituting the second-stage solutions gives welfare as a function of parameters and trade policies:

$$W_x = \left(\frac{a - 2c_x + c_y - s_x - s_y}{3} \right) \left(\frac{a - 2c_x + 2s_x + c_y - s_y}{3} \right). \qquad \textbf{(20.12)}$$

A similar derivation gives the welfare function for country y:

$$W_y = \left(\frac{a - 2c_y + c_x - s_y - s_x}{3} \right) \left(\frac{a - 2c_y + 2s_y + c_x - s_x}{3} \right). \qquad \textbf{(20.13)}$$

Each country maximizes welfare by choosing its trade policy. Using the product rule we obtain the first-order condition for country x:

$$\frac{\partial W_x}{\partial s_x} = \frac{1}{9}(-1(a - 2c_x + 2s_x + c_y - s_y) + 2(a - 2c_x + c_y - s_x - s_y)) = 0$$

or

$$a - 2c_x + c_y - 4s_x - s_y = 0 \qquad \textbf{(20.14)}$$

or

$$4s_x + s_y = a - 2c_x + c_y.$$

A similar derivation allows us to write the first-order condition for country y as

$$s_x + 4s_y = a + c_x - 2c_y. \qquad \textbf{(20.15)}$$

To solve for the optimal trade policies we write the first-order conditions in matrix form as

$$\begin{pmatrix} 4 & 1 \\ 1 & 4 \end{pmatrix} \begin{pmatrix} s_x \\ s_y \end{pmatrix} = \begin{pmatrix} a - 2c_x + c_y \\ a + c_x - 2c_y \end{pmatrix},$$ (20.16)

and use Cramer's rule we get the solutions for the subsidy rates:

$$s_x^* = \frac{3a - 9c_x + 6c_y}{15} = \frac{a - 3c_x + 2c_y}{5}$$

and (20.17)

$$s_y^* = \frac{3a - 9c_y + 6c_x}{15} = \frac{a - 3c_y + 2c_x}{5}.$$

To interpret these solutions note that if both countries export positive quantities, then both subsidy rates are also positive. The method for proving this is to substitute the subsidy rates (20.17) back into the equilibrium second-stage solutions for outputs (20.10). The result is that the conditions for positive outputs and positive subsidies are identical. Since the proof involves straightforward, but tedious, algebra, it is left as an exercise for the interested reader.

Suppose, for a moment, that production costs are the same in the two countries. The positive equilibrium subsidies then reflect a prisoners' dilemma in the choice of trade policies. Without subsidies each firm would have half the market and earn positive profits. Because each firm is earning positive profits, each country has a unilateral incentive to subsidize so as to increase its market share and profits. Yet, when both countries subsidize each still ends up with half the market *but at a lower market price*. The firms earn higher private profits (due to the subsidies and the increase in exports generated by the subsidies) but, because the market price falls we know that the cost of the subsidies must exceed the private profit gains. Thus, total economic welfare in the exporting countries is lowered. The real winner in this subsidy game is country z: its consumers pay lower subsidized prices.[2]

We conclude this model by examining how subsidy rates depend on production costs. Which country will grant a larger subsidy? Suppose that

$$c_x = c$$

and (20.18)

$$c_y = c + \Delta.$$

where $\Delta = c_y - c_x > 0$ represents country x's production-cost advantage. We can then write the subsidy rates as

$$s_x^* = \frac{a - 3c + 2(c + \Delta)}{5} = \frac{a - c + 2\Delta}{5}$$

and (20.19)

$$s_y^* = \frac{a - 3(c + \Delta) + 2c}{5} = \frac{a - c - 3\Delta}{5}.$$

The subsidy rate is higher in the country with the competitive advantage so that the net result of the combined subsidies is to reinforce competitive advantage. Higher

---

[2]The net world-welfare effects of the subsidies are positive; the gains in the importing country outweigh the losses to the exporting countries. This overall gain in welfare occurs because the increase in market quantity (decrease in market price) moves the equilibrium closer to the perfectly competitive outcome.

relative subsidies are justified (i.e., rational from the point of view of an individual country) when a country enjoys lower production costs than other exporters: export policy should promote winners rather than protect losers.

With different production costs the prisoners' dilemma nature of the first-stage game may be altered. When the production-cost difference is sufficiently large the country with the comparative advantage may experience a net welfare gain from the subsidy game. In contrast, the country with the higher cost will always experience a decline in welfare. The exact conditions under which the low-cost country earns higher welfare are derivable but complicated. The economic intuition here is that a large enough competitive advantage, reinforced by the subsidies, puts the low-cost country in a more favorable position in the final equilibrium.[3]

This two-stage model has many possible extensions. We might, for example, examine trade policy options for country z; the equilibrium if there are more than two firms; or the optimal policies in a Bertrand model. Furthermore, the basic two-stage model can be extended to other trade configurations and other trade policies. Some of these scenarios are considered in the end-of-chapter problems.

20.4   A TWO-STAGE DUOPOLY GAME

Our next application explores the incentives for firms to adopt certain types of production technologies, with lower marginal, but higher fixed costs, and examines the profit consequences of those choices.[4] We model this as a two-stage game. In the first stage each firm chooses between two available technologies. Once technology and production costs are chosen, the second stage is a Cournot duopoly game.

We will label our firms as firm 1 and firm 2. Each firm operates with a constant marginal cost, c, and some level of fixed cost, f. Using subscripts to indicate that costs need not be identical for the two firms, we can write the two firms' profit functions as

$$\Pi_1 = Pq_1 - c_1q_1 - f_1$$

and

(20.20)

$$\Pi_2 = Pq_2 - c_1q_2 - f_2.$$

The second-stage solutions (assuming linear demand with slope set equal to one) are simply the Cournot equilibrium for two duopolists with different costs. Since we have just gone through this type of model in the previous section, we won't repeat the derivations here. Instead we simply write the solutions as

$$q_1^* = \frac{a - 2c_1 + c_2}{3}$$

$$q_2^* = \frac{a - 2c_2 + c_1}{3}$$

(20.21)

$$Q^* = q_1^* + q_1^* = \frac{2a - c_1 - c_2}{3}$$

$$P^* = a - Q^* = \frac{a + c_1 + c_2}{3}.$$

---

[3]As a simple example of this result the reader can check the case of $c_x = 0$, $c_y = a/3$. With these parameter values both firms would export positive quantities in a game without subsidies. With subsidies, however, firm y drops out of the market, $q_y = 0$. Fairly straightforward calculations show that country x earns higher welfare in the game with subsidies.

[4]This section is based on Baldani and Michl, "Technical Change and Profits: The Prisoner's Dilemma," *Review of Radical Political Economics*, v. 32 #1 (2000), pp. 104–120.

Substituting the equilibrium quantities and market price into equation 20.20 allows us to write profits contingent on technology choices and costs as

$$\Pi_1 = (P - c_1)q_1 - f_1 = \frac{(a - 2c_1 + c_2)^2}{9} - f_1$$

$$\Pi_2 = (P - c_2)q_2 - f_2 = \frac{(a - 2c_2 + c_1)^2}{9} - f_2.$$

(20.22)

We now have the second-stage solutions and the firms' payoffs contingent on the solution to the first stage. We will assume that in the first stage there are two choices for production technology. Each firm will choose one technology or the other and the first stage will be static, i.e., the firms choose simultaneously. We will write a firm's total costs for the first technology as

$$TC_i = cq_i + f.$$

(20.23)

With the second technology total costs will be

$$TC_i = \hat{c}q_i + \hat{f}.$$

(20.24)

By assumption, the relationship between the two technologies will be that the first has a higher marginal cost and a lower fixed cost, i.e.,

$$c > \hat{c}$$

and

(20.25)

$$f < \hat{f}.$$

We can write the payoffs for the first-stage game in matrix form. This payoff matrix is

Firm 2

$$\text{Firm 1} \begin{array}{c} T1 \\ T2 \end{array} \begin{pmatrix} \dfrac{(a-c)^2}{9} - f, \dfrac{(a-c)^2}{9} - f & \dfrac{(a-2c+\hat{c})^2}{9} - f, \dfrac{(a-2\hat{c}+c)^2}{9} - \hat{f} \\ \dfrac{(a-2\hat{c}+c)^2}{9} - \hat{f}, \dfrac{(a-2c+\hat{c})^2}{9} - f & \dfrac{(a-\hat{c})^2}{9} - \hat{f}, \dfrac{(a-\hat{c})^2}{9} - \hat{f} \end{pmatrix},$$

with $T1$ and $T2$ column headers over Firm 2.

(20.26)

where Ti represents the choice of technology i.

Rather than trying to give a full theoretical analysis of the first-stage game, we will finish this application by assigning values to the parameters. Let the parameters take on the following values $a = 12$, $c = 6$, $f = 0$, $\hat{c} = 3$, $\hat{f} = 6$. Note that at an output of two units (the Cournot equilibrium for technology 1) both technologies have the same total and average cost. With these parameter values the payoff matrix is

$$\begin{array}{c} \\ T1 \\ T2 \end{array} \begin{array}{cc} T1 & T2 \end{array} \begin{pmatrix} 4, 4 & 1, 10 \\ 10, 1 & 3, 3 \end{pmatrix}$$

(20.27)

This is a prisoners' dilemma where the Nash equilibrium is for both firms to choose technology 2.

The economic explanation for what is happening here is that each firm has an incentive to pick the technology with a lower marginal, but higher fixed, cost. From an individual firm's point of view the lower marginal cost will result in higher output, market share, and profits. When each firm responds to its individually rational incentives and both firms adopt technology 2, the firms do end up with lower average production costs. Nevertheless, the higher outputs lead to a lower market price, which more than offsets the lower average cost, and reduces profits.

20.5 REPEATED GAMES AND OLIGOPOLY

Until now we have considered the Cournot model, where firms choose outputs, and the Bertrand model, where firms choose prices, as static models. Yet in most markets the same firms interact repeatedly over time. We model this as a repeated game. We will start with the simplest case: the Bertrand model with two firms; extend this to n firms; and, finally, consider the effects of market growth. After the Bertrand model we derive results for the Cournot duopoly model.

20.5.1 The Bertrand Model

Suppose that the market demand for a product is $Q = Q(p)$ where p is the lowest price being charged in the market. Firm 1 and firm 2 choose prices p_1 and p_2. The firm with the lowest price sells the entire market quantity, or, if prices are equal, the firms split market demand evenly. We will also assume that the firms have identical cost functions with a constant marginal cost of c and zero fixed costs.

The Nash equilibrium (see Section 18.4.1) for the static one-period pricing game is for both firms to charge a price equal to marginal cost and earn zero profits. This is also an equilibrium when we consider a repetition of the static game. The question we will address here is whether, in the infinitely repeated version of the game, there also exist trigger strategy equilibria in which the firms earn positive profits.

Let p_{it} be the price charged by firm i in period t. Let $\hat{p} > c$ be some (cooperative) price above marginal cost, c. A trigger strategy for firm i takes the form

$$t = 0 : p_{it} = \hat{p} > c$$
$$t > 0 : p_{it} = \hat{p} \quad \text{if for all time periods } s < t \text{ and for every} \qquad \text{(20.28)}$$
$$\text{player } j \ \ p_{js} = \hat{p}; \text{ otherwise } \ p_{it} = c .$$

The essence of the trigger strategy is to cooperate initially and continue to cooperate as long as all moves by all players are cooperative, but to return to the static price-equal-marginal-cost behavior forever if at any time any player makes a choice other than cooperation.

If both firms adopt this trigger strategy, then they charge equal prices and split the market evenly. Each earns a profit of $(1/2)\hat{\Pi}$, where

$$\hat{\Pi} = \hat{p}Q(\hat{p}) - cQ(\hat{p}). \qquad \text{(20.29)}$$

The present value of the trigger strategy for player i is therefore

$$V_i^C = \sum_{t=0}^{\infty} \delta^t \frac{\hat{\Pi}}{2} = \frac{\hat{\Pi}}{2(1 - \delta)}, \qquad \text{(20.30)}$$

where δ is the discount rate.

To see whether the trigger strategy is an equilibrium we need to define what happens if a firm deviates from the trigger strategy. The optimal action for defection is to charge a price just below $\hat{p}$, i.e., $p = \hat{p} - \varepsilon$ where ε is arbitrarily close to zero. The defecting firm captures the entire market and essentially earns the entire profit $\hat{\Pi}$ instead of half of this profit. Following any defection the firms revert to marginal cost pricing and earn zero profits in all subsequent periods. Thus, the present value for a firm that defects is

$$V_i^D = \hat{\Pi}. \tag{20.31}$$

The trigger strategy is an equilibrium if

$$V_i^C \geq V_i^D$$

or

$$\frac{\hat{\Pi}}{2(1 - \delta)} \geq \hat{\Pi} \tag{20.32}$$

or

$$\delta \geq \frac{1}{2}.$$

Since $\delta = 1/(1 + r)$ the trigger strategy is an equilibrium whenever the interest rate is less than 100%. Note that although we might expect the firms to pick the monopoly price, there are actually an infinite number of equilibria here. Any $\hat{p} > c$ can be an equilibrium price.

We next extend the model to the case of n firms. With n firms in the market each will have a market share of $(1/n)$ in the trigger strategy equilibrium. The present value of this equilibrium for firm i will be

$$V_i^C = \sum_{t=0}^{\infty} \delta^t \frac{\hat{\Pi}}{n} = \frac{\hat{\Pi}}{n(1 - \delta)}. \tag{20.33}$$

As before, a firm which defects from this equilibrium will charge a price $p = \hat{p} - \varepsilon$ and earn an immediate profit $\hat{\Pi}$ followed by zero profits in later time periods. The requirement for a trigger strategy equilibrium is therefore

$$V_i^C \geq V_i^D$$

or

$$\frac{\hat{\Pi}}{n(1 - \delta)} \geq \hat{\Pi} \tag{20.34}$$

or

$$\delta \geq \frac{n - 1}{n}.$$

To interpret the result here note that the right-hand side of the last line of (20.34) approaches one as n increases; the range of discount factors for which the trigger strategy is sustainable shrinks as n increases. Put differently, for a given value of the discount factor $\bar{\delta}$ we have the market price as

$$p = \hat{p} \quad \text{if} \quad \frac{n-1}{n} \leq \bar{\delta}$$

or (20.35)

$$p = c \quad \text{if} \quad \frac{n-1}{n} > \bar{\delta}.$$

Collusion (price above marginal cost) requires that the number of firms not be too large.

Our final extension of this model is to consider the effects of market growth. For simplicity we revert to the duopoly version of the model. Suppose that the quantity demanded and potential profits change at a rate g. Positive values of g indicate a growing market and negative values a shrinking market. The present value of the trigger strategy is

$$V_i^C = \sum_{t=0}^{\infty} (1+g)^t \delta^t \frac{\hat{\Pi}}{2} = \frac{\hat{\Pi}}{2\,(1-\delta(1+g))}. \tag{20.36}$$

For V_i^C to be defined, the last step of this formula requires that market growth not be too high, i.e., $\delta(1+g) < 1$.

As before, the trigger strategy is an equilibrium if

$$V_i^C \geq V_i^D$$

or

$$\frac{\hat{\Pi}}{2\,(1-\delta(1+g))} \geq \hat{\Pi} \tag{20.37}$$

or

$$\delta(1+g) \geq \frac{1}{2}.$$

The last inequality in equation (20.37) is more likely to be satisfied when the market is growing, $g > 0$, and less likely to be satisfied when the market is shrinking, $g < 0$. Intuitively, defection from the trigger strategy causes a loss of future profits, and this loss is larger (smaller) in a growing (shrinking) market.

An interesting, but messy, extension is to consider a product life cycle. The life cycle model of a market posits growth when a product is introduced; then maturation of the market; and, finally, an eventual decline in the market. This implies that g is not constant but instead is a function of time. From the conclusions above it is evident that even if collusion could be sustained initially, it would eventually break down as the market declined in the long run.

20.5.2   The Cournot Model

The repeated game version of the Cournot model is more complicated than the repeated game version of the Bertrand model. In the Cournot model firms earn positive profits in the static game; the defection strategy is therefore somewhat more difficult to derive.

We will make several assumptions that simplify the algebra of the model. First, let us assume a duopoly. Second, let us assume that demand is linear with the demand slope normalized to one, i.e.,

$$P = a - Q \quad \text{where } Q = q_1 + q_2. \tag{20.38}$$

Finally, we will assume that production costs are zero.[5]

In Section 16.2 we solved for outputs, price, and the individual firms' profit levels in a Cournot-Nash equilibrium. These solutions with (with $n = 2$, $b = 1$, and $c = 0$) are

$$q_i^{CN} = \frac{a}{3}$$

$$Q^{CN} = \frac{2a}{3}$$

$$P^{CN} = \frac{a}{3} \tag{20.39}$$

$$\Pi_i^{CN} = \frac{a^2}{9}$$

where the superscript CN stands for Cournot-Nash.

We will also simplify by focusing on a specific trigger strategy in which the two firms collude by each producing half of the monopoly output so that the market price is at the monopoly level.[6] These cooperative (denoted by a C superscript) solutions are[7]

$$q_i^C = \frac{a}{4}$$

$$Q^C = \frac{a}{2}$$

$$P^C = \frac{a}{2} \tag{20.40}$$

$$\Pi_i^C = \frac{a^2}{8}.$$

Let V_i^C be the present value of a cooperative trigger strategy. We have

$$V_i^C = \sum_{t=0}^{\infty} \delta^t \Pi_i^C = \sum_{t=0}^{\infty} \delta^t \frac{a^2}{8} = \frac{a^2}{8(1 - \delta)}. \tag{20.41}$$

To check whether trigger strategies constitute an equilibrium we must first define what happens when a firm deviates from the trigger strategy. Suppose that firm 1 defects by choosing some quantity q_1^D. With firm 2 still playing the trigger strategy firm 1's

---

[5]In the Cournot model all results depend on the difference between the demand intercept and marginal cost. We can set marginal cost to zero by using the "a" term above to represent the difference between the demand intercept and marginal cost.

[6]As in the Bertrand model there are many possible trigger strategies. The collusive, or monopoly, trigger strategy has the feature of being the "best" strategy in the sense that it yields the highest combined profits for the firms.

[7]See Chapter 2 for a derivation of the monopoly outcomes.

profits in the defection period will be given by the equation

$$\Pi_1^D = Pq_1^D = (a - q_1^D - q_2^C)\, q_1^D = \left(a - q_1^D - \frac{a}{4}\right) q_1^D$$

or   (20.42)

$$\Pi_1^D = \frac{3a}{4}q_1^D - (q_1^D)^2.$$

In a trigger strategy equilibrium the punishment for defection, which is reversion to the static Cournot-Nash equilibrium, is the same for any defection quantity. Therefore if the firm is going to defect, then the best quantity choice is that which maximizes Π_1^D. The first-order condition and the solution for the best defection quantity are

$$\frac{\partial \Pi_1^D}{\partial q_1^D} = \frac{3a}{4} - 2q_1^D = 0$$

and   (20.43)

$$q_1^D = \frac{3a}{8}.$$

Substituting this quantity into the profit function gives firm 1's profits in the period that it defects as

$$\Pi_1^D = \frac{3a}{4}q_1^D - (q_1^D)^2 = \left(\frac{3a}{4}\right)\left(\frac{3a}{8}\right) - \left(\frac{3a}{8}\right)^2 = \frac{9a^2}{64}. \qquad \textbf{(20.44)}$$

Finally, we can define the present value of defection followed by reversion to Cournot-Nash behavior. This present value is

$$V_i^D = \Pi_i^D + \sum_{t=1}^{\infty} \delta^t \Pi_i^{CN} = \frac{9a^2}{64} + \frac{\delta}{(1-\delta)}\frac{a^2}{9}. \qquad \textbf{(20.45)}$$

The trigger strategy is an equilibrium if it yields a present value greater than or equal to the present value of defection. This weak inequality is

$$V_i^C \geq V_i^D$$

or   (20.46)

$$\frac{a^2}{8(1-\delta)} \geq \frac{9a^2}{64} + \frac{\delta}{(1-\delta)}\frac{a^2}{9}.$$

Cross-multiplying, to get rid of denominators and the a^2 terms, and then solving for δ yields

$$72 \geq 81(1-\delta) + 64\delta$$

or   (20.47)

$$\delta \geq \frac{9}{17}.$$

For discount rates above this critical value the trigger strategy is an equilibrium.

The algebraically messier case of an n-firm Cournot oligopoly is left as an end-of-chapter exercise. Here, we will simply point out that the basic conclusion of the Bertrand model, that collusive trigger strategies are more likely to fail when n is large, also holds for Cournot oligopoly. Thus, although the static Bertrand and Cournot models differ in their predictions, the dynamic models are more similar in that the ability to sustain collusive prices depends on the number of competitors in the market.

The problem with these trigger-strategy models is that they yield a plethora of equilibria. It may seem natural to ask whether monopoly outcomes are an equilibrium to a repeated game. Yet, an affirmative answer implies that an infinity of other outcomes (those that lie between the static Nash equilibrium price and the monopoly price) are also equilibria. Thus, one criticism of game theory is that it can explain any type of behavior but is unable to make specific predictions about market outcomes.

Problems

20.1 Consider a variation of the bargaining model in which there is no discounting. Instead, after each time period the original one dollar prize is reduced by a fixed-amount c (this might be the case in professional sports labor negotiations where each delay wipes out games and revenue). Suppose that $c = 1/3$ and that players may alternate offers (which are shares of the remaining prize) until the prize becomes zero. Find the equilibrium to this bargaining model.

20.2 Consider a two-period bargaining model, as in Section 20.2.2. Suppose that the two players have different discount rates of δ_x for player x and δ_y for player y. Suppose that the exogenous final split is one-half for each player. How does the existence of different discount rates alter (relative to the single common discount rate case) the subgame-perfect Nash equilibrium?

20.3 Consider a Bertrand model variation with two firms (potentially) selling identical products. Each firm must choose whether to enter a market. If a firm enters, it incurs a fixed and sunk entry cost of f and can produce at a constant marginal cost c. The sequence of the game is in two stages. In stage 1, firm 1 chooses whether to enter (and incur the fixed-cost f), firm 2 observes this decision and then decides whether it will enter (and incur the fixed cost f). In stage two of the game the firm(s) choose prices simultaneously. Solve for the equilibrium.

20.4 Consider a version of the trade model in Section 20.3. Suppose that the structure of the model is identical to that in the text except that:

    (i) the two exporting firms are choosing prices (the Bertrand model) and products are differentiated. Let the demand equation for firm i be $q_i = a - 2p_i + p_j$.

    (ii) production costs are zero for each firm.

    (a) Find the Nash equilibrium prices for the second-stage game as functions of the first-stage trade policies.

    (b) Find the optimal trade policies. Are these trade policies subsidies or tariffs?

20.5 Consider a version of the trade model in Section 20.3. Suppose that the structure of the model is identical to that in the text except that only country x has an export subsidy policy. Find the optimal export subsidy for country x. Compare the equilibrium outputs with this subsidy to the Stackelberg equilibrium outputs of Section 19.3.4. Explain your conclusions.

20.6 Consider a single good, two-country trade model in which firms choose quantities. The good is produced in country y and consumed in country x (the good is not consumed in y nor is it produced in x). Suppose that there are n firms located in country y that export all their output to country x. Also assume that each firm operates with zero production costs, but that exports are taxed at a rate of $\$t$ per unit. Finally, assume that demand in country x is $P = a - Q$, where Q is the total level of imports into country x from country y.

   (a) Write the expression for a typical firm's profits. Find the first-order condition. Solve for the level of a typical firm's output, the level of total exports, and price.

   (b) Suppose that welfare in country y is $W =$ total profits $+$ tax revenue, and that the government of y chooses t so as to maximize welfare. Solve for the optimal level of t. Is it true that t is chosen so as to generate a "collusive," i.e., monopoly, price for exports? Explain.

20.7 Consider a Cournot model of international trade: There are n firms producing a good. All of the firms are located in country x. All firms operate with zero production costs and all output is sold to country y. Demand in country y is $P = a - Q$, where Q is total output. The government of country y taxes imports at a rate of $\$t$ per unit.

   (a) Write the expression for a typical firm's profits. What is the first-order condition for profit maximization? Check the second-order condition.

   (b) Solve for q^*, Q^* and P^*.

   (c) Now suppose that the government of country y chooses an optimal tariff rate, i.e., t is chosen to maximize the sum of tariff revenue plus consumer surplus. (Consumer surplus is the area of the triangle between the demand curve and the line showing the price paid by consumers.) Is t^* an increasing or decreasing function of n? Explain.

20.8 Suppose that country x produces but does not consume some product, and suppose that country y consumes but does not produce the product. Assume that demand in country y is linear and that production costs in country x are zero. Suppose that there is a single monopoly firm in country x, and that both countries choose per-unit trade policies (e.g., export subsidy for x and import tariff for y).

   (a) Find the Nash equilibrium for the case where trade policies are chosen simultaneously.

   (b) Find the subgame-perfect Nash equilibrium for the sequential game, where country x commits itself to an export subsidy level before country y chooses a tariff level.

20.9 Solve the two-agent rent-seeking model from Chapter 18 under the assumption that choices are sequential, i.e., player 1 announces and commits to a level of rent-seeking expenditures before player 2 makes a decision. [Hint: Be prepared for a surprise when you find the equilibrium.]

20.10 Consider a variation of the Stackelberg model. Demand is $P = a - Q$ and marginal costs are zero. Fixed costs are zero for firm 1, but firm 2 (if it chooses to enter the market) incurs a fixed cost of f. The sequence of the game is: (1) firm 1 enters and chooses a quantity; (2) firm 2 observes this quantity and chooses whether to enter the market; (3) if firm 2 enters it chooses its own output; and (4) the firms receive profits based on their choices. Solve for the equilibrium of the model. (Note: The nature of the equilibrium will depend on the values of the parameters.)

20.11 Consider a variation of the Stackelberg model. Demand is $P = a - Q$ and costs are zero. In this variation there is one leader firm that moves first and chooses output. In the second stage of this game, n follower firms simultaneously choose outputs in response to the leader's decision. Solve for the equilibrium. How do the leader's profits depend on the number of followers?

20.12 Consider a Bertrand duopoly with firms selling identical products. Firm 1 is the leader (it picks a price first) and firm 2 is a follower (it chooses price after observing firm 1's choice). Show that firm 1 will necessarily earn zero profits in equilibrium. Show that there exist equilibria where firm 1 has zero sales and profits, while firm 2 has positive sales and profits.

20.13 Consider a two-firm differentiated-products Bertrand price leadership model. Assume demand for firm i is $q_i = 7 - 2p_i + p_j$ and that production costs are zero. The sequence of this game is that firm 1 chooses p_1; firm 2 observes this price and then chooses p_2.

    (a)   Solve for the subgame-perfect Nash equilibrium prices and quantities.

    (b)   Compare the firms' profits in this sequential game to the profits that the firms would earn if price choices were made simultaneously in a static game. [Hint: You will probably want to use a calculator.]

20.14 Consider an infinite time horizon, trigger strategy model. Suppose that there is an n firm oligopoly, that the cooperative strategy is for each firm to produce its share of the monopoly quantity, and that the contingent strategy (if defection occurs) is for each firm to revert to its Cournot-Nash quantity solution. Assume that demand is linear, and let $r > 0$ be the interest rate.

    (a)   What is the optimal quantity choice for a firm that defects (i.e., ceases to play cooperatively)? What is the profit level for this defecting firm?

    (b)   Let r^* be the interest rate that leaves a firm indifferent between cooperating and defecting. Solve for r^* as a function of n. [Hint: This is a bit messy.]

    (c)   Graph the equilibrium price as a function of n, treating r as fixed. Interpret.

20.15 Consider a duopoly where the firms produce identical products. Suppose that firm 1 can produce at zero cost, but that firm 2 operates with a marginal cost of c. Assume that demand is linear.

    (a)   Explain why the (approximate) static Bertrand pricing equilibrium is for firm 2 to charge a price of c (and sell nothing) and firm 1 to charge a price ε below c, where ε is arbitrarily small. For part (b) you may ignore ε and treat this equilibrium as exact.

    (b)   Now suppose that we extend the model to an infinite number of time periods with a discount factor δ. Let the cooperative element of a trigger strategy be to charge the low-cost firm's monopoly price and to divide market shares on the basis of a share s for firm 1 and a share $1 - s$ for firm 2. Show that the value of s that will support a trigger price strategy is an increasing function of c. Explain this result.

20.16 Consider a duopoly where the firms produce identical products. Suppose both firms produce at zero cost.

(a)   Describe the static Bertrand pricing equilibrium.

(b)   Now suppose that we extend the model to an infinite number of time periods with a discount factor δ. Let a cooperative strategy be to charge the monopoly price. Suppose, however, that deviations from cooperation are detected and punished with a time lag of τ periods (in the model in Chapter 20, τ equaled one). When will a trigger strategy be an equilibrium (this answer will depend on both δ and τ. How does the viability of a trigger strategy depend on τ?

Index